Basics of Engineering Economy

Relations for Discrete Cash Flows with End-of-Period Compounding

Type	Find/Given	Factor Notation and Formula	Relation	Sample Cash Flow Diagram
Single Amount	F/P Compound amount	$(F/P,i,n) = (1+i)^n$	$F = P(F/P,i,n)$	
	P/F Present worth	$(P/F,i,n) = \dfrac{1}{(1+i)^n}$	$P = F(P/F,i,n)$	
Uniform Series	P/A Present worth	$(P/A,i,n) = \dfrac{(1+i)^n - 1}{i(1+i)^n}$	$P = A(P/A,i,n)$	
	A/P Capital recovery	$(A/P,i,n) = \dfrac{i(1+i)^n}{(1+i)^n - 1}$	$A = P(A/P,i,n)$	
	F/A Compound amount	$(F/A,i,n) = \dfrac{(1+i)^n - 1}{i}$	$F = A(F/A,i,n)$	
	A/F Sinking fund	$(A/F,i,n) = \dfrac{i}{(1+i)^n - 1}$	$A = F(A/F,i,n)$	
Arithmetic Gradient	P_G/G Present worth	$(P/G,i,n) = \dfrac{(1+i)^n - in - 1}{i^2(1+i)^n}$	$P_G = G(P/G,i,n)$	
	A_G/G Uniform series	$(A/G,i,n) = \dfrac{1}{i} - \dfrac{n}{(1+i)^n - 1}$ (Gradient only)	$A_G = G(A/G,i,n)$	
Geometric Gradient	P_g/A_1 and g Present worth	$P_g = \begin{cases} \dfrac{A_1\left[1 - \left(\dfrac{1+g}{1+i}\right)^n\right]}{i - g} & g \neq i \\[2em] A_1\dfrac{n}{1+i} & g = i \end{cases}$ (Gradient and base)		

Basics of Engineering Economy

Second Edition

Leland Blank, P. E.

Industrial and Systems Engineering
Texas A&M University
and
Dean Emeritus
American University of Sharjah, UAE

Anthony Tarquin, P. E.

Civil Engineering
University of Texas—El Paso

The McGraw-Hill Companies

Connect
Learn
Succeed™

BASICS OF ENGINEERING ECONOMY, SECOND EDITION

Published by McGraw-Hill, a business unit of The McGraw-Hill Companies, Inc., 1221 Avenue of the Americas, New York, NY 10020. Copyright © 2014 by The McGraw-Hill Companies, Inc. All rights reserved. Printed in the United States of America. Previous edition © 2008. No part of this publication may be reproduced or distributed in any form or by any means, or stored in a database or retrieval system, without the prior written consent of The McGraw-Hill Companies, Inc., including, but not limited to, in any network or other electronic storage or transmission, or broadcast for distance learning.

Some ancillaries, including electronic and print components, may not be available to customers outside the United States.

This book is printed on acid-free paper.

3 4 5 6 7 8 9 0 DOC/DOC 1 0 9 8 7 6 5

ISBN 978-0-07-337635-6
MHID 0-07-337635-3

Senior Vice President, Products & Markets: *Kurt L. Strand*
Vice President, General Manager: *Marty Lange*
Vice President, Content Production and Technology Services: *Kimberly Meriwether David*
Director: *Michael Lange*
Managing Director: *Thomas Timp*
Publisher: *Raghothaman Srinivasan*
Developmental Editor: *Lorraine Buczek*
Marketing Manager: *Curt Reynolds*
Director, Content Production: *Terri Schiesl*
Lead Project Manager: *Jane Mohr*
Buyer: *Jennifer Pickel*
Media Project Manager: *Prashanthi Nadipalli*
Cover Designer: *Studio Montage, St. Louis, MO*
Cover Image: *Background: Getty Images; Inset: Mike Kemp/Rubberball/Getty Images*
Compositor: *MPS Limited*
Typeface: *10/12 Times*
Printer: *R.R. Donnelley*

All credits appearing on page or at the end of the book are considered to be an extension of the copyright page.

Library of Congress Cataloging-in-Publication Data

Cataloging-in-Publication Data has been requested from the Library of Congress.

The Internet addresses listed in the text were accurate at the time of publication. The inclusion of a website does not indicate an endorsement by the authors or McGraw-Hill, and McGraw-Hill does not guarantee the accuracy of the information presented at these sites.

www.mhhe.com

Dedication

To our grandchildren Abigail, Benjamin, Grace, Taryn, and Tyler

May they lead us in successfully meeting the challenges of tomorrow

Brief Contents

Chapters

1 Foundations of Engineering Economy   1

2 Factors: How Time and Interest Affect Money   33

3 Nominal and Effective Interest Rates   71

4 Present Worth Analysis   94

5 Annual Worth Analysis   123

6 Rate of Return Analysis   141

7 Benefit/Cost Analysis and Public Sector Projects   181

8 Breakeven, Sensitivity, and Payback Analysis   206

9 Replacement and Retention Decisions   241

10 Effects of Inflation   264

11 Estimating Costs   289

12 Depreciation Methods   317

13 After-Tax Economic Analysis   344

14 Alternative Evaluation Considering Multiple Attributes and Risk   378

Appendices

A Using Spreadsheets and Microsoft Excel©   403

B Accounting Reports and Business Ratios   418

C Final Answers to Selected Problems   425

Reference Materials   437

Interest Factor Tables   439

Index   465

Contents

Preface   xi

Chapter 1

Foundations of Engineering Economy   1

Purpose and Learning Outcomes   2

1.1 What is Engineering Economy?   3

1.2 Performing an Engineering Economy Study   3

1.3 Interest Rate, Rate of Return, and MARR   5

1.4 Equivalence   8

1.5 Simple and Compound Interest   9

1.6 Terminology and Symbols   14

1.7 Cash Flows: Their Estimation and Diagramming   16

1.8 Introduction to Spreadsheet and Calculator Functions   20

1.9 Ethics and Economic Decisions   23

Summary   28

Problems   28

Additional Problems and FE Exam Review Questions   32

Chapter 2

Factors: How Time and Interest Affect Money   33

Purpose and Learning Outcomes   34

2.1 Single-Payment Formulas (F/P and P/F)   35

2.2 Uniform Series Formulas (P/A, A/P, A/F, F/A)   40

2.3 Gradient Formulas   44

2.4 Calculations for Cash Flows That are Shifted   47

2.5 Using Spreadsheets and Calculators   52

Summary   58

Problems   58

Additional Problems and FE Exam Review Questions   68

Chapter 3

Nominal and Effective Interest Rates   71

Purpose and Learning Outcomes   72

3.1 Nominal and Effective Interest Rate Statements   73

3.2 Effective Interest Rate Formulation   75

3.3 Reconciling Compounding Periods and Payment Periods   77

3.4 Equivalence Calculations Involving Only Single-Amount Factors   78

3.5 Equivalence Calculations Involving Series with PP ≥ CP   80

3.6 Equivalence Calculations Involving Series with PP < CP   82

3.7 Using Spreadsheets for Effective Interest Rate Computations   84

Summary   86

Problems   87

Additional Problems and FE Exam Review Questions   92

Chapter 4

Present Worth Analysis   94

Purpose and Learning Outcomes   95

4.1 Formulating Alternatives   96

4.2 Present Worth Analysis of Equal-Life Alternatives   98

4.3 Present Worth Analysis of Different-Life Alternatives   100

4.4 Capitalized Cost Analysis   104

4.5 Evaluation of Independent Projects   108

4.6 Using Spreadsheets for PW Analysis   110

Summary   112

Problems   113

Additional Problems and FE Exam Review Questions   121

Chapter 5

Annual Worth Analysis   123

Purpose and Learning Outcomes   124

5.1 AW Value Calculations   125

5.2 Evaluating Alternatives Based on Annual Worth   127

5.3 AW of a Long-Life or Infinite-Life Investment   130

5.4 Using Spreadsheets for AW Analysis   132

Summary   134

Problems   135

Additional Problems and FE Exam Review Questions   139

Chapter 6

Rate of Return Analysis   141

Purpose and Learning Outcomes   142

6.1 Interpretation of ROR Values   143

6.2 ROR Calculation   145

6.3 Cautions when Using the ROR Method   148

6.4 Understanding Incremental ROR Analysis   148

6.5 ROR Evaluation of Two or More Mutually Exclusive Alternatives   152

6.6 Multiple ROR Values   156

6.7 Techniques to Remove Multiple ROR Values   160

6.8 Using Spreadsheets and Calculators to Determine ROR Values   166

Summary   170

Problems   170

Additional Problems and FE Exam Review Questions   178

Chapter 7

Benefit/Cost Analysis and Public Sector Projects   181

Purpose and Learning Outcomes   182

7.1 Public Sector Projects: Description and Ethics   183

7.2 Benefit/Cost Analysis of a Single Project   188

7.3 Incremental B/C Evaluation of Two or More Alternatives   191

7.4 Using Spreadsheets for B/C Analysis   197

Summary   199

Problems   199

Additional Problems and FE Exam Review Questions   204

Chapter 8

Breakeven, Sensitivity, and Payback Analysis   206

Purpose and Learning Outcomes   207

8.1 Breakeven Analysis for a Single Project   208

8.2 Breakeven Analysis between Two Alternatives   213

8.3 Sensitivity Analysis for Variation in Estimates   216

8.4 Sensitivity Analysis of Multiple Parameters for Multiple Alternatives   221

8.5 Payback Period Analysis   223

8.6 Using Spreadsheets for Sensitivity or Breakeven Analysis   225

Summary   230

Problems   231

Additional Problems and FE Exam Review Questions   238

Chapter 9

Replacement and Retention Decisions   241

Purpose and Learning Outcomes   242

9.1 Basics of a Replacement Study   243

9.2 Economic Service Life   244

9.3 Performing a Replacement Study 246

9.4 Defender Replacement Value 250

9.5 Replacement Study Over a Specified Study Period 250

9.6 Using Spreadsheets for a Replacement Study 254

Summary 257

Problems 257

Additional Problems and FE Exam Review Questions 262

Chapter **10**

Effects of Inflation 264

Purpose and Learning Outcomes 265

10.1 Understanding the Impact of Inflation 266

10.2 PW Calculations Adjusted for Inflation 269

10.3 FW Calculations Adjusted for Inflation 274

10.4 AW Calculations Adjusted for Inflation 278

10.5 Using Spreadsheets to Adjust for Inflation 279

Summary 282

Problems 283

Additional Problems and FE Exam Review Questions 287

Chapter **11**

Estimating Costs 289

Purpose and Learning Outcomes 290

11.1 How Cost Estimates are Made 291

11.2 Unit Method 294

11.3 Cost Indexes 296

11.4 Cost-Estimating Relationships: Cost-Capacity Equations 299

11.5 Cost-Estimating Relationships: Factor Method 301

11.6 Cost-Estimating Relationships: Learning Curve 303

11.7 Indirect Cost Estimation and Allocation 305

Summary 311

Problems 312

Additional Problems and FE Exam Review Questions 315

Chapter **12**

Depreciation Methods 317

Purpose and Learning Outcomes 318

12.1 Depreciation Terminology 319

12.2 Straight Line (SL) Depreciation 321

12.3 Declining Balance Depreciation 323

12.4 Modified Accelerated Cost Recovery System (MACRS) 325

12.5 Tax Depreciation System in Canada 329

12.6 Switching Between Classical Methods; Relation to MACRS Rates 330

12.7 Depletion Methods 332

12.8 Using Spreadsheets for Depreciation Computations 334

Summary 337

Problems 338

Additional Problems and FE Exam Review Questions 342

Chapter **13**

After-Tax Economic Analysis 344

Purpose and Learning Outcomes 345

13.1 Income Tax Terminology and Relations 346

13.2 Before-Tax and After-Tax Alternative Evaluation 349

13.3 How Depreciation Can Affect an After-Tax Study 352

13.4 After-Tax Replacement Study 358

13.5 Capital Funds and the Cost of Capital 360

13.6 Using Spreadsheets for After-Tax Evaluation 364

13.7 After-Tax Value-Added Analysis 367

Summary 370

Problems 370

Additional Problems and FE Exam Review Questions 376

Chapter **14**

Alternative Evaluation Considering Multiple Attributes and Risk   378

Purpose and Learning Outcomes   379

14.1 Multiple Attribute Analysis   380

14.2 Economic Evaluation with Risk Considered   385

14.3 Alternative Evaluation Using Sampling and Simulation   394

Summary   398

Problems   398

Additional Problems   401

Appendix **A**

Using Spreadsheets and Microsoft Excel©   403

A.1 Introduction to Using Excel   403

A.2 Organization (Layout) of the Spreadsheet   406

A.3 Excel Functions Useful in Engineering Economy (alphabetical order)   407

A.4 Goal Seek—A Spreadsheet Tool for Breakeven and Sensitivity Analyses   416

A.5 Error Messages   417

Appendix **B**

Accounting Reports and Business Ratios   418

B.1 The Balance Sheet   418

B.2 Income Statement and Cost of Goods Sold Statement   419

B.3 Business Ratios   421

Appendix **C**

Final Answers to Selected Problems   425

Reference Materials   437

Interest Factor Tables   439

Index   465

Preface

All of the basic principles, techniques, and tools of undergraduate engineering economics are covered in this second edition. The textual material, examples, and problems are designed to meet the needs of a two- or three-semester/quarter credit hour *service course* for all disciplines of engineering, engineering technology, and engineering management. The printed and electronic versions are suitable for different course formats. Especially helpful are the *website-based podcasts,* which incorporate voice-over animated and annotated PPT slides. These podcasts serve as supplemental and support materials for students in any course format—resident, online, or distance education.

FEATURES — NEW AND OLD

Notable enhancements for this edition are in the areas of new or upgraded topics, teaching and learning aids, and website materials.

New Topics

- *Ethics* and its important connection to economic decisions are introduced in Chapter 1 and discussed further in Chapter 7 on public-sector projects.
- *Risk analysis* is expanded with its own new Chapter 14, which also introduces simulation with random sampling utilizing simple spreadsheet functions.
- *External rate of return* material is expanded with the modified ROR method (MIRR) and return on invested capital (ROIC) methods covered, along with hand solution and spreadsheet illustrations.

New Teaching and Learning Aids

- The *final answer* to every third end-of-chapter problem is presented in a new Appendix C. Full solutions are available on the website.
- *Financial calculator usage* is included throughout the formative chapters, with an introduction to calculator-based solutions presented in Chapter 2 in parallel with spreadsheet functions.
- *Tax tables* for corporations and individuals utilize the same format as IRS tables.
- *Spreadsheet screen shots* are streamlined and more colorful for clearer understanding of content and the approach to problem solution.
- *Using Spreadsheets* (Appendix A) is updated to Excel$^{\copyright}$ 2010.
- *End-of-chapter problems* number in excess of 870 with approximately 2/3 of them new or rewritten for this edition.

- *Solutions* using factors, calculator, and spreadsheet functions are presented for selected examples and end-of-chapter problems throughout the formative chapters.

New Website Materials

- *Podcasts,* incorporating voice-over animated and annotated PPT slides that summarize the essential topics are available on the website (www.mhhe.com/ blank). (An icon in the margin of the text identifies material included in these podcasts.)
- A *detailed solution* is available in open access form for every end-of-chapter problem that has its final answer in Appendix C.
- *Live spreadsheets* are available for all examples, plus an *image library* of all tables and figures in the text.

The familiar features that make this an easy-to-use and quick-to-learn-from text continue to be included.

- *Purpose statement* and *learning outcomes* at the beginning of each chapter with outcomes tied to individual sections.
- *Examples* in each section, taken from different engineering disciplines, are maintained from the first edition, plus several new examples to better illustrate current topics and solution approaches.
- *Spreadsheet and calculator applications* are primarily concentrated in a final section of each chapter, allowing incorporation of electronic solutions or omission of this technology, at the discretion of the professor.
- Large number of *end-of-chapter problems* cover all aspects of the text's material in each section.
- *Multiple-choice questions* for each chapter are useful in a review for the FE Exam. Alternatively, these problems can be used as additional problems or for review prior to a course exam. These can be easily incorporated into auto-grade systems for online and distance-learning course structures through course management systems such as Blackboard.
- *Solutions manual, lecture slides,* and *image library of figures* are available online for each chapter with password protection for adopters.

USES OF TEXT

The writing style emphasizes brief, crisp coverage of principles, techniques, and alternative selection guidelines based on time-value-of-money computations. This book is developed in order to reduce the time necessary to present, grasp, and apply the essentials of engineering economic analysis. Most chapters that cover the fundamentals of the subject include hand, calculator, and spreadsheet solutions. More complex solutions that utilize a spreadsheet are separately shown in a final section of each chapter.

Students should have attained a sophomore or higher level standing to thoroughly understand the engineering context of the techniques and problems addressed. A background in calculus is not necessary; however, a basic familiarity with engineering terminology in a student's own engineering discipline makes the material more meaningful and, therefore, easier to learn and apply.

The text may be used in a wide variety of ways in an undergraduate course—from a few weeks that introduce the basics of engineering economics, to a full two- or three-semester/quarter credit hour course. For senior students who have little or no background in engineering economic analysis in earlier courses, this text provides an excellent senior-level introduction as the *senior project* is designed and developed.

Engineering economy is one of the few engineering topics that is equally applicable to both individuals and corporate and government employees. It can analyze personal finances and investments in a fashion similar to corporate project finances. Students will find that this text serves well as a reference throughout their courses and senior design projects, and especially after graduation as a reference source in engineering project work.

Because various engineering curricula concentrate on different aspects of engineering economics, sections and chapters can be covered or skipped to tailor the text's usage in print or electronic forms. For example, cost estimation that is often of more importance to *chemical engineering* is concentrated in a special chapter. Public sector economics for *civil engineering* is discussed separately. After-tax analysis, cost of capital, and decision-making under risk are introduced for *industrial and systems engineering* and *engineering management* curricula that include a shortened course in engineering economy. Examples treat areas for *electrical, petroleum, mechanical,* and other engineering disciplines.

ACKNOWLEDGEMENTS

We would like to thank the following individuals who completed the review used in the development of this edition

Richard H. Bernhard, *North Carolina State University*
John Callister, *Cornell University*
Kamran Abedini, *California State Polytechnic University, Pomona*
Charles R. Glagola, *University of Florida*
Ronald Mackinnon, *University of British Columbia*
Peter Ouchida, *California State University, Sacramento*
Ken S. Sivakumaran, *McMaster University*
Matthew Sanders, *Kettering University*
Marylee Southard, *University of Kansas-Lawrence*

We thank Jack Beltran for his diligence and accuracy in checking the examples and problems. We thank our families for their patience during the preparation of the manuscript and website materials.

We welcome comments and corrections that will improve this text or its online learning materials. Our e-mail addresses are lelandblank@yahoo.com and atarquin@utep.edu.

Lee Blank and *Tony Tarquin*

Foundations of Engineering Economy

Mike Kemp/Rubberball/Getty Images

The need for engineering economy is primarily motivated by the work that engineers do in performing analysis, synthesizing, and coming to a conclusion as they work on projects of all sizes. In other words, engineering economy is at the heart of *making decisions.* These decisions involve the fundamental elements of *cash flows of money, time,* and *interest rates.* This chapter introduces the basic concepts and terminology necessary for an engineer to combine these three essential elements in organized, mathematically correct ways to solve problems that will lead to better decisions.

Purpose: Understand and apply the fundamental concepts and terminology of engineering economy.

LEARNING OUTCOMES

1. Determine the role of engineering economy in the decision-making process.

   Definition and role

2. Identify what is needed to successfully perform an engineering economy study.

   Study approach and terms

3. Perform calculations about interest rate and rate of return.

   Interest rate

4. Understand what equivalence means in economic terms.

   Equivalence

5. Calculate simple interest and compound interest for one or more interest periods.

   Simple and compound interest

6. Identify and use engineering economy terminology and symbols.

   Symbols

7. Understand cash flows, their estimation, and how to graphically represent them.

   Cash flows

8. Formulate spreadsheet and calculator functions used in engineering economy.

   Spreadsheets/Calculators

9. Describe the terms universal and personal morals, and professional ethics; understand the Code of Ethics for Engineers.

   Ethics and economics

1.1 WHAT IS ENGINEERING ECONOMY?

Before we begin to develop the fundamental concepts upon which engineering economy is based, it is appropriate to define the term engineering economy. In the simplest of terms, *engineering economy* is a collection of techniques that simplify comparisons of alternatives on an *economic* basis. In defining what engineering economy is, it might also be helpful to define what it is not. Engineering economy is not a method or process for determining what the alternatives are. On the contrary, engineering economy begins only after the alternatives have been identified. If the best alternative is actually one that the engineer has not even recognized as an alternative, then all of the engineering economic analysis tools in this book will not result in its selection.

Engineering economic analysis is able to answer professional and personal financial questions. If you wish to evaluate the economics of purchasing a new home or leasing versus buying a new automobile for yourself, the techniques of engineering economy covered in this text are just as applicable as they are for determining if a replacement piece of equipment should be purchased by your employer.

While economics will be the sole criterion for selecting the best alternatives in this book, real-world decisions usually include many other factors in the decision-making process. For example, in determining whether to build a nuclear-powered, gas-fired, or coal-fired power plant, factors such as safety, air pollution, public acceptance, water demand, waste disposal, global warming, and many others would be considered in identifying the best alternative. The inclusion of other factors (besides economics) in the decision-making process is called multiple attribute analysis. This topic is introduced in Chapter 14.

1.2 PERFORMING AN ENGINEERING ECONOMY STUDY

In order to apply economic analysis techniques, it is necessary to understand the basic terminology and fundamental concepts that form the foundation for engineering-economy studies. Some of these terms and concepts are described below.

1.2.1 Alternatives

An *alternative* is a stand-alone solution for a given situation. We are faced with alternatives in virtually everything we do, from selecting the method of transportation we use to get to work every day to deciding between buying a house or renting one. Similarly, in engineering practice, there are always several ways of accomplishing a given task, and it is necessary to be able to compare them in a rational manner so that the most economical alternative can be selected. The alternatives in engineering considerations usually involve such items as purchase cost (first cost), anticipated useful life, yearly costs of maintaining assets (annual maintenance and operating costs), anticipated resale value (salvage value), and the interest rate. After the facts and all the relevant estimates have been collected, an engineering economy analysis can be conducted to determine which is best from an economic point of view.

1.2.2 Cash Flows

The estimated inflows (revenues and savings) and outflows (costs) of money are called cash flows. These estimates are truly the heart of an engineering economic analysis. They also represent the weakest part of the analysis, because most of the numbers are judgments about what is going to happen in the *future*. After all, who can accurately predict the price of oil next week, much less next month, next year, or next decade? Thus, no matter how sophisticated the analysis technique, the end result is only as reliable as the accuracy of the data that it is based on. This means that economic decisions about proposed alternatives are made under *risk*, that is, without certainty. Techniques that utilize sensitivity analysis, risk analysis, and multiple attribute analysis assist in understanding the consequences of variation in cash flow estimates.

1.2.3 Alternative Selection

Every situation has at least two alternatives. In addition to the one or more formulated alternatives, there is always the alternative of inaction, called the *do-nothing (DN)* alternative. This is the *as-is* or *status quo* condition. In any situation, when one consciously or subconsciously does not take any action, he or she is actually selecting the DN alternative. Of course, if the status quo alternative *is* selected, the decision-making process should indicate that doing nothing is the most favorable economic outcome at the time the evaluation is made. The procedures developed in this book will enable you to consciously identify the alternative(s) that is (are) economically the best.

1.2.4 Evaluation Criteria

Whether we are aware of it or not, we use criteria every day to choose between alternatives. For example, when you drive to campus, you decide to take the "best" route. But how did you define *best?* Was the best route the safest, shortest, fastest, cheapest, most scenic, or what? Obviously, depending upon which criterion or combination of criteria is used to identify the best, a different route might be selected each time. In economic analysis, *financial units* (dollars or other currency) are generally used as the tangible basis for evaluation. Thus, when there are several ways of accomplishing a stated objective, the alternative with the lowest overall cost or highest overall net income is selected.

1.2.5 Intangible Factors

In many cases, alternatives have noneconomic or intangible factors that are difficult to quantify. When the alternatives under consideration are hard to distinguish economically, intangible factors may tilt the decision in the direction of one of the alternatives. A few examples of noneconomic factors are safety, customer acceptance, reliability, convenience, and goodwill.

1.2.6 Time Value of Money

It is often said that money makes money. The statement is indeed true, for if we elect to invest money today, we inherently expect to have more money in the future.

If a person or company borrows money today, by tomorrow more than the original loan principal will be owed. This fact is also explained by the time value of money.

> **The change in the amount of money over a given time period is called the** *time value of money;* **it is the most important concept in engineering economy.**

The time value of money can be taken into account by several methods in an economy study, as we will learn. The method's final output is a *measure of worth*, for example, rate of return. This measure is used to accept or reject an alternative.

1.3 INTEREST RATE, RATE OF RETURN, AND MARR

Interest is the manifestation of the time value of money, and it essentially represents "rent" paid for use of the money. Computationally, interest is the difference between an ending amount of money and the beginning amount. If the difference is zero or negative, there is no interest. There are always two perspectives to an amount of interest—interest paid and interest earned. Interest is *paid* when a person or organization borrows money (obtains a loan) and repays a larger amount. Interest is *earned* when a person or organization saves, invests, or lends money and obtains a return of a larger amount. The computations and numerical values are essentially the same for both perspectives, but they are interpreted differently.

Interest paid or earned is determined by using the relation

$$\text{Interest} = \text{end amount} - \text{original amount} \qquad \textbf{[1.1]}$$

When interest over a *specific time unit* is expressed as a percentage of the original amount (principal), the result is called the *interest rate* or *rate of return (ROR)*.

$$\textbf{Interest rate or rate of return} = \frac{\textbf{interest accrued per time unit}}{\textbf{original amount}} \times \textbf{100\%} \quad \textbf{[1.2]}$$

The time unit of the interest rate is called the *interest period*. By far the most common interest period used to state an interest rate is 1 year. Shorter time periods can be used, such as, 1% per month. Thus, the interest period of the interest rate should always be included. If only the rate is stated, for example, 8.5%, a 1-year interest period is assumed.

The term *return on investment (ROI)* is used equivalently with ROR in different industries and settings, especially where large capital funds are committed to engineering-oriented programs. The term *interest rate paid* is more appropriate from the borrower's perspective, while *rate of return earned* is better from the investor's perspective.

An employee at LaserKinetics.com borrows $10,000 on May 1 and must repay a total of $10,700 exactly 1 year later. Determine the interest amount and the interest rate paid.   **EXAMPLE 1.1**

Solution

The perspective here is that of the borrower since $10,700 repays a loan. Apply Equation [1.1] to determine the interest paid.

Interest paid = $10,700 − 10,000 = $700

Equation [1.2] determines the interest rate paid for 1 year.

$$\text{Percent interest rate} = \frac{\$700}{\$10,000} \times 100\% = 7\% \text{ per year}$$

EXAMPLE 1.2   **a.** Calculate the amount deposited 1 year ago to have $1000 now at an interest rate of 5% per year.
b. Calculate the amount of interest earned during this time period.

Solution

a. The total amount accrued ($1000) is the sum of the original deposit and the earned interest. If X is the original deposit,

Total accrued = original amount + original amount (interest rate)
$$\$1000 = X + X(0.05) = X(1 + 0.05) = 1.05X$$

The original deposit is

$$X = \frac{1000}{1.05} = \$952.38$$

b. Apply Equation [1.1] to determine interest earned.

Interest = $1000 − 952.38 = $47.62

    In Examples 1.1 and 1.2 the interest period was 1 year, and the interest amount was calculated at the end of one period. When more than one interest period is involved (e.g., if we wanted the amount of interest earned after 3 years in Example 1.2), it is necessary to state whether the interest is accrued on a *simple* or *compound* basis from one period to the next. Simple and compound interest will be discussed in Section 1.5.

    Engineering alternatives are evaluated upon the prognosis that a reasonable rate of return (ROR) can be realized. A reasonable rate must be established so that the accept/reject decision can be made. This reasonable rate, called the *minimum attractive rate of return* (MARR), is the lowest interest rate that will induce companies or individuals to invest their money. The MARR must be higher than the cost of money used to finance the alternative, as well as higher than the rate that would be expected from a bank or safe (minimal risk) investment. Figure 1.1 indicates the relations between different rates of return. In the United States, the current U.S. Treasury bill rate of return is sometimes used as the benchmark safe rate.

    For a corporation, the MARR is always set above its *cost of capital,* which is the interest rate a company must pay for capital funds needed to finance projects.

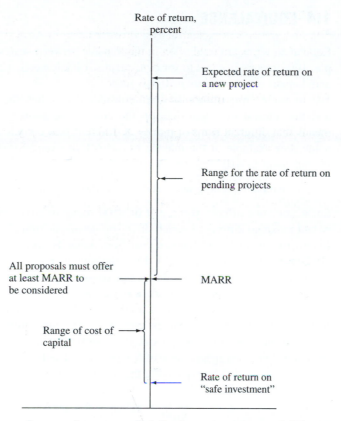

For example, if a corporation can borrow capital funds at an average of 5% per year and expects to clear at least 6% per year on a project, the MARR will be at least 11% per year.

The MARR is also referred to as the *hurdle rate;* that is, a financially viable project's expected ROR must meet or exceed the hurdle rate. Note that the MARR is not a rate calculated like the ROR; MARR is established by financial managers and is used as a criterion for accept/reject decisions. The following inequality must be correct for any accepted project.

$$\text{ROR} \geq \text{MARR} > \text{cost of capital}$$

Descriptions and problems in the following chapters use stated MARR values with the assumption that they are set correctly relative to the cost of capital and the expected rate of return. If more understanding of capital funds and the establishment of the MARR is required, refer to Section 13.5 for further detail.

An additional economic consideration for any engineering economy study is *inflation*. In simple terms, bank interest rates reflect two things: a so-called real rate of return *plus* the expected inflation rate. The safest investments (such as government bonds) typically have a 3% to 4% real rate of return built into their overall interest rates. Thus, an interest rate of, say, 9% per year on a government bond means that investors expect the inflation rate to be in the range of 5% to 6% per year. Clearly, then, inflation causes interest rates to rise. Inflation is discussed in detail in Chapter 10.

1.4 EQUIVALENCE

Podcast available

Equivalent terms are used often in the transfer between scales and units. For example, 1000 meters is equal to (or equivalent to) 1 kilometer, 12 inches equals 1 foot, and 1 quart equals 2 pints or 0.946 liter.

In engineering economy, when considered together, the time value of money and the interest rate help develop the concept of *economic equivalence,* which means that different sums of money at different times would be equal in economic value. For example, if the interest rate is 6% *per year,* $100 today (present time) is equivalent to $106 one year from today.

$$\text{Amount in one year} = 100 + 100(0.06) = 100(1 + 0.06) = \$106$$

So, if someone offered you a gift of $100 today or $106 one year from today, it would make no difference which offer you accepted from an economic perspective. In either case you have $106 one year from today. However, the two sums of money are equivalent to each other *only* when the interest rate is 6% per year. At a higher or lower interest rate, $100 today is not equivalent to $106 one year from today.

In addition to future equivalence, we can apply the same logic to determine equivalence for previous years. A total of $100 now is equivalent to $100/1.06 = $94.34 one year ago at an interest rate of 6% per year. From these illustrations, we can state the following: $94.34 last year, $100 now, and $106 one year from now are equivalent at an interest rate of 6% per year. The fact that these sums are equivalent can be verified by computing the two interest rates for 1-year interest periods.

$$\frac{\$6}{\$100} \times 100\% = 6\% \text{ per year}$$

and

$$\frac{\$5.66}{\$94.34} \times 100\% = 6\% \text{ per year}$$

Figure 1.2 indicates the amount of interest each year necessary to make these three different amounts equivalent at 6% per year.

FIGURE 1.2
Equivalence of three amounts at a 6% per year interest rate.

6% per year interest rate

$94.34 $5.66 $6.00

$100.00 $6.00

$106.00

1 year ago   Now   1 year from now

EXAMPLE 1.3

AC-Delco makes auto batteries available to General Motors dealers through privately owned distributorships. In general, batteries are stored throughout the year, and a 5% cost increase is added each year to cover the inventory carrying charge for the distributorship owner. Assume you own the City Center Delco facility. Make the calculations necessary to show which of the following statements are true and which are false about battery costs.

a. The amount of $98 now is equivalent to a cost of $105.60 one year from now.
b. A truck battery cost of $200 one year ago is equivalent to $205 now.
c. A $38 cost now is equivalent to $39.90 one year from now.
d. A $3000 cost now is equivalent to $2887.14 one year ago.
e. The carrying charge accumulated in 1 year on an investment of $2000 worth of batteries is $100.

Solution
a. Total amount accrued $= 98(1.05) = \$102.90 \neq \$105.60$; therefore, it is false. Another way to solve this is as follows: Required original cost is $105.60/1.05 = \$100.57 \neq \$98$.
b. Required old cost is $205.00/1.05 = \$195.24 \neq \$200$; therefore, it is false.
c. The cost 1 year from now is $\$38(1.05) = \$39.90$; true.
d. Cost one year ago is $3000/1.05 = \$2857.14 \neq 2887.14$; false.
e. The charge is 5% per year interest, or $\$2000(0.05) = \$100$; true.

1.5 SIMPLE AND COMPOUND INTEREST

The terms *interest, interest period,* and *interest rate* were introduced in Section 1.3 for calculating equivalent sums of money for one interest period in the past and one period in the future. However, for more than one interest period, the terms *simple interest* and *compound interest* become important.

Podcast available

Simple interest is calculated using the principal only, ignoring any interest accrued in preceding interest periods. The total simple interest over several periods is computed as

$$\textbf{Interest} = \textbf{(principal)(number of periods)(interest rate)} \qquad [1.3]$$

where the interest rate is expressed in decimal form. Therefore, the total (future) amount accumulated after several periods is the principal plus interest over all n periods.

EXAMPLE 1.4

HP borrowed money to do rapid prototyping for a new ruggedized computer that targets desert oilfield conditions. The loan is $1 million for 3 years at 5% per year simple interest. How much money will HP repay at the end of 3 years? Tabulate the results in $1000 units.

Solution

The interest for each of the 3 years in $1000 units is

Interest per year = 1000(0.05) = $50

Total interest for 3 years from Equation [1.3] is

Total interest = 1000(3)(0.05) = $150

The amount due after 3 years in $1000 units is

Total due = $1000 + 150 = $1150

The $50,000 interest accrued in the first year and the $50,000 accrued in the second year do not earn interest. The interest due each year is calculated only on the $1,000,000 principal.

The details of this loan repayment are tabulated in Table 1.1 from the perspective of the borrower. The year zero represents the present, that is, when the money is borrowed. No payment is made until the end of year 3. The amount owed each year increases uniformly by $50,000, since simple interest is figured on only the loan principal.

TABLE 1.1   Simple Interest Computations (in $1000 units)

(1) End of Year	(2) Amount Borrowed	(3) Interest	(4) Amount Owed	(5) Amount Paid
0	$1000			
1	—	$50	$1050	$ 0
2	—	50	1100	0
3	—	50	1150	1150

For *compound interest,* the interest accrued for each interest period is calculated on the *principal plus* the *total amount of interest accumulated in all previous periods.* Thus, compound interest means interest on top of interest. Compound interest reflects the effect of the time value of money on the interest also. Now the interest for one period is calculated as

Interest = (principal + all accrued interest)(interest rate)   **[1.4]**

EXAMPLE 1.5   If HP borrows $1,000,000 from a different source at 5% per year compound interest, compute the total amount due after 3 years. Compare the results of this and the previous example.

TABLE 1.2 **Compound Interest Computations (in $1000 units), Example 1.5**

(1) End of Year	(2) Amount Borrowed	(3) Interest	(4) Amount Owed	(5) Amount Paid
0	$1000			
1	—	$50.00	$1050.00	$ 0
2	—	52.50	1102.50	0
3	—	55.13	1157.63	1157.63

Solution

The interest and total amount due each year are computed separately using Equation [1.4]. In $1000 units,

$$\text{Year 1 interest:} \quad \$1000(0.05) = \$50.00$$
$$\text{Total amount due after year 1:} \quad \$1000 + 50.00 = \$1050.00$$
$$\text{Year 2 interest:} \quad \$1050(0.05) = \$52.50$$
$$\text{Total amount due after year 2:} \quad \$1050 + 52.50 = \$1102.50$$
$$\text{Year 3 interest:} \quad \$1102.50(0.05) = \$55.13$$
$$\text{Total amount due after year 3:} \quad \$1102.50 + 55.13 = \$1157.63$$

The details are shown in Table 1.2. The repayment plan is the same as that for the simple interest example—no payment until the principal plus accrued interest is due at the end of year 3. An extra $1,157,630 − 1,150,000 = $7,630 of interest is paid compared to simple interest over the 3-year period.

Comment: The difference between simple and compound interest grows significantly each year. If the computations are continued for more years, for example, 10 years, the difference is $128,894; after 20 years compound interest is $653,298 more than simple interest.

Another and shorter way to calculate the total amount due after 3 years in Example 1.5 is to combine calculations rather than perform them on a year-by-year basis. The total due each year is as follows:

$$\text{Year 1:} \quad \$1000(1.05)^1 = \$1050.00$$
$$\text{Year 2:} \quad \$1000(1.05)^2 = \$1102.50$$
$$\text{Year 3:} \quad \$1000(1.05)^3 = \$1157.63$$

The year 3 total is calculated directly; it does not require the year 2 total. In general formula form,

Total due after a number of years = principal (1 + interest rate)$^{\text{number of years}}$

This fundamental relation is used many times in upcoming chapters.

We combine the concepts of interest rate, simple interest, compound interest, and equivalence to demonstrate that different loan repayment plans may be equivalent, but they may differ substantially in monetary amounts from one year to another. This also shows that there are many ways to take into account the time value of money. The following example illustrates equivalence for five different loan repayment plans.

EXAMPLE 1.6

a. Demonstrate the concept of equivalence using the different loan repayment plans described below. Each plan repays a $5000 loan in 5 years at 8% interest per year.
  - **Plan 1: Simple interest, pay all at end.** No interest or principal is paid until the end of year 5. Interest accumulates each year on the *principal only*.
  - **Plan 2: Compound interest, pay all at end.** No interest or principal is paid until the end of year 5. Interest accumulates each year on the total of principal *and* all accrued interest.
  - **Plan 3: Simple interest paid annually, principal repaid at end.** The accrued interest is paid each year, and the entire principal is repaid at the end of year 5.
  - **Plan 4: Compound interest and portion of principal repaid annually.** The accrued interest and one-fifth of the principal (or $1000) is repaid each year. The outstanding loan balance decreases each year, so the interest for each year decreases.
  - **Plan 5: Equal payments of compound interest and principal made annually.** Equal payments are made each year with a portion going toward principal repayment and the remainder covering the accrued interest. Since the loan balance decreases at a rate slower than that in plan 4 due to the equal end-of-year payments, the interest decreases, but at a slower rate.

b. Make a statement about the equivalence of each plan at 8% simple or compound interest, as appropriate.

Solution

a. Table 1.3 presents the interest, payment amount, total owed at the end of each year, and total amount paid over the 5-year period (column 4 totals). The amounts of interest (column 2) are determined as follows:

  Plan 1 Simple interest = (original principal)(0.08)
  Plan 2 Compound interest = (total owed previous year)(0.08)
  Plan 3 Simple interest = (original principal)(0.08)
  Plan 4 Compound interest = (total owed previous year)(0.08)
  Plan 5 Compound interest = (total owed previous year)(0.08)

Note that the amounts of the annual payments are different for each repayment schedule and that the total amounts repaid for most plans are different, even though each repayment plan requires exactly 5 years. The difference in the total amounts repaid can be explained (1) by the time value of money, (2) by simple or compound interest, and (3) by the partial repayment of principal prior to year 5.

TABLE 1.3 Different Repayment Schedules Over 5 Years for $5000 at 8% Per Year Interest

(1) End of Year	(2) Interest Owed for Year	(3) Total Owed at End of Year	(4) End-of-Year Payment	(5) Total Owed after Payment
Plan 1: Simple Interest, Pay All at End				
0				$5000.00
1	$400.00	$5400.00	—	5400.00
2	400.00	5800.00	—	5800.00
3	400.00	6200.00	—	6200.00
4	400.00	6600.00	—	6600.00
5	400.00	7000.00	$7000.00	
Totals			$7000.00	
Plan 2: Compound Interest, Pay All at End				
0				$5000.00
1	$400.00	$5400.00	—	5400.00
2	432.00	5832.00	—	5832.00
3	466.56	6298.56	—	6298.56
4	503.88	6802.44	—	6802.44
5	544.20	7346.64	$7346.64	
Totals			$7346.64	
Plan 3: Simple Interest Paid Annually; Principal Repaid at End				
0				$5000.00
1	$400.00	$5400.00	$ 400.00	5000.00
2	400.00	5400.00	400.00	5000.00
3	400.00	5400.00	400.00	5000.00
4	400.00	5400.00	400.00	5000.00
5	400.00	5400.00	5400.00	
Totals			$7000.00	
Plan 4: Compound Interest and Portion of Principal Repaid Annually				
0				$5000.00
1	$400.00	$5400.00	$1400.00	4000.00
2	320.00	4320.00	1320.00	3000.00
3	240.00	3240.00	1240.00	2000.00
4	160.00	2160.00	1160.00	1000.00
5	80.00	1080.00	1080.00	
Totals			$6200.00	

TABLE 1.3 (Continued)

(1) End of Year	(2) Interest Owed for Year	(3) Total Owed at End of Year	(4) End-of-Year Payment	(5) Total Owed after Payment
Plan 5: Equal Annual Payments of Compound Interest and Principal				
0				$5000.00
1	$400.00	$5400.00	$1252.28	4147.72
2	331.82	4479.54	1252.28	3227.25
3	258.18	3485.43	1252.28	2233.15
4	178.65	2411.80	1252.28	1159.52
5	92.76	1252.28	1252.28	
Totals			$6261.41	

b. Table 1.3 shows that $5000 at time 0 is equivalent to each of the following:

Plan 1 $7000 at the end of year 5 at 8% simple interest.

Plan 2 $7346.64 at the end of year 5 at 8% compound interest.

Plan 3 $400 per year for 4 years and $5400 at the end of year 5 at 8% simple interest.

Plan 4   Decreasing payments of interest and partial principal in years 1 ($1400) through 5 ($1080) at 8% compound interest.

Plan 5 $1252.28 per year for 5 years at 8% compound interest.

Beginning in Chapter 2, we will make many calculations like plan 5, where interest is compounded and a constant amount is paid each period. This amount covers accrued interest and a partial principal repayment.

1.6 TERMINOLOGY AND SYMBOLS

The equations and procedures of engineering economy utilize the following terms and symbols. Sample units are indicated.

P = value or amount of money at a time designated as the present or time 0. Also, P is referred to as present worth (PW), present value (PV), net present value (NPV), discounted cash flow (DCF), and capitalized cost (CC); dollars

F = value or amount of money at some future time. Also, F is called future worth (FW) and future value (FV); dollars

A = series of consecutive, equal, end-of-period amounts of money. A is also called the annual worth (AW), equivalent uniform annual worth (EUAW), and equivalent annual cost (EAC); dollars per year, dollars per month

n = number of interest periods; years, months, days

i = interest rate or rate of return per time period; percent per year, percent per month, percent per day

t = time, stated in periods; years, months, days

The symbols P and F represent one-time occurrences: A occurs with the same value each interest period for a specified number of periods. It should be clear that a present value P represents a single sum of money at some time prior to a future value F or prior to the first occurrence of an equivalent series amount A.

It is important to note that the symbol A always represents a uniform amount (i.e., the same amount each period) that extends through *consecutive* interest periods. Both conditions must exist before the series can be represented by A.

The interest rate i is assumed to be a compound rate, unless specifically stated as simple interest. The rate i is expressed in percent per interest period, for example, 12% per year. Unless stated otherwise, assume that the rate applies throughout the entire n years or interest periods. The decimal equivalent for i is always used in engineering economy computations.

All engineering economy problems involve the element of time n and interest rate i. In general, every problem will involve at least four of the symbols P, F, A, n, and i, with at least three of them estimated or known.

A new college graduate has a job with Boeing Aerospace. She plans to borrow $10,000 now to help in buying a car. She has arranged to repay the entire principal plus 8% per year interest after 5 years. Identify the engineering economy symbols involved and their values for the total owed after 5 years.   **EXAMPLE 1.7**

Solution

In this case, P and F are involved, since all amounts are single payments, as well as n and i. Time is expressed in years.

$P = \$10,000 \qquad i = 8\% \text{ per year} \qquad n = 5 \text{ years} \qquad F = ?$

The future amount F is unknown.

Assume you borrow $2000 now at 7% per year for 10 years and must repay the loan in equal yearly payments. Determine the symbols involved and their values.   **EXAMPLE 1.8**

Solution

Time is in years.

$P = \$2000$

$A = ?$ per year for 10 years

$i = 7\%$ per year

$n = 10$ years

In Examples 1.7 and 1.8, the P value is a receipt *to* the borrower, and F or A is a disbursement *from* the borrower. It is equally correct to use these symbols in the reverse roles.

EXAMPLE 1.9   On July 1, 2008, your new employer Ford Motor Company deposits $5000 into your money market account, as part of your employment bonus. The account pays interest at 5% per year. You expect to withdraw an equal annual amount each year for the following 10 years. Identify the symbols and their values.

Solution

Time is in years.

$$P = \$5000$$
$$A = ? \text{ per year}$$
$$i = 5\% \text{ per year}$$
$$n = 10 \text{ years}$$

EXAMPLE 1.10   You plan to make a lump-sum deposit of $5000 now into an investment account that pays 6% per year, and you plan to withdraw an equal end-of-year amount of $1000 for 5 years, starting next year. At the end of the sixth year, you plan to close your account by withdrawing the remaining money. Define the engineering economy symbols involved.

Solution

Time is expressed in years.

$$P = \$5000$$
$$A = \$1000 \text{ per year for 5 years}$$
$$F = ? \text{ at end of year 6}$$
$$i = 6\% \text{ per year}$$
$$n = 5 \text{ years for the } A \text{ series and 6 for the } F \text{ value}$$

1.7 CASH FLOWS: THEIR ESTIMATION AND DIAGRAMMING

Cash flows are inflows and outflows of money. These cash flows may be estimates or observed values. Every person or company has cash receipts—revenue and income (inflows); and cash disbursements—expenses and costs (outflows). These receipts and disbursements are the cash flows, with a plus sign representing cash inflows and a minus sign representing cash outflows. Cash flows occur during specified periods of time, such as 1 month or 1 year.

Of all the elements of an engineering economy study, cash flow estimation is likely the most difficult and inexact. Cash flow estimates are just that—estimates about an uncertain future. Once estimated, the techniques of this book guide the decision-making process. But the time-proven accuracy of an alternative's estimated cash inflows and outflows clearly dictates the quality of the economic analysis and conclusion.

Cash inflows, or receipts, may be comprised of the following, depending upon the nature of the proposed activity and the type of business involved.

Samples of Cash Inflow Estimates
Revenues (from sales and contracts)
Operating cost reductions (resulting from an alternative)
Salvage value
Construction and facility cost savings
Receipt of loan principal
Income tax savings
Receipts from stock and bond sales

Cash outflows, or disbursements, may be comprised of the following, again depending upon the nature of the activity and type of business.

Samples of Cash Outflow Estimates
First cost of assets
Engineering design costs
Operating costs (annual and incremental)
Periodic maintenance and rebuild costs
Loan interest and principal payments
Major expected/unexpected upgrade costs
Income taxes

Background information for estimates may be available in departments such as accounting, finance, marketing, sales, engineering, design, manufacturing, production, field services, and computer services. The accuracy of estimates is largely dependent upon the experiences of the person making the estimate with similar situations. Usually *point estimates* are made; that is, a single-value estimate is developed for each economic element of an alternative. If a statistical approach to the engineering economy study is undertaken, a *range estimate* or *distribution estimate* may be developed. Though more involved computationally, a statistical study provides more complete results when key estimates are expected to vary widely. Though we use point estimates throughout most of this book, Chapter 8 discusses sensitivity analysis for estimates that vary over a specific range. Additionally, Chapter 14 introduces risk analysis using probability distributions, sampling, and spreadsheet-based simulation to understand the economic consequences of estimate variation.

Once the cash inflow and outflow estimates are developed, the net cash flow can be determined.

$$\textbf{Net cash flow = receipts} - \textbf{disbursements} \qquad \textbf{[1.5]}$$
$$= \textbf{cash inflows} - \textbf{cash outflows}$$

Since cash flows normally take place at varying times within an interest period, a simplifying end-of-period assumption is made.

FIGURE 1.3
A typical cash flow
time scale for 5 years.

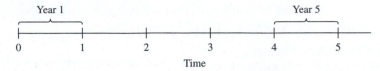

The *end-of-period convention* means that all cash flows are assumed to occur at the end of an interest period. When several receipts and disbursements occur within a given interest period, the *net* cash flow is assumed to occur at the *end* of the interest period.

It is important to understand that, although F or A amounts are located at the end of the interest period by convention, the end of the period is not necessarily December 31. In Example 1.9 the deposit took place on July 1, 2008, and the withdrawals will take place on July 1 of each succeeding year for 10 years. *Thus, end of the period means end of interest period, not end of calendar year.*

The *cash flow diagram* is a very important tool in an economic analysis, especially when the cash flow series is complex. It is a graphical representation of cash flows drawn on a time scale. The diagram includes what is known, what is estimated, and what is needed. That is, once the cash flow diagram is complete, another person should be able to work the problem by looking at the diagram.

Cash flow diagram time $t = 0$ is the present, and $t = 1$ is the end of time period 1. We assume that the periods are in years for now. The time scale of Figure 1.3 is set up for 5 years. Since the end-of-year convention places cash flows at the ends of years, the "1" marks the end of year 1.

While it is not necessary to use an exact scale on the cash flow diagram, you will probably avoid errors if you make a neat diagram to approximate scale for both time and relative cash flow magnitudes.

The direction of the arrows on the cash flow diagram is important. A vertical arrow pointing up indicates a positive cash flow. Conversely, an arrow pointing down indicates a negative cash flow. Figure 1.4 illustrates a receipt (cash inflow) at the end of year 1 and equal disbursements (cash outflows) at the end of years 2 and 3.

The perspective or vantage point must be determined prior to placing a sign on each cash flow and diagramming it. As an illustration, if you borrow $2500 to buy a $2000 used Harley-Davidson for cash, and you use the remaining $500 for a new paint job, there may be several different perspectives taken. Possible perspectives, cash flow signs, and amounts are as follows.

Perspective	Cash Flow, $
Credit union	−2500
You as borrower	+2500
You as purchaser,	−2000
and as paint customer	−500
Used cycle dealer	+2000
Paint shop owner	+500

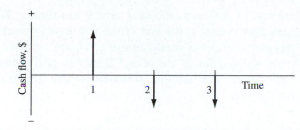

FIGURE 1.4
Example of positive
and negative cash
flows.

EXAMPLE 1.11

Reread Example 1.7, where $P = \$10,000$ is borrowed at 8% per year and F is sought after 5 years. Construct the cash flow diagram.

Solution

Figure 1.5 presents the cash flow diagram from the vantage point of the borrower. The present sum P is a cash inflow of the loan principal at year 0, and the future sum F is the cash outflow of the repayment at the end of year 5. The interest rate should be indicated on the diagram.

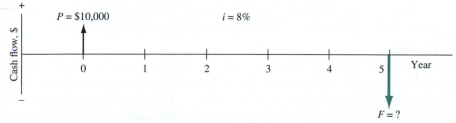

FIGURE 1.5 Cash flow diagram, Example 1.11.

EXAMPLE 1.12

Each year Exxon-Mobil expends large amounts of funds for mechanical safety features throughout its worldwide operations. Carla Ramos, a lead engineer for Mexico and Central American operations, plans expenditures of $1 million now and each of the next 4 years just for the improvement of field-based pressure-release valves. Construct the cash flow diagram to find the equivalent value of these expenditures at the end of year 4, using a cost of capital estimate for safety-related funds of 12% per year.

Solution

Figure 1.6 indicates the uniform and negative cash flow series (expenditures) for five periods, and the unknown F value (positive cash flow equivalent) at exactly the same time as the fifth expenditure. Since the expenditures start immediately, the first $1 million is shown at time 0, not time 1. Therefore, the last negative cash flow occurs at the end of the fourth year, when F also occurs. To make this diagram appear similar to that of Figure 1.5 with a full 5 years on the time scale, the addition of the year -1 prior to year 0 completes the diagram for a full 5 years. This addition demonstrates that year 0 is the end-of-period point for the year -1.

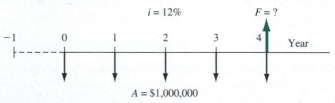

FIGURE 1.6   Cash flow diagram, Example 1.12.

EXAMPLE 1.13   A father wants to deposit an unknown lump-sum amount into an investment opportunity 2 years from now that is large enough to withdraw $4000 per year for state university tuition for 5 years starting 3 years from now. If the rate of return is estimated to be 15.5% per year, construct the cash flow diagram.

Solution

Figure 1.7 presents the cash flows from the father's perspective. The present value P is a cash outflow 2 years hence and is to be determined ($P = ?$). Note that this present value does not occur at time $t = 0$, but it does occur one period prior to the first A value of $4000, which is the cash inflow to the father.

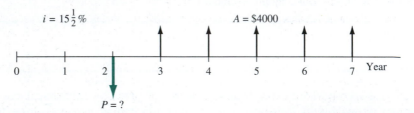

FIGURE 1.7   Cash flow diagram, Example 1.13.

1.8 INTRODUCTION TO SPREADSHEET AND CALCULATOR FUNCTIONS

There are four different ways to perform an equivalence calculation: equation, factor, spreadsheet, and financial calculator. The first two methods are discussed and illustrated in Chapter 2. Spreadsheet and calculator functions are briefly outlined in this section and utilized throughout the text in conjunction with factors and equations.

The functions on a computer spreadsheet can greatly reduce the amount of hand and calculator work for equivalency computations involving *compound interest* and the terms P, F, A, i, and n. Often a predefined function can be entered into one cell and we can obtain the final answer immediately. Any spreadsheet system can be used; Excel is used throughout this book because it is readily available and easy to use.

Appendix A is a primer on using spreadsheets and Excel. The functions used in engineering economy are described there in detail, with explanations of all the parameters (also called arguments) placed between parentheses after the function identifier. The online help system provides similar information. Appendix A also includes a section on spreadsheet layout that is useful when the economic analysis is presented to someone else—a coworker, a boss, or a professor.

A total of seven spreadsheet functions can perform most of the fundamental engineering economy calculations. However, these functions are no substitute for knowing how the time value of money and compound interest work. The functions are great supplemental tools, but they do not replace the understanding of engineering economy relations, assumptions, and techniques.

Using the symbols P, F, A, i, and n exactly as defined in Section 1.6, the spreadsheet functions most used in engineering economic analysis are formulated as follows.

To find the present value P of an A series: $= \text{PV}(i\%,n,A,F)$

To find the future value F of an A series: $= \text{FV}(i\%,n,A,P)$

To find the equal, periodic value A: $= \text{PMT}(i\%,n,P,F)$

To find the number of periods n: $= \text{NPER}(i\%,A,P,F)$

To find the compound interest rate i: $= \text{RATE}(n,A,P,F)$

To find the compound interest rate i: $= \text{IRR}(\text{first\_cell:last\_cell})$

To find the present value P of any series: $= \text{NPV}(i\%, \text{second\_cell:last\_cell}) + \text{first\_cell}$

If some of the parameters don't apply to a particular problem, they can be omitted and zero is assumed. If the parameter omitted is an interior one, the comma must be entered. The last two functions require that a series of numbers be entered into adjacent spreadsheet cells, but the first five can be used with no supporting data. In all cases, the function must be preceded by an equals sign ($=$) in the cell where the answer is to be displayed.

Most *calculators* that have financial functions include time value of money (TVM) functions. These are very useful for in-class and exam work for cash flows that are not too complex. On many calculators there are keys labeled PV, FV, PMT, i and n, or similar terms. When four of the five values are entered, the remaining parameter is calculated by pressing the appropriate key. The procedures to input the four known values and to obtain the fifth value vary slightly between calculators manufactured by different companies. For our notation purposes, each calculator function is identified in a manner similar to the first five spreadsheet functions, without an = sign. The interest rate is entered as a percent on most calculators. The notation is as follows:

Find present value P of an A series: $PV(i,n,A,F)$

Find future value F of an A series: $FV(i,n,A,P)$

Find A series of P and/or F values: $PMT(i,n,P,F)$

Find number of periods n: $n(i,A,P,F)$

Find compound interest rate i: $i(n,A,P,F)$

Some calculators use different letters than spreadsheet functions and those used in this text. For example, the letter P may represent a uniform series that we label A. Others use B to represent a present worth amount, which is P in this text. Check the user's manual to be sure you understand how to correctly apply a TVM function.

Several spreadsheet and financial calculator functions have optional parameters that can be entered as the last argument in the parenthesis. One example is "type," which identifies the cash flows as occurring at the beginning or end of each period. The default is end-of-period (type = 0); however, many finance applications require beginning-of-period cash flows, thus requiring an entry of type = 1.

There are no built-in functions for spreadsheets or calculators that accommodate *gradients* where the cash flows increase or decrease by a constant amount or constant percentage each time period. As we will learn in Chapter 2, equivalence computations involving gradients are best handled using a spreadsheet, first, with tabulated factors or equivalence equations as alternatives.

Spreadsheet and calculator functions will be introduced and illustrated at the point in this text where they are most useful. However, to get an idea of how they work, look back at Examples 1.7 and 1.8. In Example 1.7, the future amount F is unknown, as indicated by $F = ?$ in the solution. In Chapter 2, we will learn how the time value of money is used to find F, given P, i, and n. To find F in this example using a calculator or spreadsheet, simply enter the FV function. The

FIGURE 1.8

Use of spreadsheet functions for (a) Example 1.7 and (b) Example 1.8.

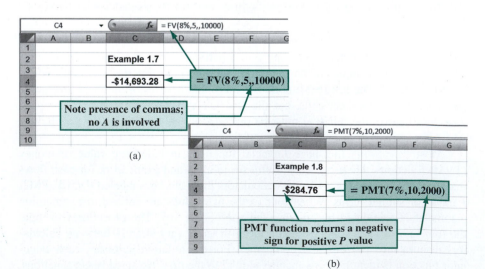

one-cell spreadsheet format is $=\text{FV}(i\%,n,,P)$ or $=\text{FV}(8\%,5,,10000)$. The extra comma is entered because there is no A involved. Figure 1.8a is a screen image of the spreadsheet with the FV function entered into cell C4. The answer of $\$-14{,}693.28$ is displayed. The answer is a negative amount from the borrower's perspective to repay the loan after 5 years. The FV function is shown in the formula bar above the worksheet, and a cell tag shows the format of the FV function.

In Example 1.8, the uniform annual amount A is sought, and P, i, and n are known. Find A using the function $=\text{PMT}(7\%,10,2000)$. Figure 1.8b shows the result.

1.9 ETHICS AND ECONOMIC DECISIONS

Initially, you may not think that engineering economic decisions are closely allied with ethics. However, economics and ethically good or poor actions can be closely connected for some individuals who work for corporations, who own their own businesses, and who serve the public in appointed or elected government positions. The fundamental cause for ethically bad decisions is commonly *money*, in the form of increased profit, lower costs, favors, or kickbacks.

Let's begin an examination of the connections between economics and ethics through an understanding of the underlying tenets that form the character and conduct of a person when judging right from wrong. These tenets are called *morals;* there are morals for society in general and for one individual. The term *ethics* is used when examining decisions and actions based on morals that can be evaluated using a measure, such as a *code of ethics*. This code forms the standards to guide a person's decisions and actions in a particular profession, for example, engineering, medicine, law, education, etc.

Universal or common morals: Fundamental moral beliefs that are held by virtually all people. Most people believe that injurious actions taken against another person are morally wrong. These include lying, stealing, physical harm, and murder.

It is possible for actions and intentions to come into conflict with a common moral. Consider the World Trade Center buildings in New York City. After their collapse on September 11, 2001, it was apparent that the design was not sufficient to withstand the heat generated in the fire storm caused by the impact of an aircraft. The structural engineers who worked on the building's design surely did not have the intent to harm occupants; however, their design did not foresee this outcome as a measurable possibility. Did they violate the common moral belief of not doing physical harm to or murdering others? Likely, they did not.

Individual or personal morals: The moral beliefs that a person has and maintains over time. These usually parallel the common morals in that stealing, lying, murdering, etc. are immoral acts.

It is quite possible that an individual strongly supports the common morals and has excellent personal morals, but these can conflict when tough decisions must be made. Consider the university student who genuinely believes that lying in the form of cheating on an exam is wrong. Assume he or she is in the last semester before graduation and does not know how to work several problems on a take-home final exam, yet a minimum grade on the exam is necessary to graduate. The decision to cheat or not on the final exam by copying from a friend is an exercise in adhering to or violating a personal moral. Likewise, if to cheat is the decision, the friend who provides the answers also has to make the decision to adhere to or violate his or her own personal moral about cheating.

Professional ethics: People working in specific disciplines are usually guided in their decision making and work activities by a formal standard or code. The code states the commonly accepted standards of honesty and integrity that each individual is expected to demonstrate. There are codes of ethics for medical doctors, attorneys, elected officials, and, of course, engineers.

Though each engineering profession has its own code of ethics, the *Code of Ethics for Engineers* published by the National Society of Professional Engineers (www.nspe.org/ethics) is commonly applied. This code, reprinted in Figure 1.9, includes numerous sections that have direct or indirect economic and financial impact upon the designs, actions, and decisions that engineers make in their professional activities. Some examples are:

- Section I.5: Avoid deceptive acts.
- Section III.3.a: Don't make statements that contain misrepresentation of fact or that omit material facts.
- Section III.5.a: Do not accept financial consideration from suppliers for specifying their product.

In the everyday practice of engineering, there are numerous money-related situations that can involve ethical dimensions. Here are a few samples:

- Compromising safety in the design of a product or service so that the price comes in low enough to win a contract when the bids are evaluated.
- Delaying regularly scheduled equipment maintenance to save money, thus placing the safety of people and product in jeopardy.
- Substituting cheaper, poorer-quality materials (chemicals, parts, etc.) to reduce cost, resulting in increased profit, and violating company health standards for employees and customers.
- Mislabeling to the public in that a much cheaper, similar-functioning product is substituted while the price charged is that of the product identified. (Example 1.14 is an illustration of this type of unethical practice.)

Many ethical questions arise when corporations operate in international settings where the corporate rules, worker incentives, cultural practices, and costs in the home country differ from those in the host country. Often these ethical dilemmas are fundamentally based in economics which provide cheaper labor, reduced raw material costs, less government oversight, and a host of other cost-reducing factors. When an engineering economy study is performed, the engineer performing the study must consider all ethically related matters to ensure that cost estimates reflect what is likely to happen once the project or system is operating.

National Society of Professional Engineers®

Code of Ethics for Engineers

Preamble

Engineering is an important and learned profession. As members of this profession, engineers are expected to exhibit the highest standards of honesty and integrity. Engineering has a direct and vital impact on the quality of life for all people. Accordingly, the services provided by engineers require honesty, impartiality, fairness, and equity, and must be dedicated to the protection of the public health, safety, and welfare. Engineers must perform under a standard of professional behavior that requires adherence to the highest principles of ethical conduct.

I. Fundamental Canons

Engineers, in the fulfillment of their professional duties, shall:
1. Hold paramount the safety, health, and welfare of the public.
2. Perform services only in areas of their competence.
3. Issue public statements only in an objective and truthful manner.
4. Act for each employer or client as faithful agents or trustees.
5. Avoid deceptive acts.
6. Conduct themselves honorably, responsibly, ethically, and lawfully so as to enhance the honor, reputation, and usefulness of the profession.

II. Rules of Practice

1. Engineers shall hold paramount the safety, health, and welfare of the public.
    a. If engineers' judgment is overruled under circumstances that endanger life or property, they shall notify their employer or client and such other authority as may be appropriate.
    b. Engineers shall approve only those engineering documents that are in conformity with applicable standards.
    c. Engineers shall not reveal facts, data, or information without the prior consent of the client or employer except as authorized or required by law or this Code.
    d. Engineers shall not permit the use of their name or associate in business ventures with any person or firm that they believe is engaged in fraudulent or dishonest enterprise.
    e. Engineers shall not aid or abet the unlawful practice of engineering by a person or firm.
    f. Engineers having knowledge of any alleged violation of this Code shall report thereon to appropriate professional bodies and, when relevant, also to public authorities, and cooperate with the proper authorities in furnishing such information or assistance as may be required.
2. Engineers shall perform services only in the areas of their competence.
    a. Engineers shall undertake assignments only when qualified by education or experience in the specific technical fields involved.
    b. Engineers shall not affix their signatures to any plans or documents dealing with subject matter in which they lack competence, nor to any plan or document not prepared under their direction and control.
    c. Engineers may accept assignments and assume responsibility for coordination of an entire project and sign and seal the engineering documents for the entire project, provided that each technical segment is signed and sealed only by the qualified engineers who prepared the segment.
3. Engineers shall issue public statements only in an objective and truthful manner.
    a. Engineers shall be objective and truthful in professional reports, statements, or testimony. They shall include all relevant and pertinent information in such reports, statements, or testimony, which should bear the date indicating when it was current.
    b. Engineers may express publicly technical opinions that are founded upon knowledge of the facts and competence in the subject matter.
    c. Engineers shall issue no statements, criticisms, or arguments on technical matters that are inspired or paid for by interested parties, unless they have prefaced their comments by explicitly identifying the interested parties on whose behalf they are speaking, and by revealing the existence of any interest the engineers may have in the matters.

4. Engineers shall act for each employer or client as faithful agents or trustees.
    a. Engineers shall disclose all known or potential conflicts of interest that could influence or appear to influence their judgment or the quality of their services.
    b. Engineers shall not accept compensation, financial or otherwise, from more than one party for services on the same project, or for services pertaining to the same project, unless the circumstances are fully disclosed and agreed to by all interested parties.
    c. Engineers shall not solicit or accept financial or other valuable consideration, directly or indirectly, from outside agents in connection with the work for which they are responsible.
    d. Engineers in public service as members, advisors, or employees of a governmental or quasi-governmental body or department shall not participate in decisions with respect to services solicited or provided by them or their organizations in private or public engineering practice.
    e. Engineers shall not solicit or accept a contract from a governmental body on which a principal or officer of their organization serves as a member.
5. Engineers shall avoid deceptive acts.
    a. Engineers shall not falsify their qualifications or permit misrepresentation of their or their associates' qualifications. They shall not misrepresent or exaggerate their responsibility in or for the subject matter of prior assignments. Brochures or other presentations incident to the solicitation of employment shall not misrepresent pertinent facts concerning employers, employees, associates, joint venturers, or past accomplishments.
    b. Engineers shall not offer, give, solicit, or receive, either directly or indirectly, any contribution to influence the award of a contract by public authority, or which may be reasonably construed by the public as having the effect or intent of influencing the awarding of a contract. They shall not offer any gift or other valuable consideration in order to secure work. They shall not pay a commission, percentage, or brokerage fee in order to secure work, except to a bona fide employee or bona fide established commercial or marketing agencies retained by them.

III. Professional Obligations

1. Engineers shall be guided in all their relations by the highest standards of honesty and integrity.
    a. Engineers shall acknowledge their errors and shall not distort or alter the facts.
    b. Engineers shall advise their clients or employers when they believe a project will not be successful.
    c. Engineers shall not accept outside employment to the detriment of their regular work or interest. Before accepting any outside engineering employment, they will notify their employers.
    d. Engineers shall not attempt to attract an engineer from another employer by false or misleading pretenses.
    e. Engineers shall not promote their own interest at the expense of the dignity and integrity of the profession.
2. Engineers shall at all times strive to serve the public interest.
    a. Engineers are encouraged to participate in civic affairs; career guidance for youths; and work for the advancement of the safety, health, and well-being of their community.
    b. Engineers shall not complete, sign, or seal plans and/or specifications that are not in conformity with applicable engineering standards. If the client or employer insists on such unprofessional conduct, they shall notify the proper authorities and withdraw from further service on the project.
    c. Engineers are encouraged to extend public knowledge and appreciation of engineering and its achievements.
    d. Engineers are encouraged to adhere to the principles of sustainable development[1] in order to protect the environment for future generations.

FIGURE 1.9 *Code of Ethics for Engineers* from the National Society of Professional Engineers

3. Engineers shall avoid all conduct or practice that deceives the public.
   a. Engineers shall avoid the use of statements containing a material misrepresentation of fact or omitting a material fact.
   b. Consistent with the foregoing, engineers may advertise for recruitment of personnel.
   c. Consistent with the foregoing, engineers may prepare articles for the lay or technical press, but such articles shall not imply credit to the author for work performed by others.
4. Engineers shall not disclose, without consent, confidential information concerning the business affairs or technical processes of any present or former client or employer, or public body on which they serve.
   a. Engineers shall not, without the consent of all interested parties, promote or arrange for new employment or practice in connection with a specific project for which the engineer has gained particular and specialized knowledge.
   b. Engineers shall not, without the consent of all interested parties, participate in or represent an adversary interest in connection with a specific project or proceeding in which the engineer has gained particular specialized knowledge on behalf of a former client or employer.
5. Engineers shall not be influenced in their professional duties by conflicting interests.
   a. Engineers shall not accept financial or other considerations, including free engineering designs, from material or equipment suppliers for specifying their product.
   b. Engineers shall not accept commissions or allowances, directly or indirectly, from contractors or other parties dealing with clients or employers of the engineer in connection with work for which the engineer is responsible.
6. Engineers shall not attempt to obtain employment or advancement or professional engagements by untruthfully criticizing other engineers, or by other improper or questionable methods.
   a. Engineers shall not request, propose, or accept a commission on a contingent basis under circumstances in which their judgment may be compromised.
   b. Engineers in salaried positions shall accept part-time engineering work only to the extent consistent with policies of the employer and in accordance with ethical considerations.
   c. Engineers shall not, without consent, use equipment, supplies, laboratory, or office facilities of an employer to carry on outside private practice.
7. Engineers shall not attempt to injure, maliciously or falsely, directly or indirectly, the professional reputation, prospects, practice, or employment of other engineers. Engineers who believe others are guilty of unethical or illegal practice shall present such information to the proper authority for action.
   a. Engineers in private practice shall not review the work of another engineer for the same client, except with the knowledge of such engineer, or unless the connection of such engineer with the work has been terminated.
   b. Engineers in governmental, industrial, or educational employ are entitled to review and evaluate the work of other engineers when so required by their employment duties.
   c. Engineers in sales or industrial employ are entitled to make engineering comparisons of represented products with products of other suppliers.
8. Engineers shall accept personal responsibility for their professional activities, provided, however, that engineers may seek indemnification for services arising out of their practice for other than gross negligence, where the engineer's interests cannot otherwise be protected.
   a. Engineers shall conform with state registration laws in the practice of engineering.
   b. Engineers shall not use association with a nonengineer, a corporation, or partnership as a "cloak" for unethical acts.

9. Engineers shall give credit for engineering work to those to whom credit is due, and will recognize the proprietary interests of others.
   a. Engineers shall, whenever possible, name the person or persons who may be individually responsible for designs, inventions, writings, or other accomplishments.
   b. Engineers using designs supplied by a client recognize that the designs remain the property of the client and may not be duplicated by the engineer for others without express permission.
   c. Engineers, before undertaking work for others in connection with which the engineer may make improvements, plans, designs, inventions, or other records that may justify copyrights or patents, should enter into a positive agreement regarding ownership.
   d. Engineers' designs, data, records, and notes referring exclusively to an employer's work are the employer's property. The employer should indemnify the engineer for use of the information for any purpose other than the original purpose.
   e. Engineers shall continue their professional development throughout their careers and should keep current in their specialty fields by engaging in professional practice, participating in continuing education courses, reading in the technical literature, and attending professional meetings and seminars.

Footnote 1 "Sustainable development" is the challenge of meeting human needs for natural resources, industrial products, energy, food, transportation, shelter, and effective waste management while conserving and protecting environmental quality and the natural resource base essential for future development.

As Revised July 2007

"By order of the United States District Court for the District of Columbia, former Section 11(c) of the NSPE Code of Ethics prohibiting competitive bidding, and all policy statements, opinions, rulings or other guidelines interpreting its scope, have been rescinded as unlawfully interfering with the legal right of engineers, protected under the antitrust laws, to provide price information to prospective clients; accordingly, nothing contained in the NSPE Code of Ethics, policy statements, opinions, rulings or other guidelines prohibits the submission of price quotations or competitive bids for engineering services at any time or in any amount."

Statement by NSPE Executive Committee
In order to correct misunderstandings which have been indicated in some instances since the issuance of the Supreme Court decision and the entry of the Final Judgment, it is noted that in its decision of April 25, 1978, the Supreme Court of the United States declared: "The Sherman Act does not require competitive bidding."

It is further noted that as made clear in the Supreme Court decision:
1. Engineers and firms may individually refuse to bid for engineering services.
2. Clients are not required to seek bids for engineering services.
3. Federal, state, and local laws governing procedures to procure engineering services are not affected, and remain in full force and effect.
4. State societies and local chapters are free to actively and aggressively seek legislation for professional selection and negotiation procedures by public agencies.
5. State registration board rules of professional conduct, including rules prohibiting competitive bidding for engineering services, are not affected and remain in full force and effect. State registration boards with authority to adopt rules of professional conduct may adopt rules governing procedures to obtain engineering services.
6. As noted by the Supreme Court, "nothing in the judgment prevents NSPE and its members from attempting to influence governmental action . . ."

Note: In regard to the question of application of the Code to corporations vis-a-vis real persons, business form or type should not negate nor influence conformance of individuals to the Code. The Code deals with professional services, which services must be performed by real persons. Real persons in turn establish and implement policies within business structures. The Code is clearly written to apply to the Engineer, and it is incumbent on members of NSPE to endeavor to live up to its provisions. This applies to all pertinent sections of the Code.

National Society of Professional Engineers®

1420 King Street
Alexandria, Virginia 22314-2794
703/684-2800 • Fax:703/836-4875
www.nspe.org
Publication date as revised: July 2007 • Publication #1102

FIGURE 1.9 *Code of Ethics for Engineers* from the National Society of Professional Engineers (continued)

In October 2011, *The Boston Globe* newspaper presented a series of articles and a video indicating that a large percentage of the fish served at restaurants in the Boston area were mislabeled. ("Seafood mislabeled at many local restaurants," LaPierre, Scott; Abelson, Jenn; and Daley, Beth; October 23, 2011; video at www.bostonglobe.com/business/specials/fish). The investigation clearly showed that the promise of larger profits, fueled by the scarcity of more expensive fish varieties, motivated restaurant owners and fish suppliers to substitute cheap fish varieties for similar looking and tasting, but more expensive fish listed on their menus. For example, catfish marketed by international suppliers were found to be purchased for $1.50 per pound, labeled as Atlantic cod, and sold for $12 per pound. DNA tests showed that some 48% of the samples from restaurants were mislabeled, all using significantly cheaper fish for more pricey varieties listed on the menu.

EXAMPLE 1.14

In one case, a restaurant owner knowingly advertised and charged the price of the expensive red snapper fish, but substituted ocean perch, a much cheaper fish. He stated that the red snapper was hard to find on the market, and that ocean perch cooks and tastes very similar to red snapper. His claim was that he wanted to offer his customers what they wanted, that is, red snapper. When interviewed, he stated that he did not even think of the substitution of perch for snapper as a price saver for him. In his mind, he was simply providing a way for customers to purchase what they wanted in fish variety. However, the price for perch averaged $4.25 per pound, while snapper cost about $8.95 a pound, when available. Yet, his menu did not state anything about the substitution or that snapper may be unavailable. Finally, when exposed, he said he would change his menu to be more truthful.

Use the NSPE Code of Ethics for Engineers as a model code to identify some of the ethically-questionable practices of the restaurant owner in mislabeling fish and selling them for the price of a more expensive variety. Describe some actions that the owner could have taken to demonstrate honestly and integrity in operating his business when the red snapper variety was unavailable.

Solution

As the NSPE Code of Ethics for Engineers is applied to this situation, consider the restaurant owner to be the engineer referenced in the code and the clients to be customers of the restaurant. Sample sections of the code that identify unethical practices are listed below; refer to the code sections for details of each violation (Figure 1.9).

Fundamental Canons: violation of *Sections I.5* and *I.6*

Rules of Practice: violation of *Section II.3.a*

Professional Obligations: violation of *Section III.3.a*

To ensure honesty and integrity in his business practices, the restaurant owner could have done several things. It could have been noted on the menu that red snapper is not always available and that substitution of a cheaper fish at the price of the snapper is never made by the restaurant. Since red snapper is one of the "big sellers," the owner could place a separate note in the menu when snapper is not available and mention that ocean perch is available, has the same taste and texture, and sells at a significantly lower price than snapper.

The biggest lesson for the owner is to realize that the "rationalization" of doing customers a service by deceptive acts is totally wrong and unethical. The owner's future practices need to be based on honesty in advertising and pricing for all of his menu items.

SUMMARY

Engineering economy is the application of economic factors to evaluate alternatives by considering the time value of money. The engineering economy study involves computing a specific economic measure of worth for estimated cash flows over a specific period of time.

The concept of *equivalence* helps in understanding how different sums of money at different times are equal in economic terms. The differences between simple interest (based on principal only) and compound interest (based on principal and interest upon interest) have been described in formulas, tables, and graphs.

The MARR is a reasonable rate of return established as a hurdle rate to determine if an alternative is economically viable. The MARR is always higher than the return from a safe investment and the corporation's cost of capital.

Also, this chapter introduced the estimation, conventions, and diagramming of cash flows.

PROBLEMS

Definitions and Basic Concepts

1.1 Discuss the importance of alternative identification in the engineering economic process.

1.2 Which of the following would be considered noneconomic factors in deciding which type of power plant to build: (a) equipment cost; (b) morale; (c) goodwill; (d) salvage value; (e) public acceptance; and (f) aesthetics?

1.3 In economic analysis, revenues and costs are examples of what?

1.4 The analysis techniques that are used in engineering economic analysis are only as good as what?

1.5 What is meant by the *do-nothing alternative*?

1.6 What is meant by the term *evaluation criterion*?

1.7 What evaluation criterion is used in economic analysis?

1.8 What is meant by the term *intangible factor*?

1.9 Give three examples of intangible factors.

1.10 What is meant by the term *time value of money*?

1.11 Interest is a manifestation of what general concept in engineering economy?

1.12 Of the fundamental dimensions length, mass, time, and electric charge, which one is the most important in economic analysis? Why?

Interest Rate and Equivalence

1.13 The term that describes compensation for "renting" money is what?

1.14 When an interest rate statement does not include a time period, e.g., 3%, the time period is assumed to be what?

1.15 The original amount of money in a loan transaction is known as what?

1.16 What is meant by the term *minimum attractive rate of return (MARR)*?

1.17 When the yield on a U.S. Government Bond is 3% per year, investors expect the inflation rate to be approximately what?

1.18 In order to build a new warehouse facility, the regional distributor for Valco Multi-position Valves borrowed $1.6 million at 10% per year interest. If the company repaid the loan in a lump sum amount after two years, what was (a) the amount of the payment, and (b) the amount of interest?

1.19 A medium-size consulting engineering firm is trying to decide whether it should remodel its office now or wait and do it one year from now. If the firm does it now, the cost will be $38,000. The interest rate is 10% per year.

   a. What would the cost have to be one year from now to render the decision indifferent?

   b. If the cost one year from now is $41,600, should the firm remodel now or later?

1.20 At an interest rate of 8% per year, $50,000 one year hence is equivalent to how much now?

1.21 Bennett Industries invested $10,000,000 in a co-venture one year ago. One year later, Bennett realized a profit of $1,450,000. What annual rate of return does this represent?

1.22 Trucking giant Yellow Corp agreed to purchase rival Roadway for $966 million in order to reduce so-called back-office costs, that is, payroll and insurance, by $45 million per year. If the savings are realized as planned, what is the rate of return on the investment?

1.23 If Ford Motor Company's profits increased from 22 cents per share to 29 cents per share in the April-June quarter compared to the previous quarter, what was the rate of increase in profits for that quarter?

1.24 A broadband service company borrowed $2 million for new equipment and repaid the principal of the loan plus $275,000 interest after 1 year. What was the interest rate on the loan?

1.25 A design-build engineering firm completed a pipeline project wherein the company realized a profit of $2.3 million in one year. If the amount of money the company invested was $6 million, what was the rate of return on the investment?

1.26 A sum of $2 million now is equivalent to $2.36 million one year from now at what interest rate?

1.27 Last year, Lee Industries decided to restructure some of its debt by paying off one of its short-term loans. To do so, the company borrowed the money one year ago at an interest rate of 10% per year. If

the total cost of repaying the loan was $53 million, what was the amount of the original loan?

1.28 A start-up company with multiple nanotechnology products established a goal of making a rate of return of at least 25% per year on its investments for the first five years. If the company acquired $400 million in venture capital, how much did it have to earn in the first year?

1.29 How many years does it take for an investment of $280,000 to accumulate to at least $425,000 at 15% per year interest?

1.30 The MARR used for a project's acceptance or rejection is set relative to what cost?

1.31 An engineer told you that a project is economically acceptable when it's rate of return equals or exceeds the corporation's cost of capital. Is this correct? Explain your answer.

Simple and Compound Interest

1.32 Valley Rendering, Inc. is considering the purchase of a new flotation system for recovering more grease. The company can finance a $150,000 system at 5% per year compound interest or 5.5% per year simple interest.

   a. If the total amount owed is due in a single payment at the end of 3 years, which interest rate should the company select?

   b. How much is the difference in interest between the two schemes?

1.33 Valtro Electronic Systems, Inc. set aside a lump sum investment four years ago in order to finance a plant expansion now. The money returned 10% per year simple interest. How much did the company set aside if the investment is now worth $850,000?

1.34 At 10% per year simple interest, how long will it take for a deposit of $1000 now to accumulate to $100,000?

1.35 Fill in the missing values A through D in the table for a loan of $10,000, if the interest rate is compounded at 10% per year.

End of Year	Interest for Year	Total Owed at End of Year	End-of-Year Payment	Total Owed after Payment
0	—	—	—	10,000
1	1000	11,000	2000	9,000
2	900	9,900	2000	A
3	B	C	2000	D

1.36 An engineer who was in the business of customizing software for small construction companies repaid a loan that she got 3 years ago at 7% per year simple interest. If the amount she repaid was $35,000, what was the principal amount of the loan?

1.37 How much can you borrow today if you promise to repay $20,000 two years from today at a compound interest rate of 20% per year?

1.38 A design-build-operate engineering company borrowed $6 million for 3 years so that it can purchase new equipment. The interest is compounded and the total amount owed will be paid in a single lump sum amount at the end of the 3-year period. The interest at the end of the first year will be $900,000.
   a. What is the interest rate on the loan?
   b. How much interest will be charged at the end of the second year?

1.39 A company that manufactures general-purpose transducers invested $2 million four years ago in high-yield junk bonds. If the bonds are now worth $2.8 million, what rate of return per year did the company make on the basis of (a) simple interest, and (b) compound interest?

1.40 How many years would it take for money to triple in value at 20% per year simple interest?

1.41 If Aquatronics Inc. wants its investments to double in value in 4 years, what rate of return would it have to make on the basis of (a) simple interest, and (b) compound interest?

1.42 Companies frequently borrow money under an arrangement that requires them to make periodic payments of "interest only" and then pay the principal all at once. If Cisco International borrowed $500,000 (identified as loan A) at 10% per year simple interest and another $500,000 (identified as loan B) at 10% per year compound interest and paid *only the interest* at the end of *each year* for three years on both loans, (a) on which loan did the company pay more interest, and (b) what was the difference in interest paid between the two loans?

Terminology and Symbols

1.43 *All* engineering economy problems will involve which two of the following symbols: P, F, A, i, n?

1.44 Define the economy symbols to determine how many years it would take for $13,000 to double in size at a compound interest rate of 6.8% per year.

1.45 At 9% per year simple interest, $1000 is equivalent to $1270 in three years. Define the symbols for the compound interest rate per year that would make this equivalence correct, i.e., $1000 now and $1270 in 3 years.

1.46 Define the symbols involved when a construction company wants to know how much money it can spend 3 years from now to purchase a new truck in lieu of spending $50,000 now. The compound interest rate is 15% per year.

1.47 Identify the symbols involved if a pharmaceutical company wants to have a liability fund worth $200 million five years from now. Assume the company will invest an equal amount of money *each year* beginning one year from now and that the investments will earn 12% per year.

1.48 Vision Technologies, Inc. is a small company that uses ultra-wideband technology to develop devices that can detect objects (including people) inside of buildings, behind walls, or below ground. The company expects to spend $100,000 per year for labor and $125,000 per year for supplies before a product can be marketed. If the company wants to know the total equivalent future amount of the company's expenses at the end of 3 years at 15% per year interest, identify the engineering economy symbols involved and the values for the ones that are given.

1.49 Corning Ceramics expects to spend $400,000 to upgrade equipment two years from now. If the company wants to know the equivalent value now of the planned expenditure, identify the symbols and their values, assuming Corning's minimum attractive rate of return is 20% per year.

1.50 Sensotech, Inc., a maker of microelectromechanical systems, believes it can reduce product recalls by 10% if it purchases new software for detecting faulty parts. The cost of the new software is $225,000. Identify the symbols involved and the values for the symbols that are given in determining how much the company would have to save each year to recover its investment in 4 years at a minimum attractive rate of return of 15% per year.

1.51 Atlantic Metals and Plastics uses austenitic nickel-chromium alloys to manufacture resistance heating wire. The company is considering a new annealing-drawing process to reduce costs. The new process will cost $1.8 million dollars now. Identify the symbols that are involved and the values of those that are given, if the company wants to know how much it must save each year to recover the investment in 6 years at an interest rate of 12% per year.

1.52 Phillips Refining plans to expand capacity by purchasing equipment that will provide additional smelting capacity. The cost of the initial investment is expected to be $16 million. The company expects revenue to increase by $3.8 million per year after the expansion. If the company's MARR is 18% per year, how long will it take for the company to recover its investment? Identify the engineering economy symbols involved and their values.

Cash Flows

1.53 What does the term "end-of-period convention" mean? What does it not mean?

1.54 In the phrase "end-of-period convention," the word "period" refers to what?

1.55 The difference between cash inflows and cash outflows is known as what?

1.56 Identify the following as cash inflows or outflows to Anderson and Dyess Design-Build Engineers: office supplies, GPS surveying equipment, auctioning of used earth-moving equipment, staff salaries, fees for services rendered, interest from bank deposits.

1.57 Identify the following as cash inflows or outflows to Honda: income taxes, loan interest, salvage value, rebates to dealers, sales revenues, accounting services, and cost reductions by subcontractors.

1.58 Construct a cash flow diagram to find the present worth of a future outflow of $40,000 in year 5 at an interest rate of 15% per year.

1.59 Construct a cash flow diagram for the following cash flows: $10,000 outflow at time zero, $3000 per year inflow in years 1 through 5 at an interest rate of 10% per year, and an unknown future amount in year 5.

1.60 Kennywood Amusement Park spends $75,000 each year in consulting services for ride inspection and maintenance recommendations. New actuator element technology enables engineers to simulate complex computer-controlled movements in any direction. Construct a cash flow diagram for determining how much the park could afford to spend now on the new technology if the cost of annual consulting services will be reduced to $30,000 per year? Assume the park uses an interest rate of 15% per year and it wants to recover its investment in 5 years.

Spreadsheet and Financial Calculator Functions

1.61 Write the engineering economy symbol that corresponds to each of the following spreadsheet functions.
- **a.** FV
- **b.** PMT
- **c.** NPER
- **d.** IRR
- **e.** PV

1.62 What are the values of the engineering economy symbols P, F, A, i, and n in the following spreadsheet or TVM calculator functions? Use a "?" for the symbol that is to be determined.
- **a.** FV(8%,10,2000,−10000)
- **b.** PMT(12%,30,16000)
- **c.** PV(9%,15,1000,700)
- **d.** n(8.5,5000,−50000,20000)

1.63 State the purpose for each of the following built-in spreadsheet functions:
- **a.** FV(i%,n,A,P)
- **b.** IRR(first_cell:last_cell)
- **c.** PMT(i%,n,P,F)
- **d.** PV(i%,n,A,F)

1.64 In a built-in spreadsheet function, if a certain parameter does not apply, under what circumstances can it be left blank and when must a comma be entered in its place?

Ethics and Economics

1.65 Explain the relation between a common moral and a personal moral.

1.66 What is one primary use of a *code of ethics* for a specific discipline of professional practice?

1.67 Yesterday, Carol, an engineer with Hancock Enterprises, was at lunch with several work friends. Joe, a person Carol has known for a year or so from similar lunches, proudly mentioned that he got a free flight and tickets to a major league playoff game two weeks from now in a distant city. Joe happened to also mention the company; it is Dryer. Carol is aware that Dryer is one of the prime bidders on a major contract to be evaluated by Hancock next month. Upon inquiry, Carol learned that both she and Joe are on the bid evaluation committee. Carol suspects that someone in Dryer has offered Joe the tickets as a consideration for Joe's favorable evaluation of their bid.
- **a.** Carol has determined that she could do one of several things about the situation: recommend

to Joe directly that he refuse the tickets; show Joe the NSPE Code of Ethics for Engineers and let him make his own decision; go to Joe's supervisor and tell her of the situation; go to her own supervisor and inform him of her suspicion; write an e-mail to Joe with a copy to her supervisor recommending that Joe consider the ethical dilemma involved for him; do nothing. Considering only these actions, select the one you think is the best and explain why you chose it.

b. Identify other options for Carol's response at this time and determine if one of them is better than her options outlined above.

1.68 Watch the video mentioned in Example 1.14. Give an example of a situation that you have experienced at a restaurant, retailer, or some other business that sells to the public. In your view, what should the business have done to be more ethical in their dealings with the public?

ADDITIONAL PROBLEMS AND FE EXAM REVIEW QUESTIONS

1.69 At a compound interest rate of 10% per year, $10,000 one year ago is equivalent to this amount now:
   a. $8,264
   b. $9,091
   c. $11,000
   d. $12,100

1.70 All of the following are intangible factors except:
   a. Taxes
   b. Goodwill
   c. Morale
   d. Convenience

1.71 Amounts of $1000 one year ago and $1345.60 one year hence are equivalent at the following compound interest rate per year:
   a. 12.5% per year
   b. 14.8% per year
   c. 17.2% per year
   d. None of the above

1.72 The simple interest rate per year required to accumulate the same amount of money in 2 years at 20% per year compound interest is:
   a. 20.5%
   b. 21%
   c. 22%
   d. 23%

1.73 An investment of $8,000 nine years ago has accumulated to $16,000 now. The compound rate of return earned on the investment is closest to:
   a. 6%
   b. 8%
   c. 10%
   d. 12%

1.74 In most engineering economy studies, the best alternative is the one which:
   a. Will last the longest time
   b. Is the easiest to implement
   c. Costs the least
   d. Is the most politically correct

1.75 The cost of tuition at a public university was $200 per credit hour 5 years ago. The cost today (exactly 5 years later) is $268. The annual rate of increase is closest to:
   a. 4%
   b. 6%
   c. 8%
   d. 10%

1.76 The time it would take for money to double at a simple interest rate of 5% per year is closest to:
   a. 10 years
   b. 12 years
   c. 15 years
   d. 20 years

1.77 For the spreadsheet built-in function PV(i%,n,A,F), the only parameter that can be completely omitted is:
   a. i%
   b. n
   c. A
   d. F

Factors: How Time and Interest Affect Money

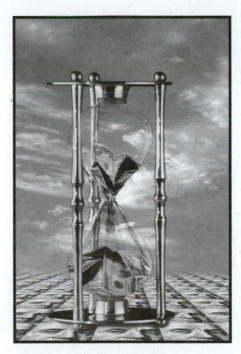

Royalty-Free/CORBIS

In Chapter 1 we learned the basic concepts of engineering economy and their role in decision making. The cash flow is fundamental to every economic study. Cash flows occur in many configurations and amounts—isolated single values, series that are uniform, and series that increase or decrease by constant amounts or constant percentages. This chapter develops the commonly used engineering economy factors that consider the time value of money for cash flows.

The application of factors is illustrated using their mathematical forms and a standard notation format. Spreadsheet and calculator functions are illustrated.

Purpose: Use tabulated factors or spreadsheet/calculator functions to account for the time value of money.

LEARNING OUTCOMES

1. Use the compound amount factor and present worth factor for single payments.

> F/P and P/F factors

2. Use the uniform series factors.

> P/A, A/P, F/A and A/F factors

3. Use the arithmetic gradient factors and the geometric gradient formula.

> Gradients

4. Use uniform series and gradient factors when cash flows are shifted.

> Shifted cash flows

5. Use a spreadsheet or calculator to make equivalency calculations.

> Spreadsheets/Calculators

2.1 SINGLE-PAYMENT FORMULAS (*F/P* AND *P/F*)

Podcast available

The most fundamental equation in engineering economy is the one that determines the amount of money F accumulated after n years (or periods) from a *single* present worth P, with interest compounded one time per year (or period). Recall that compound interest refers to interest paid on top of interest. Therefore, if an amount P is invested at time $t = 0$, the amount F_1 accumulated 1 year hence at an interest rate of i percent per year will be

$$F_1 = P + Pi$$
$$= P(1 + i)$$

where the interest rate is expressed in decimal form. At the end of the second year, the amount accumulated F_2 is the amount after year 1 plus the interest from the end of year 1 to the end of year 2 on the entire F_1.

$$F_2 = F_1 + F_1 i$$
$$= P(1 + i) + P(1 + i)i$$

The amount F_2 can be expressed as

$$F_2 = P(1 + i + i + i^2)$$
$$= P(1 + 2i + i^2)$$
$$= P(1 + i)^2$$

Similarly, the amount of money accumulated at the end of year 3 will be

$$F_3 = P(1 + i)^3$$

By mathematical induction, the future worth F can be calculated for n years using

$$F = P(1 + i)^n \qquad \text{[2.1]}$$

The term $(1 + i)^n$ is called a factor and is known as the *single-payment compound amount factor* (SPCAF), but it is usually referred to as the *F/P factor*. This is the conversion factor that yields the future amount F of an initial amount P after n years at interest rate i. The cash flow diagram is seen in Figure 2.1*a*.

   Reverse the situation to determine the P value for a stated amount F. Simply solve Equation [2.1] for P.

$$P = F\left[\frac{1}{(1 + i)^n}\right] \qquad \text{[2.2]}$$

The expression in brackets is known as the *single-payment present worth factor* (SPPWF), or the *P/F factor*. This expression determines the present worth P of a given future amount F after n years at interest rate i. The cash flow diagram is shown in Figure 2.1*b*.

   Note that the two factors derived here are for *single payments;* that is, they are used to find the present or future amount when only one payment or receipt is involved.

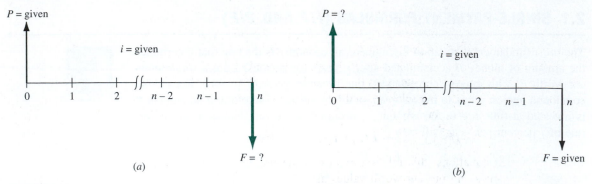

FIGURE 2.1   Cash flow diagrams for single-payment factors: (*a*) find *F* and (*b*) find *P*.

A standard notation has been adopted for all factors. It is always in the general form $(X/Y,i,n)$. The letter X represents what is sought, while the letter Y represents what is given. For example, F/P means *find F when given P*. The i is the interest rate in percent, and n represents the number of periods involved. Thus, $(F/P,6\%,20)$ represents the factor that is used to calculate the future amount F accumulated in 20 periods if the interest rate is 6% per period. The P is given. The standard notation, simpler to use than formulas and factor names, will be used hereafter. Table 2.1 summarizes the standard notation and equations for the F/P and P/F factors.

To simplify routine engineering economy calculations, tables of factor values have been prepared for a wide range of interest rates and time periods from 1 to large n values, depending on the i value. These tables are found at the end of this book.

For a given factor, interest rate, and time, the correct factor value is found at the intersection of the factor name and n. For example, the value of the factor $(P/F,5\%,10)$ is found in the P/F column of Table 10 at period 10 as 0.6139.

When it is necessary to locate a factor value for an i or n that is not in the interest tables, the desired value can be obtained in one of several ways: (1) by using the formulas derived in Sections 2.1 to 2.3, (2) by linearly interpolating between the tabulated values, or (3) by using a spreadsheet or calculator function as discussed in Section 2.5.

For many cash flow series or for the sake of speed, a spreadsheet function may be used in lieu of the tabulated factors or their equations. For single-payment series, no annual series A is present and three of the four values of F, P, i, and n are known. When solving for a future worth by spreadsheet, the F value is calculated

TABLE 2.1   *F/P* and *P/F* Factors: Notation, Equation and Function

Notation	Factor Name	Standard Notation Equation	Equation with Factor Formula	Spreadsheet Function	Calculator Function
$(F/P,i,n)$	Single-payment compound amount	$F = P(F/P,i,n)$	$F = P(1 + i)^n$	$= FV(i\%,n,,P)$	$FV(i,n,A,P)$
$(P/F,i,n)$	Single-payment present worth	$P = F(P/F,i,n)$	$P = F[1/(1 + i)^n]$	$= PV(i\%,n,,F)$	$PV(i,n,A,F)$

by the FV function, and the present worth P is determined using the PV function. The formats are included in Table 2.1. (Refer to Section 2.5 and Appendix A for more information on the FV and PV spreadsheet functions.)

The calculator formats for FV and PV functions detailed in Table 2.1 include the annual series A, which is entered as 0 when only single-amounts are involved. In standard notation form, the relation used by calculators to solve for any one of the parameters P, F, A, i, or n is

$$A(P/A,i\%,n) + F(P/F,i\%,n) + P = 0$$

Using the factor equations, this relation expresses the present worth of uniform series, future values and present worth values as:

$$A\left(\frac{1 - (1 + i/100)^{-n}}{i/100}\right) + F(1 + (i/100)^{-n} + P = 0 \qquad [2.3]$$

When only single payments are present, the first term is 0.

EXAMPLE 2.1

An engineer received a bonus of $12,000 that he will invest now. He wants to calculate the equivalent value after 24 years, when he plans to use all the resulting money as the down payment on an island vacation home. Assume a rate of return of 8% per year for each of the 24 years. Find the amount he can pay down, using the tabulated factor, the factor formula, a spreadsheet function, and a calculator function.

Solution

The symbols and their values are

$P = \$12,000 \qquad F = ? \qquad i = 8\%$ per year $\qquad n = 24$ years

The cash flow diagram is the same as that in Figure 2.1*a*.

Tabulated: Determine F, using the F/P factor for 8% and 24 years. Table 13 provides the factor value.

$F = P(F/P,i,n) = 12,000(F/P,8\%,24)$
$\qquad = 12,000(6.3412)$
$\qquad = \$76,094.40$

Formula: Apply Equation [2.1] to calculate the future worth F.

$F = P(1 + i)^n = 12,000(1 + 0.08)^{24}$
$\qquad = 12,000(6.341181)$
$\qquad = \$76,094.17$

Spreadsheet: Use the function = FV($i\%,n,A,P$). The cell entry is = FV(8%, 24,,12000). The F value displayed is ($76,094.17) in red or $-\$76,094.17$ in black to indicate a cash outflow.

Calculator: Use the TVM function FV(i,n,A,P). The numerical values are FV(8,24,0,12000), which displays the future worth value \$−76,094.17, which indicates it is a cash outflow.

The slight difference in answers between tabulated, formula, and calculator solutions is due to round-off error and how different methods perform equivalence calculations. An equivalence interpretation of this result is that \$12,000 today is worth \$76,094 after 24 years of growth at 8% per year compounded annually.

EXAMPLE 2.2   Hewlett-Packard has completed a study indicating that \$50,000 in reduced maintenance this year (i.e., year zero) on one of its processing lines resulted from improved wireless monitoring technology.

a. If Hewlett-Packard considers these types of savings worth 20% per year, find the equivalent value of this result after 5 years.
b. If the \$50,000 maintenance savings occurs now, find its equivalent value 3 years earlier with interest at 20% per year.

Solution

a. The cash flow diagram appears as in Figure 2.1*a*. The symbols and their values are

$$P = \$50,000 \qquad F = ? \qquad i = 20\% \text{ per year} \qquad n = 5 \text{ years}$$

Use the F/P factor to determine F after 5 years.

$$\begin{aligned} F &= P(F/P,i,n) = \$50,000(F/P,20\%,5) \\ &= 50,000(2.4883) \\ &= \$124,415.00 \end{aligned}$$

The function = FV(20%,5,,50000) also provides the answer. See Figure 2.2.

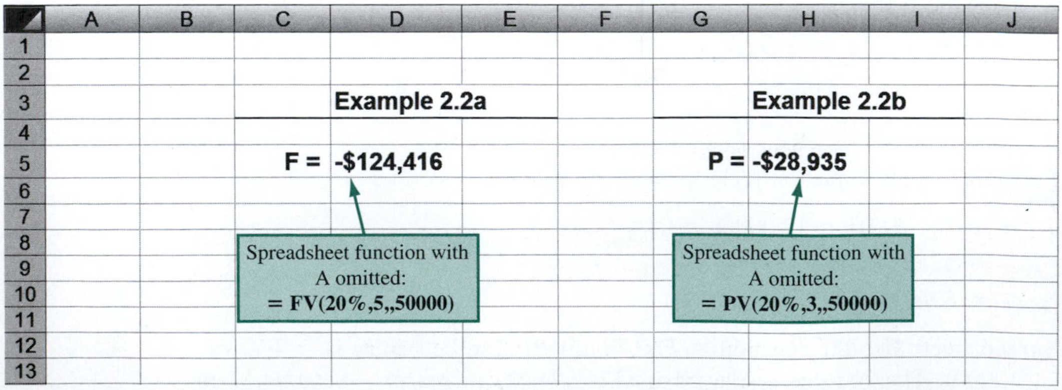

	A	B	C	D	E	F	G	H	I	J
1										
2										
3				**Example 2.2a**				**Example 2.2b**		
4										
5				F = -$124,416				P = -$28,935		
6										
7										
8				Spreadsheet function with				Spreadsheet function with		
9				A omitted:				A omitted:		
10				= FV(20%,5,,50000)				= PV(20%,3,,50000)		
11										
12										
13										

FIGURE 2.2   Use of single-cell spreadsheet functions to find F and P values, Example 2.2.

b. The cash flow diagram appears as in Figure 2.1*b* with *F* placed at time $t = 0$ and the *P* value placed 3 years earlier at $t = -3$. The symbols and their values are

$$P = ? \qquad F = \$50{,}000 \qquad i = 20\% \text{ per year} \qquad n = 3 \text{ years}$$

Use the *P/F* factor to determine *P* three years earlier.

$$P = F(P/F,i,n) = \$50{,}000\,(P/F,20\%,3)$$
$$= 50{,}000(0.5787) = \$28{,}935.00$$

Use the PV function and omit the *A* value. Figure 2.2 shows the result of entering $= \text{PV}\,(20\%,3,,50000)$ to be the same as using the *P/F* factor.

Jamie has become more conscientious about paying off his credit card bill promptly to reduce the amount of interest paid. He was surprised to learn that he paid \$400 in interest in 2007 and the amounts shown in Figure 2.3 over the previous several years. If he made his payments to avoid interest charges, he would have these funds plus earned interest available in the future. What is the equivalent amount 5 years from now that Jamie could have available had he not paid the interest penalties? Let $i = 5\%$ per year.   **EXAMPLE 2.3**

Year	2002	2003	2004	2005	2006	2007
Interest paid, $	600	0	300	0	0	400

FIGURE 2.3   Credit card interest paid over the last 6 years, Example 2.3.

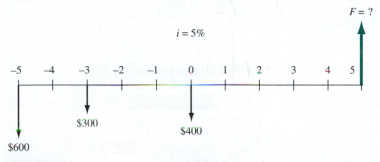

FIGURE 2.4   Cash flow diagram, Example 2.3.

Solution

Draw the cash flow diagram for the values \$600, \$300, and \$400 from Jamie's perspective (Figure 2.4). Use *F/P* factors to find *F* in the year labeled 5, which is 10 years after the first cash flow.

$$F = 600(F/P,5\%,10) + 300(F/P,5\%,8) + 400(F/P,5\%,5)$$
$$= 600(1.6289) + 300(1.4775) + 400(1.2763)$$
$$= \$1931.11$$

The problem could also be solved by finding the present worth in year -5 of the \$300 and \$400 costs using the P/F factors and then finding the future worth of the total in 10 years.

$$P = 600 + 300(P/F,5\%,2) + 400(P/F,5\%,5)$$
$$= 600 + 300(0.9070) + 400(0.7835)$$
$$= \$1185.50$$
$$F = 1185.50(F/P,5\%,10) = 1185.50(1.6289)$$
$$= \$1931.06$$

Comment: It should be obvious that there are a number of ways the problem could be worked, since any year could be used to find the equivalent total of the costs before finding the future value in year 5. As an exercise, work the problem using year 0 for the equivalent total and then determine the final amount in year 5. All answers should be the same except for round-off error.

2.2 UNIFORM SERIES FORMULAS (*P/A, A/P, A/F, F/A*)

Podcast available

There are four *uniform series* formulas that involve A, where A means that:

1. The cash flow occurs in *consecutive interest periods*, and
2. The cash flow *amount is the same* in each period.

The formulas relate a present worth P or a future worth F to a uniform series amount A. The two equations that relate P and A are as follows. (See Figure 2.5 for cash flow diagrams.)

$$P = A\left[\frac{(1 + i)^n - 1}{i(1 + i)^n}\right]$$

$$A = P\left[\frac{i(1 + i)^n}{(1 + i)^n - 1}\right]$$

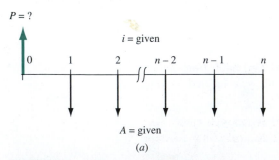

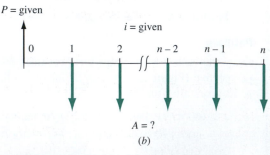

FIGURE 2.5 Cash flow diagrams used to determine (a) P of a uniform series and (b) A for a present worth.

TABLE 2.2 *P/A* and *A/P* Factors: Notation, Equation and Function

Factor Notation	Factor Name	Factor Formula	Standard Notation Equation	Spreadsheet Function	Calculator Function
$(P/A,i,n)$	Uniform-series present worth	$\dfrac{(1+i)^n - 1}{i(1+i)^n}$	$P = A(P/A,i,n)$	$= \text{PV}(i\%,n,A,F)$	$\text{PV}(i,n,A,F)$
$(A/P,i,n)$	Capital recovery	$\dfrac{i(1+i)^n}{(1+i)^n - 1}$	$A = P(A/P,i,n)$	$= \text{PMT}(i\%,n,P,F)$	$\text{PMT}(i,n,P,F)$

In standard factor notation, the equations are $P = A(P/A, i, n)$ and $A = P(A/P, i, n)$, respectively. It is important to remember that in these equations, the P and the first A value are separated by one interest period. That is, the present worth P is always located one interest period prior to the first A value. It is also important to remember that the n is always equal to the number of A values.

The factors and their use to find P and A are summarized in Table 2.2. The spreadsheet and calculator functions shown in Table 2.2 are capable of determining both P and A values in lieu of applying the *P/A* and *A/P* factors. The PV function calculates the P value for a given A over n years, and a separate F value in year n, if present. The format is

$$= \text{PV}(i\%,n,A,F)$$

Similarly, the A value is determined using the PMT function for a given P value in year 0 and a separate F, if present. The format is

$$= \text{PMT}(i\%,n,P,F)$$

In addition to i and n, the calculator functions shown in Table 2.2 include all three cash flow parameters—P, F, and A. The function uses Equation [2.3] to solve for one of the five parameters, given values for the remaining four.

How much money should you be willing to pay now for a guaranteed $600 per year for 9 years starting next year, at a rate of return of 16% per year?   **EXAMPLE 2.4**

Solution
The cash flow diagram (Figure 2.6) fits the *P/A* factor. The present worth is:

$$P = 600(P/A,16\%,9) = 600(4.6065) = \$2763.90$$

The spreadsheet PV function $= -\text{PV}(16\%,9,600)$ entered into a single spreadsheet cell will display the answer $P = \$2763.93$. Similarly, the calculator function PV(16,9,600,0) results in $P = \$-2763.93$.

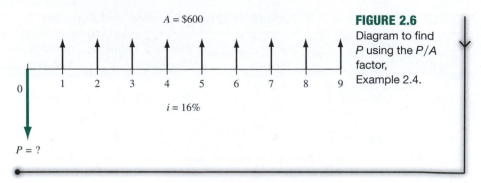

FIGURE 2.6
Diagram to find
P using the P/A
factor,
Example 2.4.

A = $600

i = 16%

P = ?

The uniform series formulas that relate A and F follow. See Figure 2.7 for the cash flow diagrams.

$$A = F\left[\frac{i}{(1 + i)^n - 1}\right]$$

$$F = A\left[\frac{(1 + i)^n - 1}{i}\right]$$

It is important to remember that these equations are derived such that the last A value occurs in the *same* time period as the future worth F, and n is always equal to the number of A values.

Standard notation follows the same form as that of other factors. They are (F/A,i,n) and (A/F,i,n). Table 2.3 summarizes the notations and equations.

If P is not present for the PMT function, the comma (spreadsheet) or a zero (calculator) must be entered to indicate that the last entry is an F value.

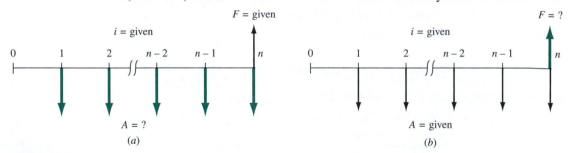

FIGURE 2.7 Cash flow diagrams to (a) find A, given F, and (b) find F, given A.

TABLE 2.3 *F/A and A/F Factors: Notation, Equation and Function*

Factor Notation	Factor Name	Factor Formula	Standard Notation Equation	Spreadsheet Function	Calculator Function
(F/A,i,n)	Uniform-series compound amount	$\frac{(1 + i)^n - 1}{i}$	F = A(F/A,i,n)	= FV(i%,n,A,P)	FV(i,n,A,P)
(A/F,i,n)	Sinking fund	$\frac{i}{(1 + i)^n - 1}$	A = F(A/F,i,n)	= PMT(i%,n,P,F)	PMT(i,n,P,F)

Formasa Plastics has major fabrication plants in Texas and Hong Kong. The president wants to know the equivalent future worth of $1 million capital investments each year for 8 years, starting 1 year from now. Formasa capital earns at a rate of 14% per year.   **EXAMPLE 2.5**

Solution

The cash flow diagram (Figure 2.8) shows the annual payments starting at the end of year 1 and ending in the year the future worth is desired. Cash flows are indicated in $1000 units. The F value in 8 years is

$$F = 1000(F/A,14\%,8) = 1000(13.2328) = \$13,232.80$$

The actual future worth is $13,232,800.

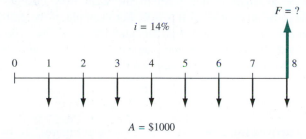

FIGURE 2.8 Diagram to find *F* for a uniform series, Example 2.5.

How much money must an electrical contractor deposit every year in her savings account starting 1 year from now at 5½% per year in order to accumulate $6000 seven years from now?   **EXAMPLE 2.6**

Solution

The cash flow diagram (Figure 2.9) fits the *A/F* factor.

$$A = \$6000(A/F,5.5\%,7) = 6000(0.12096) = \$725.76 \text{ per year}$$

The *A/F* factor value of 0.12096 was computed using the factor formula. Alternatively, use the spreadsheet function $= -\text{PMT}(5.5\%,7,,6000)$ to obtain $A = \$725.79$ per year.

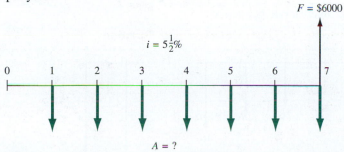

FIGURE 2.9 Cash flow diagram, Example 2.6.

When a problem involves finding i or n (instead of P, F, or A), the solution may require trial and error. Spreadsheet or calculator functions can be used to find i or n in most cases.

2.3 GRADIENT FORMULAS

Podcast available

The previous four equations involved cash flows of the *same magnitude A* in each interest period. Sometimes the cash flows that occur in consecutive interest periods are not the same amount (not an A value), but they do change in a predictable way. These cash flows are known as *gradients,* and there are two general types: arithmetic and geometric.

An *arithmetic gradient* is one wherein the cash flow changes (increases or decreases) by the same amount in each period. For example, if the cash flow in period 1 is $800 and in period 2 it is $900, with amounts increasing by $100 in each subsequent interest period, this is an arithmetic gradient G, with a value of $100.

The equation that represents the present worth of an arithmetic gradient series is:

$$P = \frac{G}{i}\left[\frac{(1 + i)^n - 1}{i(1 + i)^n} - \frac{n}{(1 + i)^n}\right] \qquad \textbf{[2.4]}$$

Equation [2.4] is derived from the cash flow diagram in Figure 2.10 by using the *P/F* factor to find the equivalent P in year 0 of each cash flow. Standard factor notation for the present worth of an arithmetic gradient is $P = G(P/G,i\%,n)$. This equation finds the present worth in year 0 of the *gradient only* (the $100 increases mentioned earlier starting in year 2). It does *not* include the base amount of money that the gradient was built upon ($800 in the example). The base amount in time period 1 must be accounted for separately as a uniform cash flow series. Thus, the *general equation* to find the present worth of an arithmetic gradient cash flow series is

P = **Present worth of base amount + present worth of gradient amount**

$= A(P/A,i\%,n) + G(P/G,i\%,n)$ \qquad **[2.5]**

where A = amount in *period* 1

G = amount of *change* in cash flow between periods 1 *and* 2

n = number of periods from 1 through n of gradient cash flow

i = interest rate per period

FIGURE 2.10
Conventional arithmetic gradient series without the base amount.

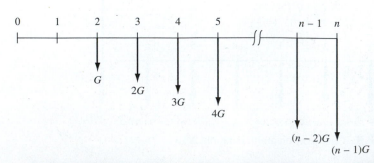

If the gradient cash flow *decreases* from one period to the next, the only change in the general equation is that the plus sign becomes a minus sign. Since the gradient G begins between years 1 and 2, this is called a *conventional gradient*.

The Highway Department expects the cost of maintenance for a piece of heavy construction equipment to be $5000 in year 1, to be $5500 in year 2, and to increase annually by $500 through year 10. At an interest rate of 10% per year, determine the present worth of 10 years of maintenance costs.   **EXAMPLE 2.7**

Solution

The cash flow includes an increasing gradient with $G = \$500$ and a base amount of $5000 starting in year 1. Apply Equation [2.5].

$$P = 5000(P/A,10\%,10) + 500(P/G,10\%,10)$$
$$= 5000(6.1446) + 500(22.8913)$$
$$= \$42,169$$

In Example 2.7 an arithmetic gradient is converted to a P value using the P/G factor. If an equivalent A value for years 1 through n is needed, the A/G factor can be used directly to convert the gradient only. The equation follows with the $(A/G,i\%,n)$ factor formula included in the brackets.

$$A = G\left[\frac{1}{i} - \frac{n}{(1 + i)^n - 1}\right] \qquad [2.6]$$

As for the P/G factor, the A/G factor converts *only the gradient* into an A value. The base amount in year 1, A_1, must be added to the Equation [2.6] result to obtain the total annual worth A_T of the cash flows.

$$A_T = A_1 \pm A_G \qquad [2.7]$$

where A_1 = cash flow (base amount) in period 1
       A_G = annual worth of gradient

Alternatively, the annual worth of the cash flow could be obtained by first finding the total present worth P_T of the cash flows and then converting it into an A value using the relation $A_T = P_T(A/P,i,n)$.

The cash flow associated with a strip mining operation is expected to be $200,000 in year 1, $180,000 in year 2, and amounts decreasing by $20,000 annually through year 8. At an interest rate of 12% per year, calculate the equivalent annual cash flow.   **EXAMPLE 2.8**

Solution

Apply Equation [2.7] and the A/G factor.

$$A_T = A_1 - A_G$$
$$= 200,000 - 20,000(A/G,12\%,8)$$
$$= 200,000 - 20,000(2.9131)$$
$$= \$141,738$$

The previous two gradient factors are for cash flows that change by a *constant amount* each period. Cash flows that change by a *constant percentage* each period are known as *geometric gradients*. The following equation is used to calculate the P value of a geometric gradient in year 0. The expression in brackets is called the $(P/A,g,i,n)$ factor.

$$P = A_1 \left[\frac{1 - \left(\frac{1+g}{1+i}\right)^n}{i - g} \right] \qquad g \neq i \qquad [2.8]$$

where A_1 = total cash flow in period 1

g = rate of change per period (decimal form)

i = interest rate per period

This equation accounts for *all* of the cash flows, including the amount in period 1. For a decreasing geometric gradient, change the sign prior to both g values. When $g = i$, the P value is

$$P = A_1[n/(1+i)] \qquad [2.9]$$

Geometric gradient factors are not tabulated; the equations are used.

There are no spreadsheet or calculator functions for arithmetic or geometric gradients that solve directly for the equivalent P or A value of the series. If the tabulated factors (P/G or A/G) for arithmetic gradients are not sufficient, the fastest approach is to use spreadsheet functions after entering the cash flow values into consecutive cells. (See Example 2.13.)

EXAMPLE 2.9   A mechanical contractor has four employees whose combined salaries through the end of this year are $250,000. If he expects to give an average raise of 5% each year, calculate the present worth of the employees' salaries over the next 5 years. Let $i = 12\%$ per year.

Solution

The cash flow at the end of year 1 is $250,000, increasing by $g = 5\%$ per year (Figure 2.11). The present worth is found using Equation [2.8].

$$P = 250,000 \left[\frac{1 - \left(\frac{1.05}{1.12} \right)^5}{0.12 - 0.05} \right]$$

$$= 250,000(3.94005)$$

$$= \$985,013$$

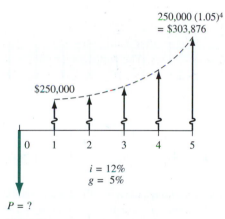

250,000 (1.05)⁴
= $303,876

$250,000

0   1   2   3   4   5

$i = 12\%$
$g = 5\%$

$P = ?$

FIGURE 2.11   Cash flow with $g = 5\%$, Example 2.9

In summary, some basics for gradients are:

- Arithmetic gradients consist of two parts: a uniform series that has an A value equal to the amount of money in period 1, and a gradient that has a value equal to the change in cash flow between periods 1 and 2.
- For arithmetic gradients, the gradient factor is preceded by a plus sign for increasing gradients and a minus sign for decreasing gradients.
- Conventional arithmetic and geometric cash flows start between periods 1 and 2, with the A value in each equation equal to the magnitude of the cash flow in period 1 and the P value in year 0.
- Geometric gradients are handled with Equation [2.8] or [2.9], which yield the present worth of *all* the cash flows.

2.4 CALCULATIONS FOR CASH FLOWS THAT ARE SHIFTED

When a uniform series begins at a time other than at the end of period 1, it is called a *shifted series*. In this case several methods based on factor equations or tabulated values can be used to find the equivalent present worth P. For example, P of the

Podcast
available

FIGURE 2.12
A uniform series that
is shifted.

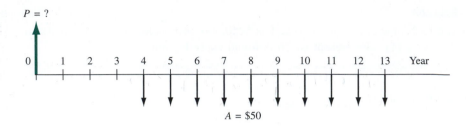

FIGURE 2.13
Location of present
worth for the shifted
uniform series in
Figure 2.12.

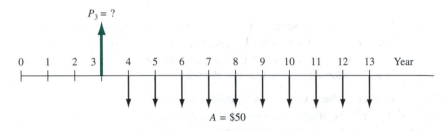

uniform series shown in Figure 2.12 could be determined by any of the following methods:

- Use the P/F factor to find the present worth of each disbursement at year 0 and add them.
- Use the F/P factor to find the future worth of each disbursement in year 13, add them, and then find the present worth of the total using $P = F(P/F,i,13)$.
- Use the F/A factor to find the future amount $F = A(F/A,i,10)$, and then compute the present worth using $P = F(P/F,i,13)$.
- Use the P/A factor to compute the "present worth" (which will be located in year 3 not year 0), and then find the present worth in year 0 by using the $(P/F,i,3)$ factor. (Present worth is enclosed in quotation marks here only to represent the present worth as determined by the P/A factor in year 3, and to differentiate it from the present worth in year 0.)

Typically the last method is used. For Figure 2.12, the "present worth" obtained using the P/A factor is located in year 3. This is shown as P_3 in Figure 2.13.

> **Remember, the present worth is always located one period prior to the first uniform-series amount when using the P/A factor.**

To determine a future worth or F value, recall that the F/A factor has the F located in the *same* period as the last uniform-series amount. Figure 2.14 shows the location of the future worth when F/A is used for Figure 2.12 cash flows.

> **Remember, the future worth is always located in the same period as the last uniform-series amount when using the F/A factor.**

It is also important to remember that the number of periods n in the P/A or F/A factor is equal to the number of uniform-series values. It may be helpful to *renumber* the cash flow diagram to avoid errors in counting. Figure 2.14 shows Figure 2.12 renumbered to determine $n = 10$.

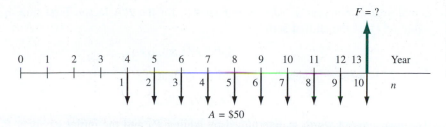

As stated above, there are several methods that can be used to solve problems containing a uniform series that is shifted. However, it is generally more convenient to use the uniform-series factors than the single-amount factors. There are specific steps that should be followed in order to avoid errors:

1. Draw a diagram of the positive and negative cash flows.
2. Locate the present worth or future worth of each series on the cash flow diagram.
3. Determine n for each series by renumbering the cash flow diagram.
4. Set up and solve the equations.

EXAMPLE 2.10

An engineering technology group just purchased new CAD software for $5000 now and annual payments of $500 per year for 6 years starting 3 years from now for annual upgrades. What is the present worth of the payments if the interest rate is 8% per year?

Solution

The cash flow diagram is shown in Figure 2.15. The symbol P_A is used throughout this chapter to represent the present worth of a uniform annual series A, and P_A' represents the present worth at a time other than period 0. Similarly, P_T represents the total present worth at time 0. The correct placement of P_A' and the diagram renumbering to obtain n are also indicated. Note that P_A' is located in

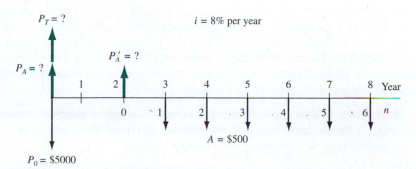

FIGURE 2.15
Cash flow diagram with placement of P values, Example 2.10.

actual year 2, not year 3. Also, $n = 6$, not 8, for the P/A factor. First find the value of P'_A of the shifted series.

$$P'_A = \$500(P/A,8\%,6)$$

Since P'_A is located in year 2, now find P_A in year 0.

$$P_A = P'_A(P/F,8\%,2)$$

The total present worth is determined by adding P_A and the initial payment P_0 in year 0.

$$
\begin{aligned}
P_T &= P_0 + P_A \\
&= 5000 + 500(P/A,8\%,6)(P/F,8\%,2) \\
&= 5000 + 500(4.6229)(0.8573) \\
&= \$6981.60
\end{aligned}
$$

To determine the present worth for a cash flow that includes both uniform series and single amounts at specific times, use the P/F factor for the single amounts and the P/A factor for the series. To calculate A for the cash flows, first convert everything to a P value in year 0, or an F value in the last year. Then obtain the A value using the A/P or A/F factor, where n is the total number of years over which the A is desired.

Many of the considerations that apply to shifted uniform series apply to arithmetic gradient series as well. Recall that a conventional gradient series starts between periods 1 and 2 of the cash flow sequence. A gradient starting at any other time is called a *shifted gradient*. The n value in the P/G and A/G factors for the shifted gradient is determined by renumbering the time scale. The period in which the *gradient first appears is labeled period 2*. The n value for the factor is determined by the renumbered period where the last gradient increase occurs. The P/G factor values and placement of the gradient series present worth P_G for the shifted arithmetic gradients in Figure 2.16 are indicated.

It is important to note that the A/G factor *cannot* be used to find an equivalent A value in periods 1 through n for cash flows involving a shifted gradient. Consider the cash flow diagram of Figure 2.16b. To find the equivalent annual series in years 1 through 10 for the arithmetic gradient series only, first find the present worth of the gradient in year 5, take this present worth back to year 0, and then annualize the present worth for 10 years with the A/P factor. If you apply the annual series gradient factor $(A/G,i,5)$ directly, the gradient is converted into an equivalent annual series over years 6 through 10 only.

Remember, to find the equivalent A series of a shifted gradient through all of the periods, first find the present worth of the gradient at actual time 0, then apply the $(A/P,i,n)$ factor.

If the cash flow series involves a *geometric gradient* and the gradient starts at a time other than between periods 1 and 2, it is a shifted gradient. The P_g is located in a manner similar to that for P_G above, and Equation [2.8] is the factor formula.

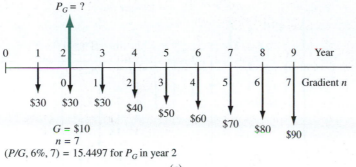

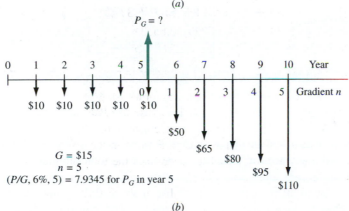

FIGURE 2.16
Determination of G
and n values used in
factors for shifted
gradients.

EXAMPLE 2.11

Chemical engineers at a Coleman Industries plant in the Midwest have determined that a small amount of a newly available chemical additive will increase the water repellency of Coleman's tent fabric by 20%. The plant superintendent has arranged to purchase the additive through a 5-year contract at $7000 per year, starting 1 year from now. He expects the annual price to increase by 12% per year starting in the sixth year and thereafter through year 13. Additionally, an initial investment of $35,000 was made now to prepare a site suitable for the contractor to deliver the additive. Use $i = 15\%$ per year to determine the equivalent total present worth for all these cash flows.

Solution

Figure 2.17 presents the cash flows. The total present worth P_T is found using $g = 0.12$ and $i = 0.15$. Equation [2.8] is used to determine the present worth P_g for the entire geometric series at actual year 4, which is moved to year 0 using $(P/F,15\%,4)$.

$$P_T = 35,000 + A(P/A,15\%,4) + A_1(P/A,12\%,15\%,9)(P/F,15\%,4)$$

$$= 35,000 + 7000(2.8550) + \left[7000\frac{1 - (1.12/1.15)^9}{0.15 - 0.12}\right](0.5718)$$

$$= 35,000 + 19,985 + 28,247$$

$$= \$83,232$$

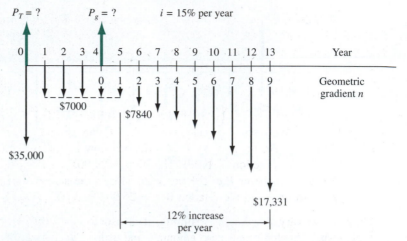

FIGURE 2.17   Cash flow diagram including a geometric gradient with $g = 12\%$, Example 2.11.

Note that $n = 4$ in the $(P/A,15\%,4)$ factor because the $7000 in year 5 is the initial amount A_1 in Equation [2.8] for the geometric gradient.

2.5 USING SPREADSHEETS AND CALCULATORS

The easiest single-cell spreadsheet functions to apply to find P, F, or A require that the cash flows exactly fit the function format. The functions apply the correct sign to the answer that would be on the cash flow diagram. That is, if cash flows are deposits (minus), the answer will have a plus sign. In order to retain the sign of the inputs, enter a minus sign prior to the function. Here is a summary and examples at 5% per year.

Present worth P: Use the PV function $= \text{PV}(i\%,n,A,F)$ if A is exactly the same for each of n years; F can be present or not. For example, if $A = \$3000$ per year deposit for $n = 10$ years, the function $= \text{PV}(5\%, 10,-3000)$ will display $P = \$23,165$. This is the same as using the P/A factor to find $P = 3000(P/A,5\%,10) = 3000(7.7217) = \$23,165$.

Future worth F: Use the FV function $= \text{FV}(i\%,n,A,P)$ if A is exactly the same for each of n years; P can be present or not. For example, if $A = \$3000$ per year deposit for $n = 10$ years, the function $= \text{FV}(5\%, 10,-3000)$ will display $F = \$37,734$. This is the same as using the F/A factor to find $F = 3000(F/A,5\%,10) = 3000(12.5779) = \$37,734$.

Annual amount A: Use the PMT function $= \text{PMT}(i\%,n,P,F)$ when there is no A present, and either P or F or both are present. For example, for

$P = \$-3000$ deposit now and $F = \$5000$ returned $n = 10$ years hence, the function $= -\text{PMT}(5\%,10,-3000,5000)$ will display $A = \$9$. This is the same as using the A/P and A/F factors to find the equivalent net $A = \$9$ per year between the deposit now and return 10 years later.

$$A = -3000\,(A/P,5\%,10) + 5000\,(A/F,5\%,10) = -389 + 398 = \$9$$

Number of periods n: Use the NPER function $= \text{NPER}(i\%,A,P,F)$ if A is exactly the same for each of n years; either P or F can be omitted, but not both. For example, for $P = \$-25,000$ deposit now and $A = \$3000$ per year return, the function $= \text{NPER}(5\%,3000,-25000)$ will display $n = 11.05$ years to recover P at 5% per year. This is the same as using trial and error to find n in the relation $0 = -25,000 + 3,000\,(P/A,5\%,n)$.

When cash flows vary in amount or timing, it is usually necessary to enter them on a spreadsheet, including all zero amounts, and utilize other functions for P, F, or A values. All spreadsheet functions allow another function to be embedded in them, thus reducing the time necessary to get final answers. Example 2.12 illustrates these functions and the embedding capability. Example 2.13 demonstrates how easily spreadsheets handle arithmetic and percentage gradients and how the IRR (rate of return) function works.

EXAMPLE 2.12

Carol just entered college and her grandparents have offered her one of two gifts. They promised to give her $25,000 toward a new car if she graduates in 4 years. Alternatively, if she takes 5 years to graduate, they offered her $5000 each year starting after her second year is complete and an extra $5000 when she graduates. Draw the cash flow diagrams first. Then, use $i = 8\%$ per year to show Carol how to use spreadsheet functions and her financial calculator TVM functions to determine the following for *each gift* offered by her grandparents.

a. Present worth P now
b. Future worth F five years from now
c. Equivalent annual amount A over a total of 5 years
d. Number of years it would take Carol to have $25,000 in hand for the new car if she were able to save $5000 each year starting next year.

Solution

Spreadsheet: The two cash flow series, labeled Gift A (lump sum) and Gift B (spread out), are in Figure 2.18. The spreadsheet in Figure 2.19a lists the cash flows (don't forget to enter the $0 cash flows so the NPV function can be used), and answers to each part using the PV, NPV, FV, or PMT functions as explained below. In some cases, there are alternative ways to obtain the answer.

Figure 2.19b shows the function formulas with some comments. Refer to Appendix A for a complete description of how each function operates. Remember that the PV, FV, and PMT functions will return an answer with the opposite

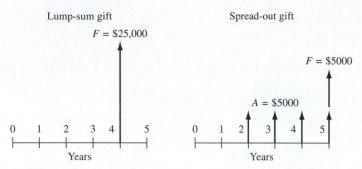

FIGURE 2.18   Cash flows for Carol's gift from her grandparents, Example 2.12.

sign from that of the cash flow entries. The same sign is maintained by entering a minus before the function name.

a. Rows 12 and 13: There are two ways to find P; either the PV or NPV function. NPV requires that the zeros be entered. (For Gift A, omitting zeros in years 1, 2, and 3 will give the incorrect answer of $P = \$23,148$, because NPV assumes the $25,000 occurs in year 1 and discounts it only one year at 8%.) The single-cell PV is hard to use for Gift B since cash flows do not start until year 2; using NPV is easier.

b. Rows 16 and 17: There are two ways to use the FV function to find F at the end of year 5. To develop FV correctly for Gift B in a single cell without listing cash flows, add the extra $5000 in year 5 separate from the FV for the four $A = \$5000$ values. Alternatively, cell D17 embeds the NPV function for the P value into the FV function. This is a very convenient way to combine functions.

c. Rows 20 and 21: There are two ways to use the PMT function to find A for 5 years; find P separately and use a cell reference, or embed the NPV function into the PMT to find A in one operation.

d. Row 24: Finding the years to accumulate $25,000 by depositing $5000 each year using the NPER function is independent of either plan. The entry $= \text{NPER}(8\%, -5000,,25000)$ results in 4.3719 years. This can be confirmed by calculating $5000(F/A,8\%,4.3719) = 5000(5.0000) = \$25,000$ (The 4.37 years is about the time it will take Carol to finish college. Of course, this assumes she can actually save $5000 a year while working on the degree.)

Calculator: Table 2.4 shows the format and completed calculator function for each gift, followed by the numerical answer below it. Minus signs on final answers have been changed to plus as needed to reflect the same sense as that in the spreadsheet solution. When calculating the values for Gift B, the functions can be performed separately, as shown, or embedded in the same way as the spreadsheet functions are embedded in Figure 2.19. In all cases, the answers are identical for the spreadsheet and calculator solutions.

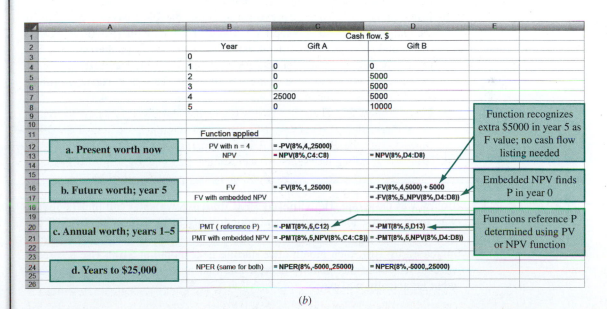

FIGURE 2.19 (a) Use of several spreadsheet functions to find P, F, A, and n values, and (b) format of functions to obtain values, Example 2.12.

TABLE 2.4 Solution Using Calculator TVM Functions, Example 2.12

Year	Cash flow, $	
	Gift A	Gift B
0		
1	0	0
2	0	5,000
3	0	5,000
4	25,000	5,000
5	0	5,000 + 5,000
	Functions applied	
a. Present worth now	$PV(i,n,A,F)$ PV(8,4,0,25000) **$18,376**	$FV(i,n,A,P) + 5,000$ FV(8,4,5000,0) + 5,000 **$27,531** $PV(i,n,A,F)$ PV(8,5,0,27531) **$18,737**
b. Future worth, year 5	$FV(i,n,A,P)$ FV(8,1,0,25000) **$27,000**	$FV(i,n,A,P) + 5,000$ FV(8,4,5000,0) + 5,000 **$27,531**
c. Annual worth, years 1–5	$PMT(i,n,P,F)$ PMT(8,5,0,27000) **$4,602**	$PMT(i,n,P,F)$ PMT(8,5,0,27531) **$4,693**
d. Years to $25,000	$n(i,A,P,F)$ n(8,−5000,0,25000) **4.37**	$n(i,A,P,F)$ n(8,−5000,0,25000) **4.37**

EXAMPLE 2.13   Bobby was desperate. He borrowed $600 from a pawn shop and understood he was to repay the loan starting next month with $100, increasing by $10 per month for a total of 8 months. Actually, he misunderstood. The repayments increased by 10% each month after starting next month at $100. Use a spreadsheet to calculate the *monthly* interest rate that he thought he was to pay, and what he actually will pay.

Solution

Figure 2.20 lists the cash flows for the assumed arithmetic gradient $G = \$10$ per month, and the actual percentage gradient $g = 10\%$ per month. Note the simple relations to construct the increasing cash flows for each type gradient. Apply the IRR function to each series using its format = IRR (first_cell: last_cell). Bobby is paying an exorbitant rate per month (and year) at 14.9% *per month*, which is higher than he expected it to be at 13.8% per month. (Interest rates are covered in detail in Chapter 3.)

	A	B	C	D	E	F	G	H
1		Cash flow, $			Cash flow, $			
2	Month	G = $10			g = 10%			
3	0	600.00			600.00			
4	1	-100.00	= B4-10		-100.00	= E4*(1.1)		
5	2	-110.00			-110.00			
6	3	-120.00			-121.00			
7	4	-130.00			-133.10			
8	5	-140.00			-146.41			
9	6	-150.00			-161.05			
10	7	-160.00			-177.16			
11	8	-170.00			-194.87	= SUM(E4:E11)		
12	Total paid back	-1080.00			-1143.59			
13	ROR per month	13.8%	= IRR(B3:B11)		14.9%	= IRR(E3:E11)		

FIGURE 2.20 Use of a spreadsheet to generate arithmetic and percentage gradient cash flows and application of the IRR function, Example 2.13.

If needed to solve a problem, the tables in the rear of this text provide the numerical value for any of the six common compound interest factors. However, the desired i or n may not be tabulated. Then the factor formula can be applied to obtain the numerical value; plus, a spreadsheet or calculator function can be used with a "1" placed in the P, A, or F location in the function. The other parameter is omitted or set to "0." For example, the P/F factor is determined using the spreadsheet's PV function with the A omitted (or set to 0) and $F = 1$, that is, $= -PV(i,n,,1)$ or $= -PV(i,n,0,1)$. The minus sign makes the result positive. If a calculator is used, the functional notation is $PV(i,n,0,1)$ for the function $PV(i,n,A,F)$. Table 2.5 summarizes the notation for spreadsheets and calculators. This information, in abbreviated form, is included inside the front cover.

When using a spreadsheet, an unknown value in one cell may be required to force the value in a different cell to equal a stated value. For example, the present worth of a given cash flow series is known to equal $10,000 and all but one of the cash flow values is known. This unknown cash flow is to be determined. The spreadsheet tool called GOAL SEEK is easily applied to find one unknown value. Refer to Section A.4 in Appendix A to learn how to use the GOAL SEEK template. This tool is applied in examples throughout the text.

TABLE 2.5
Spreadsheet and Calculator Functions That Determine Factor Values

If the Factor is:	To Do This:	The Spreadsheet Function is:	The Calculator Function is:
P/F	Find P, given F	$= -\text{PV}(i,n,,1)$	$\text{PV}(i,n,0,1)$ for $\text{PV}(i,n,A,F)$
F/P	Find F, given P	$= -\text{FV}(i,n,,1)$	$\text{FV}(i,n,0,1)$ for $\text{FV}(i,n,A,P)$
A/F	Find A, given F	$= -\text{PMT}(i,n,,1)$	$\text{PMT}(i,n,0,1)$ for $\text{PMT}(i,n,P,F)$
F/A	Find F, given A	$= -\text{FV}(i,n,1)$	$\text{FV}(i,n,1,0)$ for $\text{FV}(i,n,A,P)$
P/A	Find P, given A	$= -\text{PV}(i,n,1)$	$\text{PV}(i,n,1,0)$ for $\text{PV}(i,n,A,F)$
A/P	Find A, given P	$= -\text{PMT}(i,n,1)$	$\text{PMT}(i,n,1,0)$ for $\text{PMT}(i,n,P,F)$

SUMMARY

In this chapter, we presented formulas that make it relatively easy to account for the time value of money. In order to use the formulas correctly, certain things must be remembered.

1. When using the P/A or A/P factors, the P and the first A value are separated by one interest period.
2. When using the F/A or A/F factors, the F and the last A value are in the *same* interest period.
3. The n in the uniform series formulas is equal to the number of A values involved.
4. Arithmetic gradients change by a uniform amount from one interest period to the next, and there are two parts to the equation: a uniform series that has an A value equal to the magnitude of the cash flow in period 1 and the gradient that has the same n as the uniform series.

5. Geometric gradients involve cash flows that change by a uniform percentage from one period to the next, and the present worth of the entire cash flow sequence is determined from Equation [2.8] or [2.9].
6. For shifted gradients, the change equal to G or g occurs between renumbered periods 1 and 2. This requires that the n values be properly identified in the gradient equations.
7. For decreasing arithmetic gradients, it is necessary to change the sign in front of the P/G or A/G factors from plus to minus. For decreasing geometric gradients, it is necessary to change the sign in front of both g's in Equation [2.8].

PROBLEMS

Use of Interest Tables

2.1 Find the correct numerical value for the following factors from the interest tables:
   a. $(F/P,10\%,20)$
   b. $(A/F,4\%,8)$
   c. $(P/A,8\%,20)$
   d. $(A/P,20\%,28)$
   e. $(F/A,30\%,15)$

Determination of P, F, A, n, and i

2.2 What is the present worth of $30,000 in year 8 at an interest rate of 10% per year?

2.3 If an engineer invested $15,000 on January 1, 1991, into a retirement account that earned 6% per year interest, how much money will be in the account on January 1, 2016?

2.4 The Moller Skycar M400 is a flying car known as a personal air vehicle (PAV). The cost is $995,000, and a $100,000 deposit holds one of the first 100 vehicles. Assume a buyer pays the $885,000 balance 3 years after making the $100,000 deposit. At an interest rate of 10% per year, determine the effective total cost of the PAV in year 3 using (a) tabulated factors, and (b) a single-cell spreadsheet function.

2.5 What is the present worth of a future payment of $19,000 in year 7 if the interest rate is 10% per year using (*a*) the tabulated factor values in your book, (*b*) TVM functions on a financial calculator, and (*c*) built-in functions on a spreadsheet?

2.6 Determine the amount of money a bank will loan to a developer who repays the loan by selling 7 view lots at $120,000 each two years from now. Assume the bank's interest rate is 10% per year. Use (*a*) the tabulated factor values in your book, (*b*) TVM functions on a financial calculator, and (*c*) built-in functions on a spreadsheet.

2.7 How much will be in an investment account 12 years from now if you deposit $3000 now and $5000 four years from now and the account earns interest at a rate of 10% per year? Use (*a*) tabulated factor values, (*b*) TVM functions on a financial calculator, and (*c*) built-in functions on a spreadsheet.

2.8 Loadstar Sensors is a company that makes load/force sensors based on capacitive sensing technology. The company wants to accumulate $30 million for plant expansion five years from now. If the company already has $15 million in an investment account for the expansion, how much more must the company add to the account now so that it will have the $30 million 5 years from now? The account earns interest at 10% per year. Solve using (*a*) tabulated factors, and (*b*) a spreadsheet. Compare the answers.

2.9 Meggitt Systems, a company that specializes in extreme-high-temperature accelerometers, is investigating whether it should update equipment now or wait and do it later. If the cost now is $280,000, what is the equivalent amount 2 years from now at an interest rate of 12% per year?

2.10 By filling carbon nanotubes with miniscule wires made of iron and iron carbide, incredibly thin nanowires can be extruded by blasting the carbon nanotubes with an electron beam. If Gentech Technologies spends $12.7 million now in developing the process, how much must be received in licensing fees each year for the next 8 years (starting 1 year from now) to recover its investment at a 20% per year rate of return?

2.11 What is the present worth of a uniform series of 10 payments that begin one year from now in the amount of $6000 if the interest rate is 10% per year?

2.12 How large of a repayment must an engineer make each year starting next year if she borrows $60,000 now to start up her new consulting office and she promises to make equal annual payments for 5 years. Assume the interest rate is 8% per year. Develop the answer using (*a*) tabulated factor values, (*b*) a financial calculator, and (*c*) spreadsheet functions.

2.13 Bodine Electric makes gear motors with a three-stage, selectively hardened gearing cluster that is permanently lubricated. If the company borrows $20 million for a new warehouse/distribution facility, what is the amount to repay the loan in six equal annual payments? The loan's interest rate is 10% per year.

2.14 A small oil company wants to replace its Micro Motion Coriolis flowmeters with Emerson F-Series flowmeters in Hastelloy construction. The replacement process will cost the company $50,000 three years from now. How much money must the company set aside each year beginning one year from now in order to have the total amount in three years? Assume the company will invest its funds at 20% per year.

2.15 Ametek Technical & Industrial Products (ATIP) manufactures brushless blowers for boilers, food-service equipment, and fuel cells. The company borrowed $17,000,000 for a plant expansion and repaid the loan in eight annual payments of $2,737,680, with the first payment made one year after the company received the money. What interest rate did ATIP pay? Develop the answer using (*a*) tabulated factor values, (*b*) a financial calculator, and (*c*) spreadsheet functions.

2.16 Assume the cost of homeland security border fence is $3 million per mile. If the life of such a fence is 10 years, what is the equivalent annual cost of a 10-mile long fence at an interest rate of 8% per year? Develop the answer using: (*a*) tabulated factor values, (*b*) a financial calculator, and (*c*) spreadsheet functions.

2.17 The Border Patrol is considering the purchase of a new helicopter for aerial surveilance of the New Mexico-Texas border with Mexico. A similar helicopter was purchased 4 years ago at a cost of $1,400,000. At an interest rate of 7% per year, what would be the equivalent value today of that $1,400,000 expenditure?

2.18 Petroleum Products, Inc. is a pipeline company that provides petroleum products to wholesalers in the Northern United States and Canada. The company is considering purchasing insertion turbine flowmeters to provide better monitoring of pipeline integrity. If these meters prevent one major disruption (through early detection of product loss) valued at $600,000 four years from now, how much could the company afford to spend now at an interest rate of 12% per year?

2.19 Sensotech, Inc., a maker of microelectromechanical systems, believes it can reduce product recalls by 10% if it purchases new software for detecting faulty parts at a cost of $225,000. The minimum attractive rate of return is 15% per year.

   a. How much does the company have to save each year for 4 years to recover its investment?

   b. What was the cost of recalls per year before the software was purchased if the company exactly recovered its investment in 4 years from the 10% reduction?

2.20 Atlas Long-Haul Transportation is considering installing Valutemp temperature loggers in all of its refrigerated trucks for monitoring temperatures during transit. If the systems will reduce insurance claims by $100,000 two years from now, how much should the company be willing to spend now if it uses an interest rate of 12% per year?

2.21 The current cost of liability insurance for a consulting firm is $65,000 per year. If the cost is expected to increase by 4% each year, what will be the cost 5 years from now?

2.22 How much money could RTT Environmental Services borrow to finance a site reclamation project if it expects revenues of $280,000 per year over a 5-year cleanup period? Expenses associated with the project are expected to be $90,000 per year. Assume the interest rate is 10% per year.

2.23 Arctic and Antarctic regions are harsh environments in which to take data. A TempXZ 3000 portable temperature recorder can take and store 32,767 measurements at −40 to 150 °C. A research team from the University of Nova Scotia needs 20 of the recorders, and they are trying to decide whether they should buy them now at $649 each or purchase them 2 years from now, which is when they will be deployed. At an interest rate of 8% per year, how much can each recorder cost in 2 years to render their decision indifferent?

2.24 Since many U.S. Navy aircraft are at or near their usual retirement age of 30 years, military officials want a precise system to assess when aircraft should be taken out of service. A computational method developed at Carnegie Mellon maps in 3-D the microstructure of aircraft materials in their present state so that engineers can test them under different conditions of moisture, salt, dirt, etc. Military officials can then determine if an aircraft is fine, is in need of overhaul, or should be retired. If the 3-D system allows the Navy to use one airplane 2 years longer than it normally would have been used, thereby delaying the purchase of a $20 million aircraft for 2 years, what is the present worth of the assessment system at an interest rate of 8% per year?

2.25 GE Marine Systems is planning to supply a Japanese shipbuilder with aero-derivative gas turbines to power 11DD-class destroyers for the Japanese Self-Defense Force. The buyer can pay the total contract price of $2,100,000 two years from now, when the turbines will be needed, or an equivalent amount now. At an interest rate of 10% per year, what is the equivalent amount now?

2.26 A maker of mechanical systems can reduce product recalls by 25% if it purchases new packaging equipment. The cost of the new equipment is expected to be $40,000 four years from now. How much could the company afford to spend now, instead of 4 years from now, if it uses a minimum attractive rate of return of 12% per year?

2.27 a. How much money could Tesla-Sino Inc., a maker of superconducting magnetic energy storage systems, spend each year on new equipment in lieu of spending $850,000 five years from now, if the company's rate of return is 18% per year?

   b. What is the spreadsheet function to display an answer with the correct sign sense to the annual cash flows?

2.28 French car maker Renault signed a $95 million contract with ABB of Zurich, Switzerland, for automated underbody assembly lines, body assembly workshops, and line control systems. If ABB will be paid in 3 years, when the systems are ready, what is the present worth of the contract at 12% per year interest?

2.29 What is the future worth six years from now of a present cost of $375,000 to Corning, Inc. at an interest rate of 10% per year?

2.30 A pulp and paper company is planning to set aside $150,000 now for possibly replacing its large synchronous refiner motors. If the replacement isn't needed for 8 years, how much will the company have in the account if it earns interest at a rate of 8% per year?

2.31 Automationdirect, which makes 6-inch TFT color touch screen HMI panels, is examining its cash flow requirements for the next 5 years. The company expects to replace office machines and computer equipment at various times over the 5-year planning period. Specifically, the company expects to spend $7000 two years from now, $9000 four years from now, and $15,000 five years from now. Determine the present worth of the planned expenditures at an interest rate of 10% per year using (*a*) tabulated factors, and (*b*) a financial calculator.

2.32 A proximity sensor attached to the tip of an endoscope could reduce risks during eye surgery by alerting surgeons to the location of critical retinal tissue. If a certain eye surgeon expects that by using this technology, he will avoid lawsuits of $0.6 and $1.35 million 2 and 5 years from now, respectively, how much could he afford to spend now if his out-of-pocket costs for the lawsuits would be only 10% of the total amount of each suit? Use an interest rate of 10% per year.

2.33 Irvin Aerospace of Santa Ana, CA, just received a 5-year contract to develop an advanced space capsule airbag landing attenuation system for NASA's Langley Research Center. The company's computer system uses fluid structure interaction modeling to test and analyze each airbag design concept. What is the present worth of the contract at 10% per year interest if the annual cost for 5 years is $8 million per year?

2.34 Julong Petro Materials, Inc. ordered $10 million worth of seamless tubes for its drill collars from the Timken Company of Canton, Ohio. (A drill collar is the heavy tubular connection between a drill pipe and a drill bit.) At 10% per year interest, what is the annual worth of the purchase over a 10-year amortization period?

2.35 Improvised explosive devices (IEDs) are responsible for many deaths in times of war. Unmanned ground vehicles (robots) can be used to disarm the IEDs and to perform other tasks as well. The robots cost $140,000 each and the U.S. Army signed a contract to purchase 4000 of them now. What is the equivalent annual cost of the contract if it is amortized over a 4-year period at 10% per year interest?

2.36 The U.S. Navy's robotics lab at Point Loma Naval Base in San Diego is developing robots that will follow a soldier's command or operate autonomously. If one robot would prevent injury to soldiers or loss of equipment valued at $1.5 million per year, how much could the military afford to spend now on the robot and still recover its investment in 4 years at 8% per year?

2.37 PCM Thermal Products uses austenitic nickel-chromium alloys to manufacture resistance heating wire. The company is considering a new annealing-drawing process to reduce costs. If the new process will cost $3.25 million dollars now, how much must be saved each year to recover the investment in 6 years at an interest rate of 15% per year?

2.38 A green algae, chlamydomonas reinhardtii, can produce hydrogen when temporarily deprived of sulfur for up to 2 days at a time. How much could a small company afford to spend now to commercialize the process if the net value of the hydrogen produced is $280,000 per year? Assume the company wants to earn a rate of return of 18% per year and recover its investment in 8 years.

2.39 HydroKlean LLC, an environmental soil cleaning company, borrowed $3.5 million to finance start-up costs for a site reclamation project. How much must the company receive each year in revenue to earn a rate of return of 25% per year for the 5-year project period?

2.40 A VMB pressure regulator allows gas suppliers and panel builders to provide compact gas handling equipment, thereby minimizing the space required in clean rooms. Veritech Micro Systems is planning to expand its clean room to accommodate a new product design team. The company estimates that it can reduce the space in the clean room by 7 square meters if it uses the compact equipment. If the cost of construction for a clean room is $5000 per square meter, what is the annual worth of the savings at 10% per year interest if the cost is amortized over 10 years?

2.41 New actuator element technology enables engineers to simulate complex computer-controlled movements in any direction. If the technology results in cost savings in the design of new roller coasters, determine the future worth at 12% per year in year 6 for estimated savings of $70,000 now and $90,000 two years from now using (*a*) tabulated factors, and (*b*) spreadsheet functions.

2.42 Under an agreement with the Internet Service Providers (ISPs) Association, SBC Communications reduced the price it charges ISPs to resell its high-speed digital subscriber line (DSL) service from $458 to $360 per year per customer line for the next 5 years. A particular ISP, which has 20,000 customers, plans to pass 90% of the savings along to its customers. What is the total present worth of these savings at an interest rate of 10% per year?

2.43 Southwestern Moving and Storage wants to have enough money to purchase a new tractor-trailer in 5 years at a cost of $290,000. If the company sets aside $100,000 in year 2 and $75,000 in year 3, how much will the company have to set aside in year 4 in order to have the money it needs? Assume investments earn 9% per year. Solve using (*a*) tabulated factors, and (*b*) spreadsheet functions.

2.44 Vision Technologies, Inc., is a small company that uses ultra-wideband technology to develop devices that can detect objects (including people) inside buildings, behind walls, or below ground. The company expects to spend $100,000 per year for labor and $125,000 per year for supplies for three years before a product can be marketed. At an interest rate of 15% per year, what is the total equivalent present worth of the company's expenses?

2.45 How many years will it take Rexchem, Inc. to accumulate $400,000 for a chemical feeder if the company deposits $50,000 each year, starting one year from now, into an account that earns interest at 12% per year?

2.46 How many years will it take for money to increase to three times the initial amount at an interest rate of 10% per year?

2.47 Acceleron is planning future expansion with a new facility in Indianapolis. The company will make the move when its real estate sinking fund has a total value of $1.2 million. If the fund currently has $400,000 and the company adds $50,000 per year, how many years will it take for the account to reach the desired value? The fund earns interest at a rate of 10% per year.

2.48 The defined benefits pension fund of G-Tech Electronics has a net value of $2 billion. The company is switching to a defined contribution pension plan, but it guaranteed the current retirees that they will continue to receive their benefits as promised. If the withdrawal rate from the fund is $158 million per year starting 1 year from now, how many years will it take to completely deplete the fund if the conservatively managed fund grows at a rate of 7% per year?

Arithmetic and Geometric Gradients

2.49 Silastic-LC-50 is a liquid silicon rubber designed to provide high clarity, superior mechanical properties, and short cycle time for high speed manufacturers. One high-volume manufacturer used it to achieve smooth release from molds. The company's projected growth would result in silicon costs of $26,000 next year and costs increasing by $2000 per year through year 5. The interest rate is 10% per year. (*a*) What is the present worth of these costs using tabulated factors? (*b*) How is this problem solved using a spreadsheet? Using a financial calculator?

2.50 Calculate the equivalent annual cost of fuel for mail trucks that records indicate costs $72,000 in year one, increasing by $1000 per year through year five. Use an interest rate of 8% per year.

2.51 A company that manufactures air-operated drain valve assemblies budgeted $84,000 per year for repair components over the next five years. Assume the company uses an interest rate of 10% per year. (*a*) If the company expects to spend $15,000 in year 1, what is the annual increase (arithmetic gradient) that the company expects in the cost of the parts? (*b*) Comment on the size of the arithmetic gradient compared to the first-year cost.

2.52 A company that manufactures a revolutionary aeration system combining coarse and fine bubble aeration components had costs this year (year 1) of $9,000 for check valve components. Based on completion of a new contract with a distributor in China and volume discounts, the company expects this cost to decrease. If the cost in year 2 and each year thereafter decreases by $560, what is the equivalent annual cost for a five-year period at an interest rate of 10% per year?

2.53 For the cash flows shown, determine the value of G that makes the present worth in year 0 equal to $14,000. The interest rate is 10% per year.

Year	0	1	2	3	4
Cash flow, $ per year	—	8000	8000-G	8000-2G	8000-3G

2.54 Allen Bradley claims that its XM1Z1A and XM442 electronic over-speed detection relay modules provide customers a cost-effective monitoring and control system for turbo machinery. Estimates indicate the equipment will provide more efficient turbine performance to the extent of

$20,000 in year 1, $22,000 in year 2, and amounts increasing by $2000 per year. How much can Mountain Power and Light spend now and recover its investment in 10 years, if interest is 10% per year?

2.55 A low-cost non-contact temperature measuring tool may be able to identify railroad car wheels that are in need of repair long before a costly structural failure occurs. If the BNSF railroad saves $100,000 in year 1, $110,000 in year 2, and amounts increasing by $10,000 each year for five years, what is the future worth of the savings in year 5 at an interest rate of 10% per year?

2.56 Southwest Airlines hedged the cost of jet fuel by purchasing options that allowed the airline to purchase fuel at a fixed price for 5 years. If the market price of fuel was $0.50 per gallon higher than the option price in year 1, $0.60 per gallon higher in year 2, and amounts increasing by $0.10 per gallon higher through year 5, what was the present worth of SWA's savings per gallon? Use $i = 10\%$ per year.

2.57 NMTeX Oil company owns several gas wells in Carlsbad, NM. Income from the depleting wells has been decreasing according to an arithmetic gradient for the past five years. If the interest rate is 10% per year and income in year 1 from well no. 24 was $390,000 and it decreased by $15,000 each year thereafter, (*a*) what was the income in year 3, and (*b*) what was the equivalent annual worth of the income through year 5?

2.58 The present worth of income from an investment that follows an arithmetic gradient is projected to be $475,000. The income in year one is expected to be $25,000. What is the gradient each year through year 6 at an interest rate of 10% per year?

2.59 Very Light Jets (VLJs) are one-pilot, two-engine jets that weigh 10,000 pounds or less and have only five or six passenger seats. Since they cost half as much as the most inexpensive business jets, they are considered to be the wave of the future. MidAm Charter purchased five planes so that it can initiate service to small cities that have airports with short runways. MidAm expects revenue of $1 million in year one, $1.2 million in year two and amounts increasing by $200,000 per year thereafter. If the company's MARR is 10% per year, what is the future worth of the revenue through the end of year 5? Solve using both tabulated factors and spreadsheet functions.

2.60 The future worth in year 10 of income associated with a fixed-income investment is guaranteed to be $500,000. If the cash flow in year 1 is $20,000, how much would the arithmetic gradient have to be at the interest rate of 10% per year?

2.61 Fomguard LLC of South Korea developed a high-tech fiber-optic fencing mesh (FOM) that contains embedded sensors that can differentiate between human and animal contact. In an effort to curtail illegal entry into the United States, a FOM fence has been proposed for some sections of the U.S. border with Canada. The cost for erecting the fence in year 1 is expected to be $7 million, decreasing by $500,000 each year through year 5. At an interest rate of 10% per year, what is the equivalent uniform annual cost of the fence for years 1 to 5?

2.62 For the cash flow series shown, determine the future worth in year 5 at an interest rate of 10% per year.

Year	1	2	3	4	5
Cash Flow, $	300,000	275,000	250,000	225,000	200,000

2.63 Verizon Communications said it plans to spend $22.9 billion to expand its fiber-optic Internet and television network so that it can compete with cable-TV providers like Comcast Corp. If the company attracts 950,000 customers in year one and grows its customer base by a constant amount of 15% per year, what is the future worth of the total subscription income in year 5? Estimates indicate that income will average $800 per customer per year. Assume Verizon uses a MARR of 10% per year.

2.64 The cost for manufacturing a component used in intelligent interface converters was $23,000 the first year. The company expects the cost to increase by 2% each year. Calculate the present worth of this cost over a five-year period at an interest rate of 10% per year.

2.65 Many companies offer retirement plans wherein the company matches the contributions made by the employee up to 6% of the employee's salary. An engineer planning for her retirement expects to invest the maximum of 6% each year. Her salary in year one is $60,000 and is expected to increase by 4% each year. Including the employer's contributions, how much will she have in her account at the end of 20 years if interest accrues at 7% per year?

2.66 A concept car that will get 100 miles per gallon and carry four persons would have a carbon-fiber and aluminum composite frame with a 900 cc three-cylinder turbodiesel/electric hybrid power plant. The extra cost of these technologies is estimated to be $11,000. (*a*) If gasoline savings over a comparable conventional car would be $900 in year one, increasing by 10% each year, what is the present worth of the savings over a 10-year period at an interest rate of 8% per year? (*b*) Compare the present worth values to determine if the extra cost is recovered by fuel savings.

2.67 The National Institute on Drug Abuse has spent $15 million on clinical trials to find out whether two vaccines can end the bad habits of nicotine and cocaine addiction. A Swiss company is now testing an obesity vaccine. If the vaccines are semi-successful such that treatment costs and medical bills are reduced by an average of $15,000 per person per year, what is the annual worth of the vaccines for 10 million beneficiaries in year one and an additional 15% people each year through year 5? Use an interest rate of 8% per year.

2.68 Find the annual worth in years 1 through 10 of an investment that starts at $8000 in year 1 and increases by 10% each year. The interest rate is 10% per year.

2.69 The effort required to maintain a scanning electron microscope is known to increase by a fixed percentage each year. A high-tech equipment maintenance company has offered its services for a fee that includes automatic increases of 8% per year after year 1. A biotech company that wanted to use the service offered $65,000 as pre-payment for a 3-year contract to take advantage of a temporary tax loophole. If the biotech company used an interest rate of 10% per year in determining how much it should offer, what was the service fee amount that it assumed for year 1?

2.70 Hughes Cable Systems plans to offer its employees a salary enhancement package that has revenue sharing as its main component. Specifically, the company will set aside 1% of total sales for year-end bonuses for all its employees. The sales are expected to be $5 million the first year, $5.5 million the second year, and amounts increasing by 10% each year for the next 5 years. At an interest rate of 8% per year, what is the equivalent annual worth in years 1 through 5 of the bonus package?

2.71 Determine how much money will be in an investment account that starts at $5000 in year 1 and increases each year thereafter by 15% per year. Use an interest rate of 10% per year and a 12-year time period.

2.72 The present worth in year 10 of a decreasing geometric gradient series was calculated using tabulated factors to be $80,000. If the interest rate was 10% per year and the annual rate of decrease was 8% per year, determine the cash flow amount in year 1 using (*a*) tabulated factors, and (*b*) the Goal Seek tool in Excel. (Hint: See Appendix A, section A.4).

2.73 Altmax Ltd, a company that manufactures automobile wiring harnesses, has budgeted *P* = $400,000 *now* to pay for a certain type of wire clip over the next 5 years. If the company expects the cost of the clips to increase by 4% each year, what is the expected cost in year 3 if the company uses an interest rate of 10% per year?

2.74 Thomasville Furniture Industries offers several types of high-performance fabrics that are capable of withstanding chemicals as harsh as chlorine. A Midwestern manufacturing company that uses fabric in several products has a report showing that the present worth of fabric purchases over a specified 5-year period was $900,000. If the costs are known to have increased geometrically by 5% per year during that time and the company uses an interest rate of 15% per year for investments, what was the cost of the fabric in year 1?

2.75 A small northern California consulting firm wants to start a recapitalization pool for replacement of network servers. If the company invests $5000 at the end of year 1 but decreases the amount invested by 5% each year, how much will be in the account 5 years from now? Interest is earned at a rate of 8% per year.

2.76 A company that manufactures purgable hydrogen sulfide monitors is planning to make deposits such that each one is 5% *smaller* than the preceding one. What must be the deposit at the end of year 1 if the deposits will extend through year 10 and the fourth deposit is $1250? Use an interest rate of 10% per year.

Shifted Cash Flows

2.77 Akron Coating and Adhesives (ACA) produces a hot melt adhesive that provides a strong bond between metals and thermoplastics when used for weather stripping, seals, gaskets, hand grips for tools, appliances, etc. ACA claims that by eliminating a primer coat, manufacturers can cut costs and reduce scrap. If Porter Cable is able to save $60,000 *now* and $40,000 per year thereafter by switching to the new adhesive, what is the present worth of the savings for three years at an interest rate of 10% per year?

2.78 Attenuated Total Reflectance (ATR) is a method for looking at the surfaces of materials that are too opaque or too thick for standard transmission methods. A manufacturer of precision plastic parts estimates that ATR spectroscopy can save the company $8000 per year by reducing returns of out-of-spec parts. What is the future worth of the savings if they start now and extend through year 4? Use $i = 10\%$ per year.

2.79 To improve crack detection in aircraft, the U.S. Air Force combined ultrasonic inspection procedures with laser heating to identify fatigue cracks. Early detection of cracks may reduce repair costs by an estimated $200,000 per year. If the savings start now and continue through year 5, determine the equivalent future worth in year 5 of these savings at an interest rate of 10% per year.

2.80 An entrepreneurial new civil engineering graduate started a lab for testing endocrine disrupting compounds. The lab just broke even the first year, but in years 2 through 5, it made a profit of $97,000 each year. What is the present worth in year 0 of the profit at an interest rate of 10% per year?

2.81 A mechanical engineer who is anticipating paying for his daughter's college education plans to start depositing money *now* (year 0) and continue through year 17. If he deposits $5000 each year, how much will his daughter be able to withdraw each year starting in year 18 and continuing through year 22? Assume the account earns interest at 8% per year.

2.82 For the cash flows shown, calculate the future worth in year 8 at $i = 10\%$ per year.

Year	0	1	2	3	4	5	6
Cash flow, $	100	100	100	200	200	200	200

2.83 An Air Force Research Laboratory has spent $62 million over a decade developing and testing the Active Denial System, a device which uses energy beams to allow soldiers to break up unruly mobs without firing a shot. Another $9 million will be spent over the next two years. Assume the

$62 million was spent evenly in years 1 through 10 and that the $9 million will be expended evenly in years 11 and 12. What is the total future worth of the expenditures in year 12 at an interest rate of 8% per year? Solve using (*a*) tabulated factors, and (*b*) write the calculator function to find the future worth.

2.84 Calculate the annual cost for years 1 through 9 of the following series of disbursements. Use an interest rate of 10% per year. Solve using (*a*) factors and (*b*) a spreadsheet.

Year	Disbursement, $	Year	Disbursement, $
0	5,000	5	5,000
1	4,000	6	5,000
2	4,000	7	5,000
3	4,000	8	5,000
4	4,000	9	5,000

2.85 A company that manufactures digital pressure gauges just borrowed $10,000,000 with the understanding that it will make annual payments of $2,000,000 for three years starting next year, and then pay off the balance in year 4. If the interest rate is 8% per year, how much will the company owe at the end of year 4?

2.86 a. For the cash flow revenues shown, find the value of CF_3 that makes the equivalent annual worth in years 1 through 7 equal to exactly $200. The interest rate is 10% per year.

Year	Cash flow, $	Year	Cash flow, $
0	200	4	200
1	200	5	200
2	200	6	200
3	CF_3	7	200

b. Use a spreadsheet to find CF_3.

2.87 Some studies have shown that taller men tend to earn higher salaries than equally qualified men of shorter stature. Homotrope growth hormone can increase a child's height at a cost of $50,000 per inch. Clyde's parents wanted him to be 3" taller than he was projected to be. They paid $50,000 per year for the treatments for three years beginning on their son's 8th birthday. How much extra money would Clyde have to earn *per year* from his 26th through 60th birthdays (a total of 35 years) in order to justify their expenditure? Use $i = 8\%$ per year.

2.88 Calculate the future worth in year 7 of Merchant Trucking Company's cash flows. Use an interest rate of 10% per year.

Year	0	1	2	3	4	5	6	7
Cash Flow, $ million	450	−40	200	200	200	200	200	200

2.89 The by-product department of Iowa Packing utilizes a cooker that has the cost stream shown. Determine the annual worth for years 1 through 5 of the costs at an interest rate of 10% per year. Costs are in $1000 units.

Year	Cost, $
0	850
1	300
2	400
3	400
4	400
5	500

2.90 An entrepreneurial electrical engineer approached a large water utility with a proposal that promises to reduce the utility's power bill by at least 15% through installation of patented surge protectors. The proposal states that the engineer will not be paid for the first year, but beginning in year 2, she will receive three equal, annual payments that are equivalent to 60% of the power bill savings achieved in year 1 due to the protectors. Assuming that the utility's power bill of $1 million per year is reduced by 15% after installation of the surge protectors, what is (a) the present worth in year 0 of the uniform payments to the engineer, and (b) their future worth in year 4? Use an interest rate of 10% per year.

2.91 Metropolitan Water Utility is planning to upgrade its SCADA system for controlling well pumps, booster pumps, and disinfection equipment so that everything can be controlled from one site. The first phase will reduce labor and travel costs by an estimated $31,000 per year. The second phase will reduce costs by an estimated $20,000 per year. If phase I will occur in years 1 through 3 and phase II in years 4 through 8, what is (a) the present worth of the savings, and (b) the equivalent annual worth for years 1 through 8 of the savings? Use an interest rate of 8% per year. Solve using tabulated factors and a spreadsheet.

2.92 Infrared thermometers by Delta Thermal Products are compatible with type K thermocouples and can provide rapid non-contact measurement capabilities at a cost of $135 per unit. A small private electric utility company plans to purchase 100 of the thermometers now and 500 more 1 year from now, if the anticipated savings in labor costs can be realized. At an interest rate of 12% per year, what is the equivalent annual worth of the savings over a study period of years 1 through 5 that justifies the equipment purchases?

2.93 Encon Systems, Inc. sales revenues for a product line introduced 7 years ago is shown. Use tabulated factors, a calculator or a spreadsheet to calculate the equivalent annual worth over the 7 years using an interest rate of 10% per year.

Year	Revenue, $	Year	Revenue, $
0	4,000,000	4	5,000,000
1	4,000,000	5	5,000,000
2	4,000,000	6	5,000,000
3	4,000,000	7	5,000,000

2.94 Cisco's *gross revenue* (the percentage of total revenue left after subtracting the cost of goods sold) was 70.1% of total revenue over a 4-year period. If the *total revenue* per year was $5.8 billion for the first two years and $6.2 billion for the last 2, find the future worth of the gross revenue series in year 4 at an interest rate of 14% per year.

2.95 Calculate (a) the annual worth for years 1–8 and (b) the future worth in year 8 of the following series of incomes and expenses. The interest rate is 10% per year.

Year	Income, $	Expense, $
0		20,000,000
1–5	8,000,000	1,000,000
6–8	9,000,000	2,000,000

2.96 A supplier of certain suspension system parts for General Motors wants to have a contingency fund that it can draw on during down periods of the economy. The company wants to have $15 million in the fund 5 years from now. If the company deposits $1.5 million now, determine the uniform amount to add at the end of each of the next 5 years to reach its goal, provided the fund earns 10% per year. Solve using (*a*) tabulated factors, and (*b*) the Goal Seek tool in Excel. (Hint: See Appendix A, Section A.4.)

2.97 A rural utility company provides standby power to pumping stations using diesel-powered generators. An alternative has arisen whereby the utility could use a combination of wind and solar power to run its generators, but it will be a few years before the alternative energy systems are available. The utility estimates that the new systems will result in savings of $15,000 per year for 3 years, starting *2 years from now* and $25,000 per year for 4 more years after that, i.e., through year 8. At an interest rate of 8% per year, determine the equivalent annual worth for years 1 through 8 of the projected savings.

2.98 The cost of energy for operating high-lift pumps in a water distribution system was $1.4 million for the first three years. Beginning in year four, the cost increased by $30,000 each year for the next 12 years. What is the present worth in year 0 of these energy costs at an interest rate of 6% per year?

2.99 A chemical engineer saving for his retirement deposited $10,000 into a stock fund at $t = 0$. In year one and each year thereafter through year 20, he increased the deposit by $1500. If the rate of return on the investments is 12% per year, how much will his retirement account be worth at the end of year 20?

2.100 An industrial engineering consulting firm signed a lease agreement for simulation software. Calculate the present worth in year 0 if the lease requires a payment of $30,000 *now* and amounts increasing by 5% per year through year 7. Use an interest rate of 10% per year.

2.101 A build to operate (BTO) company signed a contract to operate several industrial wastewater treatments plants for 10 years. The contract will pay the company $2.5 million *now* and amounts increasing by $200,000 each year through year 10. At an interest rate of 10% per year, what is the present worth of the contract *now*? Solve using (*a*) tabulated factors, and (*b*) a spreadsheet.

2.102 Nippon Steel's expenses for heating and cooling one of its large manufacturing facilities are expected to increase according to an arithmetic gradient beginning in year 2. The costs are expected to be $550,000 now (year 0), $550,000 in year 1, and increases by $40,000 each year through year 12. What is the equivalent annual worth in years 1–12 of the costs at an interest rate of 10% per year?

2.103 Lifetime Savings Accounts, known as LSAs, allow people to invest after-tax money without being taxed on any of the gains. An engineer began his LSA by investing $10,000 five years ago and increased his deposit by $1000 each year, including a deposit today. How much will be in the account immediately after today's deposit (after a total of 6 deposits), if the account grew at a rate of 12% per year?

2.104 A software company that installs systems for inventory control using RFID technology spent $600,000 per year for the past 3 years in developing their latest product. The company wants to recover its investment in 5 years beginning now. If the company signed a contract that will pay $250,000 now and amounts increasing by a uniform amount each year through year 5, how much must the increase be each year? Use an interest rate of 15% per year.

2.105 The future worth in year 8 for the cash flow series shown is $20,000. At an interest rate of 10% per year, what is the value of the cash flow labeled x in year 4?

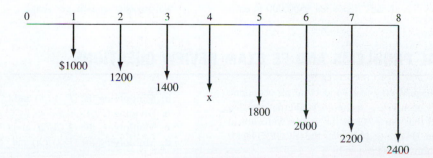

2.106 The annual worth for years 1 through 8 of the cash flows shown is $30,000. What is the amount of the cash flow labeled *x* in year 3, if *i* = 10% per year? Solve using (*a*) tabulated factors, and (*b*) a spreadsheet.

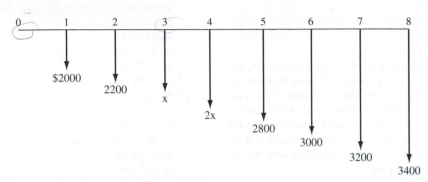

2.107 Wrangler Western has some of its jeans stone-washed under a contract with independent U.S. Garment Corp. If U.S. Garment's operating cost per machine is $22,000 per year for years 1 and 2 and then it increases by 8% per year through year 10, what is the present worth in year 0 of the machine operating cost at an interest rate of 10% per year?

2.108 Union Pacific is considering the elimination of a railroad grade crossing by constructing a dual-track overpass. The railroad subcontracts for maintenance of its crossing gates at $11,500 per year. Beginning 4 years from now, however, the costs are expected to increase by 10% per year into the foreseeable future (that is, $12,650 in year 4, $13,915 in year 5, etc.). If the railroad uses a 10-year study period and an interest rate of 15% per year, how much could the railroad afford to spend now on the overpass in lieu of the maintenance contracts?

2.109 McCarthy Construction is trying to bring the company-funded portion of its employee retirement fund into compliance with HB-301. The company has already deposited $500,000 in each of the last 5 years. Beginning in year 6, McCarthy will increase its deposits by 15% per year through year 20. How much will be in the fund immediately after the last deposit, if the fund grows at a rate of 12% per year? Solve using (*a*) tabulated factors, and (*b*) a spreadsheet.

2.110 San Antonio is considering various options for providing water in their 50-year plan, including desalting. One brackish aquifer is expected to yield desalted water that will generate revenue of $4.1 million per year for the first 4 years, after which less production will decrease revenue by 10% each year. If the aquifer will be totally depleted in 20 years, what is the present worth of the desalting option at an interest rate of 6% per year?

2.111 Revenue from gas wells that have been in production for at least 5 years tends to follow a decreasing geometric gradient. One particular rights holder received royalties of $4000 per year for years 1 through 6, but beginning in year 7, income decreased by 15% per year through year 14. What was the future value in year 14 if all of the income was invested at 10% per year?

ADDITIONAL PROBLEMS AND FE EXAM REVIEW QUESTIONS

2.112 A mechanical engineer conducting an economic analysis of wireless technology alternatives discovered that the *F/G* factor values were not in his table. He decided to create the (*F/G,i,n*) factor values himself. He did so by:

a. Multiplying the (*F/A,i,n*) and (*A/G,i,n*) values
b. Dividing (*F/A,i,n*) values by (*A/G,i,n*) values
c. Multiplying the (*F/A,i,n*) and (*P/G,i,n*) values
d. Multiplying the (*P/G,i,n*) and (*A/F,i,n*) values

2.113 The amount of money that would be accumulated in 10 years from an initial investment of $1000 at an interest rate of 8% per year is closest to:
- **a.** $2160
- **b.** $2290
- **c.** $2418
- **d.** $2643

2.114 A manufacturer of triaxial accelerometers wants to have $2,800,000 available 10 years from now so that a new product line can be initiated. If the company plans to deposit money each year, starting one year from now, the equal amount to deposit each year at 6% per year interest in order to have the $2,800,000 available immediately after the last deposit is closest to:
- **a.** $182,000
- **b.** $188,500
- **c.** $191,300
- **d.** $212,400

2.115 Yejen Industries Ltd. invested $10,000,000 in manufacturing equipment for producing small wastebaskets. If the company uses an interest rate of 15% per year, the amount of money it will have to earn each year to recover its investment in 7 years is closest to:
- **a.** $2,403,600
- **b.** $3,530,800
- **c.** $3,941,800
- **d.** $4,256,300

2.116 An engineer planning for retirement decides that she wants to have income of $100,000 per year for 20 years with the first withdrawal beginning 30 years from now. If her retirement account earns interest at 8% per year, the annual amount she would have to deposit for 29 years beginning 1 year from now is closest to:
- **a.** $7360
- **b.** $8125
- **c.** $8670
- **d.** $9445

2.117 A winner of the state lottery was given two choices: receive a single lump sum payment *now* of $50 million *or* receive 21 uniform payments, with the first payment to be made *now*, and the rest to be made at the end of each of the next 20 years. At an interest rate of 4% per year, the amount of the 21 uniform payments that would be equivalent to the $50 million lump-sum payment is closest to:
- **a.** $3,152,000
- **b.** $3,426,800
- **c.** $3,623,600
- **d.** $3,923,800

2.118 The equivalent amount of money that can be spent seven years from now in lieu of spending $50,000 now at an interest rate 18% per year is closest to:
- **a.** $15,700
- **b.** $159,300
- **c.** $199,300
- **d.** $259,100

2.119 Income from sales of an injector-cleaning gasoline additive has been averaging $100,000 per year. At an interest rate of 18% per year, the future worth of the income in years 1–5 is closest to:
- **a.** $496,180
- **b.** $652,200
- **c.** $715,420
- **d.** $726,530

2.120 The maker of a motion-sensing towel dispenser is considering adding new products to enhance offerings in the area of touchless technology. If the company does not expand its product line now, it will definitely do so in 2 years. Assume the interest rate is 10% per year. The amount the company can afford to spend *now* if the cost 2 years from now is estimated to be $100,000 is closest to:
- **a.** $75,130
- **b.** $82,640
- **c.** $91,000
- **d.** $93,280

2.121 Assume you borrow $10,000 today and promise to repay the loan in two payments, one in year 2 and the other in year 4, with the one in year 4 being only half as large as the one in year 2. At an interest rate of 10% per year, the size of the payment in year 4 will be closest to:
- **a.** $4280
- **b.** $3975
- **c.** $3850
- **d.** $2335

2.122 Chemical costs associated with a packed-bed flue gas incinerator (for odor control) have been decreasing uniformly for 5 years because of increases in efficiency. If the cost in year 1 was $100,000 and it decreased by $5,000 per year through year 5, the present worth of the costs at 10% per year is closest to:
- **a.** $344,800
- **b.** $402,200
- **c.** $485,700
- **d.** $523,300

2.123 If you borrow $24,000 now at an interest rate of 8% per year and promise to repay the loan with payments of $3000 per year starting one year from now, the number of payments that you will have to make is closest to:

    a. 8

    b. 11

    c. 14

    d. 17

2.124 You deposit $1000 now and you want the account to have a value as close to $8870 as possible in year 20. Assume the account earns interest at 10% per year. The year in which you must make another deposit of $1000 is

    a. 6

    b. 8

    c. 10

    d. 12

2.125 Maintenance costs for a regenerative thermal oxidizer have been increasing uniformly for 5 years. If the cost in year 1 was $8000 and it increased by $900 per year through year 5, the present worth of the costs at an interest rate of 10% per year is closest to:

    a. $31,670

    b. $33,520

    c. $34,140

    d. $36,500

2.126 At an interest rate of 8% per year, the present worth in year 0 of a lease that requires a payment of $9,000 *now* and amounts increasing by 8% per year through year 7 is closest to:

    a. $60,533

    b. $65,376

    c. $72,000

    d. $69,328

Nominal and Effective Interest Rates

Comstock Images/Getty Images

In all engineering economy relations developed thus far, the interest rate has been a constant, annual value. For a substantial percentage of the projects evaluated by engineers in practice, the interest rate is compounded more frequently than once a year; frequencies such as semiannual, quarterly, and monthly are common. In fact, weekly, daily, and even continuous compounding may be experienced in some project evaluations. Also, in our own personal lives, many of the financial considerations we make—loans of all types (home mortgages, credit cards, automobiles, boats), checking and savings accounts, investments, stock option plans, etc.—have interest rates compounded for a time period shorter than 1 year. This requires the introduction of two new terms—nominal and effective interest rates. This chapter explains how to understand and use nominal and effective interest rates in engineering practice and in daily life situations.

LEARNING OUTCOMES

1. Understand nominal and effective interest rate statements.

   Nominal and effective

2. Determine the effective interest rate for any time period.

   Effective interest rate

3. Determine the correct i and n values for different payment and compounding periods.

   Compare PP and CP

4. Make equivalence calculations for various payment periods and compounding periods when only single amounts occur.

   Single amounts: PP $\geq$ CP

5. Make equivalence calculations when uniform or gradient series occur for payment periods equal to or longer than the compounding period.

   Series: PP $\geq$ CP

6. Make equivalence calculations for payment periods shorter than the compounding period.

   Single and series: PP $<$ CP

7. Use a spreadsheet to perform equivalency computations involving nominal and effective interest rates.

   Spreadsheets

3.1 NOMINAL AND EFFECTIVE INTEREST RATE STATEMENTS

Podcast available

In Chapter 1, we learned that the primary difference between simple interest and compound interest is that compound interest includes interest on the interest earned in the previous period, while simple interest does not. Here we discuss *nominal and effective interest rates,* which have the same basic relationship. The difference here is that the concepts of nominal and effective are used when interest is compounded more than once each year. For example, if an interest rate is expressed as 1% per month, the terms *nominal* and *effective* interest rates must be considered. Every nominal interest rate *must* be converted into an effective rate before it can be used in formulas, factor tables, calculator, or spreadsheet functions because they are all derived using effective rates.

The term *APR (Annual Percentage Rate)* is often stated as the annual interest rate for credit cards, loans, and house mortgages. This is the same as the *nominal rate.* An APR of 15% is the same as nominal 15% per year or a nominal 1.25% per month.

Also, the term *APY (Annual Percentage Yield)* is a commonly stated annual rate of return for investments, certificates of deposit, and savings accounts. This is the same as an *effective rate.* As we will discover, the nominal rate never exceeds the effective rate, and similarly APR < APY.

Before discussing the conversion from nominal to effective rates, it is important to *identify* a stated rate as either nominal or effective. There are three general ways of expressing interest rates as shown by the three groups of statements in Table 3.1. The three statements in the top third of the table show that an interest rate can be stated over some designated time period without specifying the compounding period. Such interest rates are assumed to be effective rates with the *compounding period (CP)* the same as that of the stated interest rate.

For the interest statements presented in the middle of Table 3.1, three conditions prevail: (1) The compounding period is identified, (2) this compounding period is shorter than the time period over which the interest is stated, and (3) the interest rate is designated neither as nominal nor as effective. In such cases, the interest rate is assumed to be *nominal* and the compounding period is equal to that which is stated. (We learn how to get effective interest rates from these in the next section.)

For the third group of interest-rate statements in Table 3.1, the word *effective* precedes or follows the specified interest rate, and the compounding period is also given. These interest rates are obviously effective rates over the respective time periods stated.

The importance of being able to recognize whether a given interest rate is nominal or effective cannot be overstated with respect to the reader's understanding of the remainder of the material in this chapter and indeed the rest of the book. Table 3.2 contains a listing of several interest statements (column 1) along with their interpretations (columns 2 and 3).

TABLE 3.1 Various Interest Statements and Their Interpretations

(1) Interest Rate Statement	(2) Interpretation	(3) Comment
$i = 12\%$ per year	$i = $ *effective* 12% per year compounded yearly	When no compounding period is given, interest rate is an effective rate, with compounding period assumed to be equal to stated time period.
$i = 1\%$ per month	$i = $ *effective* 1% per month compounded monthly	
$i = 3\frac{1}{2}\%$ per quarter	$i = $ *effective* 3½% per quarter compounded quarterly	
$i = 8\%$ per year, compounded monthly	$i = $ *nominal* 8% per year compounded monthly	When compounding period is given without stating whether the interest rate is nominal or effective, it is assumed to be nominal. Compounding period is as stated.
$i = 4\%$ per quarter compounded monthly	$i = $ *nominal* 4% per quarter compounded monthly	
$i = 14\%$ per year compounded semiannually	$i = $ *nominal* 14% per year compounded semiannually	
$i = $ APY of 10% per year compounded monthly	$i = $ *effective* 10% per year compounded monthly	If interest rate is stated as an effective or APY rate, then it is an effective rate. If compounding period is not given, compounding period is assumed to coincide with stated time period.
$i = $ effective 6% per quarter	$i = $ *effective* 6% per quarter compounded quarterly	
$i = $ effective 1% per month compounded daily	$i = $ *effective* 1% per month compounded daily	

TABLE 3.2 Specific Examples of Interest Statements and Interpretations

(1) Interest Rate Statement	(2) Nominal or Effective Interest	(3) Compounding Period
15% per year compounded monthly	Nominal	Monthly
15% per year	Effective	Yearly
Effective 15% per year compounded monthly	Effective	Monthly
20% per year compounded quarterly	Nominal	Quarterly
Nominal 2% per month compounded weekly	Nominal	Weekly
2% per month	Effective	Monthly
2% per month compounded monthly	Effective	Monthly
Effective 6% per quarter	Effective	Quarterly
Effective 2% per month compounded daily	Effective	Daily
1% per week compounded continuously	Nominal	Continuously

3.2 EFFECTIVE INTEREST RATE FORMULATION

Podcast available

Understanding effective interest rates requires a definition of a nominal interest rate r as the interest rate per period times the number of periods. In equation form,

$$r = \text{interest rate per period} \times \text{number of periods} \qquad [3.1]$$

A nominal interest rate can be found for any time period that is longer than the compounding period. For example, an interest rate of 1.5% per month can be expressed as a *nominal* 4.5% per quarter (1.5% per period × 3 periods), 9% per semiannual period, 18% per year, or 36% per 2 years. Nominal interest rates obviously neglect compounding.

The equation for converting a nominal interest rate into an effective interest rate is

$$i \text{ per period} = (1 + r/m)^m - 1 \qquad [3.2]$$

where i is the *effective* interest rate for a certain period, say six months, r is the *nominal* interest rate for that *period* (six months here), and m is the number of times interest is *compounded in that same period* (six months in this case). The term m is often called the *compounding frequency*. As was true for nominal interest rates, effective interest rates can be calculated for any time period longer than the compounding period of a given interest rate. The next example illustrates the use of Equations [3.1] and [3.2].

EXAMPLE 3.1

a. A Visa credit card issued through Frost Bank carries an interest rate of 1% per month on the unpaid balance. Calculate the effective rate per semiannual and annual periods.

b. If the card's interest rate is stated as 3.5% per quarter, find the effective semiannual and annual rates.

Solution

a. The compounding period is monthly. For the effective interest rate per semiannual period, the r in Equation [3.2] must be the nominal rate per 6 months.

$$r = 1\% \text{ per month} \times 6 \text{ months per semiannual period}$$
$$= 6\% \text{ per semiannual period}$$

The m in Equation [3.2] is equal to 6, since the frequency with which interest is compounded is 6 times in 6 months. The effective semiannual rate is

$$i \text{ per 6 months} = \left(1 + \frac{0.06}{6}\right)^6 - 1$$
$$= 0.0615 \qquad (6.15\%)$$

For the effective annual rate, $r = 12\%$ per year and $m = 12$. By Equation [3.2],

$$\text{Effective } i \text{ per year} = \left(1 + \frac{0.12}{12}\right)^{12} - 1 = 0.1268 \qquad (12.68\%)$$

b. For an interest rate of 3.5% per quarter, the compounding period is a quarter. In a semiannual period, $m = 2$ and $r = 7\%$.

$$i \text{ per 6 months} = \left(1 + \frac{0.07}{2}\right)^2 - 1$$

$$= 0.0712 \quad (7.12\%)$$

The effective interest rate per year is determined using $r = 14\%$ and $m = 4$.

$$i \text{ per year} = \left(1 + \frac{0.14}{4}\right)^4 - 1$$

$$= 0.1475 \quad (14.75\%)$$

Comment: Note that the term r/m in Equation [3.2] is always the effective interest rate per compounding period. In part (a) this is 1% per month, while in part (b) it is 3.5% per quarter.

If we allow compounding to occur more and more frequently, the compounding period becomes shorter and shorter. Then m, the number of compounding periods per payment period, increases. This situation occurs in businesses that have a very large number of cash flows every day, so it is correct to consider interest as compounded continuously. As m approaches infinity, the effective interest rate in Equation [3.2] reduces to

$$i = e^r - 1 \qquad\qquad [3.3]$$

Equation [3.3] is used to compute the *effective continuous interest rate*. The time periods on i and r must be the same. As an illustration, if the nominal annual $r = 15\%$ *per year,* the effective continuous rate *per year* is

$$i\% = e^{0.15} - 1 = 16.183\%$$

For national and international chains—retailers, banks, etc.—and corporations that move thousands of items in and out of inventory each day, the flow of cash is essentially continuous. *Continuous cash flow* is a realistic model for the analyses performed by engineers and others in these organizations. The equivalence computations reduce to the use of integrals rather than summations. The topics of financial engineering analysis and continuous cash flows, coupled with continuous interest rates, are beyond the scope of this text; consult more advanced texts for formulas and procedures.

EXAMPLE 3.2   **a.** For an interest rate of 18% per year compounded continuously, calculate the effective monthly and annual interest rates.
b. An investor requires an *effective* return of at least 15%. What is the minimum annual nominal rate that is acceptable for continuous compounding?

TABLE 3.3 **Effective Annual Interest Rates for Selected Nominal Rates**

Nominal Rate $r\%$	Semiannually $(m = 2)$	Quarterly $(m = 4)$	Monthly $(m = 12)$	Weekly $(m = 52)$	Daily $(m = 365)$	Continuously $(m = \infty; e^r - 1)$
2	2.010	2.015	2.018	2.020	2.020	2.020
4	4.040	4.060	4.074	4.079	4.081	4.081
5	5.063	5.095	5.116	5.124	5.126	5.127
6	6.090	6.136	6.168	6.180	6.180	6.184
8	8.160	8.243	8.300	8.322	8.328	8.329
10	10.250	10.381	10.471	10.506	10.516	10.517
12	12.360	12.551	12.683	12.734	12.745	12.750
15	15.563	15.865	16.076	16.158	16.177	16.183
18	18.810	19.252	19.562	19.684	19.714	19.722

Solution

a. The nominal monthly rate is $r = 18\%/12 = 1.5\%$, or 0.015 per month. By Equation [3.3], the effective monthly rate is

$$i\% \text{ per month} = e^r - 1 = e^{0.015} - 1 = 1.511\%$$

Similarly, the effective annual rate using $r = 0.18$ per year is

$$i\% \text{ per year} = e^r - 1 = e^{0.18} - 1 = 19.72\%$$

b. Solve Equation [3.3] for r by taking the natural logarithm.

$$e^r - 1 = 0.15$$
$$e^r = 1.15$$
$$\ln e^r = \ln 1.15$$
$$r\% = 13.976\%$$

Therefore, a nominal rate of 13.976% per year compounded continuously will generate an effective 15% per year return.

Comment: The general formula to find the nominal rate, given the effective continuous rate i, is $r = \ln(1 + i)$

Table 3.3 summarizes the effective annual rates for frequently quoted nominal rates and various compounding frequencies.

3.3 RECONCILING COMPOUNDING PERIODS AND PAYMENT PERIODS

Now that the concepts of nominal and effective interest rates are introduced, in addition to considering the compounding period (which is also known as the interest period), it is necessary to consider the frequency of the payments or receipts within the cash-flow time interval. For simplicity, the frequency of the payments or receipts is known as the *payment period (PP)*. It is important to distinguish between

Podcast available

FIGURE 3.1 Cash-flow diagram for a monthly payment period (PP) and semiannual compounding period (CP).

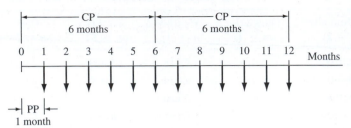

the compounding period (CP) and the payment period because in many instances the two do not coincide. For example, if a company deposited money each month into an account that pays a nominal interest rate of 6% per year compounded semi-annually, the payment period is 1 month while the compounding period is 6 months as shown in Figure 3.1. Similarly, if a person deposits money once each year into a savings account that compounds interest quarterly, the payment period is 1 year, while the compounding period is 3 months. Hereafter, for problems that involve either uniform-series or gradient cash-flow amounts, it will be necessary to determine the relationship between the compounding period and the payment period as a first step in the solution of the problem.

The next three sections describe procedures for determining the correct i and n values for use in formulas, and factor tables, as well as calculator and spreadsheet functions. In general, there are three steps:

1. Compare the lengths of PP and CP.
2. Identify the cash-flow series as involving only single amounts (P and F) or series amounts (A, G, or g).
3. Select the proper i and n values.

3.4 EQUIVALENCE CALCULATIONS INVOLVING ONLY SINGLE-AMOUNT FACTORS

Podcast available

There are many correct combinations of i and n that can be used when only single-amount factors (F/P and P/F) are involved. This is because there are only two requirements: (1) An effective rate must be used for i, and (2) the time unit on n must be the same as that on i. In standard factor notation, the single-payment equations can be generalized.

$$P = F(P/F, \text{effective } i \text{ per period, number of periods}) \qquad \textbf{[3.4]}$$

$$F = P(F/P, \text{effective } i \text{ per period, number of periods}) \qquad \textbf{[3.5]}$$

Thus, for a nominal interest rate of 12% per year compounded monthly, any of the i and corresponding n values shown in Table 3.4 could be used (as well as many others not shown) in the factors. For example, if an effective quarterly interest rate is used for i, that is, $(1.01)^3 - 1 = 3.03\%$, then the n time unit is 4 quarters in a year.

TABLE 3.4 **Various *i* and *n* Values for Single-Amount Equations Using *r* = 12% per Year, Compounded Monthly**

Effective Interest Rate, *i*	Units for *n*
1% per month	Months
3.03% per quarter	Quarters
6.15% per 6 months	Semiannual periods
12.68% per year	Years
26.97% per 2 years	2-year periods

Alternatively, it is always correct to determine the effective *i* per payment period using Equation [3.2] and to use standard factor equations to calculate *P*, *F*, or *A*.

Sherry expects to deposit $1000 now, $3000 4 years from now, and $1500 6 years from now and earn at a rate of 12% per year compounded semiannually through a company-sponsored savings plan. What amount can she withdraw 10 years from now?   **EXAMPLE 3.3**

Solution

Only single-amount *P* and *F* values are involved (Figure 3.2). Since only effective rates can be present in the factors, use an effective rate of 6% per semiannual compounding period and semiannual payment periods. The future worth is calculated using Equation [3.5].

$$F = 1000(F/P,6\%,20) + 3000(F/P,6\%,12) + 1500(F/P,6\%,8)$$
$$= \$11.634$$

An alternative solution strategy is to find the effective annual rate by Equation [3.2] and express *n* in years as determined from the problem statement.

$$\text{effective } i \text{ per year} = \left(1 + \frac{0.12}{2}\right)^2 - 1 = 0.1236 \qquad (12.36\%)$$

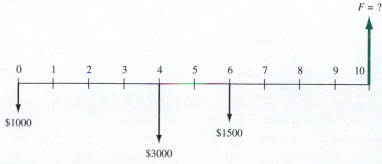

FIGURE 3.2 Cash flow diagram, Example 3.3

3.5 EQUIVALENCE CALCULATIONS INVOLVING SERIES WITH PP ≥ CP

When the cash flow of the problem dictates the use of one or more of the uniform-series or gradient factors, the relationship between the compounding period, CP, and payment period, PP, must be determined. The relationship will be one of the following three cases:

Type 1.   Payment period equals compounding period, PP = CP.
Type 2.   Payment period is longer than compounding period, PP > CP.
Type 3.   Payment period is shorter than compounding period, PP < CP.

The procedure for the first two types is the same. Type 3 problems are discussed in the following section. When PP = CP or PP > CP, the following procedure *always* applies:

Step 1.   Count the number of payments and use that number as n. For example, if payments are made quarterly for 5 years, n is 20.

Step 2.   Find the *effective* interest rate over the *same time period* as n in step 1. For example, if n is expressed in quarters, then the effective interest rate per quarter *must* be used.

Use these values for n and i (and only these!) in the factors, functions, or formulas. To illustrate, Table 3.5 shows the correct standard notation for sample cash-flow sequences and interest rates. Note in column 4 that n is always equal to the number of payments and i is an effective rate expressed over the same time period as n.

TABLE 3.5   Examples of n and i Values Where PP = CP or PP > CP

(1) Cash-flow Sequence	(2) Interest Rate	(3) What to Find; What is Given	(4) Standard Notation
$500 semiannually for 5 years	8% per year compounded semiannually	Find P; given A	$P = 500(P/A,4\%,10)$
$75 monthly for 3 years	12% per year compounded monthly	Find F; given A	$F = 75(F/A,1\%,36)$
$180 quarterly for 15 years	5% per quarter	Find F; given A	$F = 180(F/A,5\%,60)$
$25 per month increase for 4 years	1% per month	Find P; given G	$P = 25(P/G,1\%,48)$
$5000 per quarter for 6 years	1% per month	Find P; given A	$P = 5000(P/A,3.03\%,24)$

For the past 7 years, a quality manager has paid $500 every 6 months for the software maintenance contract on a laser-based measuring instrument. What is the equivalent amount after the last payment, if these funds are taken from a pool that has been returning 10% per year compounded quarterly?   **EXAMPLE 3.4**

Solution

The cash flow diagram is shown in Figure 3.3. The payment period (6 months) is longer than the compounding period (quarter); that is, PP > CP. Applying the guideline, determine an effective semiannual interest rate. Use Equation [3.2] or Table 3.3 with $r = 0.05$ per 6-month period and $m = 2$ quarters per semiannual period.

$$\text{Effective } i\% \text{ per 6-months} = \left(1 + \frac{0.05}{2}\right)^2 - 1 = 5.063\%$$

The value $i = 5.063\%$ is reasonable, since the effective rate should be slightly higher than the nominal rate of 5% per 6-month period. The number of semiannual periods is $n = 2(7) = 14$. The future worth is

$$F = A(F/A,5.063\%,14)$$
$$= 500(19.6845)$$
$$= \$9842$$

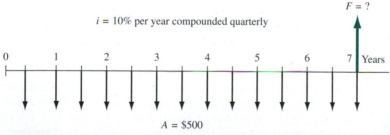

FIGURE 3.3 Diagram of semiannual payments used to determine F, Example 3.4.

ExxonMobil is using a recently installed remotely controlled system to detect underwater leakage from offshore platforms. Assume this system costs $3 million to install and an estimated $200,000 per year for all materials, operating, personnel, and maintenance costs. The expected life is 10 years. An engineer wants to estimate the total revenue requirement for each 6-month period that is necessary to recover the investment, interest, and annual costs. Find this semiannual A value if capital funds are evaluated at 8% per year compounded *semiannually*.   **EXAMPLE 3.5**

Solution

Figure 3.4 details the cash flows. There are several ways to solve this problem, but the most straightforward one is a two-stage approach. First, convert all cash

FIGURE 3.4

Cash flow diagram with different compounding and payment periods, Example 3.5.

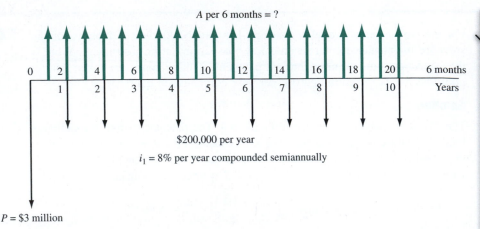

A per 6 months = ?

$200,000 per year

$i_1 = 8\%$ per year compounded semiannually

P = $3 million

flows to a P at time 0, then find the A over the 20 semiannual periods. For stage 1, recognize that PP > CP, that is, 1 year > 6 months. According to the procedure for types 1 and 2 cash flows, $n = 10$, the number of annual payments. Now, find the effective i *per year* by Equation [3.2] or Table 3.3 and use it to find P.

$$i\% \text{ per year} = (1 + 0.08/2)^2 - 1 = 8.16\%$$

$$P = 3,000,000 + 200,000(P/A,8.16\%,10)$$
$$= 3,000,000 + 200,000(6.6620)$$
$$= \$4,332,400$$

For stage 2, P is converted to a semiannual A value. Now, PP = CP = 6 months, and $n = 20$ *semiannual* payments. The effective semiannual i for use in the A/P factor is determined directly from the problem statement using r/m.

$$i\% \text{ per 6 months} = 8\%/2 = 4\%$$

$$A = 4,332,400(A/P,4\%,20)$$
$$= \$318,778$$

In conclusion, $318,778 every 6 months will repay the initial and annual costs, if money is worth 8% per year compounded semiannually.

3.6 EQUIVALENCE CALCULATIONS INVOLVING SERIES WITH PP < CP

If a person deposits money each *month* into a savings account where interest is compounded *quarterly,* do the so-called *interperiod deposits* earn interest? The usual answer is no. However, if a monthly payment on a $10 million, quarterly compounded bank loan were made early by a large corporation, the corporate financial officer would likely insist that the bank reduce the amount of interest due, based on early payment. These two are examples of PP < CP, type 3 cash flows. The timing of cash flow transactions between compounding points introduces the question of how *interperiod compounding* is handled. Fundamentally, there are two policies: interperiod cash flows earn *no interest*, or they earn *compound interest*. The only condition considered here is the first one (no interest), because many real-world transactions fall into this category.

For a no-interperiod-interest policy, deposits (negative cash flows) are all regarded as *deposited at the end of the compounding period*, and withdrawals are all regarded as *withdrawn at the beginning*. As an illustration, when interest is compounded quarterly, all monthly deposits are moved to the end of the quarter, and all withdrawals are moved to the beginning (no interest is paid for the entire quarter). This procedure can significantly alter the distribution of cash flows before the effective quarterly rate is applied to find P, F, or A. This effectively forces the cash flows into a PP = CP situation, as discussed in Section 3.5.

Rob is the on-site coordinating engineer for Alcoa Aluminum, where an under-renovation mine has new ore refining equipment being installed by a local contractor. Rob developed the cash flow diagram in Figure 3.5a in $1000 units from the project perspective. Included are payments to the contractor he has authorized for the current year and approved advances from Alcoa's home office. He knows that the interest rate on equipment "field projects" such as this is 12% per year compounded quarterly, and that Alcoa does not bother with interperiod compounding of interest. Will Rob's project finances be in the "red" or the "black" at the end of the year? By how much?

EXAMPLE 3.6

Solution

With no interperiod interest considered, Figure 3.5b reflects the moved cash flows. The future worth after four quarters requires an F at an effective rate per quarter such that PP = CP = 1 quarter, therefore, the effective $i = 12\%/4 = 3\%$. Figure 3.5b shows all negative cash flows (payments to contractor) moved to the end of the respective quarter, and all positive cash flows (receipts from home office) moved to the beginning of the respective quarter. Calculate the F value at 3%.

$$F = 1000[-150(F/P,3\%,4) - 200(F/P,3\%,3)$$
$$+ (-175 + 90)(F/P,3\%,2) + 165(F/P,3\%,1) - 50]$$
$$= \$-357,592$$

Rob can conclude that the on-site project finances will be in the red about $357,600 by the end of the year.

FIGURE 3.5
(a) Actual and
(b) moved cash
flows (in $1000)
for quarterly
compounding
periods using
no interperiod
interest,
Example 3.6.

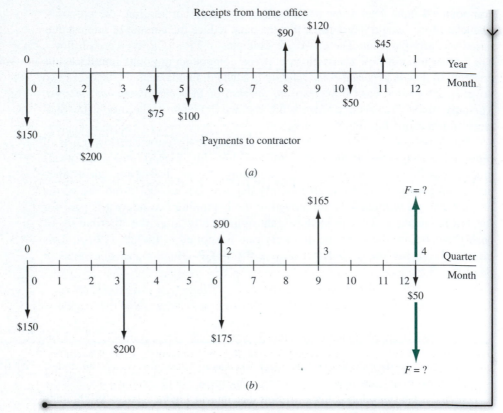

(a)

(b)

3.7 USING SPREADSHEETS FOR EFFECTIVE INTEREST RATE COMPUTATIONS

For hand calculations, converting between nominal and effective interest rates is accomplished with Equation [3.2]; for spreadsheets, the EFFECT and NOMINAL functions are applied as described below. Figure 3.6 gives two examples of each function.

Find effective rate: **EFFECT(nominal_rate, compounding frequency)**

As in Equation [3.2], the *nominal rate* is r and must be expressed over the same time period as that of the effective rate requested. The *compounding frequency* is m, which must equal the number of times interest is compounded for the period of time used in the effective rate. Therefore, in the second example of Figure 3.6 where the effective quarterly rate is requested, enter the nominal rate per quarter (3.75%) to get an effective rate per quarter, and enter $m = 3$, since monthly compounding occurs 3 times in a quarter.

Find nominal: **NOMINAL(effective_rate, compounding frequency per year)**

Example	What to find and what is given	Compounding frequency, m	Function	Result
1	Find **effective annual** rate, given 15% *nominal* compounded quarterly	4	**= EFFECT(15%,4)**	15.87%
2	Find **effective quarterly** rate, given 15% *nominal* compounded monthly	3	**= EFFECT(3.75%,3)**	3.80%
3	Find **nominal annual** rate, given 15% *effective* compounded monthy	12	**= NOMINAL(15%,12)**	14.06%
4	Find **nominal semiannual** rate, given 15% *effective* compounded monthly	12	**= NOMINAL(15%,12)/2**	7.03%

Must enter nominal rate per quarter; 15/4 = 3.75%

Must divide result by 2 to obtain semiannual nominal rate

FIGURE 3.6   Example uses of the EFFECT and NOMINAL functions to convert between nominal and effective interest rates.

This function always displays the *annual* nominal rate. Accordingly, the *m* entered must equal the number of times interest is compounded annually. If the nominal rate is needed for other than annually, use Equation [3.1] to calculate it. This is why the result of the NOMINAL function in Example 4 of Figure 3.6 is divided by 2.

If interest is compounded continuously, enter a very large value for the compounding frequency. A value of 10,000 or higher will provide sufficient accuracy. Say the nominal rate is 15% per year compounded continuously, Entering =EFFECT(15%,10000) displays 16.183%. The Equation [3.3] result is $e^{0.15} - 1 = 0.16183$, which is the same. The NOMINAL function is developed similarly to find the nominal rate per year, given the annual effective rate.

Once the effective interest rate is determined for the same timing as the cash flows, any function can be used, as illustrated in the next example.

EXAMPLE 3.7

Use a spreadsheet and a financial calculator to find the semiannual cash flow requested in Example 3.5.

Solution

Spreadsheet: This problem is important to spreadsheet use because it involves annual and semiannual cash flows as well as nominal and effective interest rates. Review the solution approach of Example 3.5 before proceeding here. Since PP > CP, the *P* of annual cash flows requires an effective interest rate for the nominal 8% per year compounded semiannually. Use EFFECT to obtain 8.16% and the PV function to obtain *P* = $-4,332,385. The spreadsheet (left side) in Figure 3.7 shows these two results.

Now refer to the right side of Figure 3.7. Determine the semiannual revenue requirement of *A* = $318,784 based on the *P* in cell B5. For the PMT function,

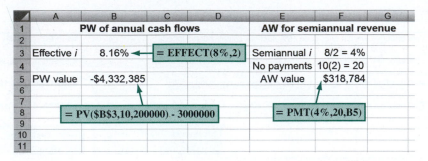

FIGURE 3.7 Use of EFFECT before applying spreadsheet functions to find P and A values over different time periods, Example 3.7.

the semiannual rate $8/2 = 4\%$ is entered. (Remember that effective rates must be used in all interest factors and spreadsheet functions. This 4% is the effective semiannual rate since interest is compounded semiannually.) Finally, note that the number of periods is 20 in the PMT function.

Calculator: First, find the effective semiannual interest rate for 8% using Equation [3.2].

$$i \text{ per 6 months} = (1 + 0.08/2)^2 - 1 = 8.16\%$$

Place an embedded $PV(i,n,A,F)$ function in a $PMT(i,n,P,F)$ function at 4% for each 6-month period and $n = 20$ periods. $F = 0$ in both functions and can be omitted. The complete function is PMT(4,20,PV(8.16,10,200000) − 3000000). The A per 6-months result is $318,784.

SUMMARY

Since many real-world situations involve cash flow frequencies and compounding periods other than 1 year, it is necessary to use nominal and effective interest rates. When a nominal rate r is stated, the effective interest rate per payment period is determined by using the effective interest rate equation.

$$\textbf{Effective } i = \left(1 + \frac{r}{m}\right)^m - 1$$

The m is the compounding frequency or number of compounding periods per payment period. If interest compounding becomes more and more frequent, the length of the CP approaches zero, continuous compounding results, and the effective i is $e^r - 1$.

All engineering economy factors require the use of an effective interest rate. The i and n values placed in a factor depend upon the type of cash flow series. If only single amounts (P and F) are present, there are several ways to perform equivalence calculations using the factors. However, when series cash flows (A, G, and g) are present, only one combination of the effective rate i and number of periods n is correct for the factors. This requires that the comparative lengths of PP and CP be considered as i and n are determined. *The interest rate and payment periods must have the same time unit* for the factors to correctly account for the time value of money.

PROBLEMS

Nominal and Effective Rates

3.1 For an interest rate of 2% per quarter, determine the nominal interest rate per (*a*) semiannual period; (*b*) year; and (*c*) 2 years.

3.2 Identify each of the following interest rate statements as either nominal or effective.
  a. 4% per year
  b. 6% per year compounded annually
  c. 10% per quarter
  d. 8% per year compounded monthly
  e. 1% per month
  f. 1% per month compounded monthly
  g. 0.1% per day compounded hourly
  h. effective 1.5% per month compounded weekly
  i. 12% per year compounded semiannually
  j. 1% per month compounded continuously

3.3 Identify the compounding period associated with each of the following interest rate statements.
  a. 2% per quarter
  b. 3% per quarter compounded monthly
  c. effective 1% per month compounded weekly
  d. nominal 6.5% per year compounded semi-annually
  e. 1% per month
  f. effective 7% per year compounded annually

3.4 Identify the compounding period for the following interest statements: (*a*) 3% per year; (*b*) 10% per year compounded quarterly; (*c*) nominal 7.2% per year compounded daily; (*d*) effective 3.4% per quarter compounded continuously; and (*e*) 0.012% per day compounded hourly.

3.5 Determine the number of times interest is compounded in 1 year for the following interest statements: (*a*) 3% per quarter; (*b*) 1% per month; and (*c*) 8% per year compounded semiannually.

3.6 Convert an interest rate of 1.5% per month into a nominal interest rate over the following time periods: (*a*) quarter; (*b*) year; (*c*) semiannual; (*d*) decade.

3.7 Convert an interest rate of 2% per quarter compounded monthly into a nominal interest rate over the following time periods: (*a*) quarter; (*b*) 2 months; (*c*) 2 years; (*d*) 1 year; (*e*) 6 months.

3.8 For an interest rate of 2% per month, find (*a*) the effective rate per quarter, and (*b*) the effective rate per year.

3.9 For an interest rate of 12% per year compounded continuously, find (*a*) the nominal rate per year, (*b*) the nominal rate per quarter; (*c*) the effective rate per quarter, and (*d*) the effective rate per month.

3.10 (*a*) An interest rate of 6.8% per semiannual period compounded weekly is equivalent to what weekly interest rate? (*b*) Is the weekly rate a nominal or effective rate? Assume 26 weeks per 6 months.

3.11 When interest is compounded monthly and single cash flows are separated by 3 years, what must the *time period* be on the interest rate if the value of n in the P/F or F/P equation is (*a*) $n = 3$; (*b*) $n = 6$; or (*c*) $n = 12$?

3.12 Deposits of $100 per week are made into an investment account that pays interest of 6% per year compounded quarterly. Identify the payment period, compounding period, and compounding frequency.

3.13 A bank advertises quarterly compounding for business checking accounts. What payment and compounding periods are associated with deposits of daily receipts?

3.14 Identify the payment period and compounding period for the following situations:
  a. Deposits are made each quarter into an account reserved for purchasing new equipment two years from now. The interest rate on the deposits is 12% per year compounded monthly.
  b. The Williams family takes a $10,000 withdrawal from a second retirement package each 6 months to pay property and income taxes. The funds are invested in a fixed return annuity that pays 7% per year.

3.15 Armstrong Fisheries took out a $400,000 loan. Clyde Armstrong wants to know the semiannual payment for the next 10 years at the loan interest rate of 8% per year compounded quarterly.
  a. Construct the cash flow diagram and indicate the compounding period and payment period. What is the compounding frequency for the quoted loan rate?
  b. What are the effective interest rate and n value that must be in the $(A/P,i\%,n)$ factor to find the semiannual payment?

3.16 Assume a problem statement involves only single amounts, that is, no series or gradients, and the interest rate is stated as 12% per year compounded quarterly. For the following n values, determine the proper interest rate to use in the factor equations: (*a*) $n = 20$ quarters; (*b*) $n = 10$ semiannual periods; (*c*) $n = 5$ years.

Equivalence for Single Amounts

3.17 A corporation deposits $20 million in a money market account for 1 year. What will be the difference in the total amount accumulated at the end of the year at 18% per year compounded monthly versus 18% per year simple interest?

3.18 The TerraMax truck currently being manufactured and field tested by Oshkosh Truck Co. is a driverless truck intended for military use. Such a truck would free personnel for non-driving tasks such as reading maps, scanning for roadside bombs, or scouting for the enemy. If such trucks would result in reduced injuries to military personnel amounting to $15 million three years from now, determine the present worth of these benefits at an interest rate of 10% per year compounded semiannually.

3.19 Federal facilities associated with an international port of entry at Tornillo, TX, are expected to cost $54 million. A six-lane, 1274-foot-long bridge is expected to cost $14 million. Land and inspection facilities, offices, and parking will add another $10 million to the cost. Construction is expected to take three years. Assuming that all of the funds are allocated at time 0, what is the future worth of the project in year 3 at an interest rate of 6% per year compounded quarterly?

3.20 Crow Corporation, a company that specializes in precision metal fabrication, is conducting a study to determine if it should update equipment now or later. If the cost 2 years from now is estimated to be $260,000, how much can the company afford to spend now if its minimum attractive rate of return is 12% per year compounded monthly?

3.21 Hydrex Mechanical Products plans to set aside $160,000 now for possibly replacing its large synchronous refiner motors whenever it becomes necessary. The replacement is expected to take place in 3-1/2 years. Determine how much the company will have in its investment set-aside account if it achieves a rate of return of 16% per year compounded quarterly. Find the amount in two different ways using each of the following: (*a*) tabulated factors, (*b*) TVM functions on a calculator, and (*c*) spreadsheet functions.

3.22 What is the future worth 4 years from now of a present cost of $242,000 to Hydron, Inc., at an interest rate of 18% per year compounded monthly?

3.23 Soil Mediators, Inc., plans to finance a site reclamation project that will require a 4-year cleanup period. If the company borrows $4.5 million now and wants to earn 16% per year compounded semiannually on its investment, how much will the company have to receive in a lump sum payment when the project is over?

3.24 A sum of $120,000 now at an interest rate of 10% per year compounded semiannually is equivalent to how much money 6 years ago? Solve using tabulated factors in two ways: using the effective annual rate and the effective semiannual rate.

3.25 Pollution control equipment for a pulverized coal cyclone furnace is expected to cost $190,000 two years from now and another $120,000 four years from now. If Monongahela Power wants to set aside enough money now to cover these costs, how much must be invested at an interest rate of 8% per year compounded quarterly?

3.26 Periodic outlays for inventory-control software at Baron Chemicals are estimated to be $120,000 next year, $180,000 in 2 years, and $250,000 in 3 years. What is the present worth of the costs at an interest rate of 10% per year compounded continuously?

3.27 For the cash flows shown, determine the future worth in year 5 at an interest rate of 10% per year compounded continuously. Solve using (*a*) tabulated factors, and (*b*) a financial calculator.

Year	1	2	3	4	5
Cash Flow, $	$300,000	0	$250,000	0	$200,000

3.28 In order to have enough money to finance his passion for top-level off-road racing, a mechanical engineer decided to sell some of his "toys." At time $t = 0$, he sold his chopper motorcycle for $18,000. Three months later, he sold his Baja prerunner SUV for $26,000. At the end of month six, he got $42,000 for his pro-light race truck. If he invested all of the money in commodity hedge funds that earned 32% per year compounded quarterly, how much did he have in the account at the end of 1-½ years?

3.29 Head & Shoulders shampoo insured spokesman Troy Polamalu's (of the Pittsburgh Steelers) head of hair for $1 million with Lloyd's of London. The insurance payout would be triggered if he lost at least 60% of his hair during an on-field event. Assume the insurance company placed the odds of a payout at 1% in year 5. If Lloyd's of London wanted a rate of return of 20% per year compounded semiannually, how much did Head and Shoulders have to pay in a lump sum amount for the insurance policy?

Equivalence when PP ≥ CP

3.30 Improvements at a Harley-Davidson Plant are estimated to be $7.8 million. Construction is expected to take three years. What is the future worth of the project in year 3 at an interest rate of 6% per year compounded quarterly, assuming the funds are allocated (a) completely at time 0, and (b) equally at the end of each year?

3.31 Hemisphere Electric Coop (HEC) is planning to outsource its 51-person information technology (IT) department to Dyonyx. HEC believes this move will allow it to have access to cutting edge technologies and skill sets that would be cost prohibitive to build on its own. If it is assumed that the loaded cost of an IT employee is $100,000 per year, and that HEC will save 25% of this cost through outsourcing, what is the present worth of the savings to HEC for a 5-year contract at an interest rate of 0.5% per month? Assume the same number of contract employees is needed as are currently employed.

3.32 McMillan Company manufactures electronic flow sensors that are designed as an alternative to ball-and-tube rotometers. The company recently spent $3 million to increase the capacity of an existing production line. If the extra revenue generated by the expansion amounts to $200,000 per month, how long will it take for the company to recover its investment at an interest rate of 12% per year compounded monthly? Solve using (a) tabulated factors, and (b) a calculator or spreadsheet.

3.33 (a) Using tabulated factors for the cash flow series shown, calculate the future worth at the end of quarter 6 using $i = 8\%$ per year compounded quarterly. (b) What are the calculator functions?

Quarter	0	1	2	3	4	5	6
Cash flow, $	100	100	300	300	300	300	300

3.34 The by-product department of Iowa Beef has the revenue stream shown. If the interest rate is an effective 12% per year compounded continuously, determine the annual worth of the revenues for years 1 through 5.

Year	Revenue, $
0	40,000
1	40,000
2	40,000
3	50,000
4	50,000
5	50,000

3.35 Video cards based on a high speed data processor typically cost $250. A recently released light version of the chip costs $150. If Karamba Videos purchases 3000 chips per quarter, what is the present worth of the savings associated with the cheaper chip over a 2-year period at an interest rate of 16% per year compounded quarterly?

3.36 The optical products division of Panasonic is planning a $3.5 million building expansion for manufacturing its powerful Lumix DMC digital zoom camera. If the company uses an interest rate of 20% per year compounded quarterly for all new investments, what is the uniform amount of revenue per quarter the company must realize to recover its investment in 3 years?

3.37 Lotus Development has a software rental plan called SmartSuite that is available on the world wide web. A number of programs are available at $2.99 for 48 hours. If a construction company uses the service an average of 48 hours per week, what is the present worth of the rental costs for 10 months at an interest rate of 1% per month compounded weekly? Assume 4 weeks per month.

3.38 Bethany Iron and Steel is evaluating possible economic advantages of being involved in electronic commerce. A modest e-commerce package is available for $30,000. If the company wants to recover the cost in 2 years, what is the equivalent amount of new income that must be realized every 6 months at an interest rate of 12% per year compounded quarterly? Solve using two of the following three approaches: factor formula, calculator, and spreadsheet.

3.39 El Paso Water Utilities purchases surface water from Elephant Butte Irrigation District at a cost of $120,000 per month in the months of February through September. Instead of paying monthly, the utility makes a single payment of $800,000 at the end of the calendar year (December) for the water it used. The delayed payment essentially represents a subsidy by the irrigation district to the water utility. At an interest rate of 0.25% per month, what is the amount of the subsidy?

3.40 Phoenix Pump and Filter projects that the cost of steel bodies for Model R910 valves will increase by $2.50 every 3 months. If the cost for the first quarter is expected to be $90 per unit, what is the present worth of a unit's costs over a 3-year time period at an interest rate of 12% per year compounded quarterly?

3.41 Spectrum Lab Products, a supplier of high quality lab equipment, borrowed $3 million to renovate one of its testing labs. The loan was repaid in 2 years through quarterly payments that increased by $50,000 each time. At an interest rate of 12% per year compounded quarterly, what was the size of the first quarterly payment? Solve using (*a*) tabulated factors, and (*b*) a spreadsheet and the GOAL SEEK tool.

3.42 According to a study by the National Association of Colleges and Employers, the average monthly salary for a recent electrical engineering graduate was $4944 versus $2795 for a graduate with a degree in liberal arts. What is the future worth in 40 years of the income difference at an interest rate of 6% per year compounded monthly?

3.43 Erbitux is a colorectal cancer treatment drug that is manufactured by ImClone Systems, Inc. If the treatment takes place over a 1 year period at a cost of $10,000 per month and the patient lives 5 years longer than he/she would have lived without the treatment, how much would that person have to earn each month beginning 1 month after treatment ends in order to earn an amount equivalent to the total treatment cost at an interest rate of 12% per year compounded monthly?

3.44 Plastics Engineering, Inc., a developer of advanced bolt drive components, recently began manufacturing sprockets made of polyacetal plastics that are approved for direct food contact. The company claims that high-precision manufacturing promotes longer belt life, thereby eliminating costs associated with belt replacement. If Kraft Foods estimates that maintenance down time costs the company an average of $12,000 per quarter, how much can the company afford to spend now if the new sprockets would cut their cost to $2000

per quarter? Assume Kraft Foods uses an interest rate of 12% per year compounded quarterly and a 2-year cost recovery period.

3.45 Linear actuators coupled to servo motors are expected to save a pneumatic valve manufacturer $19,000 per quarter through improved materials handling. If the savings begin *now*, what is the future worth of the savings through the end of year 3? The company uses an interest rate of 12% per year compounded quarterly.

3.46 An environmental soil cleaning company received a contract to remove BTEX contamination from an oil company tank farm site. The contract required the soil cleaning company to provide quarterly invoices for materials and services provided. If the material costs were $140,000 per quarter and the service charges were calculated as an additional 20% of the material costs, what is the present worth of the contract through the 3-year treatment period at an interest rate of 1% per month?

3.47 Redflex Traffic Systems manages red light camera systems that take photographs of vehicles that run red lights. Red light violations in Hereford County result in fines of $75 per incident. A two-month trial period revealed that the sheriff department could expect to issue 300 citations per month per intersection. Redflex has proposed installation of camera systems at 10 intersections. How much could the sheriff department afford to spend on the project if it wanted to recover the first cost in 2 years at an interest rate of 0.25% per month?

3.48 Anderson-McKee Construction expects to spend $835,000 for heavy equipment replacement 2 years from now and another $1.1 million 4 years from now. How much must the company deposit into a sinking fund each month to provide for the purchases, if the fund earns at a rate of return of 12% per year compounded monthly? Solve using (*a*) tabulated factors, and (*b*) a calculator.

3.49 A Hilti PP11 pipe laser can be mounted inside or on top of a pipe, a tripod, or in a manhole to provide fast and accurate positioning and alignment of pipe sections. Jordon Construction is building a 23-mile pipeline through the desert to dispose of reverse osmosis concentrate by injection into a fractured-dolomite formation. The PP11 laser can reduce construction costs through time savings by $9000 per month. What is the projected future worth of the savings over 2 years, if they start 1 month from now and the interest rate is 1% per month?

3.50 The cost of capturing and storing all CO_2 produced by coal-fired electricity generating plants during the next 200 years has been estimated to be $1.8 trillion. At an interest rate of 10% per year compounded monthly, what is the equivalent annual worth of such an undertaking over the 200 years?

3.51 Rotary encoder collars with targets on the face or outside diameter add speed sensing capability to rotating shafts. SKF Manufacturing expects sales of $150,000 in quarter one, increasing by $5,000 per quarter through year 5. Calculate the present worth of the sales through a 5-year period at an interest rate of nominal 20% per year compounded quarterly.

3.52 A company that manufactures brushless dc motors recently added a new product line of thin beam load cells. Start-up costs over the first 6 months were constant at $48,000 per month. Due to increased manufacturing efficiency, starting in month 7, costs were $46,000 and continued to decrease by $2000 per month through the end of the year. Using an interest rate of 18% per year compounded monthly, calculate the equivalent monthly cost for the first year. Solve using (*a*) tabulated factors, and (*b*) as spreadsheet.

3.53 For the cash flows shown, determine the future worth in year 5 at an interest rate of 1% per month.

Year	1	2	3	4	5
Cash Flow, $	300,000	275,000	250,000	225,000	200,000

3.54 Receipts from sales of hardened steel connectors was $50,000 in the first quarter, $51,000 in the second, and amounts increasing by $1000 per quarter through year 4. What is the equivalent uniform amount per quarter if the interest rate is 3% per quarter?

3.55 Atlas Moving and Storage wants to have enough money to purchase a new tractor-trailer in 4 years at a cost of $290,000. If the company sets aside $4000 in month 1 and plans to increase its set aside by a uniform amount each month, how much must the increase be provided the funds earn 6% per year compounded monthly?

3.56 Frontier Airlines hedged the cost of jet fuel by purchasing options that allowed the airline to purchase fuel at a fixed price for 2 years. If the savings in fuel costs were $140,000 in month 1, $141,400 in month 2, and amounts increasing by 1% per month through the 2-year option period, what was the present worth of the savings at an interest rate of 18% per year compounded monthly?

3.57 Find the present worth of an investment that starts at $8000 in year 1 and increases by 10% each year through year 7, if the interest rate is 10% per year compounded continuously.

3.58 Corrosion problems and manufacturing defects rendered a gasoline pipeline between El Paso and Phoenix subject to longitudinal weld seam failures. Therefore, pressure was reduced to 80% of the design value. If the reduced pressure results in delivery of less product valued at $100,000 per month, what will be the value of the lost revenue after a two-year period at an interest rate of 15% per year compounded continuously.

3.59 The initial cost of a pulverized coal cyclone furnace is $800,000. If the operating cost is $90,000 in year 1, $92,700 in year 2, and amounts increasing by 3% each year through a 10-year amortization period, what is the equivalent annual cost of the furnace at an interest rate of 1% per month?

Equivalence when PP < CP

3.60 In an effort to save money for early retirement, an environmental engineer plans to deposit $1200 per month starting one month from now, into a self-directed investment account that pays 8% per year compounded semiannually. How much will be in the account at the end of 25 years?

3.61 If you deposit $1500 per month into a credit union savings account that pays interest at a rate of 6% per year compounded quarterly, how much will be in the account at the end of five years? There is no interperiod compounding.

3.62 There is no interperiod compounding for the transactions shown. Your manager asks you to determine the amount of money in the account at the end of year 3 at an interest rate of 8% per year compounded semiannually. Draw the cash flow diagram and solve using (*a*) tabulated factors, and (*b*) a spreadsheet.

End of Quarter	Amount of Deposit, $	Amount of Withdrawal, $
1	900	—
2–4	700	—
7	1000	2600
11	—	1000

3.63 You plan to invest $1000 per month in a stock that pays a dividend at a rate of 4% per year compounded quarterly. How much will the stock be worth at the end of 9 years if there is no interperiod compounding?

3.64 Western Refining purchased a model MTVS peristaltic pump for injecting antiscalant at its nanofiltration water conditioning plant. The cost of the pump was $1200. If the chemical cost is $11 per day, determine the equivalent cost per month at an interest rate of 1% per month. Assume 30 days per month and a 4-year pump life.

3.65 Income from recycling the paper and cardboard generated in an office building has averaged $3000 per month for 3 years. What is the future worth of the income at an interest rate of 8% per year compounded quarterly?

3.66 Coal-fired power plants emit CO_2, which is one of the gases that is of concern with regard to global warming. A technique that power plants can adopt to keep most of the CO_2 from entering the air is called CO_2 capture and storage (CCS). If the incremental cost of the sequestration process is $0.019 per kilowatt-hour, what is the present worth of the extra cost over a 3-year period to a manufacturing plant that uses 100,000 kWh of power per month? The interest rate is 12% per year compounded quarterly.

3.67 The Autocar E3 refuse truck has an energy recovery system developed by Parker Hannifin LLC that is expected to reduce fuel consumption by 50%. Pressurized fluid flows from carbon fiber–reinforced accumulator tanks to two hydrostatic motors that propel the vehicle forward. The truck recharges the accumulators when it brakes. The fuel cost for a regular refuse truck is $800 per month. How much can a private waste hauling company afford to spend on the recovery system, if it wants to recover its investment in 3 years at an interest rate of 12% per year?

ADDITIONAL PROBLEMS AND FE EXAM REVIEW QUESTIONS

3.68 Assume you make a single deposit of $1000 into a savings account that pays interest at 0.5% per month. If you want to know how much the account will be worth in 5 years, the only values you can use for i and n are
  a. Effective i per month and $n = 60$
  b. Effective i per quarter and $n = 30$
  c. Effective i per semiannual period and $n = 10$
  d. Either (a) or (c)

3.69 An interest rate obtained by dividing the nominal rate per year by the number of compounding periods in that year is
  a. A nominal interest rate
  b. An effective interest rate
  c. An effective interest rate only if the compounding period is monthly
  d. Either (b) or (c)

3.70 If deposits are made monthly into a savings account and you want to know how much will be in the account after 5 years, the interest rate and n values you have to use are
  a. The effective rate per month and $n = 60$ months
  b. The effective rate per quarter and $n = 20$ quarters
  c. The effective rate per year and $n = 5$ quarters
  d. Either (a), (b) or (c)

3.71 An interest rate of 2% per month is the same as:
  a. 24% per year
  b. A nominal 24% per year compounded monthly
  c. An effective 24% per year compounded monthly
  d. Both (a) and (b)

3.72 The only time you change the original cash flow diagram in problems involving uniform series cash flows is when the:
  a. payment period is longer than the compounding period
  b. payment period is equal to the compounding period
  c. payment period is shorter than the compounding period
  d. it may be done in any of the cases above, depending upon how the effective interest rate is calculated

3.73 Exotic Faucets and Sinks, Ltd. guarantees that their new infra-red sensor faucet will save any household that has two or more children at least $30 per month in water costs beginning one month after the faucet is installed. If the faucet is under full warranty for five years, the minimum amount a family could afford to spend now on such a faucet at an interest rate of 6% per year compounded monthly is closest to:

a. $149
b. $1552
c. $1787
d. $1890

3.74 Border Steel invested $800,000 in a new shearing unit. At an interest rate of 12% per year compounded quarterly, the quarterly income required to recover the investment in 3 years is closest to:

a. $69,610
b. $75,880
c. $80,370
d. $83,550

3.75 An engineer analyzing cost data about hydrogen sulfide monitors discovered that the information for the first three years was missing. However, he knew the cost in year 4 was $1250 and that it increased by 5% each year thereafter. If the same trend applied to the first three years, the cost in year 1 was:

a. $1312.50
b. $1190.48
c. $1028.38
d. $1079.80

3.76 A plant manager wants to know the present worth of the maintenance costs for a reconditioned assembly line. An industrial engineer, who designed the system, estimates that the maintenance costs are expected to be zero for the first three years, $2000 in year four, $2500 in year five, and amounts increasing by $500 each year through year 10. At an interest rate of 8% per year compounded semiannually, the value of n to use in the P/G equation is:

a. 7
b. 8
c. 10
d. 14

Present Worth Analysis

Ingram Publishing; Photodisc/Getty Images

A future amount of money converted to its equivalent value now has a present worth (PW) that is less than that of the actual cash flow, because for any interest rate greater than zero, all P/F factors have a value less than 1.0. For this reason, present worth values are often referred to as *discounted cash flows (DCF)*. Similarly, the interest rate may be referred to as the *discount rate*. Besides PW, equivalent terms frequently used are present value (PV) and net present value (NPV). Up to this point, present worth computations have been made for one project. In this chapter, techniques for comparing two or more mutually exclusive alternatives by the present worth method are treated. Additionally, techniques that evaluate capitalized costs, life cycle costs, and independent projects are discussed.

LEARNING OUTCOMES

Formulating alternatives

1. Identify mutually exclusive and independent projects, and define a revenue and a cost alternative.

PW of equal-life alternatives

2. Evaluate a single alternative and select the best of equal-life alternatives using present worth analysis.

PW of different-life alternatives

3. Select the best of different-life alternatives using present worth analysis.

Capitalized cost (CC)

4. Select the best alternative using capitalized cost calculations.

Independent projects

5. Select the best independent projects with and without a budget limit.

Spreadsheets

6. Use a spreadsheet to select an alternative by PW analysis.

4.1 FORMULATING ALTERNATIVES

Alternatives are developed from project proposals to accomplish a stated purpose. The logic of alternative formulation and evaluation is depicted in Figure 4.1. Some projects are economically and technologically viable, and others are not. Once the viable projects are defined, it is possible to formulate the alternatives.

Alternatives are one of two types: mutually exclusive or independent. Each type is evaluated differently.

- **Mutually exclusive (ME).** *Only one of the viable projects can be selected.* Each viable project *is* an alternative. If no alternative is economically justifiable, do nothing (DN) is the default selection.
- **Independent.** *More than one viable project may be selected* for investment. (There may be dependent projects requiring a particular project to be selected before another, and/or contingent projects where one project may be substituted for another.)

An alternative or project is comprised of estimates for the first cost, expected life, salvage value, and annual costs. Salvage value is the best estimate of an anticipated future market or trade-in value at the end of the expected life. Salvage may be estimated as a percentage of the first cost or an actual monetary amount; salvage is often estimated as nil or zero. Annual costs are commonly termed annual operating costs (AOC) or maintenance and operating (M&O) costs. They may be uniform over the entire life, increase or decrease each year as a percentage or arithmetic gradient series, or vary over time according to some other expected pattern.

A mutually exclusive alternative selection is the most common type in engineering practice. It takes place, for example, when an engineer must select the one best diesel-powered engine from several competing models. Mutually exclusive alternatives are, therefore, the same as the viable projects; each one is evaluated, and the one best alternative is chosen. Mutually exclusive alternatives *compete with one another* in the evaluation. All the analysis techniques compare mutually exclusive alternatives. Present worth is discussed in the remainder of this chapter.

The *do-nothing (DN)* option is usually understood to be an alternative when the evaluation is performed. If it is absolutely required that one of the defined alternatives be selected, do nothing is not considered an option. (This may occur when a mandated function must be installed for safety, legal, or other purposes.) Selection of the DN alternative means that the current approach is maintained; no new costs, revenues, or savings are generated.

Independent projects are usually designed to accomplish different purposes, thus the possibility of selecting any number of the projects. These projects (or bundles of projects) do not compete with one another; each project is evaluated separately, and the *comparison is* with the MARR. Independent project selection is treated in Section 4.5.

Finally, it is important to classify an *alternative's* cash flows as revenue-based or cost-based. All alternatives evaluated in one study must be of the same type.

- **Revenue.** *Each alternative generates cost and revenue cash flow estimates, and possibly savings,* which are treated like revenues. Revenues may be different

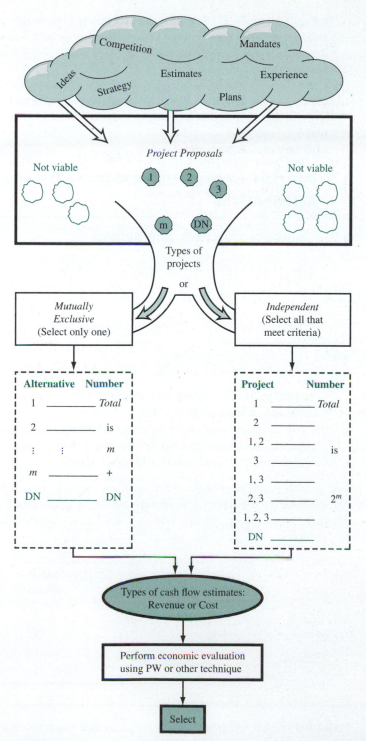

FIGURE 4.1
Logical progression from proposals to alternatives to selection.

for each alternative. These alternatives usually involve new systems, products, and services that require capital investment to generate revenues and/or savings. Purchasing new equipment to increase productivity and sales is a revenue alternative.

- **Cost.** *Each alternative has only cost cash flow estimates.* Revenues are assumed to be equal for all alternatives. These may be public sector (government) initiatives, or legally mandated or safety improvements. Cost alternatives are compared to each other; do-nothing is not an option when selecting from mutually exclusive cost alternatives.

4.2 PRESENT WORTH ANALYSIS OF EQUAL-LIFE ALTERNATIVES

4.2.1 Mutually Exclusive Alternatives

In present worth analysis, the P value, now called PW, is calculated at the MARR for each alternative. This converts all future cash flows into present dollar equivalents. This makes it easy to determine the economic advantage of one alternative over another.

The PW comparison of alternatives with equal lives is straightforward. If both alternatives are used in identical capacities for the same time period, they are termed *equal-service* alternatives.

For mutually exclusive alternatives the following guidelines are applied:

One alternative: Calculate PW at the MARR. If PW $\geq$ 0, the alternative is financially viable.

Two or more alternatives: Calculate the PW of each alternative at the MARR. *Select the alternative with the PW value that is numerically largest,* **that is, less negative or more positive.**

The second guideline uses the criterion of *numerically largest* to indicate a lower PW of costs only or larger PW of net cash flows. Numerically largest is *not the absolute value* because the sign matters here. The selections below correctly apply this guideline.

PW$_1$	PW$_2$	Selected Alternative
$-1500	$-500	2
-500	+1000	2
+2500	-500	1
+2500	+1500	1

EXAMPLE 4.1   Perform a present worth analysis of equal-service machines that have the costs shown on the next page, if the MARR is 10% per year. Revenues for all three alternatives are expected to be the same.

	Electric-Powered	Gas-Powered	Solar-Powered
First cost, $	−2500	−3500	−6000
Annual operating cost (AOC), $/year	−900	−700	−50
Salvage value, $	200	350	100
Life, years	5	5	5

Solution

These are cost alternatives. The salvage values are considered a "negative" cost, so a + sign precedes them. The PW of each machine is calculated at $i = 10\%$ for $n = 5$ years. Use subscripts E, G, and S.

$$PW_E = -2500 - 900(P/A,10\%,5) + 200(P/F,10\%,5) = \$-5788$$
$$PW_G = -3500 - 700(P/A,10\%,5) + 350(P/F,10\%,5) = \$-5936$$
$$PW_S = -6000 - 50(P/A,10\%,5) + 100(P/F,10\%,5) = \$-6127$$

The electric-powered machine is selected since the PW of its costs is the lowest; it has the numerically largest PW value.

4.2.2 Evaluation of a Bond Purchase

Often a corporation or government obtains investment capital for projects by selling *bonds*. A good application of the PW method is the evaluation of a bond purchase alternative. If PW < 0 at the MARR, the do-nothing alternative is selected. A bond is like an IOU for time periods such as 5, 10, 20, or more years. Each bond has a *face value V* of $100, $1000, $5000 or more that is fully returned to the purchaser when the bond maturity is reached. Additionally, bonds provide the purchaser with periodic *interest payments I* (also called bond dividends) using the *bond coupon* (or interest) *rate b*, and, *c*, the number of payment periods per year.

$$I = \frac{(\text{bond face value})(\text{bond coupon rate})}{\text{number of payments per year}} = \frac{Vb}{c} \qquad [4.1]$$

At the time of purchase, the bond may sell for more or less than the face value, depending upon the financial reputation of the issuer. A purchase discount is more attractive financially to the purchaser; a premium is better for the issuer. For example, suppose a person is offered a 2% discount for an 8% $10,000 20-year bond that pays the dividend quarterly. He will pay $9800 now, and, according to Equation [4.1], he will receive quarterly dividends of $I = \$200$, plus the $10,000 face value after 20 years.

To evaluate a proposed bond purchase, determine the PW at the MARR of all cash flows—initial payment and receipts of periodic dividends and the bond's face value at the maturity date. Then apply the guideline for one alternative, that is, if PW ≥ 0, the bond is financially viable. It is important to use the effective MARR rate in the PW relation that matches the time period of the payments. The simplest method is the procedure in Section 3.4 for PP = CP, as illustrated in the next example.

EXAMPLE 4.2   Marcie has some extra money that she wants to place into a relatively safe investment. Her employer is offering to employees a generous 5% discount for 10-year $5,000 bonds that carry a coupon rate of 6% paid semiannually. The expectation is to match her return on other safe investments, which have averaged 6.7% per year compounded semiannually. (This is an effective rate of 6.81% per year.) Should she buy the bond?

Solution

Equation [4.1] results in a dividend of $I = (5000)(0.06)/2 = \$150$ every 6 months for a total of $n = 20$ dividend payments. The semiannual MARR is $6.7/2 = 3.35\%$, and the purchase price now is $-5000(0.95) = \$-4750$. Using PW evaluation,

$$PW = -4750 + 150(P/A,3.35\%,20) + 5000(P/F,3.35\%,20)$$
$$= \$-2.13$$

To be correct, she should not buy the bond, because the effective rate is slightly less than 6.81% per year since PW < 0. However, if Marcie had to pay just $2.13 less for the bond, she would meet her MARR goal. She should probably purchase the bond since the return is so close to her goal.

In order to speed up a PW analysis with a spreadsheet, the PV function is utilized. If all annual amounts for AOC are the same, the PW value for year 1 to n cash flows is found by entering the function $= P - PV(i\%, n, A, F)$. In Example 4.1, $PW_E = \$-5788$ is determined by entering $= -2500 - PV(10\%,5,-900,200)$ into any cell. Spreadsheet solutions are demonstrated in detail in Section 4.6.

4.3 PRESENT WORTH ANALYSIS OF DIFFERENT-LIFE ALTERNATIVES

Present worth analysis requires an *equal service* comparison of alternatives, that is, the number of years considered must be the same for all alternatives. *If equal service is not present, shorter-lived alternatives will be favored based on lower PW of total costs,* even though they may not be economically favorable. Fundamentally, there are two ways to use PW analysis to compare alternatives with unequal life estimates; evaluate over a specific study period (planning horizon), or use the least common multiple of lives for each pair of alternatives. In both cases, the PW is calculated at the MARR, and the selection guidelines of the previous section are applied.

4.3.1 Study Period

This is a commonly used approach. Once a study period is selected, only cash flows during this time frame are considered. If an expected life is *longer* than this period, the estimated market value of the alternative is used as a "salvage value" in the last year of the study period. If the expected life is *shorter* than the study period, cash flow estimates to continue equivalent service must be made for the time period between the end of the alternative's life and the end of the study period.

In both cases, the result is an equal-service evaluation of the alternatives. As one example, assume a construction company wins a highway maintenance contract for 5 years, but plans to purchase specialized equipment expected to be operational for 10 years. For analysis purposes, the anticipated market value after 5 years is a salvage value in the PW equation, and any cash flows after year 5 are ignored. Example 4.3 illustrates a study period analysis.

4.3.2 Least Common Multiple (LCM)

The LCM approach can result in unrealistic assumptions since equal service comparison is achieved by assuming:

- The same service is needed for the LCM number of years. For example, the LCM of 5- and 9-year lives presumes the same need for 45 years!
- Cash flow estimates are initially expected to remain the same over each life cycle, which is correct *only* when changes in future cash flows exactly match the inflation or deflation rate.
- Each alternative is available for multiple life cycles, something that is usually not true.

EXAMPLE 4.3

A project engineer with EnvironCare is assigned to start up a new office in a city where a 6-year contract has been finalized to collect and analyze ozone-level readings. Two lease options are available, each with a first cost, annual lease cost, and deposit-return estimates shown below. The MARR is 15% per year.

	Location A	Location B
First cost, $	−15,000	−18,000
Annual lease cost, $ per year	−3,500	−3,100
Deposit return, $	1,000	2,000
Lease term, years	6	9

a. EnvironCare has a practice of evaluating all projects over a 5-year period. If the deposit returns are not expected to change, which location should be selected?
b. Perform the analysis using an 8-year planning horizon.
c. Determine which lease option should be selected on the basis of a present worth comparison using the LCM.

Solution

a. For a 5-year study period, use the estimated deposit returns as positive cash flows in year 5.

$$PW_A = -15,000 - 3500(P/A,15\%,5) + 1000(P/F,15\%,5)$$
$$= \$-26,236$$
$$PW_B = -18,000 - 3100(P/A,15\%,5) + 2000(P/F,15\%,5)$$
$$= \$-27,397$$

Location A is the better economic choice.

b. For an 8-year study period, the deposit return for B remains at $2000 in year 8. For A, an estimate for equivalent service for the additional 2 years is needed. Assume this is expected to be relatively expensive at $6000 per year.

$$PW_A = -15{,}000 - 3500(P/A{,}15\%{,}6) + 1000(P/F{,}15\%{,}6)$$
$$\qquad\quad - 6000(P/A{,}15\%{,}2)(P/F{,}15\%{,}6)$$
$$\qquad = \$-32{,}030$$
$$PW_B = -18{,}000 - 3100(P/A{,}15\%{,}8) + 2000(P/F{,}15\%{,}8)$$
$$\qquad = \$-31{,}257$$

Location B has an economic advantage for this longer study period.

c. Since the leases have different terms, compare them over the LCM of 18 years. For life cycles after the first, the first cost is repeated at the beginning (year 0) of each new cycle, which is the last year of the previous cycle. These are years 6 and 12 for location A and year 9 for B. The cash flow diagram is in Figure 4.2.

$$PW_A = -15{,}000 - 15{,}000(P/F{,}15\%{,}6) + 1000(P/F{,}15\%{,}6)$$
$$\qquad\quad - 15{,}000(P/F{,}15\%{,}12) + 1000(P/F{,}15\%{,}12)$$
$$\qquad\quad + 1000(P/F{,}15\%{,}18) - 3500(P/A{,}15\%{,}18)$$
$$\qquad = \$-45{,}036$$
$$PW_B = -18{,}000 - 18{,}000(P/F{,}15\%{,}9) + 2000(P/F{,}15\%{,}9)$$
$$\qquad\quad + 2000(P/F{,}15\%{,}18) - 3100(P/A{,}15\%{,}18)$$
$$\qquad = \$-41{,}384$$

Location B is selected.

FIGURE 4.2

Cash flow diagram for different-life alternatives, Example 4.3c.

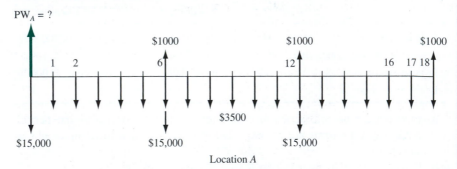

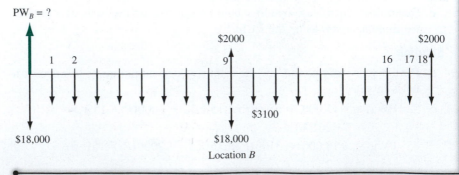

Present worth analysis illustrated in Example 4.3 using the LCM method is correct, but it is not recommended as the primary method of analysis. The same correct conclusion is easier to reach using each alternative's life and an annual worth (AW) analysis, as discussed in Chapter 5.

4.3.3 Future Worth

The future worth (FW) of an alternative may also be used to select an alternative. The FW is determined directly from the cash flows or by multiplying the PW value by the F/P factor at the established MARR. The n value in the F/P factor depends upon which time period has been used to determine PW—the LCM or a study period. Using FW values is especially applicable to large capital investment decisions when a prime goal is to maximize the *future wealth* of a corporation's stockholders. Alternatives such as electric generation facilities, toll roads, hotels, and the like can be analyzed using the FW value of investment commitments made during construction. Selection guidelines are the same as those for PW analysis.

4.3.4 Life-Cycle Cost

Life-cycle cost (LCC) is another extension of present worth analysis. The LCC method, as its name implies, is commonly applied to alternatives with cost estimates over the entire *system life span*. This means that costs from the early stage of the project (needs assessment and design), through marketing, warranty, and operation phases and through the final stage (phaseout and disposal) are estimated. Typical applications for LCC are buildings (new construction or purchases), new product lines, manufacturing plants, commercial aircraft, new automobile models, defense systems, and the like.

A PW analysis with all definable costs (and possibly incomes) estimatable are considered in a LCC analysis. However, the broad definition of the term *system life span* requires cost estimates not usually made for a regular PW analysis, such as design and development costs. *LCC is most effectively applied when a substantial percentage of the total costs over the system life span, relative to the initial investment, will be operating and maintenance costs* (postpurchase costs such as warranty, personnel, energy, upkeep, and materials). If Exxon-Mobil is evaluating the purchase of equipment for a large chemical processing plant for $150,000 with a 5-year life and annual costs of $15,000, LCC analysis is probably not justified. On the other hand, suppose Toyota is considering the design, construction, marketing, and after-delivery costs for a new automobile model. If the total start-up cost is estimated at $125 million (over 3 years) and total annual costs are expected to be 25 to 30% of this figure to build, market, and service the cars for the next 15 years (estimated life span of the model), then the logic of LCC analysis will help the decision makers understand the profile of costs and their economic consequences using PW, FW, or AW analysis. LCC is required for most defense and aerospace industries, where the approach may be called Design to Cost (see Section 11.1). LCC is usually not applied to public sector projects, because the benefits and costs are difficult to estimate with much accuracy. Benefit/cost analysis is better applied here, as discussed in Chapter 7.

4.4 CAPITALIZED COST ANALYSIS

Podcast available

Capitalized cost (CC) is the present worth of an alternative that will last "forever." Public sector projects such as bridges, dams, irrigation systems, and railroads fall into this category, since they have useful lives of 30, 40, and more years. In addition, permanent and charitable organization endowments are evaluated using capitalized cost.

The formula to calculate CC is derived from the relation $PW = A(P/A,i,n)$, where $n = \infty$. The equation can be written

$$PW = A \left[\frac{1 - \dfrac{1}{(1 + i)^n}}{i} \right]$$

As n approaches ∞, the bracketed term becomes $1/i$. The symbol CC replaces PW, and AW (annual worth) replaces A to yield

$$CC = \frac{A}{i} = \frac{AW}{i} \qquad \qquad [4.2]$$

Equation [4.2] is illustrated by considering the time value of money. If $10,000 earns 10% per year, the interest earned at the end of every year for *eternity* is $1000. This leaves the original $10,000 intact to earn more next year. In general, the equivalent A value from Equation [4.2] for an infinite number of periods is

$$A = CC(i) \qquad \qquad [4.3]$$

The cash flows (costs or receipts) in a capitalized cost calculation are usually of two types: *recurring,* also called periodic, and *nonrecurring,* also called one-time. An annual operating cost of $50,000 and a rework cost estimated at $40,000 every 12 years are examples of recurring cash flows. Examples of nonrecurring cash flows are the initial investment amount in year 0 and one-time cash flow estimates, for example, $500,000 in royalty fees 2 years hence. The following procedure assists in calculating the CC for an infinite sequence of cash flows.

1. Draw a cash flow diagram showing all nonrecurring and at least two cycles of all recurring cash flows. (Drawing the cash flow diagram is more important in CC calculations than elsewhere.)
2. Find the present worth of all nonrecurring amounts. This is their CC value.
3. Find the equivalent uniform annual worth (*A* value) through *one life cycle* of all recurring amounts. This is the same value in all succeeding life cycles. Add this to all other uniform amounts occurring in years 1 through infinity and the result is the total equivalent uniform annual worth (AW).
4. Divide the AW obtained in step 3 by the interest rate i to obtain a CC value. This is an application of Equation [4.2].
5. Add the CC values obtained in steps 2 and 4.

EXAMPLE 4.4   The property appraisal district for Marin County has just installed new software to track residential market values for property tax computations. The manager wants to know the total equivalent cost of all future costs incurred when the three county judges agreed to purchase the software system. If the new system will be used for the indefinite future, find the equivalent value (*a*) now and (*b*) for each year hereafter.

The system has an installed cost of $150,000 and an additional cost of $50,000 after 10 years. The annual software maintenance contract cost is $5000 for the first 4 years and $8000 thereafter. In addition, there is expected to be a recurring major upgrade cost of $15,000 every 13 years. Assume that $i = 5\%$ per year for county funds.

Solution

a. The detailed procedure is applied.

1. Draw a cash flow diagram for two cycles (Figure 4.3).
2. Find the present worth of the nonrecurring costs of $150,000 now and $50,000 in year 10 at $i = 5\%$. Label this CC_1.

$$CC_1 = -150,000 - 50,000(P/F,5\%,10) = \$-180,695$$

3. Convert the recurring cost of $15,000 every 13 years into an annual worth A_1 for the first 13 years.

$$A_1 = -15,000(A/F,5\%,13) = \$-847$$

The same value, $A_1 = \$-847$, applies to all the other 13-year periods as well.

4. The capitalized cost for the two annual maintenance cost series may be determined in either of two ways: (1) consider a series of $\$-5000$ from now to infinity plus a series of $\$-3000$ from year 5 on; or (2) a series of $\$-5000$ for 4 years followed by a series of $\$-8000$ from year 5 to infinity. Using the first method, the annual cost (A_2) is $\$-5000$ forever.

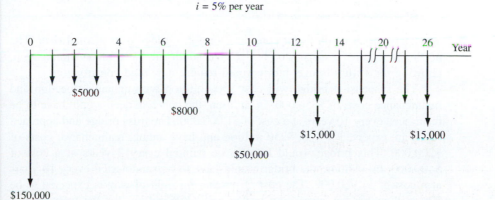

FIGURE 4.3 Cash flows for two cycles of recurring costs and all nonrecurring amounts, Example 4.4.

The capitalized cost CC_2 of $\$-3000$ from year 5 to infinity is found using Equation [4.2] times the P/F factor.

$$CC_2 = \frac{-3000}{0.05}(P/F,5\%,4) = \$-49{,}362$$

The CC_2 value is calculated using $n = 4$ because the present worth of the annual $\$3000$ cost is located in year 4, one period ahead of the first A. The two annual cost series are converted into a capitalized cost CC_3.

$$CC_3 = \frac{A_1 + A_2}{i} = \frac{-847 + (-5000)}{0.05} = \$-116{,}940$$

5. The total capitalized cost CC_T is obtained by adding the three CC values.

$$CC_T = -180{,}695 - 49{,}362 - 116{,}940 = \$-346{,}997$$

b. Equation [4.3] determines the A value forever.

$$A = CC_T(i) = \$-346{,}997(0.05) = \$-17{,}350$$

Correctly interpreted, this means Marin County officials have committed the equivalent of $\$17{,}350$ forever to operate and maintain the property appraisal software.

The CC evaluation of two or more alternatives compares them for the same number of years—infinity. The alternative with the smaller capitalized cost is the more economical one.

EXAMPLE 4.5   Two sites are currently under consideration for a bridge over a small river. The north site requires a suspension bridge. The south site has a much shorter span, allowing for a truss bridge, but it would require new road construction.

The suspension bridge will cost $\$500$ million with annual inspection and maintenance costs of $\$350{,}000$. In addition, the concrete deck would have to be resurfaced every 10 years at a cost of $\$1{,}000{,}000$. The truss bridge and approach roads are expected to cost $\$250$ million and have annual maintenance costs of $\$200{,}000$. This bridge would have to be painted every 3 years at a cost of $\$400{,}000$. In addition, the bridge would have to be sandblasted every 10 years at a cost of $\$1{,}900{,}000$. The cost of purchasing right-of-way is expected to be $\$20$ million for the suspension bridge and $\$150$ million for the truss bridge. Compare the alternatives on the basis of their capitalized cost if the interest rate is 6% per year.

Solution

Construct the cash flow diagrams over two cycles (20 years).
Capitalized cost of suspension bridge (CC_S):

$$CC_1 = \text{capitalized cost of initial cost}$$
$$= -500 - 20 = \$-520 \text{ million}$$

The recurring operating cost is $A_1 = \$-350,000$, and the annual equivalent of the resurface cost is

$$A_2 = -1,000,000\,(A/F, 6\%,10) = \$-75,870$$

$$CC_2 = \text{capitalized cost of recurring costs} = \frac{A_1 + A_2}{i}$$

$$= \frac{-350,000 + (-75,870)}{0.06} = \$-7,097,833$$

The total capitalized cost is

$$CC_S = CC_1 + CC_2 = \$-527.1 \text{ million}$$

Capitalized cost of truss bridge (CC_T):

$$CC_1 = -250 + (-150) = \$-400 \text{ million}$$
$$A_1 = \$-200,000$$
$$A_2 = \text{annual cost of painting} = -400,000(A/F,6\%,3) = \$-125,644$$
$$A_3 = \text{annual cost of sandblasting} = -1,900,000(A/F,6\%,10) = \$-144,153$$

$$CC_2 = \frac{A_1 + A_2 + A_3}{i} = \frac{\$-469,797}{0.06} = \$-7,829,950$$

$$CC_T = CC_1 + CC_2 = \$-407.83 \text{ million}$$

Conclusion: Build the truss bridge, since its capitalized cost is lower by $119 million.

If a finite-life alternative (e.g., 5 years) is compared to one with an indefinite or very long life, capitalized costs can be used. To determine CC for the finite life alternative, calculate the *A* value for *one life cycle* and divide by the interest rate (Equation [4.2]).

EXAMPLE 4.6

APSco, a large electronics subcontractor for the Air Force, needs to immediately acquire 10 soldering machines with specially prepared jigs for assembling components onto circuit boards. More machines may be needed in the future. The lead production engineer has outlined two simplified, but viable, alternatives. The company's MARR is 15% per year and capitalized cost is the evaluation technique.

Alternative LT (long-term). For $8 million now, a contractor will provide the necessary number of machines (up to a maximum of 20), now and in the future, for as long as APSco needs them. The annual contract fee is a total of $25,000 with no additional per-machine annual cost. There is no time limit placed on the contract, and the costs do not escalate.

Alternative ST (short-term). APSco buys its own machines for $275,000 each and expends an estimated $12,000 per machine in annual operating cost (AOC). The useful life of a soldering system is 5 years.

Solution

For the LT alternative, find the CC of the AOC using Equation [4.2]. Add this amount to the initial contract fee, which is already a capitalized cost.

$$CC_{LT} = \text{CC of contract fee} + \text{CC of AOC}$$
$$= -8 \text{ million} - 25,000/0.15 = \$-8,166,667$$

For the ST alternative, first calculate the equivalent annual amount for the purchase cost over the 5-year life, and add the AOC values for all 10 machines. Then determine the total CC using Equation [4.2].

$$AW_{ST} = \text{AW for purchase} + \text{AOC}$$
$$= -2.75 \text{ million}(A/P,15\%,5) - 120,000 = \$-940,380$$
$$CC_{ST} = -940,380/0.15 = \$-6,269,200$$

The ST alternative has a lower capitalized cost by approximately $1.9 million present value dollars.

4.5 EVALUATION OF INDEPENDENT PROJECTS

Podcast available

Consider a biomedical company that has a new genetics engineering product that it can market in three different countries (S, U, and R), including any combination of the three. The do-nothing (DN) alternative is also an option. All possible options are: S, U, R, SU, SR, UR, SUR, and DN. In general, for m independent projects, there are 2^m alternatives to evaluate. Selection from independent projects uses a fundamentally different approach from that for mutually exclusive (ME) alternatives. When selecting independent projects, each project's PW is calculated using the MARR. (In ME alternative evaluation, the projects compete with each other, and only one is selected.) The selection rule is quite simple for one or more *independent* projects:

Select all projects that have PW ≥ 0 at the MARR.

All projects must be developed to have revenue cash flows (not costs only) so that projects earning more than the MARR have positive PW values.

Unlike ME alternative evaluation, which assumes the need for the service over multiple life cycles, independent projects are considered one-time investments. This means the PW analysis is performed over the respective life of each project and the assumption is made that any leftover cash flows earn at the MARR when the project ends. As a result, the equal service requirement does not impose the use of a specified study period or the LCM method. The implied study period is that of the longest lived project.

There are two types of selection environments—unlimited and budget constrained.

- **Unlimited.** All projects that make or exceed the MARR are selected. Selection is made using the PW ≥ 0 guideline.
- **Budget constrained.** No more than a specified amount, b, of funds can be invested in all of the selected projects, and each project must make or exceed the MARR. Now the solution methodology is slightly more complex in that *bundles* of projects that do not exceed the investment limit b are the only ones evaluated using PW values. The procedure is:
  1. Determine all bundles that have total initial investments no more than b. (This limit usually applies in year 0 to get the project started).
  2. Find the PW value at the MARR for all projects contained in the bundles.
  3. Total the PW values for each bundle in (1).
  4. Select the bundle with the largest PW value.

EXAMPLE 4.7

Marshall Aqua Technologies has four separate projects it can pursue over the next several years. The required amounts to start each project (initial investments) now and the anticipated cash flows over the expected lives are estimated by the Project Engineering Department. At MARR = 15% per year, determine which projects should be pursued if initial funding is (*a*) not limited, and (*b*) limited to no more than $15,000.

Project	Initial Investment	Annual Net Cash Flow	Life, Years
F	$ -8,000	$3870	6
G	-15,000	2930	9
H	-6,000	2080	5
J	-10,000	5060	3

Solution

a. Determine the PW value for each independent project and select all with PW ≥ 0 at 15%.

$$PW_F = -8000 + 3870(P/A,15\%,6) = \$6646$$
$$PW_G = -15{,}000 + 2930(P/A,15\%,9) = \$-1019$$
$$PW_H = -6000 + 2080(P/A,15\%,5) = \$973$$
$$PW_J = -10{,}000 + 5060(P/A,15\%,3) = \$1553$$

Select the three projects F, H, and J for a total investment of $24,000.

TABLE 4.1 **Present Worth Analysis of Independent Projects with Investment Limited to $15,000, Example 4.7**

Bundle	Projects	Total Initial Investment	PW of Bundle at 15%
1	F	$ −8,000	$ 6646
2	G	−15,000	−1019
3	H	−6,000	973
4	J	−10,000	1553
5	FH	−14,000	7619
6	Do-nothing	0	0

b. Use the steps for a budget-constrained selection with b = $15,000.

1 and **2.** Of the $2^4 = 16$ possible bundles, Table 4.1 indicates that 6 are acceptable. These bundles involve all four projects plus do-nothing with $PW_{DN} = \$0$.

3. The PW value for a bundle is obtained by adding the respective project PW values. For example, $PW_5 = PW_F + PW_H = 6646 + 973 = \$7619$.

4. Select projects F and H, since their PW is the largest and both projects exceed the MARR, as indicated by PW > 0 at $i = 15\%$.

Comment: Budget-constrained selection from independent projects is commonly called the capital rationing or capital budgeting problem. It may be worked efficiently using a variety of techniques, one being the integer linear programming technique. Excel and its optimizing tool SOLVER handle this type of problem rather nicely.

4.6 USING SPREADSHEETS FOR PW ANALYSIS

Spreadsheet- or calculator-based evaluation of equal-life, mutually exclusive alternatives can be performed using the single-cell PV function when the annual amount A is the same. The general format to determine the PW is

$$= P - PV(i,n,A,F) \tag{4.4}$$

It is important to pay attention to the sign placed on the PV function in order to get the correct answer for the alternative's PW value. The spreadsheet function returns the opposite sign of the A series. Therefore, to retain the negative sense of a cost series A, place a minus sign immediately in front of the PV function. This is illustrated in the next example.

Cesar, a petroleum engineer, has identified two equivalent diesel-powered generators to be purchased for an offshore platform. Use $i = 12\%$ per year to determine
which is the more economic. Solve using both spreadsheet and calculator functions.

EXAMPLE 4.8

	Generator 1	Generator 2
P, $	−80,000	−120,000
S, $	15,000	40,000
n, years	3	3
AOC, $/year	−30,000	−8,000

Solution

Spreadsheet: Follow the format in Equation [4.4] in a single cell for each alternative. Figure 4.4 shows the details. Note the use of minus signs on P, the PV
function, and AOC value. Generator 2 is selected with the smaller PW of costs
(numerically larger value).

Calculator: The function and PW value for each alternative are:

Generator 1: −80000 − PV(12,3,−30000,15000) $PW_1 = \$-141{,}378$
Generator 2: −120000 − PV(12,3,−8000,40000) $PW_2 = \$-110{,}743$

As expected, the PW values and selection of Generator 2 are the same as the
spreadsheet solution.

	A	B	C D E F
2	Generator	PW value	Function to determine PW
3	1	-$141,378	= -80000 - PV(12%,3,-30000,15000)
5	2	-$110,743	= -120000 - PV(12%,3,-8000,40000)

Minus on PV function maintains correct sense of PV value

FIGURE 4.4 Equal-life alternatives evaluated using the PV
function, Example 4.8.

When different-life alternatives are evaluated, using the LCM basis, it is necessary to input all the cash flows for the LCM of the lives to ensure an equal-service
evaluation. Develop the NPV function to find PW. If cash flow is identified by CF,
the general format is

$$= P + \text{NPV}(i\%,\text{year\_1\_CF\_cell:last\_year\_CF\_cell}) \qquad [4.5]$$

It is very important that the *initial cost P not be included* in the cash flow series
identified in the NPV function. Unlike the PV function, the NPV function returns
the correct sign for the PW value.

EXAMPLE 4.9   Continuing with the previous example, once Cesar had selected generator 2 to purchase, he approached the manufacturer with the concerns that the first cost was too high and the expected life was too short. He was offered a lease arrangement for 6 years with a $20,000 annual cost and an extra $20,000 payment in the first and last years to cover installation and removal costs. Determine if generator 2 or the lease arrangement is better at 12% per year.

	A	B	C	D	E	F	G
1	Year	Generator 2	Lease				
2	0	-120,000	-40,000		Repurchase cash flow		
3	1	-8,000	-20,000		$= S - AOC - P$		
4	2	-8,000	-20,000		$= 40,000 - 8,000 - 120,000$		
5	3	-88,000	-20,000				
6	4	-8,000	-20,000				
7	5	-8,000	-20,000				
8	6	32,000	-40,000		$= -40000 + NPV(12\%,C3:C8)$		
9	PW value	-$189,568	-$132,361				
10							
11					$= -120000 + NPV(12\%,B3:B8)$		
12							

FIGURE 4.5   Different-life alternatives evaluated using the NPV function, Example 4.9.

Solution

Assuming that generator 2 can be repurchased 3 years hence and all estimates remain the same, PW evaluation over 6 years is correct. Figure 4.5 details the cash flows and NPV functions. The year 3 cash flow for generator 2 is $S - AOC - P = \$-88,000$. Note that the first costs are not included in the NPV function but are listed separately, as indicated in Equation [4.5]. The lease option is the clear winner for the next 6 years.

When evaluating alternatives for which the annual cash flows do not form an *A* series, the individual amounts must be entered on the spreadsheet and Equation [4.5] is used to find PW values. Also, remember that any zero-cash-flow year must be entered as 0 to ensure that the NPV function correctly tracks the years.

SUMMARY

This chapter explained the difference between mutually exclusive and independent alternatives, as well as revenue and cost cash flows. It discussed the use of present worth (PW) analysis to select the economically best alternative. In general, always choose an alternative with the largest PW value calculated at the MARR.

Important points to remember for mutually exclusive alternatives selection are:

1. Compare equal service alternatives over the same number of years using a specified study period. Alternatives with lives shortened by the study

period have as a "salvage value" the estimated market value. For longer lives, an equivalent-service cost must be estimated through the end of the study period.
2. PW evaluation over the least common multiple (LCM) of lives can be used to obtain equal service.
3. Capitalized cost (CC) analysis, an application of PW analysis with $n = \infty$, compares alternatives with indefinite or very long lives. In short, the CC

value is determined by dividing the equivalent A value of all cash flows by the interest rate i.

For independent projects, select all with PW ≥ 0 at the MARR if there is no budget limitation. When funds are limited, form bundles of projects that do not exceed the limit and select the bundle that maximizes PW.

PROBLEMS

Alternative Formulation

4.1 State two conditions under which the do-nothing alternative is not an option.

4.2 When evaluating projects by the present worth method, how do you know which one(s) to select, if the (a) projects are independent, and (b) alternatives are mutually exclusive?

4.3 A biomedical engineer with Johnston Implants just received estimates for replacement equipment to deliver online selected diagnostic results to doctors performing surgery who need immediate information on the patient's condition. The cost is $200,000, the annual maintenance contract costs $5000, and the useful life (technologically) is 5 years.
 a. What is the alternative if this equipment is not selected? What other information is necessary to perform an economic evaluation of the two?
 b. What type of cash flow series will these estimates form?
 c. What additional information is needed to convert the cash flow estimates to the other type of series?

4.4 The lead engineer at Bell Aerospace has $1 million in research funds to commit this year. She is considering five separate R&D projects, identified as A through E. Upon examination, she determines that three of these projects (A, B, and C) accomplish exactly the same objective using different techniques.
 a. Identify each project as mutually exclusive or independent.
 b. If the selected alternative between A, B, and C is labeled X, list all viable options (bundles) for the five projects.

4.5 List all possible bundles for the four independent projects 1, 2, 3, and 4. Projects 3 and 4 cannot both be included in the same bundle.

Evaluation of Equal-Life Alternatives

4.6 What is meant by the term *equal service*?

4.7 An irrigation return flow drain has sampling equipment that can be powered by solar cells or by running an electric line to the site and using conventional power. Solar cells will cost $14,000 to install with a useful life of 10 years. Annual costs for inspection, cleaning, etc. are expected to be $1500. A new power line will cost $12,000 to install and the power costs are estimated at $600 per year. The salvage value of the solar cells is expected to be 25% of the first cost when the sampling project ends in 4 years. The electric line will stay in place, so its salvage value is considered to be zero. At an interest rate of 10% per year, which alternative should be selected?

4.8 Oil from a particular type of marine microalgae can be converted to biodiesel that can serve as an alternate transportable fuel for automobiles and trucks. If lined ponds are used to grow the algae, the construction cost will be $13 million and the maintenance & operating (M&O) cost will be $2.1 million per year. If long plastic tubes are used for growing the algae, the initial cost will be higher at $18 million, but less contamination will render the M&O cost lower at $0.41 million per year. At an interest rate of 10% per year and a 5-year project period, which system is better, ponds or tubes? Use a present worth analysis.

4.9 American Electric Power agreed to spend $4.6 billion to clean up 46 coal-fired power plants that are believed to be contributing to acid rain. The plan is to reduce nitrogen oxide emissions by 69% by 2016 and sulfur dioxide emissions by 79% by 2018. Assume Plan A is to spend $0.575 billion per year in years 1 through 4 and an additional $0.575 billion per year in years 7 through 10. Plan B is to spend $0.46 billion in each of years 1 through 10. At an interest rate of 8% per year, which plan is more economical to the company based on a present worth analysis?

4.10 The costs associated with manufacturing a multi-function portable gas analyzer are estimated. At an interest rate of 8% per year and a present worth analysis, which method should be selected?

	Manual	**Robotic**
First cost, $	−425,000	−850,000
M&O cost, year 1, $	−90,000	−10,000
Increase in M&O, $/year	7,000	1,000
Salvage value, $	80,000	300,000
Life, years	5	5

4.11 A pilot plant for conducting research related to reverse osmosis concentrate recovery via lime softening can be leased for the 4-month project duration for $6,900 per month. Electrical work at the site will cost $8500 now and the technician to install the electrical work will charge $2000 now. Shipping will cost $2300 each way (months 0 and 4). A technician for demobilization will cost $1300 in month 4. At an interest rate of 6% per year compounded monthly, what is the present worth of the pilot plant project?

4.12 Biomet Implants is planning new online patient diagnostics for surgeons while they operate. The new system will cost $300,000 to install in an operating room, $5000 annually for maintenance, and have an expected life of 4 years. The revenue per system is estimated to be $80,000 in year 1 and to increase by $10,000 per year through year 4. Determine if the project is economically justified using PW analysis and an MARR of 10% per year.

4.13 An undergraduate engineering student and her husband operate a pet-sitting service to help make ends meet. They want to add a daily service of a photo placed online for pet owners who are traveling. The estimates are: equipment and setup cost $950; net monthly income over costs $70. Over a

period of 3 years, will the service make at least 12% per year compounded monthly?

4.14 The CFO of Marta Aaraña Cement Industries knows that many of the diesel-fueled systems in its quarries must be replaced at an estimated cost of $20 million 10 years from now. A fund for these replacements has been established with the commitment of $1 million at the end of next year (year 1) with 10% increases through the 10th year. If the fund earns at 5.25% per year, will the company have enough to pay for the replacements? Solve using (a) tabulated factors, and (b) a spreadsheet.

4.15 Burling Water Cooperative currently contracts the removal of small amounts of hydrogen sulfide from its well water using manganese dioxide filtration prior to the addition of chlorine and fluoride. Contract renewal for 5 years will cost $75,000 annually for the next 2 years and $100,000 in years 3, 4, and 5. Assume payment is made at the end of each contract year. Burling Coop can install the filtration equipment for $125,000 and perform the process for $50,000 per year. At a discount rate of 6% per year, does the contract service still save money?

4.16 Halogen-free liquid crystal polymers are used for lead-free soldering without corrosion and maintenance issues. The polymers can be produced by either of two methods. Equipment for method A costs $70,000 initially and has a $15,000 salvage value after 3 years. The operating cost with this method will be $20,000 per year. Method B will have a first cost of $140,000, an operating cost of $8000 per year, and a $40,000 salvage value after its 3-year life. At an interest rate of 12% per year, which method should be used on the basis of a present worth analysis? Solve using (a) tabulated factors, and (b) a calculator.

4.17 A software package created by Navarro & Associates can be used for analyzing and designing three-sided guyed towers and three- and four-sided self-supporting towers. A single-user license will cost $4000 per year. A site license has a one-time cost of $15,000. A structural engineering consulting company is trying to decide between 2 alternatives: first, to buy one single-user license *now* and one each year for the next four years (which will provide five years of service), or second, to buy a site license now. Determine which strategy is more economical at an interest rate of 12% per year for a 5-year planning period. Apply the present worth method of evaluation.

4.18 The Bureau of Indian Affairs provides various services to American Indians and Alaskan Natives. The Director of Indian Health Services is working with chief physicians at some of the 230 clinics nationwide to select the better of two medical X-ray system alternatives to be located at secondary-level clinics. At 5% per year, select the more economical system. Solve using (a) tabulated factors, and (b) a spreadsheet.

	Del Medical	**Siemens**
First cost, $	−250,000	−224,000
Annual operating cost, $ per year	−231,000	−235,000
Overhaul in year 3, $	—	−26,000
Overhaul in year 4, $	−140,000	—
Salvage value, $	50,000	10,000
Expected life, years	6	6

4.19 The Briggs and Stratton Commercial Division designs and manufactures small engines for golf turf maintenance equipment. A robotics-based testing system will ensure that their new signature guarantee program entitled "Always Insta-Start" does indeed work for every engine produced. Compare the two systems at MARR = 10% per year. Solve using (a) tabulated factors, and (b) single-cell spreadsheet functions.

	Pull System	**Push System**
Robot and support equipment first cost, $	−1,500,000	−2,250,000
Annual M&O cost, $ per year	−700,000	−600,000
Rebuild cost in year 3, $	0	−500,000
Salvage value, $	100,000	50,000
Estimated life, years	8	8

4.20 Chevron Corporation has a capital and exploratory budget for oil and gas production of $19.6 billion in one year. The Upstream Division has a project in Angola for which three offshore platform equipment alternatives are identified. Use the present worth method to select the best alternative at 12% per year.

	A	**B**	**C**
First cost, $ million	−200	−350	−475
Annual cost, $ million per year	−450	−275	−400
Salvage value, $ million	75	50	90
Estimated life, years	20	20	20

4.21 The TechEdge Corporation offers two forms of 4-year service contracts on its closed-loop water purification system used in the manufacture of semiconductor packages for microwave and high-speed digital devices. The Professional Plan has an initial fee of $52,000 with annual fees starting at $1000 in contract year 1 and increasing by $500 each year. Alternatively, the Executive Plan costs $62,000 up front with annual fees starting at $5000 in contract year 1 and decreasing by $500 each year. The initial charge is considered a setup cost for which there is no salvage value expected. Evaluate the plans at a MARR of 9% per year. Solve using (a) factors, and (b) a spreadsheet. (c) How is the analysis performed using a financial calculator?

4.22 Allison and Joshua are engineers at Raytheon. Each has presented a proposal to track fatigue development in composite materials installed on special-purpose aircraft. Which is the better plan economically, if i = 12% per year compounded monthly?

	Allison's Plan	**Joshua's Plan**
Initial cost, $	−40,000	−60,000
Monthly M&O costs, $ per month	−5,000	—
Semiannual M&O cost, $ per 6-month	—	−13,000
Salvage value, $	10,000	8,000
Life, years	5	5

4.23 What is the present worth of a $40,000 bond that has a bond interest rate of 6% per year, payable semiannually? The bond matures in 20 years. The interest rate in the marketplace is 8% per year compounded semiannually.

4.24 The present worth of a $10,000 municipal bond due 6 years from now is $11,000. If the bond interest is payable quarterly and the interest rate used in discounting the cash flow is 8% per year compounded quarterly, what is the bond coupon rate *b* per year?

4.25 Jamal bought a 5% $1000 20-year bond for $925. He received a semiannual dividend for 8 years, then sold it immediately after the 16th dividend for $800. Did Jamal make the return of 5% per year compounded semiannually that he wanted? Solve using (*a*) factors, and (*b*) a spreadsheet.

4.26 An investor thought that market interest rates were going to decline. He paid $19,000 for a corporate bond with a face value of $20,000. The bond has an interest rate of 10% per year payable annually. If the investor plans to sell the bond immediately after receiving the 4th interest payment, how much will he have to receive in order to make a return of 14% per year? Solve using (*a*) tabulated factors, and (*b*) the GOAL SEEK tool on a spreadsheet.

4.27 An investor pays $30,000 for a convertible bond (one that can be converted into shares of corporate common stock). The bond conversion rate is 100 shares of stock anytime within the next five years. What will the stock price have to be in year 3 in order for the investor to make 10% per year on the investment? Assume the bond interest rate is 4% per year payable annually.

4.28 Atari needs $4.5 million in new investment capital to develop and market downloadable game software for its new GPS2-ZX system. The plan is to sell $10,000 face-value corporate bonds at a discount of $9000 now. A bond pays a dividend each 6 months based on a bond interest rate of 5% per year with the $10,000 face value returned after 20 years. Will a purchase make at least 6% per year compounded semiannually?

Evaluation of Different-Life Alternatives

4.29 Heidleman Industries is considering two types of materials for roofing its warehouses. EPDM is an elastomeric polymer synthesized from ethylene, propylene, and a small amount of diene monomer, compounded with carbon black processing oils and various cross-linking and stabilizing agents. The 75 mil thickness will cost $4.10 per square foot and will last for 25 years. A thin sheet aluminum roof will cost $6.00 per square foot, but it will last for 50 years. Using an interest rate of 10% per year and a present worth comparison, determine whether the company should install the polymer or the aluminum roof.

4.30 Benjamin is an engineer with the Lego Group in Bellund, Denmark, manufacturers of Lego toy construction blocks. He is responsible for the economic analysis of a new production method of special-purpose Lego parts. Method 1 will have an initial cost of $400,000, an annual operating cost of $140,000, and a life of three years. Method 2 will have an initial cost of $600,000, an operating cost of $100,000 per year, and a six-year life. Assume 10% salvage values for both methods. If Lego Industries uses a MARR of 15% per year, which method should it select on the basis of a present worth analysis?

4.31 A mechanical engineer is considering two types of pressure sensors for the low-pressure steam lines in several of the company plants. Piezoresistive sensors use the change in conductivity of semiconductors to measure the pressure. Fiber optic sensors use the properties of fiber optic interferometers to sense nanometer scale displacement of membranes. The costs for each system are shown below. Which should be selected based on a present worth comparison at an interest rate of 1% per month?

	Piezoresistive	Fiber Optic
Purchase cost, $	−13,650	−22,900
Maintenance cost, $/month	−200	−50
Salvage value, $	0	2,000
Life, years	2	4

4.32 Two mutually exclusive projects have the estimated cash flows shown. Use a present worth analysis to determine which should be selected at an interest rate of 10% per year.

	Project P	Project Q
First cost, $	−55,000	−95,000
Annual cost, $/year	−9,000	−5,000 year 1, increasing by $1000 per year
Salvage value, $	nil	4,000
Life, years	2	4

4.33 An industrial engineer is considering two robots for improving efficiency in a fiber-optic manufacturing company. Robot X will have a first cost of $85,000, an annual maintenance and operation

(M&O) cost of $30,000, and a $35,000 salvage value after its useful life of two years. A more sophisticated model, Robot Y will cost $157,000, have an annual M&O cost of $28,000, and a $60,000 salvage value after its four-year life. Select the better robot on the basis of a future worth comparison at an interest rate of 10% per year. Solve by (a) tabulated factors, and (b) calculator.

4.34 Virgin Galactic is considering two materials for certain parts in a re-useable space vehicle: carbon fiber reinforced plastic (CFRP) and fiber reinforced ceramic (FRC). The costs are shown below. Which should be selected on the basis of a present worth comparison at an interest rate of 10% per year? Solve using (a) tabulated factors, and (b) single-cell spreadsheet functions.

	CFRP	FRC
First cost, $	−205,000	−235,000
Maintenance cost, $/year	−29,000	−27,000
Salvage value, $	2,000	20,000
Life, years	2	4

4.35 A metallurgical engineer is considering the two ceramics estimated below for use in a high-temperature annealing furnace. (a) Which should be selected on the basis of a present worth comparison at an interest rate of 12% per year? (b) If the life of material XX is increased from 3 to 4 years, determine the number of re-purchases for both alternatives necessary for a present worth analysis based on the equal-service requirement.

	Material XX	Material ZZ
First cost, $	−230,000	−380,000
Maintenance cost, $/year	−9,000	−12,000
Salvage value, $	12,000	140,000
Life, years	3	6

4.36 An environmental engineer must recommend one of two methods for monitoring high colony counts of E. coli and other bacteria in watershed area "hot spots." Estimates are tabulated and the MARR is 10% per year. Use tabulated factors or a spreadsheet for your analysis.

    a. Use present worth analysis to select the better method.

    b. For a study period of 3 years, use PW analysis to select the better method.

	Method A	Method B
Initial cost, $	−100,000	−250,000
Annual operating cost, $ per year	−30,000 in year 1, increasing by $5000 each year	−20,000
Life, years	3	6

4.37 Allen Auto Group owns corner property that can be a parking lot for customers or sold for retail sales space. The parking lot option can use concrete or asphalt. Concrete will cost $375,000 initially, last for 20 years, and have an estimated annual maintenance cost of $200 starting at the end of the eighth year. Asphalt is cheaper to install at $250,000, but it will last 10 years and cost $2500 per year to maintain starting at the end of the second year. If asphalt is replaced after 10 years, the $2500 maintenance cost will be expended in its last year. There are no salvage values to be considered. Use $i = 8\%$ per year and PW analysis to select the more economic surface, provided the property is (a) used as a parking lot for 20 years, and (b) sold after 5 years and the parking lot is completely removed.

4.38 The manager of engineering at the 900-megawatt Hamilton Nuclear Power Plant has three options to supply personal safety equipment to employees. Two are vendors who sell the items, and the third alternative is to rent the equipment for $50,000 per year, but for no more than 3 years per contract. These items have relatively short lives due to constant use. The MARR is 10% per year.

	Vendor R	Vendor T	Rental
Initial cost, $	−75,000	−125,000	0
Annual upkeep, $ per year	−27,000	−12,000	0
Annual rental, $ per year	0	0	−50,000
Salvage value, $	0	30,000	0
Estimated life, years	2	3	Maximum of 3

    a. Select from the two vendors using the LCM and PW analysis.

    b. Determine which of the three options is cheaper over a study period of 3 years.

4.39 Akash Uni-Safe in Chennai, India, makes Terminator fire extinguishers. It needs replacement equipment to form the neck at the top of each extinguisher during production. Select between two metal-constricting systems. Use the corporate MARR of 15% per year with (a) present worth analysis, and (b) future worth analysis.

	Machine D	Machine E
First cost, $	−62,000	−77,000
Annual operating cost, $ per year	−15,000	−21,000
Salvage value, $	8,000	10,000
Life, years	4	6

4.40 HJ Heinz Corporation is constructing a distribution facility in Italy for products such as Heinz Ketchup, Jack Daniel's sauces, HP steak sauce, and Lea & Perrins Worcestershire sauce. A 15-year life is expected for the structure. The exterior of the building has not yet been selected. One alternative is to use concrete walls as the facade. This will require painting now and every 5 years at a cost of $80,000 each time. Another alternative is an anodized metal exterior attached to the concrete wall. This will cost $200,000 now and require only minimal maintenance of $500 every 3 years. A metal exterior is more attractive and will have a resale value of an estimated $25,000 more than concrete 15 years from now. Assume painting (for concrete) or maintenance (for metal) will be performed in the last year of ownership to promote selling the property. Use future worth analysis and $i = 12\%$ per year to select the exterior finish.

4.41 Three types of bits can be used in an automated drilling operation. A bright high-speed steel (HSS) bit is the least expensive to buy, but it has a shorter life than either gold oxide or titanium nitride bits. The HSS bits will cost $3500 to buy and will last for 3 months under the conditions of use. The operating cost for these bits will be $2000 per month. The gold oxide bits will cost $6500 to buy and will last for 6 months with an operating cost of $1500 per month. The titanium nitride bits will cost $7000 to buy and will last 6 months with an operating cost of $1200 per month. At an interest rate of 12% per year compounded monthly, which type of drill bit should be selected? Use a future worth analysis.

Life Cycle Cost

4.42 Three different plans were presented to the GAO by a high-tech facilities manager for operating an identity-theft scanning facility. Plan A involves renewable 1-year contracts with payments of $1 million at the beginning of each year. Plan B is a 2-year contract that requires four payments of $600,000 each, with the first one made *now* and the other three at 6-month intervals. Plan C is a 3-year contract that entails a payment of $1.5 million *now* and a second payment of $0.5 million 2 years from now. Assuming that the GAO could renew any of the plans under the same payment conditions, which plan is best on the basis of a present worth analysis at an interest rate of 6% per year compounded semiannually?

4.43 The U.S. Army received two proposals for a turnkey design/build project for barracks for infantry unit soldiers in training. Proposal A involves an off-the-shelf "bare-bones" design and standard grade construction of walls, windows, doors, and other features. With this option, heating and cooling costs will be greater, maintenance costs will be higher, and replacement will occur earlier than proposal B. The initial cost for A will be $750,000. Heating and cooling costs will average $6000 per month with maintenance costs averaging $2000 per month. Minor remodeling will be required in years 5, 10, and 15 at a cost of $150,000 each time in order to render the units usable for 20 years. They will have no salvage value. Proposal B will include tailored design and construction costs of $1.1 million initially with estimated heating and cooling costs of $3000 per month and maintenance costs of $1000 per month. There will be no salvage value at the end of the 20-year life. Which proposal should be accepted on the basis of a life-cycle cost analysis at an interest rate of 0.5% per month?

4.44 A medium-size municipality plans to develop a software system to assist in project selection during the next 10 years. A life-cycle cost approach has been used to categorize costs into development, programming, operating, and support costs for each alternative. There are three alternatives under consideration, identified as A (tailored system), B (adapted system), and C (current system). The costs are summarized on the next page. Perform a life-cycle cost analysis to identify the best altérnative at 8% per year using (*a*) tabulated factors first, then (*b*) a spreadsheet to verify your selection.

Alternative	Cost Component	Cost Estimates
A	Development	$250,000 now, $150,000 years 1–4
	Programming	$45,000 now, $35,000 years 1,2
	Operation	$50,000 years 1 through 10
	Support	$30,000 years 1 through 5
B	Development	$10,000 now
	Programming	$45,000 year 0, $30,000 years 1–3
	Operation	$80,000 years 1 through 10
	Support	$40,000 years 1 through 10
C	Operation	$175,000 years 1 through 10

4.45 Recently introduced Gatorade Endurance Formula contains more electrolytes (such as calcium and magnesium) than the original sports drink formula, thus causing Endurance to taste saltier to some. It is important than the amount of electrolytes be precisely balanced in the manufacturing process. The currently installed system (called EMOST) can be upgraded to monitor the amount more precisely. It costs $12,000 per year for equipment maintenance, $45,000 a year for labor, and the upgrade will cost $25,000 now. This can serve for 10 more years, the expected remaining time the product will be financially successful. A new system (UPMOST) will also serve for the 10 years and have the following estimated costs. All costs are per year for the indicated time periods.

Equipment: $150,000 years 0 and 1
Development: $120,000 years 1 and 2
Maintain and phaseout EMOST: $20,000 years 1, 2, and 3
Maintain hardware and software: $10,000 years 3 through 10
Personnel costs: $90,000 years 3 through 10
Scrapped formula: $30,000 years 3 through 10

Sales of Gatorade Endurance with the UPMOST system installed are expected to go up by $150,000 per year beginning in year 3 and increase by $50,000 per year through year 10. Use LCC analysis at an MARR of 20% per year to select the better electrolyte monitoring system. (Choose from tabulated factors, calculator, or spreadsheet to make your evaluation.)

Capitalized Cost

4.46 The Golden Gate bridge is maintained by 17 ironworkers, who replace corroding steel and rivets, and 38 painters. If the painters have an average wage of $120,000 per year with benefits and the ironworkers get $150,000, what is the capitalized cost today of all the future wages for bridge maintenance at an interest rate of 8% per year?

4.47 Determine the capitalized cost of $100,000 now and $50,000 per year in years one through infinity at an interest rate of 10% per year compounded continuously.

4.48 Determine the capitalized cost of $1,000,000 at time 0, $125,000 in years 1 through 10, and $200,000 per year from year 11 on. Use an interest rate of 10% per year.

4.49 The cost of extending Park Road PR2 in Yellowstone National Park is $1.7 million. Resurfacing and other maintenance is expected to cost $350,000 every 3 years. What is the capitalized cost of the road extension at an interest rate of 6% per year?

4.50 John wants to have the financial ability to withdraw $80,000 per year forever beginning 30 years from now. If his retirement account earns 8% per year interest and dividends, what is the required balance in (*a*) year 29, and (*b*) year 0?

4.51 What is the capitalized cost (absolute value) of the *difference* between the following two plans at an interest rate of 10% per year? Plan A requires an expenditure of $50,000 every five years forever beginning in year 5. Plan B requires an expenditure of $100,000 every 10 years forever beginning in year 10.

4.52 An alumna of Ohio State University wants to set up an endowment fund that can award scholarships to female engineering students totalling $100,000 per year forever. The first scholarships are to be granted *now* and continue each year from now on. How much must the alumna donate now, if the endowment fund is expected to earn interest at a rate of 8% per year?

4.53 Field chlorination of reclaimed water can be accomplished via a low-cost system that uses calcium hypochlorite tables. System components include a 4 foot diameter pipe that is 4 inches long ($70), a solenoid valve ($50), a float switch ($30), a chlorine analyzer ($1500), and a programmable VFD solution pump ($1900). Assume the following component lives: pipe – 10 years; solenoid valve – 2 years; float switch – 2 years; chlorine analyzer – 5 years; solution pump – 5 years. Calculate the capitalized cost of the system at an interest rate of 8% per year.

4.54 Compare the cost of the two types of composite materials on the basis of their capitalized costs. Use an interest rate of 10% per year.

	Material J1	Material K2
First cost, $	−55,000	−325,000
Maintenance cost, $/year	−6,000	−1,000
Salvage value	2,000	200,000
Life, years	3	∞

4.55 The president of Biomed Products is considering a long-term contract to outsource maintenance and operations that will significantly improve the energy efficiency of their imaging systems. The payment schedule has two large payments in the first years with continuing payments thereafter. The proposed schedule is $200,000 now, $300,000 four years from now, $50,000 every 5 years, and an annual amount of $8000 beginning 15 years from now and continuing indefinitely. Determine the capitalized cost at 8% per year.

4.56 UPS Freight plans to spend $100 million on new long-haul tractor-trailers. Some of these vehicles will include a new shelving design with adjustable shelves to transport irregularly sized freight that requires special handling during loading and unloading. Though the life is relatively short, the director wants a capitalized cost analysis performed on the two final designs. Compare the alternatives at the MARR of 10% per year using (*a*) tabulated factors, and (*b*) a spreadsheet.

	Design A: movable shelves	Design B: adaptable frames
First cost, $	−2,500,000	−1,100,000
AOC, $ per year	−130,000	−65,000
Annual revenue, $ per year	800,000	625,000
Salvage value, $	50,000	20,000
Life, years	6	4

4.57 A water supply cooperative plans to increase its water supply by 8.5 million gallons per day to meet increasing demand. One alternative is to spend $10 million to increase the size of an existing reservoir in an environmentally acceptable way. Added annual upkeep will be $25,000 for this option. A second option is to drill new wells and provide added pipelines for transportation to treatment facilities at an initial cost of $1.5 million and annual cost of $120,000. The reservoir is expected to last indefinitely, but the productive well life is only 10 years. Compare the alternatives at 5% per year.

4.58 Three alternatives to incorporate improved techniques to manufacture computer drives to play HD DVD optical disc formats have been developed and costed. Compare the alternatives below using capitalized cost and an interest rate of 12% per year compounded quarterly.

	Alternative E	Alternative F	Alternative G
First cost, $	−2,000,000	−3,000,000	−10,000,000
Net income, $ per quarter	300,000	100,000	400,000
Salvage value, $	50,000	70,000	—
Life, years	4	8	∞

Independent Projects

4.59 A small manufacturing company is considering the addition of one or more of four new product lines. If the total amount of investment capital available for new ventures is $800,000, which one(s) should the company undertake on the basis of a present worth analysis? Assume the company uses a 5-year project recovery period and a MARR of 20% per year. All cash flows are in $1000 units.

	Product Lines			
	R1	S2	T3	U4
First cost, $	−200	−400	−500	−700
M&O cost, $/year	−50	−200	−300	−400
Revenue, $/year	150	450	520	770

4.60 Determine which of the following independent projects should be selected for investment if $240,000 is available and the MARR is 10% per year. Use the PW method to evaluate mutually exclusive bundles to make the selection.

Project	Initial Investment, $	Net Cash Flow, $/year	Life, Years
A	−100,000	50,000	8
B	−125,000	24,000	8
C	−120,000	75,000	8
D	−220,000	39,000	8
E	−200,000	82,000	8

4.61 Feng Seawater Desalination Systems has established a capital investment limit of $800,000 for next year for projects that target improved recovery of highly brackish groundwater. Select any or all of the projects using a MARR of 10% per year. All projects have a 4-year life.

Project	Initial Investment, $	Net Cash Flow, $/year	Salvage Value, $
X	−250,000	50,000	45,000
Y	−300,000	90,000	−10,000
Z	−550,000	150,000	100,000

4.62 Dwayne has four independent vendor proposals to contract the nationwide oil recycling services for the Ford Corporation manufacturing plants. All combinations are acceptable, except that vendors B and C cannot both be chosen. Revenue sharing of recycled oil sales with Ford is a part of the requirement. Develop all possible mutually exclusive bundles under the additional following restrictions and select the best projects. The corporate MARR is 10% per year.
   a. A maximum of $4 million can be spent.
   b. A larger budget of $5.5 million is allowed, but no more than two vendors can be selected.
   c. There is no limit on spending.

Vendor	Initial Investment, $	Life, Years	Annual Net Revenue, $ per Year
A	−1.5 million	8	360,000
B	−3.0 million	10	600,000
C	−1.8 million	5	620,000
D	−2.0 million	4	630,000

ADDITIONAL PROBLEMS AND FE EXAM REVIEW QUESTIONS

4.63 In the PW method of alternative evaluation, equal service means that:
   a. all projects must start at the same time.
   b. all projects are evaluated over the same time period.
   c. all projects must have the same operating cost.
   d. all projects have equal salvage values.

The following estimates are used in Problems 4.64 through 4.66.

The cost of money is 10% per year.

	Machine P	Machine Q
Initial cost, $	−35,000	−66,000
Annual cost, $/year	−20,000	−15,000
Salvage value, $	10,000	23,000
Life, years	2	4

4.64 In comparing the machines on a present worth basis, the present worth of machine P is closest to:
   a. $−82,130
   b. $−87,840
   c. $−91,568
   d. $−112,230

4.65 In comparing the machines on a present worth basis, the present worth of machine Q is closest to:
   a. $−68,445
   b. $−97,840
   c. $−125,015
   d. $−223,120

4.66 The capitalized cost of machine P is closest to:
   a. $−35,405
   b. $−97,840
   c. $−354,050
   d. $−708,095

4.67 The cost of maintaining a public monument in Washington, D.C. occurs as periodic outlays of

$10,000 every 5 years. If the first outlay is 5 years from now, the capitalized cost of the maintenance at an interest rate of 10% per year is closest to:
- **a.** $-1638
- **b.** $-16,380
- **c.** $-26,380
- **d.** $-29,360

4.68 A grateful donor wishes to start an endowment at her alma mater that will provide scholarship money of $40,000 per year beginning *now* (time 0) and continue indefinitely. If the funds earn 10% per year, the amount she must donate now is closest to:
- **a.** $340,000
- **b.** $400,000
- **c.** $440,000
- **d.** $493,800

4.69 A corporate bond has a face value of $10,000, a bond interest rate of 8% per year payable semiannually, and a maturity date of 20 years from now. If a person purchases the bond for $9000 when the interest rate in the market place is 8% per year compounded semiannually, the size and frequency of the interest payments the person will receive are:
- **a.** $270 every six months
- **b.** $300 every six months
- **c.** $360 every six months
- **d.** $400 every six months

The following estimates are used in Problems 4.70 and 4.71.

The MARR is 12% per year.

	Alternative 1	Alternative 2
First cost, $	−40,000	−65,000
Annual cost, $ per year	−20,000	−15,000
Salvage value, $	10,000	25,000
Life, years	3	4

4.70 The relation that correctly calculates the present worth of alternative 2 when comparing it to alternative 1 is:
- **a.** $-65,000 - 15,000(P/A,12\%,12) + 25,000(P/F,12\%,8) + 25,000(P/F,12\%,12)$
- **b.** $-65,000 - 15,000(P/A,12\%,4) + 25,000(P/F,12\%,4)$
- **c.** $-65,000 - 15,000(P/A,12\%,12) + 25,000(P/F,12\%,12)$
- **d.** $-65,000 - 40,000[(P/F,12\%,4) + (P/F,12\%,8)] - 15,000(P/A,12\%,12) + 25,000(P/F,12\%,12)$

4.71 The number of life cycles for each alternative when performing a present worth evaluation based on the LCM for equal service is:
- **a.** 2 for each alternative
- **b.** 1 for each alternative
- **c.** 3 for alternative 1; 4 for alternative 2
- **d.** 4 for alternative 1; 3 for alternative 2

Annual Worth Analysis

Imagestate Media (John Foxx); Ingram Publishing

A n AW analysis is commonly preferred over a PW analysis because the AW value is easy to calculate; the measure of worth—AW in monetary units (dollars) per year—is understood by most individuals; and its assumptions are essentially the same as those of the PW method.

Annual worth is known by other titles. Some are equivalent annual worth (EAW), equivalent annual cost (EAC), annual equivalent (AE), and EUAC (equivalent uniform annual cost). The alternative selected by the AW method will always be the same as that selected by the PW method, and all other alternative evaluation methods, provided they are performed correctly.

LEARNING OUTCOMES

1. Calculate capital recovery and AW over one life cycle.

AW calculation

2. Select the best alternative on the basis of an AW analysis.

Alternative selection by AW

3. Select the best long-life (infinite-life) investment alternative using AW values.

Long-life investment AW

4. Use a spreadsheet to perform an AW evaluation.

Spreadsheets

5.1 AW VALUE CALCULATIONS

The annual worth (AW) method is commonly used for comparing alternatives. All cash flows are converted to an equivalent uniform annual amount over one life cycle of the alternative. The AW value is easily understood by all since it is stated in terms of dollars per year. The major advantage over all other methods is that the equal service requirement is met without using the least common multiple (LCM) of alternative lives. The AW value is calculated over one life cycle and is assumed to be exactly the same for any succeeding cycles, provided all cash flows change with the rate of inflation or deflation. If this cannot be reasonably assumed, a study period and specific cash flow estimates are needed for the analysis. The repeatability of the AW value over multiple cycles is demonstrated in Example 5.1.

Podcast available

New digital scanning graphics equipment is expected to cost $20,000, to be used for 3 years, and to have an annual operating cost (AOC) of $8000. Determine the AW values for one and two life cycles at $i = 22\%$ per year.

EXAMPLE 5.1

Solution
First use the cash flows for one life cycle (Figure 5.1) to determine AW.

$$AW = -20,000(A/P,22\%,3) - 8000 = \$-17,793$$

For two life cycles, calculate AW over 6 years. Note that the purchase for the second cycle occurs at the end of year 3, which is year zero for the second life cycle (Figure 5.1).

$$AW = -20,000(A/P,22\%,6) - 20,000(P/F,22\%,3)(A/P,22\%,6) - 8000$$
$$= \$-17,793$$

The same AW value can be obtained for any number of life cycles, thus demonstrating that the AW value for one cycle represents the equivalent annual worth of the alternative for every cycle.

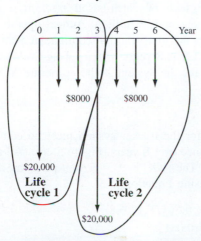

FIGURE 5.1
Cash flows over two life cycles of an alternative.

It is always possible to determine the AW, PW, and FW values from each other using the following relation.

$$AW = PW(A/P,i,n) = FW(A/F,i,n) \qquad \text{[5.1]}$$

The equal-service requirement necessary for a PW comparison means that the n value in this equation is the LCM of lives.

The AW value of an alternative is the addition of two distinct components: *capital recovery (CR)* of the initial investment and the *equivalent A value* of the annual operating costs (AOC).

$$AW = CR + A \text{ of AOC} \qquad \text{[5.2]}$$

The recovery of an amount of capital P committed to an asset, plus the time value of the capital at a particular interest rate, is a fundamental principle of economic analysis. *Capital recovery is the equivalent annual cost of owning the asset plus the return on the initial investment.* The A/P factor is used to convert P to an equivalent annual cost. If there is some anticipated positive salvage value S at the end of the asset's useful life, its equivalent annual value is removed using the A/F factor. This action reduces the equivalent annual cost of owning the asset. Accordingly, CR is

$$CR = -P(A/P,i,n) + S(A/F,i,n) \qquad \text{[5.3]}$$

The annual amount (A of AOC) is determined from uniform recurring costs (and possibly receipts) and nonrecurring amounts. The P/A and P/F factors may be necessary to first obtain a present worth amount, then the A/P factor converts this amount to the A value in Equation [5.2].

EXAMPLE 5.2   Lockheed Martin is increasing its booster thrust power in order to win more satellite launch contracts from European companies interested in new global communications markets. A piece of earth-based tracking equipment is expected to require an investment of $13 million. Annual operating costs for the system are expected to start the first year and continue at $0.9 million per year. The useful life of the tracker is 8 years with a salvage value of $0.5 million. Calculate the AW value for the system if the corporate MARR is currently 12% per year.

Solution

The cash flows (Figure 5.2*a*) for the tracker system must be converted to an equivalent AW cash flow sequence over 8 years (Figure 5.2*b*). (All amounts are expressed in $1 million units.) The AOC is $A = \$-0.9$ per year, and the capital recovery is calculated by using Equation [5.3].

$$
\begin{aligned}
CR &= -13(A/P,12\%,8) + 0.5(A/F,12\%,8) \\
&= -13(0.2013) + 0.5(0.0813) \\
&= \$-2.576
\end{aligned}
$$

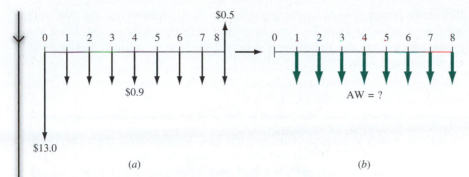

FIGURE 5.2 (*a*) Cash flow diagram for satellite tracker costs, and (*b*) conversion to an equivalent AW (in $1 million), Example 5.2.

The correct interpretation of this result is very important to Lockheed Martin. It means that each and every year for 8 years, the equivalent total revenue from the tracker must be at least $2,576,000 *just to recover the initial present worth investment plus the required return of 12% per year.* This does not include the AOC of $0.9 million each year. Total AW is found by Equation [5.2].

$$AW = -2.576 - 0.9 = \$-3.476 \text{ million per year}$$

This is the AW for all future life cycles of 8 years, provided the costs rise at the same rate as inflation, and the same costs and services apply for each succeeding life cycle.

For solution by computer, use the PMT function to determine CR only in a single spreadsheet cell. The format is = PMT($i\%,n,P,-S$). As an illustration, the CR in Example 5.2 is displayed when = PMT(12%,8,13,−0.5) is entered.

The annual worth method is applicable in any situation where PW, FW, or Benefit/Cost analysis can be utilized. The AW method is especially useful in certain types of studies: asset replacement and retention studies to minimize overall annual costs, breakeven studies and make-or-buy decisions (all covered in later chapters), and all studies dealing with production or manufacturing where cost/unit is the focus.

5.2 EVALUATING ALTERNATIVES BASED ON ANNUAL WORTH

The annual worth method is typically the easiest of the evaluation techniques to perform, when the MARR is specified. The alternative selected has the lowest equivalent annual cost (cost alternatives), or highest equivalent income (revenue alternatives). The selection guidelines for the AW method are the same as for the PW method.

One alternative: AW ≥ 0, the alternative is financially viable.

Two or more alternatives: Choose the numerically largest AW value (lowest cost or highest income).

If a study period is used to compare two or more alternatives, the AW values are calculated using cash flow estimates over only the study period. For a study period shorter than the alternative's expected life, use an estimated market value for the salvage value.

EXAMPLE 5.3   PizzaRush, which is located in the general Los Angeles area, fares very well with its competition in offering fast delivery. Many students at the area universities and community colleges work part-time delivering orders made via the web at PizzaRush.com. The owner, a software engineering graduate of USC, plans to purchase and install five portable, in-car systems to increase delivery speed and accuracy. The systems provide a link between the web order-placement software and the in-car GPS system for satellite-generated directions to any address in the Los Angeles area. The expected result is faster, friendlier service to customers, and more income for PizzaRush.

Each system costs $4600, has a 5-year useful life, and may be salvaged for an estimated $300. Total operating cost for all systems is $650 for the first year, increasing by $50 per year thereafter. The MARR is 10% per year. Perform an annual worth evaluation that answers the following questions:

a. How much new annual revenue is necessary to recover only the initial investment at an MARR of 10% per year?
b. The owner conservatively estimates increased income of $5000 per year for all five systems. Is this project financially viable at the MARR? See cash flow diagram in Figure 5.3.
c. Based on the answer in part (b), determine how much new income PizzaRush must have to economically justify the project. Operating costs remain as estimated.

FIGURE 5.3
Cash flow diagram used to compute AW, Example 5.3.

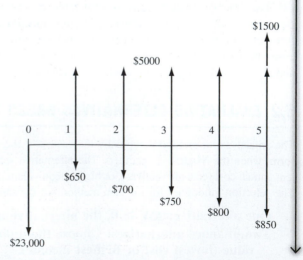

Solution

a. The CR value will answer this question. Use Equation [5.3] at 10%.

$$CR = -5(4600)(A/P,10\%,5) + 5(300)(A/F,10\%,5)$$
$$= \$-5822$$

b. The financial viability could be determined now without calculating the AW value, because the $5000 in new income is lower than the CR of $5822, which does not yet include the annual costs. So, the project is not economically justified. However, to complete the analysis, determine the total AW. The annual operating costs and incomes form an arithmetic gradient series with a base of $4350 in year 1, decreasing by $50 per year for 5 years. The AW relation is

$$AW = \text{capital recovery} + A \text{ of net income}$$
$$= -5822 + 4350 - 50(A/G,10\%,5) \qquad \text{[5.4]}$$
$$= \$-1562$$

This shows conclusively that the alternative is not financially viable at MARR = 10%.

c. An equivalent of the projected $5000 plus the AW amount are necessary to make the project economically justified at a 10% return. This is 5000 + 1562 = $6562 per year in new revenue. At this point AW will equal zero based on Equation [5.4].

A quarry outside of Austin, Texas, wishes to evaluate two similar pieces of equipment by which the company can meet new state environmental requirements for dust emissions. The MARR is 12% per year. Determine which alternative is economically better using (a) the AW method, and (b) AW method with a 3-year study period.

EXAMPLE 5.4

Equipment	X	Y
First cost, $	−40,000	−75,000
AOC, $ per year	−25,000	−15,000
Life, years	4	6
Salvage value, $	10,000	7,000
Estimated value after 3 years, $	14,000	20,000

Solution

a. Calculating AW values over the respective lives indicates that Y is the better alternative.

$$AW_X = -40{,}000(A/P{,}12\%{,}4) - 25{,}000 + 10{,}000(A/F{,}12\%{,}4)$$
$$= \$-36{,}077$$
$$AW_Y = -75{,}000(A/P{,}12\%{,}6) - 15{,}000 + 7{,}000(A/F{,}12\%{,}6)$$
$$= \$-32{,}380$$

b. All n values are 3 years and the "salvage values" become the estimated market values after 3 years. Now X is economically better.

$$AW_X = -40{,}000(A/P{,}12\%{,}3) - 25{,}000 + 14{,}000(A/F{,}12\%{,}3)$$
$$= \$-37{,}505$$
$$AW_Y = -75{,}000(A/P{,}12\%{,}3) - 15{,}000 + 20{,}000(A/F{,}12\%{,}3)$$
$$= \$-40{,}299$$

5.3 AW OF A LONG-LIFE OR INFINITE-LIFE INVESTMENT

The annual worth equivalent of a very long-lived project is the AW value of its capitalized cost (CC), discussed in Section 4.4. The AW value of the first cost, P, or present worth, PW, of the alternative uses the same relation as Equation [4.2].

$$\mathbf{AW = CC}(i) = \mathbf{PW}(i) \qquad\qquad \text{[5.5]}$$

Cash flows that occur at regular intervals are converted to AW values over one life cycle of their occurrence. All other nonregular cash flows are first converted to a P value and then multiplied by i to obtain the AW value over infinity.

EXAMPLE 5.5 If you receive an inheritance of $10,000 today, how long do you have to invest it at 8% per year to be able to withdraw $2000 every year forever? Assume the 8% per year is a return that you can depend on forever.

Solution

Cash flow is detailed in Figure 5.4. Solving Equation [5.5] for PW indicates that it is necessary to have $25,000 accumulated at the time that the $2000 annual withdrawals start.

$$PW = 2000/0.08 = \$25{,}000$$

Find $n = 11.91$ years using the relation $\$25{,}000 = 10{,}000(F/P{,}8\%{,}n)$.

Comment: It is easy to use a spreadsheet to solve this problem. In any cell write the function $= \text{NPER}(8\%,,-10000,25000)$ to display the answer of 11.91 years. The financial calculator function $n(8,0,-10000,25000)$ displays the same n value.

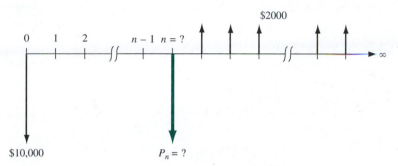

FIGURE 5.4 Diagram to determine n for a perpetual withdrawal, Example 5.5.

The state is considering three proposals for increasing the capacity of the main drainage canal in an agricultural region. Proposal A requires dredging the canal. The state is planning to purchase the dredging equipment and accessories for $650,000. The equipment is expected to have a 10-year life with a $17,000 salvage value. The annual operating costs are estimated to total $50,000. To control weeds in the canal itself and along the banks, environmentally safe herbicides will be sprayed during the irrigation season. The yearly cost of the weed control program is expected to be $120,000.

    Proposal B is to line the canal walls with concrete at an initial cost of $4 million. The lining is assumed to be permanent, but minor maintenance will be required every year at a cost of $5000. In addition, lining repairs will have to be made every 5 years at a cost of $30,000.

    Proposal C is to construct a new pipeline along a different route. Estimates are: an initial cost of $6 million, annual maintenance of $3000 for right-of-way, and a life of 50 years.

    Compare the alternatives on the basis of annual worth, using an interest rate of 5% per year.

EXAMPLE 5.6

Solution

Since this is an investment for a long-life project, compute the AW for one cycle of all recurring costs. For proposals A and C, the CR values are found using Equation [5.3], with $n_A = 10$ and $n_C = 50$, respectively. For proposal B, the CR is simply $P(i)$.

Proposal A

CR of dredging equipment:

$-650,000(A/P,5\%,10) + 17,000(A/F,5\%,10)$ $\$ -82,824$

Annual cost of dredging $-50,000$

Annual cost of weed control $-120,000$

     $\$-252,824$

Proposal B

CR of initial investment: $-4,000,000(0.05)$ $\$-200,000$

Annual maintenance cost $-5,000$

Lining repair cost: $-30,000(A/F,5\%,5)$ $-5,429$

     $\$-210,429$

Proposal C

CR of pipeline: $-6,000,000(A/P,5\%,50)$ $\$-328,680$

Annual maintenance cost $-3,000$

     $\$-331,680$

Proposal B is selected.

Comment: The A/F factor is used instead of A/P in B because the lining repair cost begins in year 5, not year 0, and continues indefinitely at 5-year intervals.

If the 50-year life of proposal C is considered infinite, CR $= P(i) =$ $\$-300,000$, instead of $\$-328,680$ for $n = 50$. This is a small economic difference. How expected lives of 40 or more years are treated economically is a matter of local practice.

5.4 USING SPREADSHEETS FOR AW ANALYSIS

Annual worth evaluation of equal or unequal life, mutually exclusive alternatives is simplified using the PMT function. The general format for determining an alternative's AW is

$$= \text{PMT}(i\%,n,P,F) - A$$

PMT determines the required capital recovery (CR). Usually F is the estimated salvage value entered as $-S$, and $-A$ is the annual operating cost (AOC). Since AW analysis does not require evaluation over the least common multiple of lives, the n values can be different for each alternative.

As with the PV function, the $+$ and $-$ signs on PMT values must be correctly entered to ensure the appropriate sense on the result. To obtain a negative AW value for a cost alternative, enter $+P$, $-S$, and $-A$. The next two examples illustrate spreadsheet-based AW evaluations for unequal life alternatives, including a long-life investment discussed in Section 5.3.

Herme, the quarry supervisor for Espinosa Stone, wants to select the more economical of two alternatives for quarry dust control. Help him perform a spreadsheet-based AW evaluation for the estimates with MARR = 12% per year.

EXAMPLE 5.7

Equipment	X	Y
First cost, $	−40,000	−75,000
AOC, $ per year	−25,000	−15,000
Life, years	4	6
Salvage value, $	10,000	7,000

Solution

Because these are the same estimates as Example 5.4*a*, you can see what is required to hand-calculate AW values by referring back. Spreadsheet evaluation is much faster and easier; it is possible to enter the two single-cell PMT functions for X and Y using the general format presented earlier. Figure 5.5 includes details. The decision is to select Y with the lower-cost AW value.

	A	B	C
1		**Equipment X**	**Equipment Y**
2			
3	AW value	-$36,077	-$32,379
4			
5	Function	= - PMT(12%,4,-40000,10000)-25000	= - PMT(12%,6,-75000,7000)-15000
6			

FIGURE 5.5 Evaluation by AW method using PMT functions, Example 5.7.

Perform a spreadsheet-based AW evaluation of the three proposals in Example 5.6. Note that the lives vary from 10 years for A, to 50 for C, to "infinite" for B.

EXAMPLE 5.8

Solution

Figure 5.6 summarizes the estimates on the left. Obtaining AW values is easy using PMT and a few computations shown on the right. The PMT function determines capital recovery of first cost at 5% per year for A and C. For B, capital recovery is simply P times i, according to Equation [5.5], because proposal B (line walls with cement) is considered permanent. (Remember to enter $-4,000,000$ for P to ensure a minus sign on the CR amount.)

All annual costs are uniform series except proposal B's periodic lining repair cost of $30,000 every 5 years, which is annualized (in cell G7) using a PMT function. Adding all costs (row 8) indicates that proposal B has the lowest AW of costs.

	A	B	C	D	E	F	G	H	I	J	K
1	**Recap of Estimates**				**Calculation of AW values**						
2	Proposal	A	B	C	Proposal	A	B	C	$= -PMT(5\%,50,-6000000)$		
3	First cost, $	-650,000	-4,000,000	-6,000,000	Capital recovery, CR	-82,826	-200,000	-328,660			
4	Life, years	10	Infinite	50					$= -4000000*(0.05)$		
5	Annual costs, $	-170,000	-5000	-3000							
6	Periodic cost, $	None	-30,000 every 5 years	None	Annual costs, A	-170,000	-5,000	-3,000	$= -PMT(5\%,10,-650000,17000)$		
7	Salvage, $	17,000	0	0				-5,429	$= -PMT(5\%,5,,-30000)$		
8					AW value, CR + A	-252,826	-210,429	-331,660			
9											

FIGURE 5.6   Use of PMT functions to perform an AW evaluation of three proposals with different lives, Example 5.8.

When estimated annual costs are not a uniform series A, the cash flows must be entered individually (using the negative sign for cash outflows) and the use of single-cell functions is somewhat limited. In these cases, first use the NPV function to find the PW value over one life cycle, followed by the PMT function to determine the AW value. Alternatively, though it is more involved, embed the NPV function directly into PMT using the general format

$$= PMT(i\%,n,P+NPV(i\%,year\_1\_cell:year\_n\_cell))$$

SUMMARY

The annual worth method of comparing alternatives is preferred to the present worth method, because the AW comparison is performed for only one life cycle. This is a distinct advantage when comparing different-life alternatives. When a study period is specified, the AW calculation is determined for that time period only, and the estimated value remaining in the alternative at the end of the study period becomes the salvage value.

For infinite-life alternatives, the initial cost is annualized by multiplying P by i. For finite-life alternatives, the AW through one life cycle is equal to the perpetual equivalent annual worth.

PROBLEMS

Capital Recovery and AW

5.1 Heyden Motion Solutions ordered $7 million worth of seamless tubes for manufacturing their high performance and precision linear motion products. If their annual operating costs are $860,000 per year, how much annual revenue is required over a 3-year planning period to recover the initial investment and operating costs at the company's MARR of 15% per year?

5.2 NRG Energy plans to construct a giant solar plant in Santa Teresa, NM to supply electricity to 30,000 southern NM and western TX homes. The plant will have 390,000 heliostats to concentrate sunlight onto 32 water towers to generate steam. NRG will spend $560 million in constructing the plant and $430,000 per year in operating it. If a salvage value of 20% of the initial cost is assumed, how much will the company have to make each year for 15 years in order to recover its investment at a MARR of 18% per year?

5.3 Environmental recovery company RexChem Partners plans to finance a site reclamation project that will require a 4-year cleanup period. The company will borrow $3.8 million now to finance the project. How much will the company have to receive in annual payments for 4 years, provided it will also receive a final lump sum payment after 4 years in the amount of $500,000? The MARR is 20% per year on this investment.

5.4 U.S. Steel is planning a plant expansion to produce austenitic, precipitation-hardened, duplex and martensitic stainless steel round bar (as well as various nickels) that is expected to cost $13 million now and another $10 million 1 year from now. If total operating costs will be $1.2 million per year starting 1 year from now, how much must the company realize in annual revenue in years 1 through 10 to recover its investment plus 15% per year?

5.5 A small metal plating company wants to become involved in electronic commerce. A modest e-commerce package is available for $20,000. Semiannual updates and site maintenance will cost $300. The salvage value of the package is estimated to be $1500 after 3 years. If the company wants to recover its total cost in 3 years, what is the equivalent semiannual amount of new income

that must be realized at an interest rate of 5% per 6-month period?

5.6 Toro Company is expanding its U.S.-based plastic molding plant as it continues to transfer work from Juarez, Mexico contractors. The plant bought a $1.1 million precision injection molding machine to make plastic parts for Toro lawn mowers, trimmers, and snow blowers. The plant also spent $275,000 for three smaller plastic injection molding machines to make plastic parts for a new line of sprinkler systems. The plant expects to hire 13 people, including some engineers for the expansion. If the average loaded cost (i.e., including benefits) of each employee is $100,000 per year, determine the annual worth of the new systems over a five-year planning period at an interest rate of 10% per year. Assume a 25% salvage value for the new equipment.

5.7 In an effort to retain troops who are proficient with weapons and who can speak the languages of Middle Eastern countries, the Pentagon offered bonuses of $150,000 to specialized personnel who were near or already eligible for retirement. If 400 enlisted personnel accepted the bonus in year one, 300 in year two, and 600 in year three, what was the equivalent annual cost of the program over the 3-year period at an interest rate of 6% per year?

5.8 A company that manufactures magnetic flow meters expects to undertake a project that will have the cash flows below. At an interest rate of 10% per year, what is the equivalent annual cost of the project? Find the AW value using (*a*) tabulated factors, (*b*) calculator functions, and (*c*) a spreadsheet. Which method did you find the easiest to use?

First cost, $	−800,000
Equipment replacement cost in year 2, $	−300,000
Annual operating cost, $/year	−950,000
Salvage value, $	250,000
Life, years	4

5.9 A small commercial building contractor purchased a used crane 2 years ago for $60,000. Its operating cost was $2500 in month one, $2550 in month two, and amounts increasing by $50 per month through the end of year two (now). If the crane was sold for $48,000 now, what was its equivalent monthly cost at an interest rate of 1% per month?

5.10 A 600-ton press used to produce composite-material fuel cell components for automobiles using proton exchange membrane (PEM) technology can reduce the weight of enclosure parts up to 75%. At MARR = 12% per year, calculate (*a*) capital recovery and (*b*) annual revenue required. (*c*) Solve using a spreadsheet.

Installed cost = $−3.8 million $n = 12$ years

Salvage value = $250,000

Annual operating costs = $−350,000 in year 1, increasing by $25,000 per year

Evaluating Alternatives Using AW

5.11 Two machines with the following cost estimates are under consideration for a dishwasher assembly process. Using an interest rate of 10% per year, determine which alternative should be selected on the basis of an annual worth analysis.

	Machine X	Machine Y
First cost, $	−300,000	−430,000
Annual operating cost, $/year	−60,000	−40,000
Salvage value, $	70,000	95,000
Life, years	4	6

5.12 A public water utility is replacing its service trucks with more fuel-efficient vehicles. Two types of trucks are under consideration: Type 1 uses "start-stop" technology that turns the engine off when the vehicle comes to a halt in traffic or at a stop light. These trucks will get 22 miles per gallon of gasoline and will cost $27,000 to buy. At the end of the truck's 5-year service life, it is expected to sell for $10,000. Type 2 trucks use a system called Variable Cylinder Management wherein the engine operates on fewer cylinders when the vehicle is cruising under light load conditions. These trucks will cost $29,500 and will get 25 miles per gallon. Their salvage value is expected to be $12,000 after 5 years. The trucks are driven an average of 21,000 miles per year and the wholesale

price of gasoline is $4.00 per gallon. Which type of truck should the utility buy on the basis of an annual worth comparison at an interest rate of 6% per year?

5.13 An engineer is considering two different liners for an evaporation pond that will receive salty concentrate from a brackish water desalting plant. A plastic liner will cost $0.90 per square foot and will have to be replaced in 20 years when precipitated solids have to be removed from the pond using heavy equipment. A rubberized elastomeric liner is tougher and, therefore, is expected to last 30 years, but it will cost $2.20 per square foot. The pond covers 110 acres (1 acre = 43,560 square feet). Which liner is more cost effective on the basis of an annual worth analysis at an interest rate of 8% per year? Solve using (*a*) tabulated factors, and (*b*) a calculator.

5.14 One of two methods will produce solar panels for electric power generation. Method 1 will have an initial cost of $550,000, an annual operating cost of $160,000 per year, and a $125,000 salvage value after its three-year life. Method 2 will cost $830,000 with an annual operating cost of $120,000, and a $240,000 salvage value after its five-year life. The company has asked you to determine which method is economically better, but it wants the analysis done over a three-year planning period. The salvage value of Method 2 will be 35% higher after 3 years than it is after 5 years. If the company's MARR is 10% per year, which method should the company select?

5.15 An environmental engineer is considering three methods for disposing of a non-hazardous chemical sludge: land application, fluidized-bed incineration, and private disposal contract. The estimates for each method are below. (*a*) Determine which has the least cost on the basis of an annual worth comparison at 10% per year. (*b*) Determine the equivalent present worth value of each alternative using its AW value.

	Land Application	Incineration	Contract
First cost, $	−150,000	−900,000	0
Annual cost, $/year	−95,000	−60,000	−170,000
Salvage value, $	25,000	300,000	0
Life, years	4	6	2

5.16 BP Oil is in the process of replacing sections of its Prudhoe Bay, Alaska oil transit pipeline. This will reduce corrosion problems, while allowing higher line pressures and flow rates to downstream processing facilities. The installed cost is expected to be about $170 million. Alaska imposes a 22.5% tax on annual profits (net revenue over costs), which are estimated to average $85 million per year for a 20-year period. Use tabulated factors and a spreadsheet to answer the following: (*a*) At a corporate MARR of 10% per year, does the project AW indicate it will make at least the MARR? (*b*) Recalculate the AW at MARR values increasing by 10% per year, that is, 20%, 30%, etc. At what required return does the project become financially unacceptable?

5.17 Equipment needed at a Valero Corporation refinery for the conversion of corn stock to ethanol, a cleaner burning gasoline additive, will cost $175,000 and have net cash flows of $35,000 the first year, increasing by $10,000 per year over the life of 5 years. (*a*) Use a spreadsheet (and tabulated factors, if instructed to do so) to calculate the AW amounts at different MARR values to determine when the project switches from financially justified to unjustified. (*b*) Develop a spreadsheet chart that plots AW versus interest rate.

5.18 The TT Racing and Performance Motor Corporation wishes to evaluate two alternative machines for NASCAR motor tune-ups. (*a*) Use the AW method at 9% per year to select the better alternative. (*b*) Use spreadsheet single-cell functions to find the better alternative.

	Machine R	Machine S
First cost, $	−250,000	−370,500
Annual operating cost, $ per year	−40,000	−50,000
Life, years	3	5
Salvage value, $	20,000	20,000

5.19 Estimates have been presented to Holly Farms, which is considering two environmental chambers for a project that will detail laboratory confirmations of on-line bacteria tests in chicken meat for the presence of *E. coli* 0157:H7 and *Listeriamonocytogenes*. (*a*) If the project will last for 6 years and *i* = 10% per year, perform an AW evaluation to determine which chamber is more economical. (*b*) Chamber D103 can be purchased with different options and, therefore, at different installed costs. They range from $300,000 to $500,000. Will the selection change if one of these other models is installed? (*c*) Use single-cell spreadsheet functions to solve part (*b*).

	Chamber D103	Chamber 490G
Installed cost, $	−400,000	−250,000
Annual operating cost, $ per year	−4000	−3000
Salvage value at 10% of *P*, $	40,000	25,000
Life, years	3	2

5.20 Blue Whale Moving and Storage recently purchased a warehouse building in Santiago. The manager has two good options for moving pallets of stored goods in and around the facility. Alternative 1 includes a 4000-pound capacity, electric forklift (*P* = $−30,000; *n* = 12 years; AOC = $−1000 per year; *S* = $8000) and 500 new pallets at $10 each. The forklift operator's annual salary and indirect benefits are estimated at $32,000.

Alternative 2 uses two electric pallet movers ("walkies") each with a 3000-pound capacity (for each mover, *P* = $−2000; *n* = 4 years; AOC = $−150 per year; no salvage) and 800 pallets at $10 each. The two operators' salaries and benefits will total $55,000 per year. For both options, new pallets are purchased now and every two years that the equipment is in use. (*a*) If the MARR is 8% per year, use tabulated factors to determine which alternative is better. (*b*) Rework using a spreadsheet solution.

Evaluating Long-Life Alternatives

5.21 Calculate the equivalent annual cost for years 1 through infinity of $1,000,000 now and $1,000,000 three years from now at an interest rate of 10% per year.

5.22 Calculate the infinite-life equivalent annual cost of $5,000,000 in year 0, $2,000,000 in year 10, and $100,000 in years 11 through infinity. The interest rate is 10% per year.

5.23 Compare the alternatives below using the annual worth method at an interest rate of 10% per year. Use (a) tabulated factors, and (b) calculator functions.

	A	B
First cost, $	−60,000	−380,000
Annual cost, $/year	−30,000	−5000
Salvage value, $	10,000	25,000
Life, years	3	∞

5.24 For the cash flows below, use an annual worth comparison to determine which alternative is best at an interest rate of 1% per month.

AW can compare	X	Y	Z
projects at offset lift cycle			
First cost, $	−90,000	−400,000	−900,000
M&O costs, $/month	−30,000	−20,000	−13,000
Overhaul every 10 years, $	—	—	−80,000
Salvage value, $	7000	25,000	200,000
Life, years	3	10	∞

5.25 Cheryl and Gunther wish to place into a retirement fund an equal amount each year for 20 consecutive years to accumulate just enough to withdraw $24,000 per year starting exactly one year after the last deposit is made. The fund has a reliable return of 8% per year. Determine the annual deposit for two withdrawal plans: (a) forever (years 21 to infinity); (b) 30 years (years 21 through 50). (c) How much less per year is needed when the withdrawal horizon decreases from infinity to 30 years?

5.26 Baker|Trimline owned a specialized tools company for a total of 12 years when it was sold for $38 million cash. During the ownership, annual net cash flow varied significantly as follows:

Year	1 2 3 4 5 6 7 8 9 10 11 12
Net Cash Flow, $ million per year	4 0 −1 −3 −3 1 4 6 8 10 12 12

The company made 12% per year on its positive cash flows and paid 10% per year on short-term loans to cover the lean years. The president wants to use the cash accumulated after 12 years to improve capital investments starting in year 13 and forward. If an 8% per year return is expected after

the sale, what annual amount can Baker|Trimline invest forever?

5.27 A major repair on the suspension system of Jane's 3-year old car cost her $2000 because the warranty expired after 2 years of ownership. Based on this experience, she will plan on additional $2000 expenses every 3 years henceforth. Also, she spends $800 every 2 years for maintenance now that the warranty is over. This is for years 2, 4, 6, 8, and 10, when she plans to donate the car to charity. Use these costs to determine Jane's equivalent annual cost for years 1 through infinity at $i = 5\%$ per year, if cars she owns in the future have the same cost pattern. Solve using tabulated factors and a spreadsheet, as requested by your instructor.

5.28 A West Virginia coal mining operation has installed an in-shaft monitoring system for oxygen tank and gear readiness for emergencies. Based on maintenance patterns for previous systems, costs are minimal for the first few years, increase for a time period, and then level off. Maintenance costs are expected to be $150,000 in year 3, $175,000 in year 4, and amounts increasing by $25,000 per year through year 6 and remain constant thereafter for the expected 10-year life of this system. If similar systems will replace the current one, determine the perpetual equivalent annual maintenance cost at $i = 10\%$ per year. Solve using tabulated factors and a spreadsheet, as requested by your instructor.

5.29 Harmony Auto Group sells and services imported and domestic cars. The owner is considering the outsourcing of all its new car warranty service work to Winslow, Inc., a private repair service that works on any make and year car. Both a 5-year contract basis or 10-year license agreement are available from Winslow. Revenue from the manufacturer will be shared with no added cost incurred by the car/warranty owner. Alternatively, Harmony can continue to do warranty work in-house. Use the estimates made by the Harmony owner to perform an annual worth evaluation at 10% per year to select the best option. All dollar values are in millions.

	Contract	License	In-house
First cost, $	0	−2	−20
Annual cost, $ per year	−1	−0.2	−4
Annual income, $ per year	2.5	1.3	8
Life, years	5	10	∞

5.30 ABC Drinks purchases its 355 ml cans in large bulk from Wald-China Can Corporation. The finish on the anodized aluminum surface is produced by mechanical finishing technology called brushing or bead blasting. Engineers at Wald are switching to more efficient, faster, and cheaper machines to supply ABC. Use the estimates and MARR = 8% per year to select between two alternatives.

Brush alternative: $P = \$-400,000$; $n = 10$ years; $S = \$50,000$; nonlabor AOC = $\$-60,000$ in year 1, decreasing by \$5000 annually starting in year 2.

Bead blasting alternative: $P = \$-400,000$; n is large, assume permanent; no salvage; nonlabor AOC = $\$-70,000$ per year.

5.31 You are an engineer with Yorkshire Shipping in Singapore. Your boss, Zul, asks you to recommend one of two methods to reduce or eliminate rodent damage to silo-stored grain as it awaits shipment. Perform an AW analysis at 10% per year compounded quarterly. Dollar values are in millions.

	Alternative A Major Reduction	Alternative B Almost Eliminate
First cost, $	−10	−35
Annual operating cost, $ per year	−1.8	−0.6
Salvage value, $	0.7	0.2
Life, years	5	Almost permanent

ADDITIONAL PROBLEMS AND FE EXAM REVIEW QUESTIONS

5.32 In comparing alternatives that have different lives by the annual worth method,
 a. the annual worth value of both alternatives must be calculated over a time period equal to the life of the shorter-lived one.
 b. the annual worth value of both alternatives must be calculated over a time period equal to the life of the longer-lived asset.
 c. the annual worth values must be calculated over a time period equal to the least common multiple of the lives.
 d. the annual worth values can be compared over one life cycle of each alternative.

5.33 If you have the present worth of an alternative with a 5-year life, you can obtain its annual worth by:
 a. multiplying the PW by i.
 b. multiplying the PW by $(A/F,i,5)$.
 c. multiplying the PW by $(P/A,i,5)$.
 d. multiplying the PW by $(A/P,i,5)$.

5.34 An automation asset with a high first cost of \$10 million has a capital recovery (CR) of \$1,985,000 per year. The correct interpretation of this CR value is that:
 a. the owner must pay an additional \$1,985,000 each year to retain the asset.
 b. each year of its expected life, a net revenue of \$1,985,000 must be realized to recover the

\$10 million first cost and the required rate of return on this investment.
 c. each year of its expected life, a net revenue of \$1,985,000 must be realized to recover the \$10 million first cost.
 d. the services provided by the asset will stop if less than \$1,985,000 in net revenue is reported in any year.

5.35 The AWs of three cost alternatives are $\$-23,000$ for Alternative A, $\$-21,600$ for B, and $\$-27,300$ for C. On the basis of AW values, the best economic choice is:
 a. select alternative A.
 b. select alternative B.
 c. select alternative C.
 d. select the do nothing alternative.

5.36 The initial cost of a packed-bed degassing reactor for removing trihalomethanes from potable water is \$84,000. The annual operating cost for power, site maintenance, etc. is \$13,000. If the salvage value of the pumps, blowers, and control systems is expected to be \$9000 at the end of 10 years, the AW of the packed-bed reactor at an interest rate of 8% per year is closest to:
 a. $\$-26,140$
 b. $\$-25,520$
 c. $\$-24,900$
 d. $\$-13,140$

5.37 The AW values of three revenue alternatives are $-23,000 for A, $-21,600 for B, and $-27,300 for C. On the basis of these AW values, the correct decision is to:
 a. select alternative A.
 b. select alternative B.
 c. select alternative C.
 d. select the do nothing alternative.

Problems 5.38 through 5.40 are based on the following estimates.

Use an interest rate of 10% per year.

Alternative	A	B
First cost, $	−50,000	−80,000
Annual cost, $/year	−20,000	−10,000
Salvage value, $	10,000	25,000
Life, years	3	6

5.38 The equivalent annual worth of alternative A is closest to:
 a. $-25,130
 b. $-37,100
 c. $-41,500
 d. $-42,900

5.39 The equivalent annual worth of alternative B is closest to:
 a. $-25,130
 b. $-28,190
 c. $-37,080
 d. $-39,100

5.40 The equivalent annual worth of alternative A over an infinite time period is closest to:
 a. $-25,000
 b. $-27,200
 c. $-31,600
 d. $-37,100

5.41 If you have the capitalized cost of an alternative that has an infinite life, you can get its annual cost over a very long number of years by:
 a. multiplying the capitalized cost by i.
 b. multiplying the capitalized cost by $(A/F,i,n)$.
 c. dividing the capitalized cost by $(P/A,i,n)$.
 d. dividing the capitalized cost by i.

5.42 If you have the annual worth of an alternative that has a 5-year life, you can obtain its perpetual annual worth by:
 a. doing no calculations, since perpetual annual worth equals the annual worth.
 b. multiplying the annual worth by $(A/P,i,5)$.
 c. dividing the annual worth by i.
 d. multiplying the annual worth by i.

Rate of Return Analysis

Flatliner/Getty Images

Although the most commonly quoted measure of economic worth for a project or alternative is the rate of return (ROR), its meaning is easily misinterpreted, and the methods to determine ROR are often applied incorrectly. This chapter presents the methods by which one, two, or more alternatives are evaluated using the ROR procedure. Though the computations vary slightly, ROR is known by several other names: internal rate of return (IRR), return on investment (ROI), and profitability index (PI), to name three.

In some cases, more than one ROR value may satisfy the rate of return equation. This chapter describes how to recognize this possibility and presents an approach to find the multiple values. Alternatively, one unique ROR value can be obtained by using information determined independently of the project's cash flows.

Purpose: Understand the meaning of rate of return (ROR) and perform ROR calculations when considering one or more alternatives.

LEARNING OUTCOMES

1. State the meaning of rate of return.

   Definition of ROR

2. Calculate the rate of return using a present worth, annual worth, or future worth equation.

   ROR using PW, AW, or FW

3. Understand the difficulties of using the ROR method, relative to PW, AW, and FW methods.

   Cautions about ROR

4. Tabulate incremental cash flows and interpret ROR on the incremental investment.

   Incremental analysis

5. Select the best of multiple alternatives using an incremental ROR analysis.

   Alternative selection

6. Determine the maximum number of possible ROR values and their values for a cash flow series.

   Multiple RORs

7. Calculate the external rate of return using reinvestment and/or borrowing rates.

   External ROR

8. Use a spreadsheet to perform ROR analysis of one or more alternatives.

   Spreadsheets

6.1 INTERPRETATION OF ROR VALUES

As stated in Chapter 1, interest rate and rate of return refer to the same thing. We commonly use the term *interest rate* when discussing borrowed money and *rate of return* when dealing with investments.

From the perspective of someone who has *borrowed money,* the interest rate is applied to the *unpaid balance* so that the total loan amount and interest due are paid in full exactly with the last loan payment. From the perspective of the *investor* (or lender) there is an *unrecovered balance* at each time period. The interest rate is the return on this unrecovered balance so that the total amount lent and interest earned are recovered exactly with the last receipt. Calculation of the *rate of return* describes both of these perspectives.

> **Rate of return (ROR) is the rate paid on the unpaid balance of borrowed money, or the rate earned on the unrecovered balance of an investment, so that the final payment or receipt brings the balance to exactly zero with interest considered.**

The rate of return is expressed as a percent per period, for example, $i = 10\%$ per year. It is stated as a positive percentage; the fact that interest paid on a loan is actually a negative rate of return from the borrower's perspective is not considered. The numerical value of i can range from -100% to infinity, that is, $-100\% < i < \infty$. In terms of an investment, a return of $i = -100\%$ means the entire amount is lost.

The definition above does *not* state that the rate of return is on the initial amount of the investment; rather it is on the *unrecovered balance,* which changes each time period. Example 6.1 illustrates this difference.

EXAMPLE 6.1

Jefferson Bank lent a newly graduated engineer $1000 at $i = 10\%$ per year for 4 years to buy home office equipment. From the bank's perspective (the lender), the investment in this young engineer (the borrower) is expected to produce an equivalent net cash flow of $315.47 for each of 4 years.

$$A = \$1000(A/P,10\%,4) = \$315.47$$

This represents a 10% per year rate of return on the bank's unrecovered balance. Compute the amount of the unrecovered investment for each of the 4 years using (*a*) the rate of return on the unrecovered balance (the correct basis) and (*b*) the return on the initial $1000 investment (the incorrect basis).

Solution

a. Table 6.1 shows the unrecovered balance at the end of each year in column 6 using the 10% rate on the *unrecovered balance at the beginning of the year.* After 4 years, the total $1000 is recovered, and the balance in column 6 is exactly zero.

TABLE 6.1 Unrecovered Balances Using a Rate of Return of 10% on the Unrecovered Balance

(1) Year	(2) Beginning Unrecovered Balance	(3) = 0.10 × (2) Interest on Unrecovered Balance	(4) Cash Flow	(5) = (4) − (3) Recovered Amount	(6) = (2) + (5) Ending Unrecovered Balance
0	—	—	$−1,000.00	—	—
1	$−1,000.00	$100.00	+315.47	$215.47	$−784.53
2	−784.53	78.45	+315.47	237.02	−547.51
3	−547.51	54.75	+315.47	260.72	−286.79
4	−286.79	28.68	+315.47	286.79	0
		$261.88		$1,000.00	

TABLE 6.2 Unrecovered Balances Using a 10% Return on the Initial Amount

(1) Year	(2) Beginning Unrecovered Balance	(3) = 0.10 × (2) Interest on Initial Amount	(4) Cash Flow	(5) = (4) − (3) Recovered Amount	(6) = (2) + (5) Ending Unrecovered Balance
0	—	—	$−1,000.00	—	—
1	$−1,000.00	$100	+315.47	$215.47	$−784.53
2	−784.53	100	+315.47	215.47	−569.06
3	−569.06	100	+315.47	215.47	−353.59
4	−353.59	100	+315.47	215.47	−138.12
		$400		$861.88	

b. Table 6.2 shows the unrecovered balance if the 10% return is always figured on the *initial $1000*. Column 6 in year 4 shows a remaining unre-covered amount of $138.12, because only $861.88 is recovered in the 4 years (column 5).

Because rate of return is the interest rate on the unrecovered balance, the com-putations in *Table 6.1 present a correct interpretation of a 10% rate of return.* An interest rate applied to the original principal represents a higher rate than stated. From the standpoint of the borrower, it is better that interest is charged on the unpaid balance than on the initial amount borrowed.

6.2 ROR CALCULATION

The basis for calculating an unknown rate of return is an *equivalence* relation in PW, AW, or FW terms. The objective is to *find the interest rate,* represented as i^*, at which the cash flows are equivalent. The calculations are the reverse of those made in previous chapters, where the interest rate was known. For example, if you invest $1000 now and are promised payments of $500 three years from now and $1500 five years from now, the rate of return relation using PW factors is

$$1000 = 500(P/F,i^*,3) + 1500(P/F,i^*,5) \qquad [6.1]$$

The value of i^* is sought (see Figure 6.1). Move the $1000 to the right side in Equation [6.1].

$$0 = -1000 + 500(P/F,i^*,3) + 1500(P/F,i^*,5)$$

The equation is solved to obtain $i^* = 16.9\%$ per year. The rate of return will always be greater than zero if the total amount of receipts is greater than the total amount of disbursements.

It should be evident that rate of return relations are merely a rearrangement of a present worth equation. That is, if the above interest rate is known to be 16.9%, and it is used to find the present worth of $500 three years from now and $1500 five years from now, the PW relation is

$$PW = 500(P/F,16.9\%,3) + 1500(P/F,16.9\%,5) = \$1000$$

This illustrates that rate of return and present worth equations are set up in exactly the same fashion. The only differences are what is given and what is sought. The PW-based ROR equation can be generalized as

$$0 = -PW_D + PW_R \qquad [6.2]$$

where PW_D = present worth of disbursements or cash outflows
PW_R = present worth of receipts or cash inflows

Annual worth or future worth values can also be used in Equation [6.2].

There are different ways to determine i^* once the PW relation is established: solution via trial and error with tabulated factors and solution by calculator or spreadsheet function. The first helps in understanding how ROR computations work; the other two are faster.

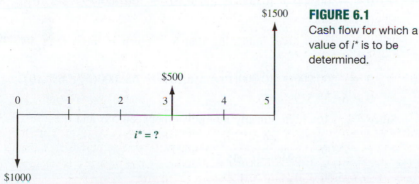

FIGURE 6.1
Cash flow for which a value of i^* is to be determined.

i* Using Tabulated Factors

The general procedure for using a PW-based equation is

1. Draw a cash flow diagram.
2. Set up the rate of return equation in the form of Equation [6.2].
3. Select values of i by trial and error until the equation is balanced.

The next two examples illustrate PW and AW equivalence relations to find i^*.

i* by Calculator or Spreadsheet

The fastest way to determine an i^* value, when there is a series of equal cash flows (A series), is to apply the i function on a calculator or the RATE function on a spreadsheet. These are powerful functions, where it is acceptable to have a separate P value in year 0 and an F value in year n that is separate from the series A amount. The spreadsheet format is $= \text{RATE}(n, A, P, F)$ and the calculator format is $i(n, A, P, F)$. If cash flows vary over the years, the IRR function on a spreadsheet is used to determine i^*; there is no similar calculator function. These functions are illustrated in the last section of this chapter.

EXAMPLE 6.2

The HVAC engineer for a company that constructed one of the world's tallest buildings (Burj Khalifa in the United Arab Emirates) requested that $500,000 be spent on software and hardware to improve the efficiency of the environmental control systems. This is expected to save $10,000 per year for 10 years in energy costs and $700,000 at the end of 10 years in equipment refurbishment costs. Find the rate of return.

Solution

For trial-and-error use the procedure based on a PW equation.

1. Figure 6.2 shows the cash flow diagram.
2. Use Equation [6.2] format for the ROR equation.

$$0 = -500,000 + 10,000(P/A,i^*,10) + 700,000(P/F,i^*,10)$$

3. Try $i = 5\%$.

$$0 = -500,000 + 10,000(P/A,5\%,10) + 700,000(P/F,5\%,10)$$
$$0 < \$6946$$

The result is positive, indicating that the return is more than 5%. Try $i = 6\%$.

$$0 = -500,000 + 10,000(P/A,6\%,10) + 700,000(P/F,6\%,10)$$
$$0 > \$-35,519$$

Since 6% is too high, linearly interpolate between 5% and 6%.

$$i^* = 5.00 + \frac{6946 - 0}{6946 - (-35,519)}(1.0)$$
$$= 5.00 + 0.16 = 5.16\%$$

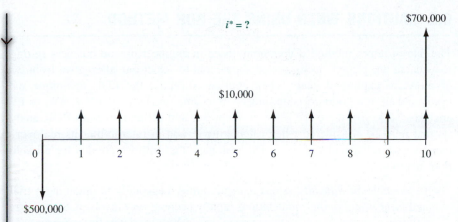

FIGURE 6.2   Cash flow diagram, Example 6.2.

Comment: For a spreadsheet solution, a single-cell entry of the function = RATE(n,A,P,F), which is = RATE(10,10000, −500000,700000) in this case, displays $i* = 5.16\%$. The i function on a calculator has the same contents.

EXAMPLE 6.3

Allied Materials needs $8 million in new capital for expanded composites manufacturing. It is offering small-denomination corporate bonds at a deep discount price of $800 for a 4% $1000 face value bond that matures in 20 years and pays the dividend semiannually. Find the nominal and effective annual rates, compounded semiannually, that Allied is paying an investor.

Solution

By Equation [4.1], the semiannual income from the bond dividend is $I = 1000(0.04)/2 = \$20$. This will be received by the investor for a total of 40 6-month periods. The AW-based relation to calculate the effective semiannual rate is

$$0 = -800(A/P,i*,40) + 20 + 1000(A/F,i*,40)$$

By trial-and-error and linear interpolation, $i^* = 2.87\%$ semiannually. The nominal annual rate is $i*$ times 2.

Nominal $i = 2.87(2) = 5.74\%$ per year compounded semiannually

Using Equation [3.2], the effective annual rate is

Effective $i = (1.0287)^2 - 1 = 0.0582$ (5.82% per year)

Comment: The calculator function $i(40,20,-800,1000)$ results in $i* = 2.84\%$ semiannually. This can be compared with the trial-and-error result of 2.87%.

6.3 CAUTIONS WHEN USING THE ROR METHOD

The rate of return method is commonly used in engineering and business settings to evaluate one project, as discussed above, and to select one alternative from two or more, as explained later. When applied correctly, the ROR technique will always result in a good decision, indeed, the same one as with a PW, AW, or FW analysis. However, there are some assumptions and difficulties with ROR analysis that must be considered when calculating i^* and in interpreting its real-world meaning for a particular project. The summary that follows applies for all solution methods.

- *Computational difficulty versus understanding.* Especially in obtaining a trial-and-error solution, the computations rapidly become very involved. Spreadsheet solution is easier; however, there are no spreadsheet functions that offer the same level of understanding to the learner as that provided by hand (or calculator) solution of PW, AW, and FW relations.
- *Special procedure for multiple alternatives.* To correctly use the ROR method to choose from two or more mutually exclusive alternatives requires an analysis procedure significantly different from that used in other methods. Section 6.5 explains this procedure.
- *Multiple i^* values.* Depending upon the sequence of cash flow disbursements and receipts, there may be more than one real-number root to the ROR equation, resulting in more than one i^* value. There are procedures to use the ROR method and obtain one unique i^* value. Sections 6.6 and 6.7 cover these aspects of ROR analysis.
- *Reinvestment at i^*.* The PW, AW, and FW methods assume that any net positive investment (i.e., net positive cash flows once the time value of money is considered) are *reinvested at the MARR*. But the ROR method assumes reinvestment at the i^* *rate*. When i^* is not close to the MARR (e.g., if i^* is substantially larger than MARR), or if multiple i^* values exist, this is an unrealistic assumption. In such cases, the i^* value is not a good basis for decision making.

In general, it is good practice to use the MARR to determine PW, AW, or FW. If the ROR value is needed, find i^* while taking these cautions into consideration. As an illustration, if a project is evaluated at MARR = 15% and has PW < 0, there is no need to calculate i^*, because $i^* < 15\%$. However, if PW > 0, then calculate the exact i^* and report it along with the conclusion that the project is financially justified.

6.4 UNDERSTANDING INCREMENTAL ROR ANALYSIS

From previous chapters, we know that the PW (or AW or FW) value calculated at the MARR identifies the one mutually exclusive alternative that is best from the economic viewpoint. The best alternative is simply the one that has the numerically largest PW value. (This represents the equivalently lowest net cost or highest net revenue cash flow.) In this section, we learn that the ROR can also be used to identify the best alternative; however, it is *not* always as simple as selecting the highest rate of return alternative.

Let's assume that a company uses a MARR of 16% per year, that the company has $90,000 available for investment, and that two alternatives (A and B) are being evaluated. Alternative A requires an investment of $50,000 and has an internal rate of return i_A^* of 35% per year. Alternative B requires $85,000 and has an i_B^* of 29% per year. Intuitively we may conclude that the better alternative is the one that has the larger ROR value, A in this case. However, this is not necessarily so. While A has the higher projected ROR, it requires an initial investment that is much less than the total money available ($90,000). What happens to the investment capital that is left over? It is generally assumed that excess funds will be invested at the company's MARR, as we learned earlier. Using this assumption, it is possible to determine the consequences of the alternative investments. If alternative A is selected, $50,000 will return 35% per year. The $40,000 left over will be invested at the MARR of 16% per year. The rate of return on the total capital available, then, will be the weighted average. Thus, if alternative A is selected,

$$\text{Overall ROR}_A = \frac{50,000(0.35) + 40,000(0.16)}{90,000} = 26.6\%$$

If alternative B is selected, $85,000 will be invested at 29% per year, and the remaining $5000 will earn 16% per year. Now the weighted average is

$$\text{Overall ROR}_B = \frac{85,000(0.29) + 5000(0.16)}{90,000} = 28.3\%$$

These calculations show that even though the i^* for alternative A is higher, alternative B presents the better overall ROR for the $90,000. If either a PW, AW, or FW comparison is conducted using the MARR of 16% per year as i, alternative B will be chosen.

This example illustrates a major dilemma of the rate of return method when comparing alternatives: Under some circumstances, alternative ROR (i^*) values do not provide the same ranking of alternatives as do the PW, AW, and FW analyses. To resolve the dilemma, conduct an *incremental analysis* between two alternatives at a time and base the alternative selection on the *ROR of the incremental cash flow series*.

A standardized format (Table 6.3) simplifies the incremental analysis. If the alternatives have *equal lives,* the year column will go from 0 to n. If the alternatives have *unequal lives,* the year column will go from 0 to the LCM (least common multiple) of the two lives. The use of the LCM is necessary because *incremental ROR analysis requires equal-service comparison* between alternatives. Therefore, all the assumptions and requirements developed earlier apply for any incremental ROR evaluation. When the LCM of lives is used, the salvage value and reinvestment in each alternative are shown at their respective times. If a study period is defined, the cash flow tabulation is for the specified period.

For the purpose of simplification, use the convention that between two alternatives, the one with the *larger initial investment* will be regarded as *alternative B.* Then, for each year in Table 6.3,

Incremental cash flow = cash flow$_B$ − cash flow$_A$ [6.3]

TABLE 6.3 Format for Incremental Cash Flow Tabulation

	Cash Flow		Incremental
Year	**Alternative A** **(1)**	**Alternative B** **(2)**	**Cash Flow** **(3) = (2) − (1)**
0			
1			
·			
·			
·			

As discussed in Chapter 4, an alternative's cash flow series is one of two types:

Revenue alternative, where there are both negative and positive cash flows.

Cost alternative, where all cash flow estimates are negative.

In either case, Equation [6.3] is used to determine the incremental cash flow series with the sign of each cash flow carefully determined. The next two examples illustrate incremental cash flow tabulation of cost alternatives of equal and different lives. A later example treats revenue alternatives.

EXAMPLE 6.4

A tool and die company in Sydney is considering the purchase of a drill press with fuzzy-logic software to improve accuracy and reduce tool wear. The company has the opportunity to buy a slightly used machine for $15,000 or a new one for $21,000. Because the new machine is a more sophisticated model, its operating cost is expected to be $7000 per year, while the used machine is expected to require $8200 per year. Each machine is expected to have a 25-year life with a 5% salvage value. Tabulate the incremental cash flow.

Solution

Incremental cash flow is tabulated in Table 6.4 using Equation [6.3]. The subtraction performed is (new − used) since the new machine has a larger initial cost. The salvage values are separated from the year 25 cash flow for clarity.

TABLE 6.4 Cash Flow Tabulation for Example 6.4

	Cash Flow		Incremental Cash Flow
Year	**Used Press**	**New Press**	**(New − Used)**
0	$ −15,000	$ −21,000	$ −6,000
1–25	−8,200	−7,000	+1,200
25	+750	+1,050	+300
Total	$−219,250	$−194,950	$+24,300

Comment: When the cash flow columns are subtracted, the difference between the totals of the two cash flow series should equal the total of the incremental cash flow column. This merely provides a check of the addition and subtraction in preparing the tabulation.

EXAMPLE 6.5

Sandersen Meat Processors has asked its lead process engineer to evaluate two different types of conveyors for the beef cutting line. Type A has an initial cost of $70,000 and a life of 3 years. Type B has an initial cost of $95,000 and a life expectancy of 6 years. The annual operating cost (AOC) for type A is expected to be $9000, while the AOC for type B is expected to be $7000. If the salvage values are $5000 and $10,000 for type A and type B, respectively, tabulate the incremental cash flow using their LCM.

Solution

The LCM of 3 and 6 is 6 years. In the incremental cash flow tabulation for 6 years (Table 6.5), note that the reinvestment and salvage value of A is shown in year 3.

TABLE 6.5 Incremental Cash Flow Tabulation, Example 6.5

Year	Cash Flow Type A	Cash Flow Type B	Incremental Cash Flow (B − A)
0	$ −70,000	$ −95,000	$−25,000
1	−9,000	−7,000	+2,000
2	−9,000	−7,000	+2,000
3	{−70,000 / −9,000 / +5,000	−7,000	+67,000
4	−9,000	−7,000	+2,000
5	−9,000	−7,000	+2,000
6	{−9,000 / +5,000	{−7,000 / +10,000	+7,000
	$−184,000	$−127,000	$+57,000

Once the incremental cash flows are tabulated, determine the incremental rate of return on the extra amount required by the larger investment alternative. This rate, termed Δi^*, represents the return over n years expected on the optional extra investment in year 0. The general selection guideline is to make the extra investment if the incremental rate of return meets or exceeds the MARR. Briefly stated,

If $\Delta i^* \geq$ MARR, select the larger investment alternative (labeled B).

Otherwise, select the lower investment alternative (labeled A).

Use of this guideline is demonstrated in Section 6.5. The best rationale for understanding incremental ROR analysis is to think of only *one alternative* under consideration, that alternative being represented by the incremental cash flow series. Only if the return on the extra investment, which is the Δi^* value, meets or exceeds the MARR is it financially justified, in which case the larger investment alternative should be selected.

As a matter of efficiency, if the analysis is between *multiple revenue alternatives,* an acceptable procedure is to initially determine each alternative's i^* and remove those alternatives with $i^* <$ MARR, since their return is too low. Then complete the incremental analysis for the remaining alternatives. If no alternative i^* meets or exceeds the MARR, the do-nothing alternative is economically the best. This initial "weeding out" can't be done for cost alternatives since they have no positive cash flows.

When *independent projects* are compared, no incremental analysis is necessary. All projects with $i^* \geq$ MARR are acceptable. Limitations on the initial investment amount are considered separately, as discussed in Section 4.5.

6.5 ROR EVALUATION OF TWO OR MORE MUTUALLY EXCLUSIVE ALTERNATIVES

Podcast available

When selecting from two or more mutually exclusive alternatives on the basis of ROR, equal-service comparison is required, and an incremental ROR analysis must be used. The incremental ROR value between two alternatives (B and A) is correctly identified as Δi^*_{B-A}, but it is usually shortened to Δi^*. The selection guideline, as introduced in Section 6.4, is:

Select the alternative that:
  1. **requires the largest initial investment, and**
  2. **has a $\Delta i^* \geq$ MARR, indicating that the extra initial investment is economically justified.**

If the higher initial investment is not justified, it should not be made as the extra funds could be invested elsewhere.

Before conducting the incremental evaluation, classify the alternatives as *cost* or *revenue* alternatives. The incremental comparison will differ slightly for each type.

Cost: Evaluate alternatives only against each other.

Revenue: First evaluate against do-nothing (DN), then against each other.

The following procedure for comparing multiple, mutually exclusive alternatives, using a PW-based equivalence relation, can now be applied.

  1. Order the alternatives by increasing initial investment. *For revenue alternatives* add DN as the first alternative.
  2. Determine the incremental cash flow between the first two ordered alternatives (B − A) over their least common multiple of lives. (For revenue alternatives, the first ordered alternative is DN.)

3. Set up a PW-based relation of this incremental cash flow series and determine Δi^*, the incremental rate of return.
4. If $\Delta i^* \geq$ MARR, eliminate A; B is the survivor. Otherwise, A is the survivor.
5. Compare the survivor to the next alternative. Continue to compare alternatives using steps (2) through (4) until only one alternative remains as the survivor.

The next two examples illustrate this procedure for cost and revenue alternatives, respectively, as well as for equal and different-life alternatives.

For completeness's sake, it is important to understand the procedural difference for comparing *independent projects*. If the projects are independent rather than mutually exclusive, the preceding procedure does not apply. As mentioned in Section 6.4, no incremental evaluation is necessary; all projects with $i^* \geq$ MARR are selected, thus comparing each project's i^* against the MARR, not each other.

As the film of an oil spill from an at-sea tanker moves ashore, great losses occur for aquatic life as well as shoreline feeders and dwellers, such as birds. Environmental engineers and lawyers from several international petroleum corporations and transport companies—Exxon-Mobil, BP, Shell, and some transporters for OPEC producers—have developed a plan to strategically locate throughout the world newly developed equipment that is substantially more effective than manual procedures in cleaning crude oil residue from bird feathers. The Sierra Club, Greenpeace, and other international environmental interest groups are in favor of the initiative. Alternative machines from manufacturers in Asia, America, Europe, and Africa are available with the cost estimates in Table 6.6. Annual cost estimates are expected to be high to ensure readiness at any time. The company representatives have agreed to use the average of the corporate MARR values, which results in MARR = 13.5%. Use incremental ROR analysis to determine which manufacturer offers the best economic choice.

EXAMPLE 6.6

Solution

Follow the procedure for incremental ROR analysis.

1. These are *cost alternatives* and are arranged by increasing first cost.
2. The lives are all the same at $n = 8$ years. The B − A incremental cash flows are indicated in Table 6.7. The estimated salvage values are shown separately in year 8.

TABLE 6.6   Costs for Four Alternative Machines, Example 6.6

	Machine A	Machine B	Machine C	Machine D
First cost, $	−5,000	−6,500	−10,000	−15,000
AOC, $/year	−3,500	−3,200	−3,000	−1,400
Salvage value, $	+500	+900	+700	+1,000
Life, years	8	8	8	8

TABLE 6.7 Incremental Cash Flow for Comparison of Machine B-to-A

Year	Cash Flow Machine A	Cash Flow Machine B	Incremental Cash Flow for (B − A)
0	$−5000	$−6500	$−1500
1–8	−3500	−3200	+300
8	+500	+900	+400

3. The following PW relation for (B − A) results in $\Delta i^* = 14.57\%$.

$$0 = -1500 + 300(P/A,\Delta i^*,8) + 400(P/F,\Delta i^*,8)$$

4. Since this return exceeds MARR = 13.5%, A is eliminated and B is the survivor.

5. The comparison of C-to-B results in the elimination of C based on $\Delta i^* = -18.77\%$ from the incremental relation

$$0 = -3500 + 200(P/A,\Delta i^*,8) - 200(P/F,\Delta i^*,8)$$

The D-to-B incremental cash flow PW relation for the final evaluation is

$$0 = -8500 + 1800(P/A,\Delta i^*,8) + 100(P/F,\Delta i^*,8)$$

With $\Delta i^* = 13.60\%$, machine D is the overall, though marginal, survivor of the evaluation; it should be purchased and located in the event of oil spill accidents.

EXAMPLE 6.7

Harold owns a construction company that subcontracts to international power equipment corporations such as GE, ABB, Siemens, and LG. For the last 4 years he has leased crane and lifting equipment for $32,000 annually. He now wishes to purchase similar equipment. Use an MARR of 12% per year to determine if any of the options detailed in Table 6.8 are financially justified.

Solution

Apply the incremental ROR procedure with MARR = 12% per year.

1. Because these are revenue alternatives, add the do-nothing option as the first alternative and order the remaining ones. The comparison order is DN, 4, 2, 1, 3.

TABLE 6.8 Estimates for Alternative Equipment, Example 6.7

Alternative	1	2	3	4
First cost, $	−80,000	−50,000	−145,000	−20,000
Annual cost, $/year	−28,000	−26,000	−16,000	−21,000
Annual revenue, $/year	61,000	43,000	51,000	29,000
Life, years	4	4	8	4

2. Each annual cash flow for the DN alternative is \$0. Therefore, the incremental cash flows for comparing 4-to-DN are the same as those for alternative 4.
3. The Δi^* for the comparison 4-to-DN is actually the project ROR. Since $n = 4$ years, the PW relation and return are

$$0 = -20,000 + (29,000 - 21,000)(P/A,\Delta i^*,4)$$
$$(P/A,\Delta i^*,4) = 2.5$$
$$\Delta i^* = 21.9\%$$

4. Since $21.9\% > 12\%$, eliminate DN and proceed with the 2-to-4 comparison.
5. Both alternatives 2 and 4 have $n = 4$. The incremental cash flows are $-30,000 in year 0 and $(43,000-26,000) - (29,000-21,000) = \$+9000$ in years 1 to 4. Incremental analysis results in $\Delta i^* = 7.7\%$ from the PW relation

$$0 = -30,000 + 9000(P/A,\Delta i^*,4)$$
$$\Delta i^* = 7.7\%$$

Alternative 4 is, again, the survivor. Continue with the comparison 1-to-4 to obtain

$$0 = -60,000 + 25,000(P/A,\Delta i^*,4)$$
$$\Delta i^* = 24.1\%$$

Now alternative 1 is the survivor. The final comparison of 3 to 1 must be conducted over the LCM of 8 years for equal service. Table 6.9 details the incremental cash flows, including the alternative 1 repurchase in year 4. The PW relation is

$$0 = -65,000 + 2000(P/A,\Delta i^*,8) + 80,000(P/F,\Delta i^*,4)$$
$$\Delta i^* = 10.1\%$$

Since $10.1\% < 12\%$, eliminate 3; declare alternative 1 the survivor and select it as the one that is economically justified.

TABLE 6.9 **Incremental Cash Flows for the Comparison of Alternatives 3-to-1, Example 6.7**

	Alternative Cash Flows		Incremental
Year	Alternative 1	Alternative 3	Cash Flow for (3–1)
0	\$-80,000	\$-145,000	\$-65,000
1	+33,000	+35,000	+2,000
2	+33,000	+35,000	+2,000
3	+33,000	+35,000	+2,000
4	-47,000	+35,000	+82,000
5	+33,000	+35,000	+2,000
6	+33,000	+35,000	+2,000
7	+33,000	+35,000	+2,000
8	+33,000	+35,000	+2,000

The previous incremental analyses were performed using the PW relations. It is equally correct to apply an AW-based or FW-based analysis; however, the LCM of lives must be used since ROR analysis requires an equal-service comparison. Consequently, there is usually no advantage to developing AW relations to find Δi^* for different-life alternatives.

The use of the IRR spreadsheet function can greatly speed up incremental ROR comparison of multiple alternatives, especially for those with unequal lives. This is fully illustrated in the last section of this chapter.

6.6 MULTIPLE ROR VALUES

Podcast available

For some cash flow series (net for one project or incremental for two alternatives) it is possible that more than one unique rate of return i^* exists. This is referred to as *multiple i^* values*. It is difficult to complete the economic evaluation when multiple i^* values are present, since none of the values may be the *correct* rate of return. The discussion that follows explains how to predict the number of i^* values in the range -100% to infinity, how to determine their values, and how to resolve the difficulty of knowing the "true" ROR value (if this is important). If using ROR evaluation is not absolutely necessary, a simple way to avoid this dilemma is to use the PW, AW, or FW evaluation method at the MARR.

In actuality, finding the rate of return is solving for the root(s) of an nth order polynomial. *Conventional* or *simple* cash flows have only one sign change over the entire series, as shown in Table 6.10. Commonly, this is negative in year 0 to positive at some time during the series. There is a unique, real number i^* value for a conventional series. A *nonconventional* series (Table 6.10) has more than one sign change and multiple roots may exist. The *cash flow rule of signs* (based upon Descartes' rule) states:

The maximum number of i^* values is equal to the number of sign changes in the cash flow series.

When applying this rule, zero cash flow values are disregarded.

TABLE 6.10 Examples of Conventional and Nonconventional Net or Incremental Cash Flows for a 6-year Period

Type of Series	Sign on Cash Flow							Number of Sign Changes
	0	1	2	3	4	5	6	
Conventional	−	+	+	+	+	+	+	1
Conventional	−	−	−	+	+	+	+	1
Conventional	+	+	+	+	+	−	−	1
Nonconventional	−	+	+	+	−	−	−	2
Nonconventional	+	+	−	−	−	+	+	2
Nonconventional	−	+	−	−	+	+	+	3

Prior to determining the multiple i^* values, a second rule can be applied to indicate that a unique nonnegative i^* value exists for a nonconventional series. It is the *cumulative cash flow test* (also called Norstom's criterion). It states:

> **There is one real-number, positive i^* value if the cumulative cash flow series $S_0, S_1, \ldots, S_n$ changes sign only once and $S_0 < 0$.**

To perform the test, observe the sign on S_0 and count the number of sign changes in the S_t series, where

$$S_t = \text{cumulative cash flows through period } t$$

More than one sign change provides no information, and the rule of signs is applied to indicate the possible number of i^* values.

Now the unique or multiple i^* value(s) can be determined using trial-and-error with tabulated factors, by graphical interpolation using a PW-based relation, or using the IRR spreadsheet function that incorporates the "guess" option to search for the multiple i^* values. A calculator function is equally useful if the cash flow series is not too complex. The next example illustrates the two rules on sign changes and solution using PW-based relations. The use of spreadsheets is shown in the last section of the chapter.

The engineering design and testing group for Honda Motor Corp. does contract-based work for automobile manufacturers throughout the world. During the last 3 years, the net cash flows for contract payments have varied widely, as shown below, primarily due to a large manufacturer's inability to pay its contract fee.

EXAMPLE 6.8

Year	0	1	2	3
Cash Flow ($1000)	+2000	−500	−8100	+6800

a. Determine the maximum number of i^* values that may satisfy the ROR relation.

b. Write the PW-based ROR relation and approximate the i^* value(s) by plotting PW versus i.

Solution

a. Table 6.11 shows the annual cash flows and cumulative cash flows. Since there are two sign changes in the cash flow sequence, the rule of signs indicates a maximum of two i^* values. The *cumulative* cash flow sequence has two sign changes and $S_0 > 0$, indicating there is not just one nonnegative root. The conclusion is that as many as two i^* values can be found.

b. The PW relation is

$$PW = 2000 - 500(P/F,i,1) - 8100(P/F,i,2) + 6800(P/F,i,3)$$

Select values of i to find the two i^* values, and plot PW versus i. The PW values are shown on the next page and plotted in Figure 6.3 (using a smooth

TABLE 6.11 **Cash Flow and Cumulative Cash Flow Sequences, Example 6.8**

Year	Cash Flow ($1000)	Sequence Number	Cumulative Cash Flow ($1000)
0	+2000	S_0	+2000
1	−500	S_1	+1500
2	−8100	S_2	−6600
3	+6800	S_3	+200

approximation) for i values of 0, 5, 10, 20, 30, 40, and 50%. The characteristic parabolic shape for a second-degree polynomial is obtained, with PW crossing the i axis at approximately $i_1^* = 8$ and $i_2^* = 41\%$.

$i\%$	0	5	10	20	30	40	50
PW ($1000)	+200	+51.44	−39.55	−106.13	−82.01	−11.83	+81.85

FIGURE 6.3 Present worth of cash flows at several interest rates, Example 6.8.

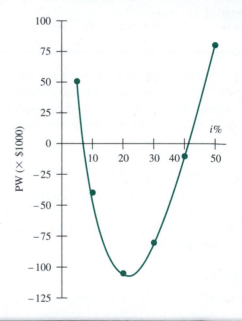

EXAMPLE 6.9   An American-Australian joint venture has been contracted to provide the train cars for a 25-mile subway system using new tunnel-boring and track-design technologies. Austin, Texas was selected as the proof-of-concept site based on its variety in landscape features (hilly terrain, lake and green space areas, and relatively low precipitation) and its public environmental mindedness. Selection of either a Swiss

or Japanese contractor to provide the gears and power components for the electric transfer motor assemblies resulted in two cost alternatives. Table 6.12 gives the incremental cash flow estimates (in $1000) over the expected 10-year life of the motors. Determine the number of i^* values and estimate them graphically.

TABLE 6.12 Incremental and Cumulative Cash Flow Series, Example 6.9

| | Cash Flow, $1000 | | | Cash Flow, $1000 | |
Year	Incremental	Cumulative	Year	Incremental	Cumulative
0	−500	−500	6	+800	+200
1	−2000	−2500	7	+400	+600
2	−2000	−4500	8	+300	+900
3	+2500	−2000	9	+200	+1100
4	+1500	−500	10	+100	+1200
5	−100	−600			

Solution

The incremental cash flows form a nonconventional series with three sign changes in years 3, 5, and 6. The cumulative series starts with $S_0 < 0$ and has one sign change in year 6. This test indicates a single nonnegative root. The incremental ROR is determined from a PW relation (in $1000).

$$0 = -500 - 2000(P/F,\Delta i^*,1) - 2000(P/F,\Delta i^*,2) \cdots + 100(P/F,\Delta i^*,10)$$

Calculation of PW at various i values is plotted (Figure 6.4) to estimate the unique Δi^* of 8% per year.

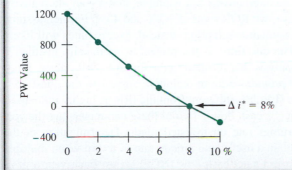

FIGURE 6.4
Graphical estimation of Δi^* using PW values, Example 6.9.

Commonly when multiple i^* or Δi^* values are indicated, there is only one realistic root. Others may be large negative or positive numbers that make no real-world sense and can be neglected. (One clear advantage of using a spreadsheet or calculator for ROR evaluation, as described in Section 6.8, is that realistic i^* values are commonly determined first by the functions.) Here are some useful guidelines for retaining and discarding multiple rates. Assume there are two $i*$ values for a cash flow series.

If $i*$ values are	Do this
Both $i* < 0$	Discard both values
Both $i* > 0$	Discard both values
One $i* > 0$; one $i* < 0$	Use $i* > 0$ as ROR

As mentioned previously, the use of a PW, AW, or FW analysis eliminates the multiple ROR dilemma since the MARR is used in all equivalence relations, and excess funds are assumed to earn the MARR. (See Section 6.4 for a quick review.) It is because of the complexities of the ROR method, namely, incremental analysis, the use of LCM for equal service, reinvestment rate assumption, and possible multiple $i*$ values, that other methods are preferred to ROR. Yet, the ROR result is important in that some people wish to know the estimated return on proposed project(s).

6.7 TECHNIQUES TO REMOVE MULTIPLE ROR VALUES

Podcast available

All ROR values calculated thus far can be termed *internal rates of return (IRR)*. As discussed in Section 6.1, the IRR guarantees that the last receipt or payment brings the balance to exactly zero with interest considered. No excess funds are generated in any year, so all funds are kept internal to the project. However, a project can generate excess funds prior to the end of the project's life when the net cash flow in any year is positive ($NCF_t > 0$). This can result in a nonconventional series, as we learned in Section 6.6. The ROR method assumes these excess funds can earn at any one of the multiple $i*$ values. This generates ambiguity when the ROR method is used to evaluate alternatives. For example, assume an alternative's nonconventional cash flows have two ROR roots at -2% and 40%. Further, assume that neither is a realistic reinvestment assumption. Instead, excess funds will likely earn at the MARR of 15%. The question "Is the project economically justified?" is not answered by the ROR analysis that resulted in the multiple IRR values. Finding the external rate of return provides a more definitive answer to the question.

The *external rate of return* (EROR), different from the IRR, is highly influenced by parameters *outside* the project's cash flows. Two of these parameters are the cost to borrow money and the earnings rate on invested funds. Determination of the EROR is the correct way to obtain a useful and unique rate of return when multiple $i*$ values are indicated by a project's net cash flow (NCF) series. Each year a project will generate excess funds (positive NCF) that can be reinvested, or negative NCF, which indicates that funds must be borrowed from a source external to the project. The value and accuracy of the EROR is a function of: (1) the reinvestment rate earned by excess funds; (2) the interest rate paid on borrowed funds; and (3) the reliability of these estimates. Two different methods that determine an EROR are discussed below. The resulting EROR is different for each method, and the EROR is *not* equal to any of the multiple i^* values determined from the cash flow series, since i^* is an *internal* ROR. The benefit is that the EROR determined by either method can be used to make a sound decision about the economic viability of a project. First, let's define the external rates necessary to apply one or both methods.

> **Reinvestment rate i_r** – The rate at which extra funds are invested in some source external to the project. Also called the *investment rate*, this rate is applied to all positive NCF. It is common that i_r is set equal to the MARR.

> **Borrowing rate i_b** – The interest rate at which funds are borrowed from an external source to provide capital to the project. This applies to all negative annual NCF. The cost of capital (CoC was introduced in Section 1.3) or weighted average cost of capital (WACC is discussed in Chapter 13) can be used for this rate.

Though the two rates can be set equal to each other, it is not a good idea. Setting $i_r = i_b$ implies that the company is willing to borrow funds and reinvest funds at the same rate; this means no profit margin over time. The company can't survive for long. Commonly MARR > CoC, which means that $i_r > i_b$.

6.7.1 MIRR—Modified ROR Method

This is the easier approach to apply and it has a spreadsheet function that can display the EROR value, which is identified by the symbol i'. The reinvestment and borrowing rates must be reliably estimated, since the resulting i' using the MIRR method may be quite sensitive to them. Figure 6.5 is a reference diagram. The net cash flows change sign several times; multiple i^* values are likely. The modified ROR method uses the following procedure to determine the unique external rate of return i'.

1. *For all negative NCF:* Determine the PW in year 0 at the *borrowing rate i_b*. (Gray shaded area and resulting PW_0 value in Figure 6.5.)
2. *For all positive NCF:* Determine the FW value in year n at the *reinvestment rate i_r*. (Green shaded area and resulting FW_n value.)
3. Determine the external rate of return i' at which PW_0 and FW_n are *equivalent* using the relation

$$FW_n = PW_0(F/P,i',n)$$ [6.4]

4. The selection guideline is the same as applied previously:

> **If $i' \geq$ MARR, the project is economically justified**
> **If $i' <$ MARR, the project is not economically justified**

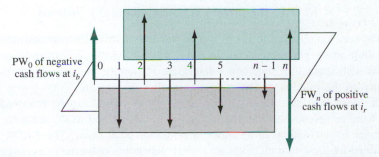

FIGURE 6.5 Sample cash flow series used to determine the external ROR by the modified ROR (MIRR) method.

The MIRR spreadsheet function displays i' directly once the NCF values are entered into adjacent cells. The borrowing rate i_b is called the finance rate in the MIRR function. The format is:

$$= \textbf{MIRR(first\_cell:last\_cell},i_b,i_r)$$

6.7.2   ROIC—Return on Invested Capital Method

From a project perspective, the return on invested capital is a measure of how effectively the project utilizes funds invested in it. From the viewpoint of an entire corporation, ROIC is a measure of how effectively all of its resources (facilities, equipment, people, systems, process, and all other assets) are utilized in the conduct of business.

The symbol for the ROIC rate is i''. The ROIC method utilizes the *reinvestment rate* i_r for excess funds generated by a project in any year. Often, i_r is set equal to the MARR. It is not necessary to estimate a borrowing rate for the ROIC method. The EROR is determined using a technique called the *net-investment procedure,* which involves developing a series of future worth relations (FW values) moving forward 1 year at a time from time $t = 0$ to $t = n$ with time value of money considered. In those years that the net balance of the project cash flows is positive (extra funds generated by the project), the funds are reinvested at the rate i_r. When the net balance is negative, the ROIC rate i'' accounts for the time value of money. The procedure follows.

1. Develop a series of future worth relations for each year t ($t = 1, 2, \ldots, n$).

$$\textbf{FW}_t = \textbf{FW}_{t-1}(1 + k) + \textbf{NCF}_t \qquad [6.5]$$

where FW_t = future worth in year t based on previous year and time value of money

         NCF_t = net cash flow in year t

$$k = \begin{cases} i_r & \text{if } FW_{t-1} > 0 \quad \text{(extra funds available)} \\ i'' & \text{if } FW_{t-1} < 0 \quad \text{(project uses all available funds)} \end{cases}$$

2. Set the future worth relation for the last year n equal to 0, that is, $FW_n = 0$, and solve for i''. The i'' value is the ROIC for the specified reinvestment rate i_r.
3. The selection guideline is the same as above.

If $i'' \geq$ MARR, the project is economically justified
If $i'' <$ MARR, the project is not economically justified

The FW_t series and solution for i'' can be become involved mathematically when hand solution is performed. Fortunately, the GOAL SEEK spreadsheet tool, coupled with logical IF statements determine i'' rapidly, because this is the only unknown in the relation and the target value is $FW_n = 0$.

The next example illustrates how multiple $i*$ values can be removed using either hand or spreadsheet solution utilizing the MIRR and ROIC methods. However, prior to presenting an example, a reminder is in order. The EROR values determined by these two methods are highly dependent upon the reinvestment rate and/or borrowing rate estimates. Additionally, remember that the EROR values determined are not the same as any multiple i^* rate, but each EROR value is unique for the estimated i_r and i_b rates, as required by the method.

Remember: The MIRR method and ROIC method are used when multiple i^* values are indicated. Multiple i^* values are present when a nonconventional cash flow series does not have a single, unique root. Finally, it is important to remember that these procedures are unnecessary if the PW, AW or FW method is used to perform the economic evaluation at the MARR.

EXAMPLE 6.10

Large oil exploration corporations are using better machinery and technology to cap offshore oil spills before they become major disasters. Marine Wells, a company experienced in providing containment response equipment, has estimated the net annual savings shown below (in $ million of cash flow) over the current and next 3 years if their equipment is contracted for by international offshore exploration corporations such as BP, Exxon-Mobil, Chevron, and Total SA. The negative amount in year 1 assumes no oil spill is experienced; the cost is that of the annual contract. Find a unique rate of return using (a) the MIRR method, and (b) the ROIC method, if the following external rates are estimated.

MARR = 12% per year

Borrowing rate for extra funds = 10% per year

Reinvestment rate for excess funds = 15% per year

Year	Cash Flow, $ million
0	50
1	−200
2	50
3	100

Solution

This is a nonconventional cash flow series and does have multiple i^* values as indicated by the cash flow sign test (2 changes) and cumulative cash flow test (inconclusive and $S_0 > 0$). Mathematically, two positive i^* values can be determined: 0% and 256%, neither of which is useful for economic decision making.

a. For the MIRR method, reinvestment is at $i_r = 15\%$ for any excess fund years, and the borrowing rate is $i_b = 10\%$. For a hand solution of the external rate of return i', use the MIRR procedure and utilize Figure 6.5 as a general reference.

1. Negative NCF in year 1 at borrowing rate:

$$PW_0 = -200(P/F,10\%,1) = \$-181.82$$

2. Positive NCF in years 0, 2 and 3 at reinvestment rate:

$$FW_3 = 50[(F/P,15\%,3) + (F/P,15\%,1)] + 100 = \$233.54$$

3. Per Equation [6.4], set $FW_3 = PW_0$ considering the time value of money and solve for i'.

$$233.54 = 181.82(F/P,i',3)$$
$$181.82(1 + i')^3 = 233.54$$
$$(1 + i')^3 = 1.2845$$
$$i' = 0.0870 \qquad (8.70\%)$$

4. The external rate of return of $i' = 8.70\%$ is less than MARR $= 12\%$. The project is not economically viable.

If a spreadsheet has the cash flows entered into cells B2 through B5, the function = MIRR(B2:B5,10%,15%) will display $i' = 8.70\%$.

b. The ROIC method uses the reinvestment rate of $i_r = 15\%$ to determine a significantly lower EROR value of $i'' = 3.13\%$. The hand and spreadsheet procedures are summarized here.

Hand solution: Apply the 3-step procedure to develop the FW series and find i''. Figure 6.6a shows the original cash flows, and the remaining diagrams track the development year-by-year.

1. Develop the FW relations for years 0 through 3 using $i_r = 15\%$ only when $FW_{t-1} > 0$.

Year 0: $FW_0 = \$50$ (Reinvest at 15%)
Year 1: $FW_1 = 50(1.15) - 200$ (Figure 6.6b; use i'' for year 2)
                $= \$-142.50$
Year 2: $FW_2 = -142.50(1 + i'') + 50$ (Figure 6.6c; use i'' for year 3)
Year 3: $FW_3 = [-142.50(1 + i'') + 50]$ (Figure 6.6d)
                $\times (1 + i'') + 100$

2. Set $FW_3 = 0$ and solve for i'' using the quadratic equation.

$$-142.50(1 + i'')^2 + 50(1 + i'') + 100 = 0$$

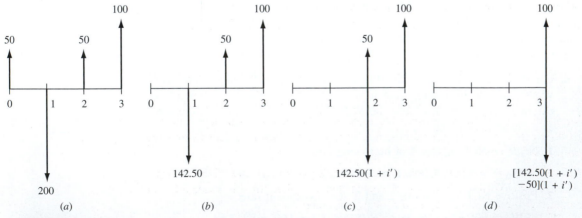

FIGURE 6.6 Cash flow series for which the external rate of return i'' is computed using the ROIC method: (a) original form; equivalent form in (b) year 1, (c) year 2, and (d) year 3.

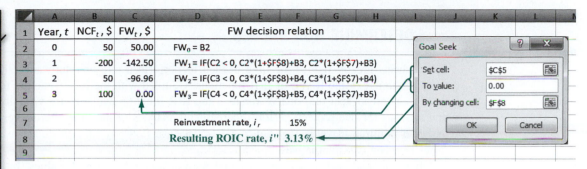

FIGURE 6.7   Application of ROIC method using logical IF statements and the GOAL SEEK tool, Example 6.10.

The two roots for $1 + i''$ are -0.68 and 1.0313. This translates into the rates of -168% and 3.13%. Discard the negative 168%, since it is below the -100% lower limit for a rate of return. We conclude that the EROR is $i'' = 3.13\%$.

3. Since 3.13% is much less than the MARR of 12%, again the project is not economically viable.

Spreadsheet solution: Figure 6.7 details the logical IF statements and resulting FW values (column C) for each year. As shown in the IF statement of column D, when $FW_{t-1} < 0$ the logic statement is "true" and the ROIC rate in cell F8 is applied to the next FW. Alternatively, when $FW_{t-1} > 0$, the reinvestment rate of 15% (cell F7) accounts for the time value of money. The GOAL SEEK tool is used to force FW_3 to equal 0 by changing the trial ROIC value, thus resulting in $i'' = 3.13\%$.

Comment

Note that the two EROR rates—8.70% by the MIRR method and 3.13% by the ROIC method—are different; plus they are both different from the multiple i^* rates (0% and 256%). This illustrates how dependent the different methods are upon the additional information provided by i_b and i_r.

The two methods discussed above remove multiple i^* values. They are very helpful when the multiple i^* values are unrealistic and the reinvestment assumption that excess cash flows are reinvested at these rates makes no sense. Here are some interesting relations between the multiple i^* values for a nonconventional cash flow series, the external reinvestment rate i_r, the borrowing rate i_b, and the resulting EROR rates i' and i''.

MIRR method – When both i_b and i_r are exactly equal to any one of the multiple i^* values, the MIRR-method rate i' equals that i^* value. In this case, all four parameters have the same value; if $i^* = i_b = i_r$, then $i' = i^*$.

ROIC method – Similarly, if i_r equals a multiple i^* value, the ROIC-method rate is $i'' = i^*$.

6.8 USING SPREADSHEETS AND CALCULATORS TO DETERMINE ROR VALUES

Spreadsheets greatly reduce the time needed to perform a rate of return analysis through the use of the RATE or IRR functions. Coupled with the NPV function to develop a spreadsheet plot of PW versus i, the IRR function can perform virtually any analysis for one project, perform incremental analysis of multiple alternatives, and find multiple i^* values for nonconventional cash flows and incremental cash flows between two alternatives.

If the annual cash flows are all equal with separate P and/or F values, find i^* using the spreadsheet function

$$= \text{RATE}(n,A,P,F)$$

For series that are not too complex, the financial calculator function $i(n,A,P,F)$ is a very rapid way to find the i^* value. It is the same as the RATE function on a spreadsheet.

If cash flows vary throughout the n years, a spreadsheet must be used to find one or multiple i^* values using

$$= \text{IRR}(\text{first\_cell:last\_cell, guess})$$

For IRR, each cash flow must be entered in succession by spreadsheet row or column. A "zero" cash flow year must be entered as "0" so the year is accounted for. "Guess," an optional entry that starts the ROR analysis, is used most commonly to find multiple i^* values for nonconventional cash flows, or if the #NUM error is displayed when IRR is initiated without a guess entry. The next two examples illustrate RATE, IRR, and NPV use as follows:

One project—RATE, IRR, and NPV for single and multiple i^* values (Example 6.11)

Multiple alternatives—IRR for incremental evaluation (Example 6.12)

EXAMPLE 6.11   Two brothers, Gerald and Henry, own the Edwards Service Company in St. Johns, Newfoundland. It provides onshore services for spent lubricants from North Atlantic offshore platforms. The company needs immediate cash flow. Because the Edwards have an excellent reputation among major oil producers, they have been offered an 8-year contract that pays $200,000 total with 50% upfront and 50% at the end of the suggested 8-year contract. The estimated annual cost for Edwards to provide the services is $30,000. Assume you are the financial person for Edwards. Is the project justified if the brothers want to make at least 8% per year? Use calculator and spreadsheet functions and charts to perform a thorough ROR analysis.

Solution

The analysis can be accomplished in several ways, some more thorough than others. The approaches illustrated here are of increasing thoroughness.

1. The $i(n,A,P,F)$ calculator function is fast and easy to use. The function $i(8,-30000,100000,100000)$ will display one of several responses for i^*, depending upon the method used by the calculator to solve for i in Equation [2.3]. The normal display is $i^* = 15.91\%$. However, some calculators return the statement "Divided by 0" and provide no numerical answer, and some display a negative rate of return of $i^* = -23.98\%$. In this case, it is better to rely on a spreadsheet, as illustrated below.

2. Refer to Figure 6.8 (left). The easiest and quickest spreadsheet approach is to develop the RATE function for $n = 8$, $A = \$-30,000$, $P = \$100,000$, and $F = \$100,000$ in a single cell. The value $i^* = 15.91\%$ is displayed. The project is definitely justified since MARR $= 8\%$.

3. Refer to Figure 6.8 (right). Alternatively, to obtain the same answer, enter the years and associated net cash flows in adjacent cells. Develop the IRR function for the 9 entries to display $i^* = 15.91\%$; the project is justified.

4. Refer to Figure 6.9 (top). The cash flow (upper left of the figure) is a non-conventional series with two sign changes. The cumulative series has one sign change, but $S_0 > 0$. However, there is one real-number, positive i^*; it is 15.91%, displayed when the function $=$ IRR(B3:B11) is input. However, as predicted by the rule of signs, there is a second, though negative, i^* value at -23.98% found by using the guess option. A variety of guess percentages can be used; Figure 6.9 shows only four, but the results will always be $i_1^* = -23.98\%$ and $i_2^* = 15.91\%$.

   Is the project justified, knowing there are two roots? If reinvestment can be assumed to be at 15.91% rather than the MARR of 8%, the project is justified. If released funds are actually expected to make 8%, the project's

	A	B	C	D	E	F	G	H	I	J
1										
2	(2) i^* using RATE is **15.91%**					Year	Cash flow, $	(3) i^* using IRR is **15.91%**		
3						0	100,000			
4						1	-30,000			
5		= RATE(8, -30000, 100000, 100000)				2	-30,000			
6						3	-30,000			
7						4	-30,000	= IRR(G3:G11)		
8						5	-30,000			
9						6	-30,000			
10						7	-30,000			
11						8	70,000			
12										
13										

FIGURE 6.8 ROR analysis of a project using the RATE and IRR functions, Example 6.11.

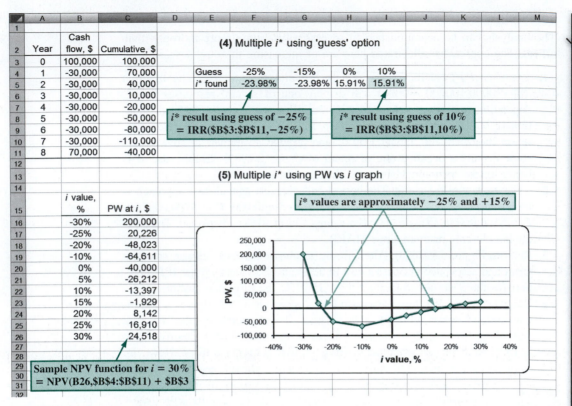

FIGURE 6.9 ROR analysis of a project with multiple *i** values using the IRR function with "guess" option and PW versus *i* graphical analysis, Example 6.11.

return is between 8 and 15.91%, and it is still justified. However, if funds are actually expected to make less than the MARR, it is not justified because the true rate is less than MARR, in fact between the MARR and −23.98%. *The conclusion is that the ROR analysis does not provide a conclusive answer.* If reinvestment is not assumed at either *i** value, one of the procedures of Section 6.7 should be performed.

5. Refer to Figure 6.9 (bottom). An excellent graphical way to approximate *i** values is to generate an *x-y* scatter chart of PW versus *i*. Use the NPV function with different *i* values to determine where the PW curve crosses the PW = 0 line. Here *i* values from −30% to +30% were chosen. For details, refer to the sample NPV function cell tag using the cash flows at the top of the figure and different *i* values. The approximate *i** values are −25% and +15%. Basically, this graphical analysis provides the same information as that of (4).

Perform a spreadsheet analysis of the four alternatives for Harold's construction company in Example 6.7.

EXAMPLE 6.12

Solution

First, review the situation in Example 6.7 for the four revenue alternatives. Figure 6.10 performs the complete analysis in one spreadsheet starting with the addition of the do-nothing alternative and the ordered alternatives DN, 4, 2, 1, and 3.

The top portion provides the estimates, including net annual cash flows in row 6. The middle portion calculates the incremental investment and cash flows for each comparison of two alternatives. Equal lives of 4 years are present for the first three comparisons. Details for the final comparison of 3-to-1 over the LCM of 8 years are included in the cell tags. (Multiple i^* values are not indicated for any incremental series.)

The bottom portion shows conclusions after using the IRR function to find the incremental ROR, comparing it to MARR = 12%, and identifying the surviving alternative. The entire logic of the analysis is identical to that of the hand solution. Alternative 1 is the best economic choice.

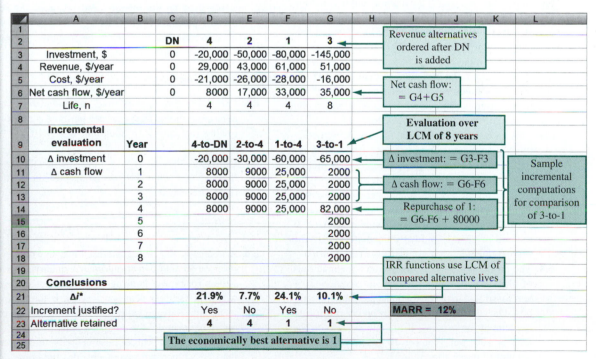

FIGURE 6.10   Incremental ROR analysis of four revenue alternatives, Example 6.12.

SUMMARY

Just as present worth, annual worth, and future worth methods find the best alternative from several, the ROR calculations serve the same purpose. With ROR analysis, however, the *incremental* cash flow series between two alternatives is evaluated. Alternatives are ordered by increasing initial investments, and the pairwise ROR evaluation proceeds from smallest to largest investment. The alternative with the larger incremental *i** value is selected as each comparison is conducted. Once eliminated, an alternative cannot be reconsidered.

Rate-of-return calculations performed by hand typically require trial-and-error solutions using tabu-lated factors. Spreadsheets and calculators greatly speed up this process. The analysis may result in more than one ROR value depending upon the number of sign changes present in the cash flow series. The cash flow rule of signs and cumulative cash flow test assist in determining if a unique ROR value does exist. The dilemma of multiple rates can be effectively dealt with by calculating the external rate of return (EROR) using the MIRR or ROIC method with either hand or spreadsheet procedures. In the end, if multiple rates are present, it is strongly recommended that the PW, AW, or FW value is determined at the MARR.

PROBLEMS

Understanding ROR

6.1 In percent, what is (*a*) the highest, and (*b*) the lowest rate of return that is possible?

6.2 When interest is charged on the unrecovered balance, if you borrow $10,000 at 10% per year interest and repay the loan in equal payments over a 5-year period, the payment amount is $2638 per year. How much will the annual payment be if the interest rate is charged on the *initial loan amount* instead of the unrecovered balance?

6.3 Spectra Scientific of Santa Clara, CA, manufactures Q-switched solid state industrial lasers for LED substrate scribing and silicon wafer dicing. The company got a $60 million loan, amortized over a 5-year period at 8% per year interest. What is the amount of the unrecovered balance (*a*) immediately *before* the payment is made at the end of year 1, and (*b*) immediately *after* the first payment?

6.4 The production of polyamide from raw materials of plant origin, such as castor oil, requires 20% less fossil fuel than conventional production methods. Darvon Chemicals borrowed $6 million to implement the process. If the interest rate on the loan is 10% per year for 10 years, what is the amount of interest for year 2?

6.5 General Dynamics obtained a 0.5%-per-month $100 million loan to be repaid over a 5-year period. (*a*) What is the difference in the amount of interest in the second month's payment if interest is charged on the original principal of the loan rather than on the unrecovered balance? (*b*) As months pass, for which basis—principal only or unrecovered balance—does the monthly interest decrease in amount?

ROR Calculation

6.6 Use tabulated factors and a spreadsheet to determine the interest rate per period for the following rate of return equation: $0 = -40,000 + 8000(P/A,i*,5) + 8000(P/F,i*,8)$.

6.7 Determine the rate of return per year for the cash flows shown below. Use (*a*) tabulated factors, and (*b*) a spreadsheet.

Year	1	2	3	4
Cash Flow, $	−80,000	9000	70,000	30,000

6.8 A company that manufactures brushless blowers invested $650,000 in an automated quality control system for blower housings. The resultant savings was $160,000 per year for 5 years. If the equipment had a salvage value of $50,000, what rate of return per year did the company make?

6.9 A University of Massachusetts study found that married women who work outside the home do about one hour less of housework per week for every $7500 they earn outside the home. Assume that they hire a housekeeper one time per week for $120; that the $7500 is received in uniform amounts of $625 per month; and, that the housekeeper is paid weekly with these payments made before the $625 is received. What rate of return are they making *per week* on their "investment" in the housekeeper?

6.10 A 473-foot, 7000-ton World War II troop carrier (once commissioned as the SS Excambion) was sunk in the Gulf of Mexico to serve as an underwater habitat and diving destination. The project took 10 years of planning and cost $4 million, which was spent equally at $400,000 in years 1 through 10. Fishing and recreation activities, estimated at $270,000 per year, will begin in year 11 and are expected to continue in perpetuity. Determine the rate of return on the venture using (*a*) tabulated factors, and (*b*) the GOAL SEEK tool.

6.11 The Closing the Gaps initiative by the Texas Higher Education Coordinating Board established the goal of increasing the number of students in higher education in Texas from 1,064,247 in 2000 to 1,694,247 in 2015. If the increase occurs uniformly and is compounded annually, what rate of increase is required each year to meet the goal?

6.12 When Hurricane Katrina struck New Orleans, there was a significant loss of aquarium fish at the Audubon Aquarium of the Americas. FEMA originally stated that the aquarium needed to buy the fish from commercial vendors, a method the agency said would cost $616,849 but would comply with disaster aid laws. FEMA later reversed their decision and allowed the aquarium staff to catch the fish themselves at a total cost of $99,766. If it is assumed that the aquarium staff spent the $99,766 equally over a 12-month period of time, what rate of return per month did their effort represent? Assume FEMA would have given the aquarium the $616,849 at the end of month 12.

6.13 Texas Governor Rick Perry promised to put hundreds of cameras on the Texas-Mexico border and broadcast the video over the Web so that anyone, anywhere could become a border patroller, helping root out border crime and illegal crossings. As part of that project, Texas secured a federal grant for $3 million that paid for 200 mobile cameras in strategic high-traffic areas. If the 200 cameras are considered to be equivalent to 20 border patrol agents, each with an annual salary of $75,000, what is the rate of return over a 3-year project period?

6.14 The University of California at San Diego is considering a plan to build a 8-megawatt cogeneration plant to provide for part of its power needs. The cost of the plant is expected to be $41 million. The university consumes 55,000 megawatt-hours per year at a cost of $120 per megawatt-hour. (*a*) If the university will be able to produce power at half the cost that it now pays, what rate of return will it make on its investment for an expected power plant life of 30 years? (*b*) If, in addition, the university can sell an average of 12,000 megawatt-hours per year back to the utility at $90 per megawatt-hour, what rate of return will it make?

6.15 The Camino Real Landfill was required to install a plastic liner to prevent leachate from migrating into the groundwater. The fill area was 50,000 m^2 and the installed liner cost was $8 per square meter. In order to recover the investment, the owner charged $10 for pick-up loads, $25 for dump truck loads, and $70 for compactor-truck loads. The annual distribution is 2400 pick-up loads, 600 dump truck loads, and 1200 compactor-truck loads. What rate of return will the landfill owner make on the investment if the fill area is adequate for 4 years?

6.16 U.S. Census Bureau statistics show that the annual earnings for a person with a high-school diploma are $35,220 versus $57,925 for someone with a bachelor's degree. If the cost of attending college is assumed to be $30,000 per year for four years and the foregone earnings during those years is assumed to be $35,220 per year, what rate of return does earning a bachelor's degree represent? Assume a 35-year study period.

6.17 Rubber sidewalks made from ground-up tires are said to be environmentally friendly and easier on peoples' knees. Rubbersidewalks, Inc. of Gardena, CA, manufactures the small rubberized squares that are installed where tree roots, freezing weather, and snow removal require sidewalk replacement or major repairs every three years. The District of Columbia spent $60,000 for a rubber sidewalk to replace broken concrete in a residential neighborhood lined with towering willow oaks. If a concrete sidewalk costs $28,000 and lasts only 3 years versus a 9-year life for the rubber sidewalks, what rate of return does this represent?

6.18 Steel cable barriers in highway medians are a low cost way to improve traffic safety without busting state department of transportation budgets. Cable barriers cost $44,000 per mile, compared with $72,000 per mile for guardrail and $419,000 per mile for concrete barriers. Furthermore, cable barriers tend to snag tractor-trailer rigs, keeping them from ricocheting back into same-direction traffic. The state of Ohio spent $4.97 million installing 113 miles of cable barriers. (a) If the cables prevent accidents totaling $1.3 million per year, determine the rate of return that this represents over a 10-year study period. Use all three methods—tabulated factors, a calculator, and a spreadsheet. (b) Now, determine the rate of return for 113 miles of guardrail if accident prevention is $1.1 million per year over a 10-year study period. To do so, first write the ROR relation and then find i^* using a single-cell spreadsheet function.

6.19 A broadband service company borrowed $2 million for new equipment and repaid the loan in amounts of $200,000 in years 1 and 2 plus a lump sum amount of $2.2 million at the end of year 3. What was the interest rate on the loan?

6.20 A new permanent endowment at the University of Alabama will award scholarships to engineering students twice per year (end of June and end of December). The first awards are to be made beginning 5-½ years after the $20 million lump sum donation is made. If the interest from the endowment is intended to fund 100 students each semester in the amount of $5000 twice per year, what semiannual rate of return must the endowment fund earn?

6.21 An Indium-Gallium-Arsenide-Nitrogen alloy developed at Sandia National Laboratory is said to have potential uses in electricity-generating solar cells. The new material is expected to have a longer life, and it is believed to have a 40% efficiency rate, which is nearly twice that of standard silicon solar cells. The useful life of a telecommunications satellite could be extended from 10 to 15 years by using the new solar cells. What rate of return could be realized if an extra investment now of $950,000 would result in extra revenues of $450,000 in year 11, $500,000 in year 12, and amounts increasing by $50,000 per year through year 15?

6.22 Barron Chemical used a thermoplastic polymer to enhance the appearance of certain RV panels. The initial cost of one process was $130,000 with annual costs of $49,000. Revenues were $78,000 in year 1, increasing by $1000 per year. A salvage value of $23,000 was realized when the process

was discontinued after 8 years. What rate of return did the company make on the process?

Incremental Analysis

6.23 Why is an incremental analysis necessary when conducting a rate of return evaluation of cost alternatives?

6.24 What is the overall rate of return on a $100,000 investment that returns 20% on the first $30,000 and 14% on the remaining $70,000?

6.25 Alternatives X and Y have rates of return of 10% and 18%, respectively. What is known about the rate of return on the increment between X and Y if the investment required in Y is (a) larger than that required for X, and (b) smaller than that required for X? (c) Develop two spreadsheet examples that illustrate your responses to parts (a) and (b).

6.26 A company that manufactures rigid shaft couplings has $600,000 to invest. The company is considering three different projects that will yield the following rates of return:

Project X $i_X = 24\%$
Project Y $i_Y = 18\%$
Project Z $i_Z = 30\%$

The initial investment required for each project is $100,000, $300,000, and $200,000, respectively. If the company's MARR is 15% per year and the company invests in all three projects, what overall rate of return will the company make?

6.27 For each of the following scenarios, state whether an incremental investment analysis is required to select an alternative and state why or why not. Assume that alternative Y requires a larger initial investment than alternative X and that the MARR is 20% per year.
   a. X has $i^* = 22\%$ per year, and Y has $i^* = 20\%$ per year.
   b. X has $i^* = 19\%$ per year, and Y has $i^* = 21\%$ per year.
   c. X has $i^* = 16\%$ per year, and Y has $i^* = 19\%$ per year.
   d. X has $i^* = 25\%$ per year, and Y has $i^* = 23\%$ per year.
   e. X has $i^* = 20\%$ per year, and Y has $i^* = 22\%$ per year.

6.28 For the cash flows shown and in preparation for a PW-based rate of return analysis, determine the incremental cash flow between machines B and A for (a) year 0, (b) year 3, and (c) year 6.

	Machine A	Machine B
First cost, $	−15,000	−25,000
Annual operating cost, $ per year	−1,600	−400
Salvage value, $	3,000	6,000
Life, years	3	6

6.29 Determine the sum of the cash flows in the incremental difference column (i.e., Y-X) for systems X and Y.

	System X	System Y
First cost, $	−45,000	−65,000
Annual operating cost, $	−21,800	−14,000
Salvage value, $	3,000	6,000
Life, years	5	5

6.30 For the alternatives shown, determine the sum of the cash flows in the Z-X difference column.

	System X	System Z
First cost, $	−40,000	−95,000
Annual operating cost, $/year	−12,000	−5,000
Salvage value, $	6,000	14,000
Life, years	3	6

ROR Evaluation of Two or More Alternatives

6.31 The incremental cash flows for alternatives P and Q are shown. Determine which should be selected using a FW-based rate of return analysis. The MARR is 15% per year and alternative Q requires the larger initial investment.

Year	Incremental Cash Flow (Q-P)
0	$−250,000
1–8	+50,000
8	+30,000

6.32 The Chem-Tex Chemical company is considering two additives for improving the dry-weather stability of its low-cost acrylic paint. Additive A will have a first cost of $110,000 and an annual operating cost of $60,000. Additive B will have a first cost of $175,000 and an annual operating cost of $35,000. If the company uses a three-year recovery period for paint products and a MARR of 20% per year, which process is favored on the basis of an incremental rate of return analysis?

6.33 Liquid Sleeve, Inc. is a company that makes a sealing solution for machine shaft surfaces that have been compromised by abrasion, high pressures, or inadequate lubrication. The manager is considering adding a metal-based nanoparticle (Type Al or Fe) to its solution to increase the product's performance at high temperatures. The costs associated with each type are estimated. If the company's MARR is 20% per year, which nanoparticle type should the company select? Utilize a rate of return analysis.

	Type Fe	Type Al
First cost, $	−150,000	−280,000
Annual operating cost, $/year	−92,000	−74,000
Salvage value, $	30,000	70,000
Life, years	2	4

6.34 A mechanical engineer is considering two robots for improving materials handling in the production of rigid shaft couplings that mate dissimilar drive components. Robot X has a first cost of $84,000, an annual maintenance and operation (M&O) cost of $31,000, a $40,000 salvage value, and will improve net revenues by $96,000 per year. Robot Y has a first cost of $146,000, an annual M&O cost of $28,000, a $47,000 salvage value, and will increase net revenues by $119,000 per year. Which one should be selected on the basis of a rate of return analysis if the company's MARR is 15% per year? Use a three-year study period.

6.35 Old Southwest Canning Co. has determined that any one of four machines can be used in a certain phase of its chili-canning operation. The first costs and annual operating costs (AOC) are estimated below, and all machines have a 5-year life. The MARR is 25% per year. (*a*) Determine which machine should be selected on the basis of a rate of return analysis. (*b*) Use a spreadsheet to perform PW analysis of each machine. Compare the machine selections with that of the ROR analysis.

Machine	First Cost, $	AOC, $
1	−28,000	−20,000
2	−51,000	−12,000
3	−32,000	−19,000
4	−33,000	−18,000

6.36 The four alternatives described below are being evaluated.

  a. If the proposals are independent, which one(s) should be selected at a MARR of 17% per year?

b. If the proposals are mutually exclusive, which one should be selected at a MARR of 14.5% per year?

c. If the proposals are mutually exclusive, which one should be selected at a MARR of 10.0% per year?

Alternative	Initial Investment, $	Overall Rate of Return, %	Incremental Rate of Return, %, When Compared with Alternative		
			A	B	C
A	−60,000	11.7			
B	−90,000	22.2	43.3		
C	−140,000	17.9	22.5	10.0	
D	−190,000	15.8	17.8	10.0	10.0

6.37 A small manufacturing company expects to expand its operation by adding new product lines. Any or all of four new lines can be added. If the company uses a MARR of 15% per year and a 5-year project period, which products, if any, should the company manufacture? Monetary terms are in $1000.

	Product			
	1	2	3	4
Initial cost, $	−340	−500	−570	−620
Annual cost, $/year	−70	−64	−48	−40
Annual savings, $/year	180	190	220	205

6.38 A WiMAX wireless network integrated with a satellite network can provide connectivity to any location within 10 km of the base station. The number of sectors per base station can be varied to increase the bandwidth. An independent cable operator is considering the three bandwidth alternatives shown below (monetary values in $1000 units). Assume a life of 20 years and a MARR of 25% per year to determine which alternative is best using an incremental ROR analysis.

Bandwidth, Mbps	First Cost, $	Operating Cost, $ per year	Annual Income, $ per year
44	−40,000	−2000	+4000
55	−46,000	−1000	+5000
88	−61,000	−500	+8000

6.39 Ashley Foods, Inc. has determined that only one of five machines can be used in one phase of its dairy products operation. The first and annual costs are estimated; all machines are expected to have a 4-year useful life. If the MARR is 20% per year, determine which machine should be selected on the basis of rate of return.

Machine	First Cost, $	AOC, $ per year
1	−31,000	−16,000
2	−29,000	−19,300
3	−34,500	−17,000
4	−49,000	−12,200
5	−41,000	−15,500

6.40 Five *revenue* projects are under consideration by General Dynamics for improving material flow through an assembly line. The initial cost (in $1000 units) and life of each project are estimated. Income estimates are not known at this point.

	Project				
	A	B	C	D	E
Initial cost, $	−700	−900	−2300	−300	−1600
Life, years	5	5	5	5	5

An engineer determined the incremental ROR (Δi^*) values. From these results, determine which project, if any, should be undertaken, provided the company's MARR is (*a*) 13.5% per year, and (*b*) 16% per year. If other calculations must be made in order to make a decision, state which ones are necessary.

Comparison	Δi*, %	Comparison	Δi*, %
C to DN	24	B to A	53
C to B	25	B to DN	23
D to DN	15	A to DN	13
E to C	59	E to B	−16
E to D	4	B to D	26
E to A	0	D to C	25
E to DN	6	A to D	12

6.41 Four different machines are under consideration for improving material flow in a production process. An engineer performed an economic analysis to select the best machine, but some of his calculations were deleted from the report by a disgruntled employee. All machines are assumed to have a 10-year life. (*a*) Fill in the missing numbers in the report. (*b*) Which machine should the company select if its MARR is 18% per year and one of the machines must be selected?

	Machine			
	1	2	3	4
Initial cost, $	?	−60,000	−72,000	−98,000
Annual cost, $ per year	−70,000	−64,000	−61,000	−58,000
Annual savings, $ per year	+80,000	+80,000	+80,000	+82,000
Overall ROR, %	18.6%	?	23.1%	20.8%
Machines compared		2 to 1	3 to 2	4 to 3
Incremental investment, $		−16,000	?	−26,000
Incremental cash flow, $ per year		+6,000	+3,000	?
ROR on increment, %		35.7%	?	?

6.42 Allstate Insurance Company is considering adopting one of five fraud detection systems, all of which can be considered to last indefinitely. If the company's MARR is 14% per year, determine which one should be selected on the basis of a rate of return analysis.

	A	B	C	D	E
First cost, $	−10,000	−25,000	−15,000	−70,000	−50,000
Annual net income, $/year	2,000	4,000	2,900	10,000	6,000
Overall ROR, %	20	20	19.3	14.3	12

6.43 The four revenue proposals described below are being evaluated.
   a. If the proposals are independent, which one(s) should be selected with MARR = 15.5% per year?
   b. If the proposals are mutually exclusive, which one should be selected with MARR = 10% per year?
   c. If the proposals are mutually exclusive, which one should be selected with MARR = 14% per year?

Proposal	Initial Investment, $	Proposal i*, %	Δi* when Compared with Proposal, %		
			A	B	C
A	−40,000	29			
B	−75,000	15	1		
C	−100,000	16	7	20	
D	−200,000	14	10	13	12

6.44 An engineer initiated a rate of return analysis for the infinite-life revenue proposals detailed below, but was unable to complete the evaluation.
   a. Fill in the eleven blanks in the table.
   b. What proposal(s) should be selected if they are independent and the MARR is 21% per year?
   c. What proposal should be selected if they are mutually exclusive and the MARR is 13% per year.

Proposal	Investment, $	Proposal i^*, %	Δi^* on Incremental Cash Flow when Compared with Proposal, %			
			X1	X2	X3	X4
X1	−20,000	?	—	?	?	?
X2	−30,000	13.33	2	—	?	?
X3	−50,000	?	14	20	—	?
X4	−75,000	12	?	?	?	—

Multiple ROR Values

6.45 For the following incremental cash flow series, what is the maximum number of i^* values according to the cash flow rule of signs?

Year	0	1	2	3	4	5	6	7	8
Cash Flow, $	−100	40	35	−15	−11	60	42	12	−10

6.46 Jenco Electric manufactures washdown adjustable speed drives in open loop, encoderless, and closed-loop servo configurations. The net cash flow associated with one phase of the production operation is shown below.
   a. How many possible rate of return values are there according to the cash flow rule of signs?
   b. How many changes of sign occur in the cumulative cash flow series? What does this mean?

Year	Net Cash Flow, $
0	−40,000
1	32,000
2	18,000
3	−2000
4	−1000

6.47 For the following incremental cash flow series, use a spreadsheet to find all rate of return values between 0% and 100%.

Year	Incremental Cash Flow, $
0	−50,000
1	+22,000
2	+38,000
3	−2000
4	−1000
5	+5000

6.48 According to Descartes' rule of signs, how many possible i^* values are there for net cash flows that have the following signs?: (a) $---+++--+$ (b) $-----+++++$ (c) $++++------$ $-+-+---$

6.49 The cash flow (in $1000 units) associated with a new method of manufacturing box cutters is shown below for a 2-year period. (a) Use Descartes' rule to determine the maximum number of possible rate of return values. (b) Use Norstrom's criterion to determine if there is only one positive rate of return value.

Quarter	Expenses, $1000	Revenue, $1000
0	−20	0
1	−20	5
2	−10	10
3	−10	25
4	−10	26
5	−10	20
6	−15	17
7	−12	15
8	−15	2

6.50 ARCI Instruments manufactures a ventilation controller designed for monitoring and controlling carbon monoxide in parking garages, boiler rooms, tunnels, etc. The net cash flow associated with one phase of the operation is shown on the next page. (a) How many possible rate of return values are there for this cash flow series? (b) Find all the rate of return values between 0 and 100% using tabulated factors and a spreadsheet.

Year	Net Cash Flow, $
0	−30,000
1	20,000
2	15,000
3	−2000

Year	Expense, $	Receipts, $
0	−3000	0
1	−1500	2900
2	−4000	5700
3	−2000	5500
4	−1300	1100

6.51 The cash flows associated with sales of handheld refractometers (instruments that measure the concentration of an aqueous solution by determining its refractive index) are shown. Determine the cumulative cash flow through year 4 and estimate the expected number of positive, real-number i^* values.

Year	Revenue, $	Costs, $
1	25,000	30,000
2	15,000	7,000
3	4,000	6,000
4	18,000	12,000

6.52 Boron nitride spray II (BNS II) from GE's Advanced Material Ceramics Division is a release agent and lubricant that prevents materials such as molten metal, rubber, plastics, and ceramic materials from sticking to or reacting with dies, molds, or other surfaces. A European distributor of BNS II and other GE products had the net cash flows shown. (a) Determine the number of possible rate of return values. (b) Find all rate of return values between −30% and 130%.

Year	Net Cash Flow, $
0	−17,000
1	−20,000
2	4,000
3	−11,000
4	32,000
5	47,000

6.53 Faro laser trackers are portable contact measurement systems that use laser technology to measure large parts and machinery to accuracies of 0.0002 inches across a wide range of industrial applications. A customer that manufactures and installs cell phone relay dishes and satellite receiving stations reported the cash flows (in $1000 units) for one of its product lines. (a) Determine the number of possible rate of return values. (b) Find all rate of return values between 0 and 150%.

Removing Multiple ROR Values

6.54 For the cash flow series shown, find the external rate of return using a reinvestment rate of 15% per year, using (a) the manual ROIC method, and (b) a spreadsheet to verify the answer.

Year	Incremental Cash Flow, $
0	+48,000
1	+20,000
2	−90,000
3	+50,000
4	−10,000

6.55 Carl, an engineer working for GE, invested his bonus money each year in company stock. His bonus at the end of each year 1 through 6 has been $5000. At the end of year 7, Carl received no bonus and he sold $9000 worth of stock to remodel his kitchen. In years 8 through 10, he again received a bonus and invested the $5000. Carl sold all the remaining stock for $50,000 immediately after the last investment at the end of year 10.
a. Determine the expected number of positive rate of return values.
b. Find the internal rate(s) of return.
c. Use hand solution and the MIRR spreadsheet function to determine the external rate of return using the modified rate of return approach with a borrowing rate of 8% and a reinvestment rate of 20% per year
d. Determine the external rate of return using the ROIC approach with a reinvestment rate of 20% per year. Apply both the net-investment procedure and spreadsheet functions to obtain the EROR.

6.56 A company that makes clutch disks for race cars had the cash flows shown for one department.
a. Calculate the internal rate of return.
b. Calculate the external rate of return using the ROIC method with a reinvestment rate of 15% per year.

c. Calculate the external rate of return using the MIRR method with a reinvestment rate of 15% per year and a borrowing rate of 8% per year.
d. Rework parts (*b*) and (*c*) using a spreadsheet.

Year	Cash Flow, $1000
0	−65
1	30
2	84
3	−10
4	−12

6.57 Gemini Products makes vitamin-enriched cereal products for large supermarket chains. They have forecasted the cash flows (in $1000 units) for this and the next 4 years. Develop hand and spreadsheet solutions that determine the EROR for (*a*) the ROIC method with $i_r = 14\%$ per year, and (*b*) the MIRR method with $i_r = 14\%$ and $i_b = 8\%$ per year. (*c*) Comment on these two EROR values compared to the IRR value(s) obtained using the IRR-function "guess" option to discover multiple roots.

Year	Cash Flow, $
0	3000
1	−2000
2	1000
3	−6000
4	3800

6.58 A public-private initiative in Florida will significantly expand the wind-generated energy throughout the state. The cash flow for one phase of the project involving Central Point Energy, a transmission utility company, is shown. Calculate the external rate of return (*a*) using the ROIC method and a reinvestment rate of 14% per year, and (*b*) using the modified ROR approach with a reinvestment rate of 14% and a borrowing rate of 7% per year. Solve by hand or spreadsheet, as requested by your instructor.

Year	Cash Flow, $1,000
0	5000
1	−2000
2	−1500
3	−7000
4	4000

6.59 A new advertising campaign by a company that manufactures products that rely on biometrics, surveillance, and satellite technologies resulted in the cash flows shown (in $1000 units). Develop one spreadsheet that displays the following: external rate of return using both the ROIC method with $i_r = 30\%$ per year, and the modified ROR approach with $i_r = 30\%$ and $i_b = 10\%$ per year; and the unique or multiple internal rate of return value(s) indicated by the two multiple-root sign tests.

Year	Cash Flow, $1,000
0	2000
1	1200
2	−4000
3	−3000
4	2000

ADDITIONAL PROBLEMS AND FE EXAM REVIEW QUESTIONS

6.60 The lowest rate of return possible is:
a. 0%
b. −∞
c. −100%
d. the company's MARR
6.61 When calculating an i^* value, all net positive cash flows are assumed to be reinvested at:
a. the current market interest rate.
b. the i^* rate.
c. the company's MARR.
d. the company's cost of capital.

6.62 Alternative A has a rate of return of 14% and alternative B has a rate of return of 17%. If the investment required in B is larger than that required for A, the rate of return on the increment of investment between A and B is:
a. larger than 14%
b. larger than 17%
c. between 14% and 17%
d. smaller than 14%
6.63 A small manufacturing company borrowed $1 million and repaid the loan through monthly

payments of $20,000 for 2 years plus a single lump-sum payment of $1 million at the end of 2 years. The interest rate on the loan was closest to:
a. 0.5% per month
b. 2% per month
c. 2% per year
d. 8% per year

6.64 An investment of $60,000 resulted in uniform income of $10,000 per year for 10 years. The rate of return on the investment was closest to:
a. 10.6% per year
b. 14.2% per year
c. 16.4% per year
d. 18.6% per year

6.65 Assume you are told that by investing $100,000 now, you will receive $10,000 per year *starting in year 5* and continuing forever. If you accept the offer, the rate of return on the investment is:
a. 4% per year
b. between 6% and 7% per year
c. between 7% and 10% per year
d. over 12% per year

6.66 A chemical engineer working for a large chemical products company was asked to make a recommendation about which of three mutually exclusive revenue alternatives should be selected for improving the marketability of personal care products used for conditioning hair, cleansing skin, removing wrinkles, etc. The alternatives (X, Y, and Z) were ranked in order of increasing initial investment and then compared by incremental rate of return analysis. The rate of return on each increment of investment was less than the company's MARR of 17% per year. The alternative to select is:
a. DN
b. alternative X
c. alternative Y
d. alternative Z

Problems 6.67 and 6.68 are based on the following data.
The five alternatives are being evaluated by the rate of return method.

| Proposal | Initial Investment, $ | Overall i*, % | Δi^*, %, when Compared with Proposal | | | | |
			A	B	C	D	E
A	−25,000	9.6	—	27.3	19.4	35.3	25.0
B	−35,000	15.1		—	0	38.5	24.4
C	−40,000	13.4			—	46.5	27.3
D	−60,000	25.4				—	26.8
E	−75,000	20.2					—

6.67 If the alternatives are independent and the MARR is 15% per year, the one(s) that should be selected is (are):
a. only D
b. only D and E
c. only B, D, and E
d. only E

6.68 If the alternatives are mutually exclusive and the MARR is 15% per year, the alternative to select is:
a. either B, C, D or E
b. only B
c. only D
d. only E

6.69 Jewel-Osco evaluated three different pay-by-touch systems that identify customers by a finger scan and then deduct the amount of the bill directly from their checking accounts. The alternatives were ranked according to increasing initial investment and identified as alternatives A, B, and C.

Comparison	Δi^*, %
DN to A	23.4
DN to B	8.1
DN to C	16.6
B to A	−5.1
C to A	12.0
C to B	83.9

Based on the incremental rates of return and the company's MARR of 16% per year, the alternative that should be selected is:

a. alternative A
b. alternative B
c. alternative C
d. alternative DN

6.70 For the cash flows shown, the correct equation for FW_2 using the ROIC method at the reinvestment rate of 20% per year is:

a. $[10,000(1 + i'') + 6000](1.20) - 8000$
b. $[10,000(1.20) + 6000(1 + i'')](1.20) - 8000$
c. $[10,000(1.20) + 6000](1.20) - 8000$
d. $[10,000(1.20) + 6000](1 + i'') - 8000$

Year	Cash Flow, $
0	10,000
1	6,000
2	−8,000
3	−19,000

Benefit/Cost Analysis and Public Sector Projects

PhotoLink/Getty Images

The evaluation methods of previous chapters are usually applied to alternatives in the private sector, that is, for-profit and not-for-profit corporations and businesses. This chapter introduces *public sector alternatives* and their economic consideration. Here the owners and users (beneficiaries) are the citizens of the government unit—city, county, state, province, or nation. There are substantial differences in the characteristics of public and private sector alternatives and their economic evaluation. Partnerships of the public and private sector have become increasingly common, especially for large infrastructure construction projects such as major highways, power generation plants, water resource developments, and the like.

The benefit/cost (B/C) ratio was developed, in part, to introduce objectivity into the economic analysis of public sector evaluation, thus reducing the effects of politics and special interests. The different formats of B/C analysis are discussed. The B/C analysis can use equivalency computations based on PW, AW, or FW values.

Purpose: Understand public versus private sector projects; evaluate alternatives using the benefit/cost ratio method.

LEARNING OUTCOMES

1. Identify fundamental differences between public and private sector alternatives; understand ethical aspects of the public sector.

| Public sector |

2. Use the benefit/cost ratio to evaluate a single project.

| B/C for single project |

3. Select the best of two or more alternatives using the incremental B/C ratio method.

| Alternative selection |

4. Use a spreadsheet to perform B/C analysis of one or more alternatives.

| Spreadsheets |

7.1 PUBLIC SECTOR PROJECTS: DESCRIPTION AND ETHICS

7.1.1 Public Sector Project Description

Podcast available

Public sector projects are owned, used, and financed by the citizenry of any government level, whereas projects in the private sector are owned by corporations, partnerships, and individuals. Virtually all the examples in previous chapters have been from the private sector. Notable exceptions occur in Chapters 4 and 5 where capitalized cost was introduced for long-life alternatives and infinite-life (perpetual) investments.

Public sector projects have a primary purpose to provide services to the citizenry for the public good at no profit. Areas such as health, transportation, safety, economic development, and utilities comprise a majority of alternatives that require engineering economic analysis. Some public sector examples are

Hospitals and clinics	Transportation: highways, bridges,
Parks and recreation	waterways
Utilities: water, electricity, gas,	Police and fire protection
sewer, sanitation	Courts and prisons
Schools: primary, secondary,	Food stamp and rent relief programs
community colleges, universities	Job training
Economic development	Public housing
Convention centers	Emergency relief
Sports arenas	Codes and standards

There are significant differences in the characteristics of private and public sector alternatives.

Characteristic	Public Sector	Private Sector
Size of investment	Larger	Some large; more medium to small

Often alternatives developed to serve public needs require large initial investments, possibly distributed over several years. Modern highways, public transportation systems, airports, and flood control systems are examples.

Life estimates	Longer (30–50+ years)	Shorter (2–25 years)

The long lives of public projects often prompt the use of the capitalized cost method, where infinity is used for n and annual costs are calculated as $A = P(i)$.

Annual cash flow estimates	No profit; costs, benefits, and disbenefits are estimated	Revenues and savings contribute to profits; costs are estimated

Public sector projects (also called publicly owned) do not have profits; they do have costs paid by the appropriate government unit; and they benefit the citizenry. Public sector projects often have undesirable consequences (*disbenefits*). It is these consequences that can cause public controversy about the projects, because benefits to one group of taxpayers might be disbenefits to other taxpayers, as discussed more fully below. Due to the subjective nature of public project data, it is relatively easy to manipulate estimates; however, to perform an economic analysis of public alternatives, the costs (initial and annual), the benefits, and the disbenefits, if considered, must be estimated as accurately as possible in *monetary units*.

> **Costs**—estimated expenditures *to the government entity* for construction, operation, and maintenance of the project, less any expected salvage value.
>
> **Benefits**—advantages to be experienced *by the owners, the public*. Benefits can include revenue and savings.
>
> **Disbenefits**—expected undesirable consequences that are indirect economic disadvantages *to the owners* if the alternative is implemented.

In many cases, it is difficult to estimate and agree upon the economic impact of benefits and disbenefits for a public sector alternative. For example, assume a short bypass around a congested traffic area is recommended. How much will it benefit a driver in *dollars per driving minute* to be able to bypass five traffic lights, as compared to stopping at an average of two lights for 45 seconds each? The bases and standards for benefits estimation are always difficult to establish and verify. Relative to revenue cash flow estimates in the private sector, benefit estimates are much harder to make, and they vary more widely around uncertain averages. And the disbenefits that accrue from an alternative are harder to estimate.

The examples in this chapter include straightforward identification of benefits, disbenefits, and costs. However, in actual situations, judgments are subject to interpretation, particularly in determining which elements of cash flow should be included in the economic evaluation. For example, improvements to the condition of pavement on city streets might result in fewer accidents, an obvious benefit to the taxpaying public. But fewer damaged cars and personal injuries mean less work and money for auto repair shops, towing companies, car dealerships, doctors and hospitals—also part of the taxpaying public. It may be necessary to take a limited viewpoint, because in the broadest viewpoint benefits are usually offset by approximately equal disbenefits.

Funding	Taxes, fees, bonds, private funds	Stocks, bonds, loans, individual owners

The capital used to finance public sector projects is commonly acquired from taxes, bonds, fees, and gifts from private donors. Taxes are collected from those who are the owners—the citizens (e.g., gasoline taxes for highways are paid by all gasoline users). This is also the case for fees, such as toll road fees for drivers. Bonds are often issued: municipal bonds and special-purpose bonds, such as utility district bonds.

Interest rate	Lower	Higher, based on market cost of capital

The interest rate for public sector projects, also called the *discount rate,* is virtually always lower than for private sector alternatives. Government agencies are exempt from taxes levied by higher-level government units. For example, municipal projects do not have to pay state taxes. Also, many loans are government subsidized and carry low interest rates.

Selection criteria	Multiple criteria	Primarily based on MARR

Multiple categories of users, economic as well as noneconomic interests, and special-interest political and citizen groups make the selection of one alternative over another much more difficult in public sector economics. Seldom is it possible to select an alternative on the sole basis of a criterion such as PW or ROR. Multiple attribute evaluation is discussed further in Chapter 14.

Environment of the evaluation	Politically inclined	Primarily economic

There are often public meetings and debates associated with public sector projects. Elected officials commonly assist with the selection, especially when pressure is brought to bear by voters, developers, environmentalists, and others. The selection process is not as "clean" as in private sector evaluation.

The *viewpoint* of the public sector analysis must be determined before cost, benefit, and disbenefit estimates are made. There are always several viewpoints that may alter how a cash flow estimate is classified. Some example viewpoints are the citizen; the tax base; number of students in the school district; creation and retention of jobs; economic development potential; or a particular industry interest. Once established, the viewpoint assists in categorizing the costs, benefits, and disbenefits of each alternative.

EXAMPLE 7.1

The citizen-based Capital Improvement Projects (CIP) Committee for the city of Dundee has recommended a $25 million bond issue for the purchase of greenbelt/floodplain land to preserve low-lying green areas and wildlife habitat. Developers oppose the proposal due to the reduction of available land for commercial development. The city engineer and economic development director have made preliminary estimates for some obvious areas over a projected 15-year planning horizon. The inaccuracy of these estimates is made very clear in a report to the Dundee City Council. The estimates are not yet classified as costs, benefits, or disbenefits.

Economic Dimension	Estimate
1. Annual cost of $5 million in bonds over 15 years at a 6% bond interest rate	$300,000 (years 1–14) $5,300,000 (year 15)
2. Annual maintenance, upkeep, and program management	$75,000 + 10% per year increase
3. Annual parks development budget	$500,000 (years 5–10)
4. Annual loss in commercial development	$2,000,000 (years 8–10)
5. State sales tax rebates not realized	$275,000 + 5% per year (years 8 on)
6. Annual municipal income from park use and regional sports events	$100,000 + 12% per year (years 6 on)
7. Savings in flood control projects	$300,000 (years 3–10) $1,400,000 (years 11–15)
8. Property damage (personal and city) not incurred due to flooding	$500,000 (years 10 and 15)

Identify different viewpoints for an economic analysis of the proposal, and classify the estimates accordingly.

Solution

There are many perspectives to take; three are addressed here. The viewpoints and goals are identified and each estimate is classified as a cost, benefit, or disbenefit. (How the classification is made will vary depending upon who does the analysis. *This solution offers only one logical answer.*)

Viewpoint 1: *Citizen of the city.* Goal: Maximize the quality and wellness of citizens with family and neighborhood as prime concerns.

    Costs: 1, 2, 3   Benefits: 6, 7, 8   Disbenefits: 4, 5

Viewpoint 2: *City budget.* Goal: Ensure the budget is balanced and of sufficient size to fund rapidly growing city services.

    Costs: 1, 2, 3, 5   Benefits: 6, 7, 8   Disbenefits: 4

Viewpoint 3: *Economic development.* Goal: Promote new commercial and industrial economic development for creation and retention of jobs.

    Costs: 1, 2, 3, 4, 5   Benefits: 6, 7, 8   Disbenefits: none

If the analyst favors the economic development goals of the city, commercial development losses (4) are considered real costs, whereas they are undesirable consequences (disbenefits) from the citizen and budget viewpoints. Also, the loss of sales tax rebates from the state (5) is interpreted as a real cost from the budget and economic development perspectives, but only as a disbenefit from the citizen viewpoint.

Many large public sector projects are developed through public-private partnerships (PPP). This approach works because of the greater efficiency of the private sector and because full funding is not possible using traditional means of government financing—fees, taxes, and bonds. This is an especially useful approach for projects in international settings and in developing countries. Examples of these projects are major highways, tunnels, airports, water resources, and public transportation. In these joint ventures, the government cannot make a profit, but the corporate partner can realize a reasonable profit.

Historically, public projects were designed for and financed by a government unit with a contractor doing the construction under either a *fixed price* (lump sum) or *cost plus* (cost reimbursement) contract with a profit margin specified. Here the contractor did not share the risk of success with the government unit. More recently, when a PPP is developed, the project is commonly arranged using a *design-build contract*. One common form of this contract is a BOT-administered project, which stands for build-operate-transfer. This type of contract requires that the contractor be responsible partially or completely for design and completely responsible for the construction (the build element), operation (operate), and maintenance for a specified number of years. After this time period, the title of ownership is transferred (transfer) at no or very low cost to the government. A turnkey approach to contracting is a DBOMF(design-build-operate-maintain-finance) contract, in which the contractor is responsible for managing the cash flows for the project, but obtaining capital and operating funds remains the government's responsibility through loans, bonds, taxes, grants and gifts.

7.1.2 Ethics Considerations

Engineers are involved in a wide range of public sector activities, which may be generally classified into two categories—policy and planning.

Policy making—These activities include *strategy development* based upon feasibility studies, surveys, historical precedent, legal requirement, current data, and hypothesis testing. Examples are highway and air transportation management and healthcare systems policies. In the case of highway transportation, contract and government-employed engineers make most of the recommendations on elements such as highway expansion, routing, capacity, zoning, speed limits, and traffic signaling systems.

Planning—This includes *project development and oversight* that implement approved policies and strategies. These projects affect people, the environment, and finances. The planning level of examples mentioned above could be healthcare delivery methods, highway traffic control, and air traffic control projects. In the case of highway traffic control, engineers implement policy through plans for commercial and residential corridors, speed control, and monitoring projects (e.g., camera surveillance), parking restrictions, traffic light placement, toll roads, etc.

In all of these types of activities, the public expects its engineers, like elected officers and politicians, to have good morals and ethics (See Section 1.9 for a review.). Among other things, they are expected to:

- Demonstrate high standards
- Make realistic assumptions and conclusions
- Gather and use data and information fairly
- Be impartial in decision making
- Consider a wide range of circumstances before deciding on a particular strategy or plan

In other words, it is assumed that public servants of all types (elected, employed, and contracted) will have and demonstrate *integrity and fairness* in all of their dealings. Most citizens become very disappointed and discouraged in the public leadership, including engineers, when these qualities are compromised.

It is vitally important that engineers maintain adherence to the Code of Ethics for Engineers (Figure 1.9) and stay clear of potentially unethical practices. However, it is quite easy to become involved in situations that involve a public sector project that offers ethical challenges. Here is one simple example.

The situation – Joe, a city of Vickor engineer employed by the energy department, is currently full time on a project that involves the conversion of all residential electrical meters from read-on-site to read-remotely. Some 95,000 meters will be replaced at an expenditure of several million dollars by a contractor to be chosen.

Over the last few weeks, Joe has become a close friend of Lisa, who, he has learned, works as a proposal writer for Lange Contractors, one of the two or three companies expected to bid on the meter replacement contract.

The temptation – In his efforts to impress Lisa, Joe has thought of mentioning to Lisa that his boss will cast a major vote when the meter-replacement contract is awarded, and that he, Joe's manager, is very positively impressed with the meters manufactured by Hammond Industries. In fact, Joe's manager has stated privately to Joe that a proposing contractor will have an edge for his vote if Hammond meters are specified in the equipment portion of the proposal.

The ethics dilemma – As an experienced engineer, Joe realizes he must not provide Lisa with this sort of advantage when writing the proposal. Though he may be tempted to provide "hints" if the conversation turns to work topics, to remain professionally ethical, Joe must refrain from any sort of information that will favor the Lange proposal. This is ethically correct even if Joe learns that Lange is already planning to propose meters manufactured by Hammond Industries. On the other side, Lisa should not ask Joe about the proposal or share any of its contents during their conversations.

7.2 BENEFIT/COST ANALYSIS OF A SINGLE PROJECT

The benefit/cost ratio, a fundamental analysis method for public sector projects, was developed to introduce more objectivity into public sector economics. It was developed in response to the U.S. Flood Control Act of 1936. There are several variations of the B/C ratio; however, the basic approach is the same. All cost and benefit estimates must be converted to a common equivalent monetary unit (PW, AW, or FW) at the discount rate (interest rate). The B/C ratio is then calculated using one of these relations:

$$B/C = \frac{\text{PW of benefits}}{\text{PW of costs}} = \frac{\text{AW of benefits}}{\text{AW of costs}} = \frac{\text{FW of benefits}}{\text{FW of costs}} \qquad [7.1]$$

The sign convention for B/C analysis is positive signs, so *costs are preceded by a + sign.* Salvage values, when they are estimated, are subtracted from costs. Disbenefits are considered in different ways depending upon the model used. Most commonly, disbenefits are subtracted from benefits and placed in the numerator. The different formats are discussed below. The decision guideline for a single project is simple:

> **If B/C ≥ 1.0, accept the project as economically acceptable for the estimates and discount rate applied.**
>
> **If B/C < 1.0, the project is not economically acceptable.**

The *conventional B/C ratio* is the most widely used. It subtracts disbenefits from benefits.

$$\mathbf{B/C} = \frac{\textbf{benefits} - \textbf{disbenefits}}{\textbf{costs}} = \frac{B - D}{C} \qquad [7.2]$$

The B/C value would change considerably were disbenefits added to costs. For example, if the numbers 10, 8, and 5 are used to represent the PW of benefits, disbenefits, and costs, respectively, Equation [7.2] results in B/C = (10 − 8)/5 = 0.40. The incorrect placement of disbenefits in the denominator results in B/C = 10/(8 + 5) = 0.77, which is approximately twice the correct B/C value. Clearly, then, the method by which disbenefits are handled affects the magnitude of the B/C ratio. However, it does not matter whether disbenefits are (correctly) subtracted from the numerator or (incorrectly) added to costs in the denominator, *a B/C ratio of less than 1.0 by the first method will always yield a B/C ratio less than 1.0 by the second method, and vice versa.*

The *modified B/C ratio* places benefits (including revenues and savings), disbenefits, and maintenance and operation (M&O) costs in the numerator. The denominator includes only the equivalent PW, AW, or FW of the *initial investment.*

$$\textbf{Modified B/C} = \frac{\textbf{benefits} - \textbf{disbenefits} - \textbf{M\&O costs}}{\textbf{initial investment}} \qquad [7.3]$$

Salvage value is included in the denominator with a negative sign. The modified B/C ratio will obviously yield a different value than the conventional B/C method. However, as discussed above with disbenefits, *the modified procedure can change the magnitude of the ratio but not the decision to accept or reject the project.* The decision guideline remains the same; if the modified B/C ratio exceeds or equals 1.0, the project is justified.

The *benefit and cost difference* measure of worth, which does not involve a ratio, is based on the difference between the PW, AW, or FW of benefits (including income and savings) and costs, that is, $B - C$. If $(B - C) \geq 0$, the project is acceptable. This method has the advantage of eliminating the discrepancies noted above when disbenefits are regarded as costs, because B represents *net benefits*. Thus, for the numbers 10, 8, and 5 the same result is obtained regardless of how disbenefits are treated.

Subtracting disbenefits from benefits: $B - C = (10 - 8) - 5 = -3$

Adding disbenefits to costs: $B - C = 10 - (8 + 5) = -3$

EXAMPLE 7.2

The Ford Foundation expects to award $15 million in grants to public high schools to develop new ways to teach the fundamentals of engineering that prepare students for university-level material. The grants will extend over a 10-year period and will create an estimated savings of $1.5 million per year in faculty salaries and student-related expenses. The Foundation uses a discount rate of 6% per year.

   This grants program will share Foundation funding with ongoing activities, so an estimated $200,000 per year will be removed from other program funding. To make this program successful, a $500,000 per year operating cost will be incurred from the regular M&O budget. Use the B/C method to determine if the grants program is economically justified.

Solution

Use annual worth as the common monetary equivalent. For illustration only, all three B/C models are applied.

   AW of investment cost. $15,000,000(A/P,6\%,10) = \$2,038,050$ per year

   AW of M&O cost. $500,000 per year

   AW of benefit. $1,500,000 per year

   AW of disbenefit. $200,000 per year

Use Equation [7.2] for conventional B/C analysis, where M&O is placed in the denominator as an annual cost. The project is not justified, since B/C < 1.0.

$$B/C = \frac{1,500,000 - 200,000}{2,038,050 + 500,000} = \frac{1,300,000}{2,538,050} = 0.51$$

   By Equation [7.3] the modified B/C ratio treats the M&O cost as a reduction to benefits.

$$\text{Modified B}/C = \frac{1,500,000 - 200,000 - 500,000}{2,038,050} = 0.39$$

   For the $(B - C)$ model, B is the net benefit, and the annual M&O cost is included with costs.

$$B - C = (1,500,000 - 200,000) - (2,038,050 - 500,000)$$
$$= \$-1.24 \text{ million}$$

7.3 INCREMENTAL B/C EVALUATION OF TWO OR MORE ALTERNATIVES

Incremental B/C analysis of two or more alternatives is very similar to that for incremental ROR analysis in Chapter 6. The incremental B/C ratio, ΔB/C, between two alternatives is based upon the PW, AW, or FW equivalency of costs and benefits. Selection of the survivor of pairwise comparison is made using the following guideline:

Podcast available

If ΔB/C $\geq$ 1, select the larger-cost alternative.

Otherwise, select the lower-cost alternative.

Note that the decision is based upon incrementally justified *total* costs, not incrementally justified *initial* cost.

There are several special considerations for B/C analysis of multiple alternatives that make it slightly different from ROR analysis. As mentioned earlier, all costs have a positive sign in the B/C ratio. Also, the *ordering of alternatives is done on the basis of total costs* in the denominator of the ratio. Thus, if two alternatives have equal initial investments and lives, but 2 has a larger equivalent annual cost, then 2 must be incrementally justified against 1. If this convention is not correctly followed, it is possible to get a negative cost value in the denominator, which can incorrectly make ΔB/C $<$ 1 and reject a higher-cost alternative that is actually justified. In the unusual circumstance that the two alternatives have equal costs (yielding a ΔB/C of infinity), the alternative with the larger benefits is selected by inspection.

Like the ROR method, B/C analysis requires *equal-service comparison* of alternatives. Usually, the expected useful life of a public project is long (25 or 30 or more years), so alternatives generally have equal lives. However, when alternatives do have unequal lives, the use of PW to determine the equivalent costs and benefits requires that the LCM of lives be used.

There are two types of benefits. Before conducting the incremental evaluation, classify the alternatives as *usage cost* estimates or *direct benefit* estimates. Usage cost estimates have implied benefits based on the difference in costs between alternatives. Direct benefit alternatives have benefit amounts estimated. The incremental comparison differs slightly for each type; direct benefit alternatives are initially compared with the DN alternative. (This is the same treatment made for revenue alternatives in an ROR evaluation, but the term *revenue* is not used for public projects.)

Usage cost: Evaluate alternatives only against each other.

Direct benefit: First evaluate against do-nothing, then against each other.

Apply the following procedure for comparing multiple, mutually exclusive alternatives using the conventional B/C ratio:

1. For each alternative, determine the equivalent PW, AW, or FW values for costs C, benefits B, and disbenefits D, if considered.
2. Order the alternatives by increasing total equivalent cost. For *direct benefit* alternatives, add DN as the first alternative.

3. Determine the incremental costs and benefits between the first two ordered alternatives (that is, 2 − 1) over their least common multiple of lives. For *usage cost* alternatives, incremental benefits are determined as the difference in usage costs.

$$\Delta B = \text{usage cost of 2} - \text{usage cost of 1} \qquad [7.4]$$

4. Calculate the incremental conventional B/C ratio using Equation [7.2]. If disbenefits are considered, this is

$$\Delta B/C = \Delta(B - D)/\Delta C \qquad [7.5]$$

5. If $\Delta B/C \geq 1$, eliminate 1; 2 is the survivor; otherwise 1 is the survivor.
6. Continue to compare alternatives using steps 2 through 5 until only one alternative remains as the survivor.

In step 3, prior to calculating the $\Delta B/C$ ratio, visually check the PW, AW, or FW values to ensure that the larger cost alternative also yields larger benefits. If benefits are not larger, the comparison is unnecessary.

The next two examples illustrate this procedure; the first for two direct benefit estimate alternatives, and the second for four usage fee estimate alternatives.

EXAMPLE 7.3   The city of Garden Ridge, Florida, has received two designs for a new wing to the municipal hospital. The costs and benefits are the same in most categories, but the city financial manager decided that the following estimates should be considered to determine which design to recommend at the city council meeting next week.

	Design 1	Design 2
Construction cost, $	10,000,000	15,000,000
Building maintenance cost, $/year	35,000	55,000
Patient benefits, $/year	800,000	1,050,000

The patient benefit is an estimate of the amount paid by an insurance carrier, not the patient, to occupy a hospital room with the features included in the design of each room. The discount rate is 5% per year, and the life of the addition is estimated at 30 years.

a. Use conventional B/C ratio analysis to select design 1 or 2.
b. Once the two designs were publicized, the privately owned hospital in the adjacent city of Forest Glen lodged a complaint that design 1 will reduce its own municipal hospital's income by an estimated $600,000 per year because some of the day-surgery features of design 1 duplicate its services. Subsequently, the Garden Ridge merchants' association argued that design 2 could reduce its annual revenue by an estimated $400,000 because it will eliminate an entire parking lot used for short-term parking. The city financial manager stated that these concerns would be entered into the evaluation. Redo the B/C analysis to determine if the economic decision is still the same.

Solution

a. Apply the incremental B/C procedure with no disbenefits included and direct benefits estimated.

1. Since most of the cash flows are already annualized, ΔB/C is based on AW values. The AW of costs is the sum of construction and maintenance costs.

$$AW_1 = 10,000,000(A/P,5\%,30) + 35,000 = \$685,500$$
$$AW_2 = 15,000,000(A/P,5\%,30) + 55,000 = \$1,030,750$$

2. Since the alternatives have direct benefits estimated, the DN option is added as the first alternative with AW of costs and benefits of $0. The comparison order is DN, 1, 2.
3. The comparison 1–to–DN has incremental costs and benefits exactly equal to those of alternative 1.
4. Calculate the incremental B/C ratio.

$$\Delta B/C = 800,000/685,500 = 1.17$$

5. Since $1.17 > 1.0$, design 1 is the survivor over DN.

The comparison continues for 2-to-1 using incremental AW values.

$$\Delta B = 1,050,000 - 800,000 = \$250,000$$
$$\Delta C = 1,030,750 - 685,500 = \$345,250$$
$$\Delta B/C = 250,000/345,250 = 0.72$$

Since $0.72 < 1.0$, design 2 is eliminated, and design 1 is the selection for the construction bid.

b. The revenue loss estimates are considered disbenefits. Since the disbenefits of design 2 are $200,000 less than those of 1, this positive difference is added to the $250,000 incremental benefits of 2 to give it a total ΔB of $450,000. Now,

$$\Delta B/C = \frac{\$450,000}{\$345,250} = 1.30$$

Design 2 is now favored. The inclusion of disbenefits reversed the decision.

EXAMPLE 7.4

The Economic Development Corporation (EDC) for the city of Bahia, California, and Moderna County is operated as a not-for-profit corporation. It is seeking a developer that will place a major water park in the city or county area. Financial incentives will be awarded. In response to a request for proposal (RFP) to the major water park developers in the country, four proposals have been received. Larger and more intricate water rides and increased size of the park will attract more customers, thus different levels of initial incentives are requested in the proposals.

Approved and in-place economic incentive guidelines allow entertainment industry prospects to receive up to $1 million cash as a first-year incentive award and 10% of this amount each year for 8 years in property tax reduction. Each proposal includes a provision that residents of the city or county will benefit from reduced entrance (usage) fees when using the park. This fee reduction will be in effect as long as the property tax reduction incentive continues. The EDC has estimated the annual total entrance fees with the reduction included for local residents. Also, EDC estimated the benefits of extra sales tax revenue. These estimates and the costs for the initial incentive and annual 10% tax reduction are summarized in the top section of Table 7.1.

Perform an incremental B/C study to determine which park proposal is the best economically. The discount rate is 7% per year.

Solution

The viewpoint taken for the economic analysis is that of a resident of the city or county. The first-year cash incentives and annual tax reduction incentives are real costs to the residents. Benefits are derived from two components: the decreased entrance fee estimates and the increased sales tax receipts. These will benefit each citizen indirectly through the increase in money available to those who use the park and through the city and county budgets where sales tax receipts are deposited. Since benefits must be calculated indirectly from

TABLE 7.1 Estimates of Costs and Benefits, and the Incremental B/C Analysis for Four Water Park Proposals, Example 7.4

	Proposal 1	Proposal 2	Proposal 3	Proposal 4
Initial incentive, $	250,000	350,000	500,000	800,000
Tax incentive cost, $/year	25,000	35,000	50,000	80,000
Resident entrance fees, $/year	500,000	450,000	425,000	250,000
Extra sales taxes, $/year	310,000	320,000	320,000	340,000
Study period, years	8	8	8	8
AW of total costs, $/year	66,867	93,614	133,735	213,976
Alternatives compared		2-to-1	3-to-2	4-to-2
Incremental costs ΔC, $/year		26,747	40,120	120,362
Entrance fee reduction, $/year		50,000	25,000	200,000
Extra sales tax, $/year		10,000	0	20,000
Incremental benefits ΔB, $/year		60,000	25,000	220,000
Incremental B/C ratio		2.24	0.62	1.83
Increment justified?		Yes	No	Yes
Alternative selected		2	2	4

the two components mentioned, the alternatives are classified as usage cost estimates.

Table 7.1 includes the results of applying an AW-based incremental B/C procedure.

1. For each alternative, the capital recovery amount over 8 years is determined and added to the annual property tax incentive cost. For proposal #1,

$$\text{AW of total costs} = \text{initial incentive } (A/P,7\%,8) + \text{tax cost}$$
$$= \$250,000 \, (A/P,7\%,8) + 25,000 = \$66,867$$

2. The usage-cost alternatives are ordered by the AW of total costs in Table 7.1.
3. Table 7.1 shows incremental costs. For the 2-to-1 comparison,

$$\Delta C = \$93,614 - 66,867 = \$26,747$$

Incremental benefits for an alternative are the sum of the resident entrance fees compared to those of the next-lower-cost alternative, plus the increase in sales tax receipts over those of the next-lower-cost alternative. Thus, the benefits are determined incrementally for each pair of alternatives. For the 2-to-1 comparison, resident entrance fees decrease by \$50,000 per year and the sales tax receipts increase by \$10,000. The total benefit is the sum, $\Delta B = \$60,000$ per year.

4. For the 2-to-1 comparison, Equation [7.5] results in

$$\Delta B/C = \$60,000/\$26,747 = 2.24$$

5. Alternative #2 is clearly justified. Alternative #1 is eliminated.
6. This process is repeated for the 3-to-2 comparison, which has $\Delta B/C < 1.0$ because the incremental benefits are substantially less than the increase in costs. Proposal #3 is eliminated, and the 4-to-2 comparison results in

$$\Delta B = 200,000 + 20,000 = \$220,000$$
$$\Delta C = 213,976 - 93,614 = \$120,362$$
$$\Delta B/C = \$220,000/\$120,362 = 1.83$$

Since $\Delta B/C > 1.0$, proposal #4 survives as the one remaining alternative.

When the lives of alternatives are so long that they can be considered infinite, the capitalized cost is used to calculate the equivalent PW or AW values. Equation [5.5], $\text{AW} = \text{PW}(i)$, determines the equivalent AW values in the incremental B/C analysis.

If two or more *independent projects* are evaluated using B/C analysis and there is no budget limitation, no incremental comparison is necessary. The only comparison is between each project separately with the do-nothing alternative. The project B/C values are calculated, and those with B/C ≥ 1.0 are accepted.

EXAMPLE 7.5   The Army Corps of Engineers wants to construct a dam on a flood-prone river. The estimated construction cost and average annual dollar benefits are listed below. A 6% per year rate applies and dam life is infinite for analysis purposes. (*a*) Select the one best location using the B/C method. (*b*) If the sites are now considered independent projects, which sites are acceptable?

Site	Construction Cost, $ millions	Annual Benefits, $
A	6	350,000
B	8	420,000
C	3	125,000
D	10	400,000
E	5	350,000
F	11	700,000

Solution

a. The capitalized cost AW = PW(*i*) is used to obtain AW values of the construction cost, as shown in the first row of Table 7.2. Since benefits are estimated directly, initial comparison with DN is necessary. For the analysis, sites are ordered by increasing AW-of-cost values. This is DN, C, E, A, B, D, and F. The analysis between the ordered mutually exclusive alternatives is detailed in the lower portion of Table 7.2. Since only site E is incrementally justified, it is selected.

b. The dam site proposals are now independent projects. The site B/C ratio is used to select from none to all six sites. In Table 7.2, third row, B/C > 1.0 for sites E and F only; they are acceptable.

TABLE 7.2   Use of Incremental B/C Ratio Analysis for Example 7.5 (Values in $1000)

	DN	C	E	A	B	D	F
AW of cost, $/year	0	180	300	360	480	600	660
Annual benefits, $/year	0	125	350	350	420	400	700
Site B/C	—	0.69	1.17	0.97	0.88	0.67	1.06
Comparison		C-to-DN	E-to-DN	A-to-E	B-to-E	D-to-E	F-to-E
Δ Annual cost, $/year		180	300	60	180	300	360
Δ Annual benefits, $/year		125	350	0	70	50	350
Δ B/C ratio		0.69	1.17	—	0.39	0.17	0.97
Increment justified?		No	Yes	No	No	No	No
Site selected		DN	E	E	E	E	E

7.4 USING SPREADSHEETS FOR B/C ANALYSIS

Formatting a spreadsheet to apply the incremental B/C procedure for mutually exclusive alternatives is basically the same as that for an incremental ROR analysis (Section 6.8). Once all estimates are expressed in terms of either PW, AW, or FW equivalents using spreadsheet functions, the alternatives are *ordered by increasing total equivalent cost.* Then, the incremental B/C ratios from Equation [7.5] assist in selecting the one best alternative.

A significant new application of nanotechnology is the use of thin-film solar panels applied to houses to reduce the dependency on fossil-fuel generated electrical energy. A community of 400 new all-electric public housing units will utilize the technology as anticipated proof that significant reductions in overall utility costs can be attained over the expected 15-year life of the housing. Table 7.3 details the three bids received. Also included are estimated annual electricity usage costs for the community with the panels in use, and the PW of backup systems required in case of panel failures. Use a spreadsheet, the conventional B/C ratio, and $i = 5\%$ per year to select the best bid.

EXAMPLE 7.6

Solution

Benefits are estimated from utility bill differences, so the alternatives are classified as usage cost. Initial comparison to DN is unnecessary. The step-by-step procedure (Section 7.3) is included in the Figure 7.1 solution.

1. Base the analysis on present worth. With the use of PV functions to determine PW of construction costs and utility bills (rows 9 and 10), all benefit (grey-screened rows), disbenefit (white rows), and cost (green-screened rows) terms are prepared. (Note the inclusion of the minus sign on the PV function to ensure that cost terms remain positive.)

TABLE 7.3   Estimates for Alternatives Using Nanocrystal-Layered Thin-Film Solar Panel Technology for Energy, Example 7.6

Bidder	Geyser, Inc.	Harris Corp.	Quimbley, LLP
Bid identification	G	H	Q
PW of initial cost, $	2,400,000	1,850,000	6,150,000
Annual maintenance, $/year	500,000	650,000	450,000
Annual utility bill, $/year	960,000	1,000,000	550,000
PW of backup systems, $	650,000	750,000	950,000

	A	B	C	D	E	F	G	H
1								
2			Geyser	Harris	Quimbley			
3		**Estimate**	**G**	**H**	**Q**			
4	PW of initial cost, $	**Cost**	2,400,000	1,850,000	6,150,000			
5	Annual maintenance, $/year	**Cost**	500,000	650,000	450,000			
6	Annual utility bill, $/year	**Benefit (usage cost)**	960,000	1,000,000	550,000			
7	PW of back-up system, $	**Disbenefit**	650,000	750,000	950,000			
8								
9	PW of total costs, $	**Cost**	7,589,829	8,596,778	10,820,846			
10	PW of utility bills, $	**Benefit (implied)**	9,964,472	10,379,658	5,708,812			
11								
12	Incremental comparison			H-to-G	Q-to-G			
13	PW of ΔB, $			-415,186	4,255,660			
14	PW of ΔD, $			100,000	300,000			
15	PW of ΔC, $			1,006,949	3,231,017			
16	$\Delta(B\text{-}D)/C$			-0.51	1.22			
17	Increment justified?			No	Yes			
18	Selected bid			G	Q			
19								
20								
21								
22								

Annotations:
- Estimate of costs, benefits, disbenefits
- = −PV(5%,15,500000) + 2400000
- = − (D10-C10)
- = − (E10-C10)
- = E9-C9
- = (E13-E14)/E15
- Bidder Q is the best economically

FIGURE 7.1 Spreadsheet evaluation of multiple alternatives using incremental B/C analysis, Example 7.6.

2. Based on PW of costs (row 9), the order of evaluation is G, H, and finally Q. It is important to realize that even though H has a smaller initial cost, G has a smaller equivalent total cost.
3. The first comparison is H-to-G. The form of Equation [7.4] helps determine ΔB with the minus sign on utility bill costs changed to a plus for the $\Delta B/C$ computation.

$$\Delta B = \text{utility bills for H} - \text{utility bills for G}$$
$$= -(10,379,658 - 9,964,472)$$
$$= \$-415,186$$

In this case, $\Delta B < 0$ due to the larger PW of utility bills for H. There is no need to complete the comparison, since $\Delta(B - D)/C$ will be negative.
4. The value $\Delta(B - D)/C = -0.51$ is determined for illustration only.
5. G is the survivor; now compare Q-to-G.

The comparison in column E results in $\Delta(B - D)/C = 1.22$, indicating that the Quimbley bid is incrementally justified and is the selected bid. This is the most expensive bid, especially in initial cost, but the savings in utility bills help make it the most economic over the 15-year planning horizon.

SUMMARY

The benefit/cost method is used primarily to evaluate alternatives in the public sector. When one is comparing mutually exclusive alternatives, the incremental B/C ratio must be greater than or equal to 1.0 for the incremental equivalent total cost to be economically justified. The PW, AW, or FW of the initial costs and estimated benefits can be used to perform an incremental B/C analysis.

Public sector economics are substantially different from those of the private sector. For public sector projects, the initial costs are usually large, the expected life is long (25, 35, or more years), and the sources for capital are usually a combination of taxes levied on the citizenry, user fees, bond issues, and private lenders. *It is very difficult to make accurate estimates of benefits and disbenefits for a public sector project.* The discount rates in the public sector are lower than those for corporate projects.

PROBLEMS

B/C Considerations

7.1 In conducting a B/C analysis, why is it best to take a limited viewpoint in determining benefits and disbenefits?

7.2 Identify the following as primarily public or private sector projects.
   a. Bridge across Ohio River
   b. Coal mine
   c. Baja 1000 race team
   d. Consulting engineering firm
   e. County courthouse
   f. Flood control project
   g. Endangered species designation
   h. Freeway lighting
   i. Antarctic cruise
   j. Crop dusting

7.3 Identify the following as primarily public or private sector undertakings: eBay, farmer's market, state police department, car racing facility, Social Security, EMS, ATM, travel agency, amusement park, gambling casino, and swap meet.

7.4 What are the primary financial responsibilities of a contractor and the government when operating under a DBOMF contract?

7.5 Identify the following funding sources as primarily public or private.
   a. Municipal bonds
   b. Retained earnings
   c. Sales taxes
   d. Automobile license fees
   e. Bank loans
   f. Savings accounts
   g. Engineer's retirement plan
   h. State fishing license revenues
   i. Entrance fees to Disneyland
   j. State park entrance fees

7.6 Identify the following as primarily private or public sector characteristics:
   a. Large investment
   b. No profits
   c. Funding from fees
   d. MARR-based selection criteria
   e. Low interest rate
   f. Short project life estimate
   g. Disbenefits

Single Project B/C

7.7 Calculate the conventional B/C ratio for a county government project that is projected to have the following cash flows: costs of $2,000,000 per year; benefits of $2,740,000 per year; disbenefits of $380,000 per year.

7.8 The Hawaii Department of Transportation has planned a bypass loop that is expected to cost $9,000,000 and save motorists $820,000 per year in gasoline and other automobile-related expenses. However, local businesses will suffer sales losses estimated at $135,000 each year. (*a*) Calculate the conventional B/C ratio using a discount rate of 6% per year and a 20-year study period. (*b*) Is the project economically justified if disbenefits are considered? If disbenefits are not considered?

7.9 A southwestern city that has 170,000 households is required to install treatment systems for the removal of arsenic from drinking water. The annual cost is projected to be $50 per household per year. Assume that one life will be saved every three years as a result of the arsenic removal system and that the EPA values a human life at $4.8 million. Use a discount rate of 8% per year and assume the life is saved at the end of each three-year period. Utilize a conventional B/C ratio to determine if the project is economically justified.

7.10 The following estimates (in $1000 units) have been developed for a new security system at Chicago's O'Hare Airport.

First cost, $	13,000
AW of benefits, $/year	3,800
FW (in year 20) of disbenefits, $	6,750
M&O costs, $/year	400
Life of project, years	20

a. Calculate the conventional B/C ratio at a discount rate of 10% per year.

b. Determine the minimum first cost necessary to make the project economically *unjustified*.

7.11 An Army Corps of Engineers project for improving navigation on the Ohio River will have an initial cost of $6,500,000 and annual maintenance of $130,000. Benefits for barges and paddle wheel touring boats are estimated at $820,000 per year. If the project is assumed to be permanent, use the conventional B/C ratio to determine if it is economically justified at 8% per year.

7.12 The sheriff of Los Lunas county along the Arizona-Mexico border asked the county to build a new minimum security detention facility for persons caught while attempting to enter the United States illegally. The construction cost will be $22 million, with annual operating costs of $2.1 million. The new facility will create jobs that produce benefits for many local businesses including realtors, restaurants, etc. The benefits are estimated to be $5 million in years 1 and 2, $2.8 million in year 3, and $1.12 million per year beginning in year 4 and continuing through the 30-year life of the facility. At a discount rate of 8% per year, does the conventional B/C ratio indicate that the project is economically justified?

7.13 The cash flows associated with a public works project in Buffalo, NY, are shown. Calculate the modified B/C ratio at a discount rate of 5% per year.

First cost, $	35,000,000
AW of benefits, $/year	6,500,000
AW of disbenefits, $/year	1,700,000
M&O costs, $/year	900,000
Life of project, years	30

7.14 Determine the B/C ratio for a project that has an infinite life and the following estimates. Use an interest rate of 8% per year.

To the people	To the government
Annual benefits = $180,000/year	First cost = $950,000
Annual disbenefits = $30,000/year	Annual cost = $60,000/year
	Annual savings = $25,000/year

7.15 As part of the rehabilitation of the downtown area of a northwestern Florida city, the Parks and Recreation Department is planning to develop the space below several overpass bridges into basketball, handball, miniature golf, and tennis courts. The initial cost is expected to be $150,000 for improvements which are expected to have a 20-year life. Annual maintenance costs are projected to be $12,000. The department expects 20,000 people per year to use the facilities for an average of 2 hours per person. The value of the recreation has been conservatively set at $0.50 per hour. At a discount rate of 6% per year, does the conventional B/C ratio indicate economic justification?

7.16 The conventional B/C ratio for a flood control project along the Mississippi River was calculated to be 1.3. The benefits were $500,000 per year and the maintenance costs were $200,000 per year. What was the initial cost of the project if a discount rate of 7% per year was used and the project was assumed to have a 50-year life?

7.17 The *modified* B/C ratio for a city-owned hospital heliport project is 1.7. The initial cost is $1 million, annual benefits are $150,000, and the estimated life is 30 years. What is the amount of the annual M&O costs used in the calculation at a discount rate of 6% per year?

7.18 The conventional B/C ratio estimate of 2.1 was reported to the County Commissioners for a proposed mosquito control program. The person who prepared the report stated that the health benefits were estimated to be $400,000 per year, and that disbenefits of $25,000 per year were used in the calculation. She also stated that the costs for chemicals, machinery, maintenance, and labor were estimated at $150,000 per year. However, she forgot to list the cost for initiating the program (trucks, pumps, tanks, etc.). If the project has a 10-year study period and an 8%-per-year discount rate, determine the estimated initial cost.

7.19 The cash flows associated with a Death Valley County arroyo improvement project are as follows: initial cost $650,000; life 20 years; annual maintenance cost $150,000 per year; benefits $600,000 per year; disbenefits $190,000 per year. The discount rate is 6% per year. Determine if the project is justified using (*a*) the conventional B/C ratio, and (*b*) the modified B/C ratio.

7.20 The Parks and Recreation Department of Burkett County estimates that the initial cost of a "bare-bones" permanent river park will be $2.3 million. Annual upkeep costs are estimated at $120,000. Benefits of $340,000 per year and disbenefits of $40,000 per year have also been identified. Using a discount rate of 6% per year, calculate (*a*) the conventional B/C ratio and (*b*) the modified B/C ratio.

7.21 From the following data, calculate the (*a*) conventional and (*b*) modified benefit/cost ratios using a discount rate of 6% per year and a very long (infinite) project life.

	To the People	To the Government
Benefits:	$300,000 now and $100,000 per year thereafter	Costs: $1.5 million now and $200,000 three years from now
Disbenefits:	$40,000 per year	Savings: $70,000 per year

7.22 When red light cameras are installed at high-risk intersections, rear-end collisions go up, but all other types of accidents go down, including those involving pedestrians. Analysis of traffic accidents in a northwestern city revealed that the total number of collisions at photo-controlled intersections decreased from 33 per month to 18. At the same time, the number of traffic tickets issued for red light violations averaged 1100 per month at an average cost to violators of $85 per citation. The cost to install the basic camera system at selected

intersections was $750,000. If the cost of a collision is estimated at $41,000 and traffic ticket costs are considered disbenefits, calculate the B/C ratio for the camera system. Use a discount rate of 0.5% per month and a 3-year study period.

Alternative Comparison Using B/C

7.23 A state agency is considering two mutually exclusive alternatives for upgrading the skills of its technical staff. Alternative 1 involves purchasing software that will reduce the time required to collect background information on each client. The total cost for the purchase, installation, and training associated with the new software is $840,900. The present worth of the benefits from increased efficiency is expected to be $1,020,000. Alternative 2 involves multimedia training to improve the performance of the staff technicians. The total cost to develop, install, and train the technicians will be $1,780,000. The present worth of the benefits from increased performance due to training is expected to be $1,850,000. Use a B/C analysis to determine which alternative, if any, the agency should undertake.

7.24 A project to control flooding from rare, but sometimes heavy, rainfalls in the arid southwest will have the following cash flow estimates. Determine which project should be selected on the basis of a B/C analysis using an interest rate of 8% per year and a 20-year study period.

	Sanitary Sewers	Open Channels
First cost, $ (millions)	26	53
M&O cost, $/year	400,000	30,000
Homeowner cleanup costs, $/year	60,000	0

7.25 In order to safeguard the public health, environment, public beaches, water quality, and economy of south San Diego County, California, and Tijuana, Mexico, federal agencies in the United States and Mexico developed four alternatives for treating wastewater prior to discharge into the ocean. The project will minimize untreated wastewater flows that have caused chronic and substantial pollution in the Tijuana River valley, Tijuana River National Estuarine Research Reserve, coastal areas used for agriculture and public recreation, and areas designated as critical habitat for federal and state-listed endangered species. If the costs and benefits are as shown, which alternative should be selected on the basis of a B/C analysis at 6% per year and a 40-year project period? All monetary amounts are in $ million units.

	Pond System	Expand Plant	Advanced Primary	Partial Secondary
Capital cost, $	58	76	2	48
M&O cost, $/year	5.5	5.3	2.1	4.4
Benefits, $/year	11.1	12.0	2.7	8.3

7.26 The present worth of cash flows associated with developing oceanfront property for commercial use are estimated. Determine which plan, if any, should be selected on the basis of a B/C analysis at 8% per year.

	Present Worth ($1000)	
Plan	Costs	Benefits
A	1400	1246
B	2220	2560
C	4680	4710

7.27 Use the B/C method to compare four mutually exclusive alternatives for recycling plastic bottles. Make any additional calculations necessary to determine which alternative should be selected.

	PW of Total Costs, $ Millions	Overall B/C Ratio	ΔB/C when Compared with Alternative			
Alternative			M	N	O	P
M	10	0.91	—	1.69		0.96
N	21	1.32	1.69	—		
O	44	1.25			—	
P	52	0.95	0.96	0.80	0.08	—

7.28 Two sites for suspension bridges are under consideration for crossing the Allegheny River in Pittsburgh, PA. Use the B/C ratio method at a discount rate of 6% per year to determine which bridge, if either, should be built.

	Site N	Site S
Initial Cost, $	11×10^6	27×10^6
Annual M&O, $/year	100,000	90,000
Benefits, $/year	990,000	2,100,000
Disbenefits, $/year	120,000	300,000
Life, years	∞	∞

7.29 Two alternatives are under consideration for providing energy at a remote research station, one of which must be selected: goethermal and solar. The benefits and costs associated with each alternative are shown. Use the B/C method to determine which should be selected at an interest rate of 6% per year over a 20-year study period.

	Geothermal	Solar
Initial cost, $	200,000	50,000
Annual M&O costs, $/year	31,000	9,000
Annual benefits, $/year	39,500	16,000

7.30 Two relatively inexpensive alternatives are available to reduce potential earthquake damage at a top secret government research site. The cash flow estimates for each alternative are shown. At an interest rate of 8% per year and a 20-year study period, apply the B/C ratio method to select an alternative. Assume the damage costs would occur in the middle of the study period, that is, year 10.

	Alternative 1	Alternative 2
Initial cost, $	600,000	1,100,000
Annual maintenance, $ per year	50,000	90,000
Potential damage costs, $	950,000	350,000

7.31 There are two methods under consideration for improving security at a county jail. Select one based on a B/C analysis at an interest rate of 7% per year and a 10-year study period.

	Extra Cameras (EC)	New Sensors (NS)
First cost, $	38,000	87,000
Annual M&O, $ per year	49,000	74,000
Benefits, $ per year	110,000	130,000
Disbenefits, $ per year	26,000	18,000

7.32 The public utility in a medium-sized city is considering two cash rebate programs to improve water conservation. Program 1, expected to cost an average of $60 per household, offers a rebate of 75% of the purchase and installation costs of an ultra low-flush toilet ($100 maximum). This program is projected to achieve a 5% reduction in overall household water use over a 5-year evaluation period. This will benefit the citizenry to the extent of $1.25 per household per month.

Program 2 is the replacement of turf grass with xeriscape landscaping. This is expected to cost $500 per household, but it will result in an average reduced water cost estimated at $8 per household per month. At a discount rate of 0.5% per month, use the B/C method to determine which program(s) the utility should undertake if the programs are (a) mutually exclusive, and (b) independent.

7.33 A consulting engineer is currently evaluating 4 projects for the U.S. government. The present worth values of the costs, benefits, disbenefits, and savings are shown. Assuming the discount rate is 10% per year compounded continuously, determine which projects, if any, should be selected if they are (a) independent, and (b) mutually exclusive.

	Fair	Good	Better	Best
PW of costs, $	10,000	8,000	20,000	14,000
PW of benefits, $	15,000	11,000	25,000	42,000
PW of disbenefits, $	6,000	1,000	20,000	31,000
PW of cost savings, $	1,500	2,000	16,000	3,000

7.34 From the data below for projects regarding campgrounds and lodging at a national park, determine which one, if any, should be selected from the 6 mutually exclusive projects. If the proper comparisons have not been made, state which one(s) should be performed.

	Project Identification					
	G	H	I	J	K	L
Cost, $	20,000	45,000	50,000	35,000	85,000	70,000
Project B/C	1.15	0.89	1.10	1.11	0.94	1.06

Selected incremental B/C ratios:

Comparison	ΔB/C	Comparison	ΔB/C
H to G	0.68	L to H	1.36
I to G	0.73	K to J	0.82
J to H	0.10	L to J	1.00
I to J	1.07	K to L	0.40
J to G	1.07	L to G	1.02
K to H	1.00		

7.35 The 4 mutually exclusive alternatives shown below are compared by the incremental B/C method. What alternative, if any, should be selected?

	First Cost,	Overall	ΔB/C when Compared with Alternative			
Alternative	$ Millions	B/C Ratio	X	Y	Z	ZZ
X	20	0.75	—			
Y	30	1.07	1.70	—		
Z	50	1.20	1.50	1.40	—	
ZZ	90	1.11	1.21	1.13	1.00	—

ADDITIONAL PROBLEMS AND FE EXAM REVIEW QUESTIONS

7.36 All of the following are primarily associated with public sector projects, except:
  a. profits
  b. taxes
  c. disbenefits
  d. infinite life

7.37 All of the following cash flows should be identified as benefits, except:
  a. longer tire life because of smooth pavement
  b. $200,000 annual income to local businesses because of tourism created by a water reservoir
  c. expenditure of $20 million for construction of a highway
  d. fewer highway accidents because of improved lighting.

7.38 In a modified B/C ratio:
  a. disbenefits and M&O costs are subtracted from benefits
  b. disbenefits are subtracted from benefits, and M&O costs are added to costs
  c. disbenefits and M&O costs are added to costs
  d. disbenefits are added to costs, and M&O costs are subtracted from benefits

7.39. If two mutually exclusive alternatives have B/C ratios of 1.4, and 1.5 for the lower and higher-cost ones, respectively,
  a. the B/C ratio on the increment between them is equal to 1.5
  b. the B/C ratio on the increment between them is between 1.4 and 1.5
  c. the B/C ratio on the increment between them is less than 1.5
  d. the higher-cost alternative is definitely the better one

7.40 From the PW, AW, and FW values below, the conventional B/C ratio is closest to:
  a. 1.27
  b. 1.33
  c. 1.54
  d. 2.76

	PW, $	AW, $/year	FW, $
First cost	100,000	16,275	259,370
M&O cost	68,798	11,197	178,441
Benefits	245,784	40,000	637,496
Disbenefits	30,723	5,000	79,687

7.41 The first cost of grading and spreading gravel on a short rural road is expected to be $700,000. The road will have to be maintained at a cost of $25,000 per year. Even though the new road is not very smooth, it allows access to an area that previously could be reached only with off-road vehicles. This improved accessibility has increased the property values along the road from $400,000 to $700,000. The conventional B/C ratio at a discount rate of 6% per year for a 10-year study period is closest to:
  a. 2.5
  b. 3.3
  c. 4.1
  d. 5.8

7.42 A permanent flood control dam is expected to have an initial cost of $2.8 million and an annual upkeep cost of $20,000. In addition, minor reconstruction will be required every 5 years at a cost of $200,000. As a result of the dam, flood damage will be reduced by an average of $180,000 per year. Using an interest rate of 6% per year, the conventional B/C ratio will be closest to:
  a. 0.46
  b. 0.81
  c. 0.97
  d. 1.06

7.43 If the two *independent* projects shown are evaluated using an interest rate of 10% per year and the B/C ratio method, the decision is to:
  a. select do-nothing
  b. select X
  c. select Y
  d. select X and Y

	Project X	Project Y
Annualized first cost, $/year	60,000	90,000
Annual M&O cost, $/year	45,000	35,000
Annual benefits, $/year	110,000	150,000
Annual disbenefits, $/year	20,000	45,000
Life, years	∞	∞

7.44 The four *mutually exclusive* alternatives shown on the next page are compared using the B/C method. The alternative to select is:
  a. J
  b. K
  c. L
  d. M

Alternative	Cost, $ millions	B/C Ratio Compared to DN	ΔB/C when Compared with Alternative			
			J	K	L	M
J	20	1.1	—			
K	25	0.96	0.40	—		
L	33	1.22	1.42	2.14	—	
M	45	0.89	0.72	0.80	0.08	—

The following information is used for Problems 7.45 through 7.47

The Corps of Engineers compiled the following data to determine which one of two flood control dams should be constructed in a flood-prone residential area.

	Mountain	Valley
Flood damage, $/year	220,000	140,000
Disbenefits, $/year	30,000	10,000
Costs, $/year	300,000	450,000

7.45 In conducting an incremental B/C analysis of this data:
   a. the DN alternative is not an option
   b. the DN alternative is an option
   c. there is not enough information given to know if DN is an option or not
   d. DN is an option only if the alternatives are mutually exclusive

7.46 The *conventional* ΔB/C ratio between alternatives Mountain and Valley is closest to:
   a. 0.33
   b. 0.40
   c. 0.53
   d. 0.73

7.47 If the flood damage estimates for both alternatives were reduced to zero and replaced with benefits of $310,000 per year and $470,000 per year for the two alternatives, respectively, the B/C ratio for the Valley alternative is closest to:
   a. 0.93
   b. 1.02
   c. 1.23
   d. 1.56

7.48 If benefits are $10,000 per year forever, starting in year 1, and costs are $50,000 at time zero and $50,000 at the end of year 2, the B/C ratio at $i = 10\%$ per year is closest to:
   a. 1.1
   b. 1.8
   c. 0.90
   d. less than 0.75

7.49 If the first cost of a permanent national monument is $2 million with annual benefits and disbenefits estimated to be $360,000 and $42,000, respectively, the B/C ratio at 6% per year is closest to:
   a. 0.16
   b. 0.88
   c. 1.73
   d. 2.65

Chapter

Breakeven, Sensitivity, and Payback Analysis

Randy Allbritton/Getty Images

This chapter covers several related topics that assist in evaluating the effects of varying estimated values for one or more parameters present in an economic study. Since all estimates are for the future, it is important to understand which parameter(s) may make a significant impact on the economic justification of a project.

Breakeven analysis is performed for one project or two alternatives. For a single project, it determines a parameter value that makes revenue equal cost. Two alternatives break even when they are equally acceptable based upon a calculated value of one parameter common to both alternatives. *Make-or-buy decisions*, also called inhouse-outsource decisions, for most subcontractor services, manufactured components, or international contracts, are routinely based upon the outcome of a breakeven analysis.

Sensitivity analysis is a technique that determines the impact on a measure of worth—most commonly PW, AW, ROR, or B/C—caused by varying estimated values. In short, it answers the question "What if?" It can be applied to one project, to two or more alternatives, and to one or more parameters. Since every study depends upon good estimation of costs (and revenues), sensitivity analysis is a vital tool to learn and utilize.

Payback period is a good technique to initially determine if a project may be financially acceptable. The technique determines the amount of time necessary for a proposed project to do two things—generate enough net cash flow to recover its initial investment (first cost) and to meet or exceed the required MARR. As discussed in the chapter, there are a couple of drawbacks to payback analysis that must be remembered when the technique is used.

Purpose: Determine the value of a parameter to break even or to pay back the initial investment with the time value of money considered.

LEARNING OUTCOMES

Breakeven point	1. Determine the breakeven value for a single project.
Two alternative breakeven	2. Calculate the breakeven value between two alternatives and use it to select one alternative.
Sensitivity to estimates	3. Evaluate the sensitivity of one or more parameters to variation in the estimates.
Sensitivity of alternatives	4. Evaluate the sensitivity of mutually exclusive alternative selection to variation in estimates.
Payback period	5. Calculate the payback period for a project at $i = 0\%$ and $i > 0\%$, and state the cautions on using payback analysis.
Spreadsheets	6. Use a spreadsheet to perform sensitivity and breakeven analyses.

8.1 BREAKEVEN ANALYSIS FOR A SINGLE PROJECT

Breakeven analysis determines the value of a parameter or decision variable that makes two relations equal. For example, breakeven analysis can determine the required years of use to recover the initial investment and annual operating costs. There are many forms of breakeven analysis; some equate PW or AW equivalence relations, some involve equating revenue and cost relations, others may equate demand and supply relations. However, they all have a common approach, that is, to equate two relations, or to set their difference equal to zero, and solve for the breakeven value of one variable that makes the equation true.

The need to determine the *breakeven value of a decision variable* without including the time value of money is common. For example, the variable may be a design capacity that will minimize cost, or the sales volume necessary to cover costs, or the cost of fuel to maximize revenue from electricity generation.

Figure 8.1a presents different shapes of a revenue relation identified as R. A linear revenue relation is commonly assumed, but a nonlinear relation is often more realistic with increasing per unit revenue for larger volumes (curve 1), or decreasing per unit revenue (curve 2).

FIGURE 8.1
Linear and nonlinear revenue and cost relations.

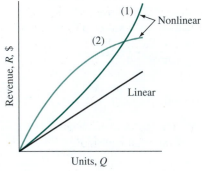

(*a*) Revenue relations—linear, increasing (1)
and decreasing (2) per unit

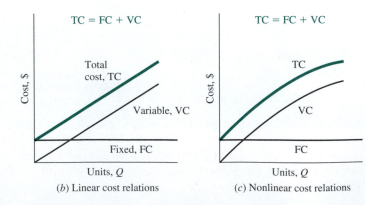

(*b*) Linear cost relations   (*c*) Nonlinear cost relations

Costs, which may be linear or nonlinear, usually include two components—fixed and variable—as indicated in Figures 8.1*b* and *c*.

> **Fixed cost (FC).** Includes costs such as buildings, insurance, fixed overhead, some minimum level of labor, equipment capital recovery, and information systems.
>
> **Variable cost (VC).** Includes costs such as direct labor, subcontractors, materials, indirect costs, marketing, advertisement, legal, and warranty.

The fixed cost component is essentially constant for all values of the variable, so it does not vary significantly over a wide range of operating parameters. Even if no output is produced, fixed costs are incurred at some threshold level. (Of course, this situation cannot last long before the operation must shut down.) Fixed costs are reduced through improved equipment, information systems and workforce utilization, less costly fringe benefit packages, subcontracting specific functions, and so on.

A simple VC relation is vQ, where v is the variable cost per unit and Q is the quantity. Variable costs change with output level, workforce size, and many other parameters. It is usually possible to decrease variable costs through improvements in design, efficiency, automation, materials, quality, safety, and sales volume.

When FC and VC are added, they form the total cost relation TC. Figure 8.1*b* illustrates linear fixed and variable costs. Figure 8.1*c* shows TC for a nonlinear VC in which unit variable costs decrease as Q rises.

At some value of Q, the revenue and total cost relations will intersect to identify the breakeven point Q_{BE} (Figure 8.2*a*). If $Q > Q_{BE}$, there is a profit; but if $Q < Q_{BE}$, there is a loss. For linear R and TC, the greater the quantity, the larger the profit. Profit is calculated as

$$\textbf{Profit} = \textbf{revenue} - \textbf{total cost} = \textbf{\textit{R}} - \textbf{TC} \qquad \textbf{[8.1]}$$

A closed-form solution for Q_{BE} may be derived when revenue and total cost are linear functions of Q by equating the relations, indicating a profit of zero.

$$R = TC$$
$$rQ = FC + VC = FC + vQ$$

where r = revenue per unit

         v = variable cost per unit

Solve for Q to obtain the breakeven quantity.

$$Q_{BE} = \frac{FC}{r - v} \qquad \textbf{[8.2]}$$

The breakeven graph is an important management tool because it is easy to understand. For example, if the variable cost per unit is reduced, the linear TC line has a smaller slope (Figure 8.2*b*), and the breakeven point decreases. This is an advantage because the smaller the value of Q_{BE}, the greater the profit for a given amount of revenue.

FIGURE 8.2
(a) Breakeven point and (b) effect on breakeven point when the variable cost per unit is reduced

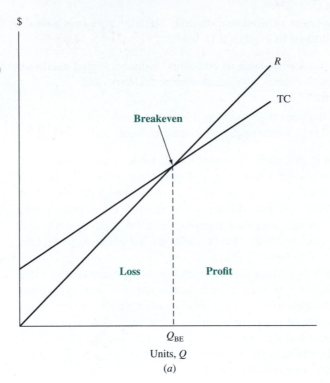

(a)

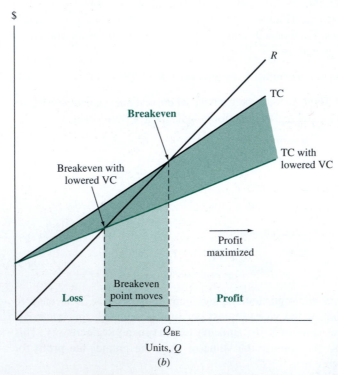

(b)

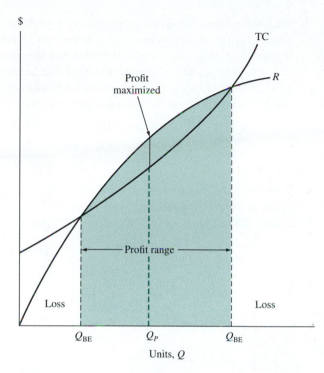

FIGURE 8.3
Breakeven points and maximum profit point for a nonlinear analysis.

If nonlinear R or TC models are used, there may be more than one breakeven point. Figure 8.3 presents this situation for two breakeven points. The maximum profit occurs at Q_P where the distance between the R and TC curves is greatest.

EXAMPLE 8.1

Nicholea Water LLC dispenses its product Nature's Pure Water via vending machines with most current locations at food markets and pharmacy or chemist stores. The average monthly fixed cost per site is $900, while each gallon costs 18¢ to purify and sells for 30¢. (*a*) Determine the monthly sales volume needed to break even. (*b*) Nicholea's president is negotiating for a sole-source contract with a municipal government where several sites will dispense larger amounts. The fixed cost and purification costs will be the same, but the sales price per gallon will be 30¢ for the first 5000 gallons per month and 20¢ for all above this threshold level. Determine the monthly breakeven volume at each site.

Solution

a. Use Equation [8.2] to determine the monthly breakeven quantity of 7500 gallons.

$$Q_{BE} = \frac{900}{0.30 - 0.18} = 7500$$

b. At 5000 gallons, the profit is negative at $\$-300$, as determined by Equation [8.1]. The revenue curve has a lower slope above this threshold gallonage level. Since Q_{BE} can't be determined directly from Equation [8.2], it is found by equating revenue and total cost relations with the threshold level of 5000 included. If Q_U is termed the breakeven quantity above threshold, the equated R and TC relations are

$$0.30(5000) + 0.20(Q_U) = 900 + 0.18(5000 + Q_U)$$

$$Q_U = \frac{900 + 900 - 1500}{0.20 - 0.18} = 15{,}000$$

Therefore, the required volume per site is 20,000 gallons per month, the point at which revenue and total cost break even at \$4500. Figure 8.4 details the relations and points.

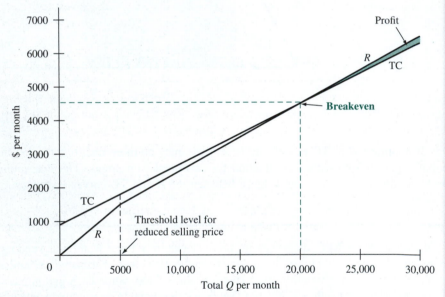

FIGURE 8.4   Breakeven graph with a volume discount for sales, Example 8.1*b*.

In some circumstances, breakeven analysis of revenue and cost is better if performed on a per unit basis. The revenue per unit is $R/Q = r$, and the TC relation is divided by Q to obtain cost per unit, also termed *average cost per unit* C_u.

$$C_u = \frac{TC}{Q} = \frac{FC + vQ}{Q} = \frac{FC}{Q} + v \qquad [8.3]$$

The relation $R/Q = TC/Q$ is solved for Q. The result for Q_{BE} is the same as Equation [8.2].

When an engineering economy study for a single project includes a P, F, A, i, or n value that cannot be reliably estimated, a breakeven quantity for one of the parameters can be determined by setting an equivalence relation for PW, FW, or AW equal to zero and solving for the unknown variable, as presented in Chapter 6 to determine the breakeven rate of return i^*. As an illustration, assume a professional musician who specializes in jazz fests pays $20,000 for new equipment that will have a 10% salvage value after five years. Further, assume that his costs are $100 per day and that he charges $300 per day for an appearance. We can find X, the number of "gigs" per year to breakeven at an interest rate of 5% per year by setting the AW-based relation equal to zero and solving for X.

$$0 = -20{,}000(A/P,5\%,5) + 0.10(20{,}000)(A/F,5\%,5) - 100X + 300X$$
$$= -20{,}000(0.23097) + 2000(0.18097) + 200X$$
$$200X = 4257.46$$
$$X = 21.3 \text{ gigs/year}$$

This means, for example, that 22 gigs per year will recover the investment with slightly more than a 5% per year rate of return.

8.2 BREAKEVEN ANALYSIS BETWEEN TWO ALTERNATIVES

Breakeven analysis is an excellent technique with which to determine the value of a parameter that is common to two alternatives. The parameter can be the interest rate, capacity per year, first cost, annual operating cost, or any parameter. We have already performed breakeven analysis between alternatives in Chapter 6 for the incremental ROR value (Δi^*).

Podcast available

Breakeven analysis usually involves revenue or cost variables common to both alternatives. Figure 8.5 shows two alternatives with linear total cost (TC) relations. The fixed cost of alternative 2 is greater than that of alternative 1. However, alternative 2 has a smaller variable cost, as indicated by its lower slope. If the number of units of the common variable is greater than the breakeven amount, alternative 2 is selected, since the total cost will be lower. Conversely, an anticipated level of operation below the breakeven point favors alternative 1.

It is common to find the breakeven value by *equating PW or AW equivalence relations*. The AW is preferred when the variable units are expressed on a yearly

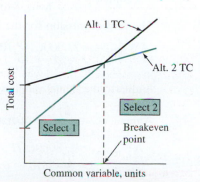

FIGURE 8.5
Breakeven between two alternatives with linear cost relations.

basis or when alternatives have unequal lives. The following steps determine the breakeven point of the common variable.

1. Define the common variable and its dimensional units.
2. Use AW or PW analysis to express the total cost of each alternative as a function of the common variable.
3. Equate the two relations and solve for the breakeven value.
4. The selection guideline is based on the expected level of the common variable and the size of the variable cost. Refer to Figure 8.5.

   Expected level < breakeven value: Select alternative with higher variable cost   (larger slope on TC line)
   Expected level > breakeven value: Select alternative with lower variable cost   (smaller slope on TC line)

EXAMPLE 8.2   A small aerospace company is evaluating two alternatives: the purchase of an automatic feed machine and a manual feed machine for a finishing process. The auto-feed machine has an initial cost of $23,000, an estimated salvage value of $4000, and a predicted life of 10 years. One person will operate the machine at a rate of $24 per hour. The expected output is 8 tons per hour. Annual maintenance and operating cost is expected to be $3500.

   The alternative manual feed machine has a first cost of $8000, no expected salvage value, a 5-year life, and an output of 6 tons per hour. However, three workers will be required at $12 per hour each. The machine will have an annual maintenance and operation cost of $1500. All projects are expected to generate a return of 10% per year. How many tons per year must be finished in order to justify the higher purchase cost of the auto-feed machine?

Solution

Use the steps above to calculate the breakeven point between the two alternatives.

1. Let x represent the number of tons per year.
2. For the auto-feed machine the annual variable cost is

$$\text{Annual VC} = \frac{\$24}{\text{hour}} \; \frac{1 \text{ hour}}{8 \text{ tons}} \; \frac{x \text{ tons}}{\text{year}} = 3x$$

   The AW expression for the auto-feed machine is

$$\text{AW}_{\text{auto}} = -23{,}000(A/P,10\%,10) + 4000(A/F,10\%,10) - 3500 - 3x$$
$$= \$-6992 - 3x$$

   Similarly, the annual variable cost and AW for the manual feed machine are

$$\text{Annual VC} = \frac{\$12}{\text{hour}} (3 \text{ operators}) \frac{1 \text{ hour}}{6 \text{ tons}} \frac{x \text{ tons}}{\text{year}} = 6x$$
$$\text{AW}_{\text{manual}} = -8000(A/P,10\%,5) - 1500 - 6x$$
$$= \$-3610 - 6x$$

3. Equate the two cost relations and solve for x.

$$\text{AW}_{auto} = \text{AW}_{manual}$$
$$-6992 - 3x = -3610 - 6x$$
$$x = 1127 \text{ tons per year}$$

4. If the output is expected to exceed 1127 tons per year, purchase the auto-feed machine, since its VC slope of 3 is smaller.

The breakeven approach is commonly used for make-or-buy decisions. The alternative to buy (or subcontract) often has no fixed cost, but has a larger variable cost. Where the two cost relations cross is the make-buy decision quantity. Amounts above this indicate that the item should be made, not outsourced.

Guardian is a national manufacturing company of home health care appliances. It is faced with a make-or-buy decision. A newly engineered lift can be installed in a car trunk to raise and lower a wheelchair. The steel arm of the lift can be purchased for $0.60 per unit or made inhouse. If manufactured on site, two machines will be required. Machine A is estimated to cost $18,000, have a life of 6 years, and a $2000 salvage value; machine B will cost $12,000, have a life of 4 years, and a $−500 salvage value (carry-away cost). Machine A will require an overhaul after 3 years costing $3000. The AOC for A is expected to be $6000 per year and for B $5000 per year. A total of four operators will be required for the two machines at a rate of $12.50 per hour per operator. One thousand units will be manufactured in a normal 8-hour period. Use an MARR of 15% per year to determine the following:

EXAMPLE 8.3

a. Number of units to manufacture each year to justify the inhouse (make) option.
b. The maximum capital expense justifiable to purchase machine A, assuming all other estimates for machines A and B are as stated. The company expects to produce 125,000 units per year.

Solution

a. Use steps 1 to 3 stated previously to determine the breakeven point.
  1. Define x as the number of lifts produced per year.
  2. There are variable costs for the operators and fixed costs for the two machines for the make option.

$$\text{Annual VC} = (\text{cost per unit})(\text{units per year})$$
$$= \frac{4 \text{ operators}}{1000 \text{ units}} \frac{\$12.50}{\text{hour}} (8 \text{ hours})x$$
$$= 0.4x$$

The annual fixed costs for machines A and B are the AW amounts.

$$AW_A = -18,000(A/P,15\%,6) + 2000(A/F,15\%,6)$$
$$-6000 - 3000(P/F,15\%,3)(A/P,15\%,6)$$
$$AW_B = -12,000(A/P,15\%,4) - 500(A/F,15\%,4) - 5000$$

Total cost is the sum of AW_A, AW_B, and VC.

3. Equating the annual costs of the buy option ($0.60x$) and the make option yields

$$-0.60x = AW_A + AW_B - VC$$
$$= -18,000(A/P,15\%,6) + 2000(A/F,15\%,6) - 6000$$
$$-3000(P/F,15\%,3)(A/P,15\%,6) - 12,000(A/P,15\%,4)$$
$$-500(A/F,15\%,4) - 5000 - 0.4x$$
$$-0.2x = -20,352.43$$
$$x = 101,762 \text{ units per year}$$

A minimum of 101,762 lifts must be produced each year to justify the make option, which has the lower variable cost of $0.40x$.

b. The production level is above breakeven. To find the maximum justifiable P_A, substitute 125,000 for x and P_A for the first cost of machine A. Solution yields $P_A = \$35,588$. This means that approximately twice the estimated first cost of $18,000 could be spent on A.

8.3 SENSITIVITY ANALYSIS FOR VARIATION IN ESTIMATES

There is always some degree of risk in undertaking any project, often due to uncertainty and variation in parameter estimates. The effect of variation may be determined by using *sensitivity analysis*. Usually, one factor at a time is varied, and independence with other factors is assumed. This assumption is not completely correct in real-world situations, but it is practical since it is difficult to accurately account for dependencies among parameters.

Sensitivity analysis determines how a measure of worth—PW, AW, ROR, or B/C—and the alternative may be altered if a particular parameter varies over a stated range of values. For example, some variation in MARR will likely not alter the decision when all alternatives have $i^* > \text{MARR}$; thus, the decision is relatively insensitive to MARR. However, variation in the estimated P value may make selection from the same alternatives sensitive.

There are several types of sensitivity analyses. It is possible to examine the sensitivity to variation for one, two, or more parameters for one alternative, as well as evaluating the impact on selection between mutually exclusive alternatives. *Breakeven analysis is actually a form of sensitivity analysis* in that the accept/reject decision changes depending upon where the parameter's most likely estimate lies relative to the breakeven point. There are three types of sensitivity analyses covered in this and the next section:

- Variation of one parameter at a time for a single project (Example 8.4) or for selecting between mutually exclusive alternatives
- Variation of more than one parameter for a single project (Example 8.5)
- Sensitivity of selection from multiple mutually exclusive alternatives to variation of more than one parameter (Section 8.4).

In all cases, the targeted parameter(s) and measure of worth must be selected prior to initiating the analysis. A general procedure to conduct a sensitivity analysis follows these steps:

1. Determine which parameter(s) of interest might vary from the most likely estimated value.
2. Select the probable range (numerical or percentage) and an increment of variation for each parameter.
3. Select the measure of worth.
4. Compute the results for each parameter using the measure of worth.
5. To better interpret the sensitivity, graphically display the parameter versus the measure of worth.

It is best to routinely use the AW or PW measure of worth. The ROR should not be used for multiple alternatives because of the incremental analysis required (discussed in Section 6.4). Spreadsheets are very useful in performing sensitivity analysis. One or more values can be easily changed to determine the effect. See Section 8.6 for a complete example.

Hammond Watches, Inc. is about to purchase equipment that will significantly improve the light reflected by the watch dial, thus making it easier to read the time in very low light environments. Most likely estimates are $P = \$80,000$, $n = 10$ years, $S = 0$, and increased net revenue after expenses of \$25,000 the first year decreasing by \$2000 per year thereafter. The VP of Manufacturing is concerned about the project's economic viability if the equipment life varies from the 10-year estimate, and the VP of Marketing is concerned about sensitivity to the revenue estimate if the annual decrease is larger or smaller than the \$2000. Use MARR = 15% per year and PW equivalence to determine sensitivity to (*a*) variation in life for 8, 10, and 12 years, and (*b*) variation in revenue decreases from 0 to \$3000 per year. Plot the resulting PW values for each parameter.

EXAMPLE 8.4

Solution

a. Follow the procedure outlined above for sensitivity to life estimates.
   1. Asset life is the parameter of interest.
   2. Range for n is 8 to 12 in 2-year increments.

3. PW is the measure of worth.

4. Use the most likely estimate of $G = \$-2000$ in revenue, and set up the PW relation for varying n values. Then insert values of $n = 8$, 10, and 12 to obtain the PW values.

$$PW = -80,000 + 25,000(P/A,15\%,n) - 2000(P/G,15\%,n)$$

n	PW
8	$ 7,221
10	11,511
12	13,145

5. Figure 8.6a plots PW versus n. The nonlinear result indicates minimal sensitivity to this range of life estimates. All PW values are clearly positive, indicating that the MARR of 15% is well exceeded.

b. Now the revenue gradient is examined using the same PW relation with $n = 10$. Set the increment at $1000 to obtain sensitivity to variation in G.

$$PW = -80,000 + 25,000(P/A,15\%,10) - G(P/G,15\%,10)$$

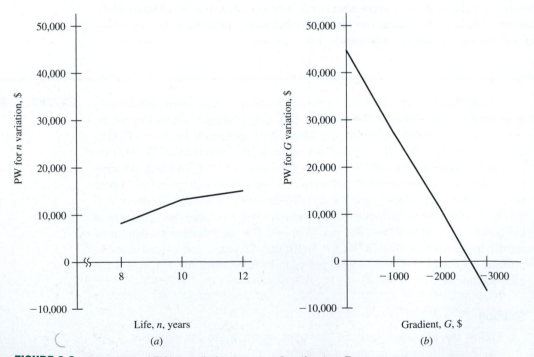

FIGURE 8.6   Sensitivity of PW to variation in n and G estimates, Example 8.4.

G	PW
$ 0	$45,470
−1000	28,491
−2000	11,511
−3000	−5,469

Figure 8.6*b* graphs PW versus *G*. The resulting linear plot indicates high sensitivity over the range. In fact, if revenue does drop by $3000 per year, and other estimates prove to be accurate, the project will not make the 15% return.

Comments: It is important to recognize a significant limitation of one-at-a-time analysis. All other parameters are fixed at their most likely estimate when the PW value is calculated. This is a good approach when one parameter contributes most of the sensitivity, but not so good when several parameters are expected to contribute significantly to the sensitivity.

Sensitivity plots involving *n* and *i* will not be linear because of the mathematical form of the factor. Those involving *P*, *G*, and *S* will normally be linear since *n* and *i* are fixed. In Figure 8.6*b*, the decrease in PW is $16,980 per $1000 change in *G*. Also, the breakeven point can be estimated from the plot at about $−2700. Solving for *G* in the PW relation results in the exact value of $−2678. If the annual decrease in revenue is expected to exceed this amount, the project will not be economically justifiable at MARR = 15%.

As a parameter estimate varies, selection between alternatives can change. To determine if the possible parameter range can alter the selection, calculate the PW, AW, or other measure of worth at different parameter values. This determines the sensitivity of the selection to the variation. Alternatively, the two PW or AW relations can be set equal and the parameter's breakeven value determined. If breakeven is in the range of the parameter's expected variation, the decision is considered sensitive. If this sensitivity/breakeven analysis of two alternatives (A and B) is performed using a spreadsheet, the SOLVER tool can be used to equate the two equivalency relations by setting up the constraint $AW_A = AW_B$. Reference the Excel help function for details on SOLVER.

When the sensitivity of *several parameters* is considered for *one project* using a *single measure of worth,* it is helpful to graph *percentage change* for each parameter versus the measure of worth. The variation in each parameter is indicated as a percentage deviation from the most likely estimate on the horizontal axis. If the response curve is flat and approaches horizontal over the expected range of variation, there is little sensitivity. As the curve increases in slope (positive or negative), the sensitivity increases. Those with large slopes indicate parameters that may require further study.

EXAMPLE 8.5   Janice is a process engineer with Upland Chemicals currently working in the research and planning department. She has performed an economic analysis for a process involving alternative fuels that is to be undertaken within the next 12 months. An ROR of 10% balanced the PW equivalence relation for the project. The required MARR for similar projects has been 5%, since some government subsidy is available for new fuels research. In an effort to evaluate the sensitivity of ROR to several variables, she constructed the graph in Figure 8.7 for variation ranges of selected parameters from the single-point estimates provided to her.

Sales price, $ per gallon	±20%
Materials cost, $ per ton	−10% to +30%
Labor cost, $ per hour	±30%
Equipment maintenance cost, $ per year	±10%

What should Janice observe and do about the sensitivity of each parameter studied?

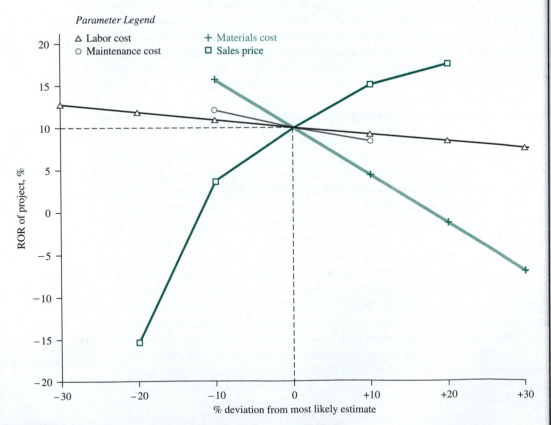

FIGURE 8.7   Sensitivity of ROR to four parameters, Example 8.5.

Solution

First of all, the range studied for each parameter varies from as little as 10% to as much as 30%. Therefore, sensitivity outside these ranges cannot be determined by this analysis. For the ranges studied, *labor and annual maintenance costs do not cause much change in the ROR* of the project. In fact, a 30% increase in labor cost will decrease the ROR by only an estimated 2 to 3%.

An increase in materials cost can cause a significant decrease in ROR. If raw materials cost increases by 10%, the 5% MARR will not be met, and a 20% increase will likely drive the ROR negative. Knowledge about and control of these costs are essential, especially in a cost inflationary environment.

Sales price can also have a large effect on the project's profitability. A relatively small change of $\pm 10\%$ from the estimated price can make ROR swing from about 3% to as much as 15%. Decreases beyond -10% can force the project into a negative return situation, and price increases above 10% should have a diminishing effect on the overall ROR. Sales price is a very powerful parameter; efforts by Upland's management to reduce predictable variation are clearly needed.

Janice needs to ensure that the results of her analysis are brought to the attention of appropriate management personnel. She should be sure that the sample data she used to perform the analysis are included in her report so the results can be verified and/or expanded. If no response is received in a couple of weeks, she should do a personal followup regarding reactions to her analysis.

8.4 SENSITIVITY ANALYSIS OF MULTIPLE PARAMETERS FOR MULTIPLE ALTERNATIVES

The economic advantages and disadvantages among two or more mutually exclusive alternatives can be studied by making three estimates for each parameter expected to vary enough to affect the selection: *a pessimistic, a most likely, and an optimistic estimate.* Depending upon the nature of a parameter, the pessimistic estimate may be the lowest value (alternative life is an example) or the largest value (such as asset first cost). This approach analyzes the measure of worth and alternative selection sensitivity within a predicted range of variation for each parameter. Usually the most likely estimate is used for all other parameters when the measure of worth is calculated for each alternative.

EXAMPLE 8.6

An engineer is evaluating three alternatives for which he has made three estimates for salvage value, annual operating cost, and life (Table 8.1). For example, alternative B has pessimistic estimates of $S = \$500$, AOC $= \$-4000$, and $n = 2$ years. The first costs are known, so they have the same value. Perform a sensitivity analysis and determine the most economic alternative using AW analysis at an MARR of 12% per year.

TABLE 8.1 Competing Alternatives with Three Estimates for Selected Parameters

Strategy		First Cost, $	Salvage Value, $	AOC, $	Life n, Years
Alternative A					
	P	−20,000	0	−11,000	3
Estimates	ML	−20,000	0	−9,000	5
	O	−20,000	0	−5,000	8
Alternative B					
	P	−15,000	500	−4,000	2
Estimates	ML	−15,000	1000	−3,500	4
	O	−15,000	2000	−2,000	7
Alternative C					
	P	−30,000	3000	−8,000	3
Estimates	ML	−30,000	3000	−7,000	7
	O	−30,000	3000	−3,500	9

P = pessimistic; ML = most likely; O = optimistic.

Solution

For each alternative in Table 8.1, calculate the AW value. For example, the AW relation for A, pessimistic estimates, is

$$AW_A = -20,000(A/P,12\%,3) - 11,000 = \$-19,327$$

There are a total of nine AW relations for the three alternatives and three estimates. Table 8.2 presents all AW values. Figure 8.8 is a plot of AW versus the three estimates of life for each alternative (plots of AW for any parameter with variation, that is n, AOC or S, give the same conclusion.) Since the AW calculated using the ML estimates for B ($\$-8229$) is economically better than even the optimistic AW values for A and C, alternative B is clearly favored.

TABLE 8.2 Annual Worth Values for Varying Parameters, Example 8.6

Estimates	Alternative AW Values		
	A	B	C
P	$−19,327	$−12,640	$−19,601
ML	−14,548	−8,229	−13,276
O	−9,026	−5,089	−8,927

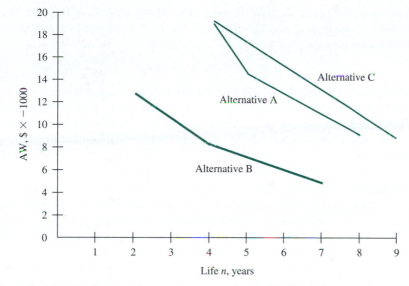

FIGURE 8.8   Plot of AW sensitivity for different life estimates, Example 8.6.

Comment: While the alternative that should be selected here is quite obvious, this is not always the case. Then it is necessary to select one set of estimates (P, ML, or O) upon which to base the selection.

8.5 PAYBACK PERIOD ANALYSIS

Payback analysis (also called *payout analysis*) is another form of sensitivity analysis that uses a PW equivalence relation. Payback can take two forms: one for $i > 0\%$ (also called *discounted payback*) and another for $i = 0\%$ (also called *no-return payback*). *The payback period n_p is the time, usually in years, it will take for estimated revenues and other economic benefits to recover the initial investment P and a specific rate of return i%.* The n_p value is generally not an integer.

The payback period should be calculated using a required return that is greater than 0%. In practice, however, the payback period is often determined with a no-return requirement ($i = 0\%$) to initially screen a project and determine whether it warrants further consideration.

To find the discounted payback period at a stated rate $i > 0\%$, calculate the years n_p that make the following expression correct.

$$0 = -P + \sum_{t=1}^{t=n_p} \text{NCF}_t(P/F,i,t) \qquad [8.4]$$

As discussed in Chapter 1, NCF is the estimated net cash flow for each year t, where NCF = receipts − disbursements. If the NCF values are equal each year, the P/A factor may be used to find n_p.

$$0 = -P + \text{NCF}(P/A,i,n_p) \qquad \text{[8.5]}$$

After n_p years, the cash flows will recover the investment and a return of $i\%$. If, in reality, the asset or alternative is used for more than n_p years, a larger return may result; but if the useful life is less than n_p years, there is not enough time to recover the initial investment and the $i\%$ return. It is very important to realize that in payback analysis *all net cash flows occurring after n_p years are neglected*. This is significantly different from the approach of all other evaluation methods (PW, AW, ROR, B/C) where all cash flows for the entire useful life are included. As a result, payback analysis can unfairly bias alternative selection. The payback period n_p *should not be used as the primary measure of worth to select an alternative*. It provides initial screening or supplemental information in conjunction with an analysis performed using another measure of worth.

No-return payback analysis determines n_p at $i = 0\%$. This n_p value serves merely as an initial indicator that a proposal is viable and worthy of a full economic evaluation. To determine the payback period, substitute $i = 0\%$ in Equation [8.4] and find n_p.

$$0 = -P + \sum_{t=1}^{t=n_p} \text{NCF}_t \qquad \text{[8.6]}$$

For a uniform net cash flow series, Equation [8.6] is solved for n_p directly.

$$n_p = \frac{P}{\text{NCF}} \qquad \text{[8.7]}$$

An example use of *no-return payback* as an initial screening of proposed projects is a corporation president who absolutely insists that every project must recover the investment in 3 years or less. Therefore, no proposed project with $n_p > 3$ at $i = 0\%$ should be considered further.

As with n_p for $i > 0\%$, it is incorrect to use the no-return payback period to make final alternative selections. It *neglects any required return*, since the time value of money is omitted when $i = 0\%$, and it *neglects all net cash flows after time n_p*, including positive cash flows that may contribute to the return on the investment.

EXAMPLE 8.7   This year the owner/founder of J&J Health allocated a total of $18 million to develop new treatment techniques for sickle cell anemia, a blood disorder that primarily affects people of African ancestry and other ethnic groups, including people who are of Mediterranean and Middle Eastern descent. The results are

estimated to positively impact net cash flow starting 6 years from now and for the foreseeable future at an average level of $6 million per year.

a. As an initial screening for economic viability, determine both the no-return and $i = 10\%$ payback periods.
b. Assume that any patents on the process will be awarded during the sixth year of the project. Determine the project ROR if the $6 million net cash flow were to continue for a total of 17 years (through year 22), when the patents legally expire.

Solution

a. The NCF for years 1 through 5 is $0 and $6 million thereafter. Let $x =$ number of years beyond 5 when NCF > 0. For no-return payback, apply Equation [8.6], and for $i = 10\%$, apply Equation [8.4]. In $ million units,

$$i = 0\%: \quad 0 = -18 + 5(0) + x(6)$$
$$n_p = 5 + x = 5 + 3 = 8 \text{ years}$$
$$i = 10\%: \quad 0 = -18 + 5(0) + 6(P/A,10\%,x)(P/F,10\%,5)$$

$$(P/A,10\%,x) = \frac{18}{6(0.6209)} = 4.8317$$

$$x = 6.9$$
$$n_p = 5 + x = 5 + 7 = 12 \text{ years (rounded up)}$$

b. The PW relation to determine i^* over the 22 years is satisfied at $i^* = 15.02\%$. In $ million units, PW is

$$PW = -18 + 6(P/A,i\%,17)(P/F,i\%,5)$$

The conclusions are: a 10% return requirement increases payback from 8 to 12 years; and when cash flows expected to occur after the payback period are considered, project return increases to 15% per year.

If two or more alternatives are evaluated using payback periods to indicate initially that one may be better than the other(s), the primary shortcoming of payback analysis (neglect of cash flows after n_p) may lead to an economically incorrect decision. When cash flows that occur after n_p are neglected, it is possible to favor short-lived assets when longer-lived assets produce a higher return. In these cases, PW or AW analysis should always be the primary selection method.

8.6 USING SPREADSHEETS FOR SENSITIVITY OR BREAKEVEN ANALYSIS

Spreadsheets are excellent tools to perform sensitivity, breakeven, and payback analyses. Estimates can be varied repeatedly one at a time to determine how PW, AW, or ROR changes. The GOAL SEEK tool is a great advantage when solving for breakeven values. The following examples illustrate spreadsheet development for both breakeven and sensitivity analyses.

To include a spreadsheet chart, care must be taken to arrange the data of interest in rows or columns. The *x-y* scatter chart is a commonly used tool for engineering economic analyses. See Appendix A and Excel© help system for details on constructing charts and using the GOAL SEEK tool.

EXAMPLE 8.8 Find the breakeven point in Example 8.3*a* between the make and buy alternatives using a spreadsheet and the GOAL SEEK tool.

Solution

Let the breakeven point be identified as BE. Figure 8.9 details the cash flows and AW values for machines A and B over their lives of 6 and 4 years, respectively. Column D shows the annual variable cost expression of 0.4 times BE with the same expression developed in the hand solution. Rows 13 and 14 show the complete AW expression using the BE value in cell E10. The solution is initiated with a test BE value; we used 25,000. GOAL SEEK calculated the correct value of 101,762 units per years by setting the difference of AW values to zero (cell E15). If more than 101,762 units are produced each year, the make alternative is selected based on its smaller slope.

	A	B	C	D	E	F
1		Machine A cash flow, $	Machine B cash flow, $	Annual VC, $/unit	Breakeven (BE), units/year	
2	Year					
3	0	-18,000	-12,000			
4	1	-6,000	-5,000			
5	2	-6,000	-5,000	VC function = 4*12.5*8 1000	Breakeven value using GOAL SEEK	
6	3	-9,000	-5,000			
7	4	-6,000	-5,500			
8	5	-6,000				
9	6	-4,000				
10	AW @ 15%	-11,049	-9,303	0.4	101,762	
11						
12			Breakeven relations			
13	Make cost/year: AW_A + AW_B + 0.4*BE				-61,057	
14	Buy cost/year: 0.6*BE				-61,057	
15			AW difference between make and buy:		0	
16						
17			= E13-E14			

Goal Seek — Set cell: E15 — To value: 0 — By changing cell: E10

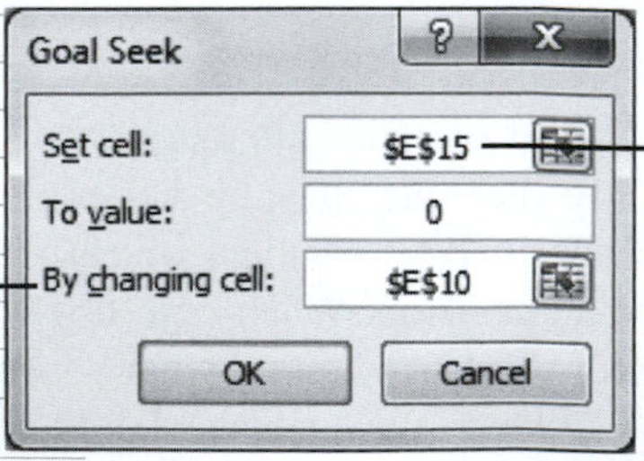

FIGURE 8.9 Spreadsheet solution for breakeven of make-buy alternatives using GOAL SEEK, Example 8.3*a*.

EXAMPLE 8.9

Halcrow, Inc., Division of Road and Highway Consultancy, is anticipating the purchase of concrete strength test equipment for use in highway construction. Estimates are:

First cost, $P = \$-100,000$ Annual operating costs, $AOC = \$-20,000$
Life, $n = 5$ years   Annual revenue, $R = \$50,000$

The Halcrow MARR for such projects is 10% per year. With an annual net cash flow (NCF) of $30,000, the IRR function indicates the project is economically justified with $i^* = 15.2\%$ for these most likely estimates. Use spreadsheets to perform a sensitivity analysis of ROR to variation in first cost and revenue.

a. Depending upon the model purchased, P may vary as much as $\pm25\%$ from $\$-75,000$ to $\$-125,000$.

b. Revenue variations up to $+20\%$ are possible, but R may go as low as $\$25,000$ per year in the worst case.

Solution

a. Figure 8.10a details the ROR values using the IRR function for first cost values from $\$-75,000$ to $\$-125,000$ with the most likely estimate of

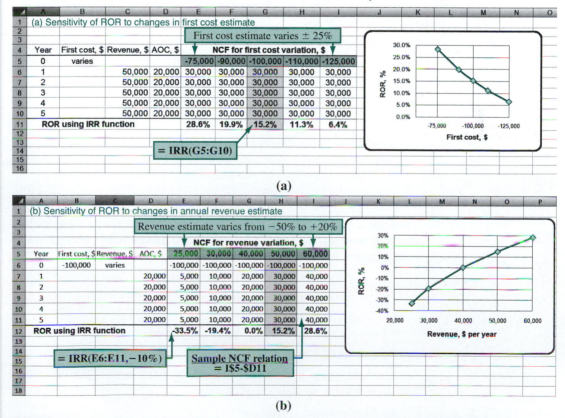

FIGURE 8.10 Sensitivity analysis of ROR to changes in (a) first cost estimate and (b) annual revenue estimate, Example 8.9.

$-100,000$ and $i* = 15.2\%$ in column G. (P and ROR values are placed in rows to facilitate plotting the x-y scatter chart.) The plot of rows 5 and 11 shows the significant reduction in ROR from about 28% to 6% over the possible range for P. From a sensitivity perspective, if P increases by 25%, a project return of considerably less than 10% is predicted.

b. Figure 8.10b uses the same spreadsheet format to evaluate ROR sensitivity of -50% to $+20\%$ variation in annual revenue. (The IRR function's 'guess' entry of -10% avoids a #NUM error when $i* = -33.5\%$ is determined at $R = \$25,000$.) A reduction of 20% to $40,000, as indicated on the chart, reduces ROR to 0%, which is much less than MARR $= 10\%$. A drop in R to $25,000 per year indicates a large negative return of -33.5%. The analysis indicates that ROR is very sensitive to revenue variation.

EXAMPLE 8.10   The Helstrom Corp. is a subcontractor in field operations to Boeing Company, manufacturer of the F18 Super Hornet carrier-based multirole fighter aircraft. Helstrom is about to purchase diagnostics equipment to test on-board electronics systems. Estimates are:

First cost	$8,000,000
Annual operating cost	$100,000
Life	5 years
Variable cost per test	$800 (most likely)
Fixed contract income per test	$1600

Use GOAL SEEK to determine the breakeven number of tests per year if the variable cost per test (*a*) equals $800, the most likely estimate, and (*b*) varies from $800 to $1400. For simplicity purposes only, utilize a 0% interest rate.

Solution

a. There are a couple of ways to approach breakeven problems using spreadsheets. Equation [8.2] finds Q_{BE} directly. The annual fixed cost and breakeven relations are

$$FC = 8,000,000\,(A/P,0\%,5) + 100,000$$
$$= 8,000,000\,(0.2) + 100,000$$
$$= \$1,700,000$$

$$Q_{BE} = \frac{1,700,000}{1600 - 800} = 2125 \text{ tests}$$

A second approach, used in the spreadsheets here, applies Equation [8.1] with $Q =$ number of tests per year.

$$\text{Profit} = \text{revenue} - \text{total cost}$$
$$= (1600 - 800)Q - (8,000,000(0.2) + 100,000)$$

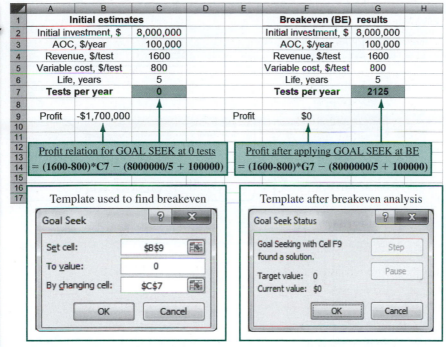

FIGURE 8.11 Determination of breakeven using GOAL SEEK, Example 8.10*a*.

Figure 8.11 (left side) shows all estimated parameters and the profit of $-1.7 million at 0 tests. The GOAL SEEK tool on the left side is set up to change the negative profit (cell B9) to $0 by increasing Q (cell C7). The right side presents the results with breakeven at 2125 tests per year. (When applying GOAL SEEK, the breakeven value 2125 will be displayed in cell B9; the results are shown separately here for cell F9 only for illustration.)

b. To evaluate breakeven's sensitivity to a range of variable costs, the spreadsheet is reconfigured to Figure 8.12. It now includes values from $800 to $1400 per test in the profit relations in columns C and E. Column D displays breakeven values as GOAL SEEK is applied repeatedly for each variable cost. The template in the inset will find breakeven for $1400 per test. (As before, in routine applications the results from GOAL SEEK would be in columns B and C; we have added columns D and E only to indicate before and after values.)

Breakeven is sensitive to test cost. Breakeven tests increase fourfold (2125 to 8500) while the cost per test increases less than twofold ($800 to $1400).

Comment: GOAL SEEK is excellent when only one parameter is varied. The more powerful spreadsheet tool SOLVER is better when multiple values (cells) are to be examined and when equality and inequality constraints are placed on parameter values. Consult the Excel© help system for details.

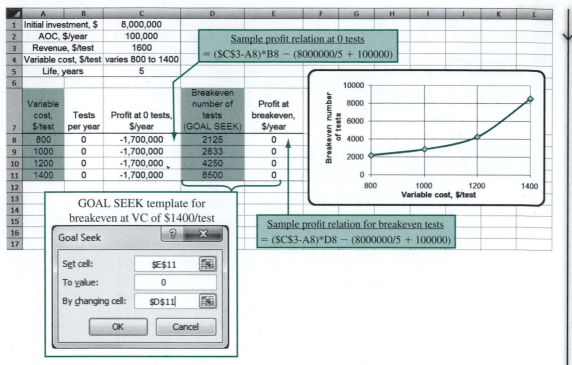

FIGURE 8.12 Determination of breakeven using GOAL SEEK repeatedly for changing variable cost estimates, Example 8.10*b*.

SUMMARY

This chapter treated sensitivity analysis and the related topics of breakeven analysis and payback period. Sensitivity to changing estimates for a parameter or for competing alternatives is determined via changes in a measure of worth. The variation in PW, AW, ROR, or B/C value for one or more parameters over a range of values is calculated (and may be plotted). Alternative selection sensitivity can be determined using three estimates of a parameter—optimistic, most likely, and pessimistic. In each sensitivity study, one parameter at a time is varied and independence between parameters is assumed.

The breakeven value Q_{BE} identifies a point of economic indifference. The acceptance criterion for one project is:

Accept the project if the estimated quantity exceeds Q_{BE}

For two alternatives determine the breakeven amount for a common parameter. The selection guideline is:

Select the alternative with the lower variable cost if the estimated quantity exceeds the breakeven amount.

Payback analysis estimates the number of years necessary to recover the initial investment plus a stated MARR. This is a supplemental analysis tool, best used for initial screening prior to full economic analysis. The technique should be applied with caution, as it has several drawbacks, especially for no-return payback where the MARR is set to 0%. The main drawback is that payback analysis does not consider cash flows that may occur after the payback period has expired.

PROBLEMS

Single Project Breakeven

8.1 The fixed costs at Harley Motors are $1 million annually. The main product has revenue of $9.90 per unit and $4.50 variable cost. Determine the following:

    a. Breakeven quantity per year

    b. Annual profit if 150,000 units are sold

    c. Annual profit if 480,000 units are sold

8.2 A professional photographer who specializes in wedding-related activities paid $16,000 for equipment that has a $2000 estimated salvage value after five years. He estimates that his costs associated with each event amount to $65 per day. If he charges $300 per day for his services, how many days per year must he be employed in order to break even at an interest rate of 8% per year?

8.3 An independent over-the-road (OTR) truck driver-owner paid $98,000 for a used tractor-trailer. The salvage value of the rig after five more years of use is expected to be $66,000. The operating cost is $0.60 per mile and the base mileage rate (revenue) is $0.71 per mile.

    a. How many miles per year must the owner drive just to break even at an interest rate of 10% per year?

    b. If the owner drives 550 miles per day, how many days per year will be required for breakeven?

8.4 A small consulting engineering company bought an office building for $900,000. The company has ten engineers and eight support staff. Monthly expenses for salaries, utilities, grounds maintenance, etc. are $1.1 million. If the average billing rate per engineer is $90 per hour, how many hours per month must be billed in order for the company to make a profit of $15,000 per month? Use an interest rate of 1% per month and assume the building will have a market value of $1.5 million after 10 years.

8.5 How long will it take to recover an investment of $245,000 in enhanced CNC controls that include axis control to 8 axes (on the milling model) if the associated income is $92,000 per year, expenses are $38,000 per year, and the salvage value is assumed to be 15% of the first cost? Use a MARR of 15% per year.

8.6 A call center in India used by U.S. and U.K. credit card holders has a capacity of 1,500,000 calls annually. The fixed cost of the center is $850,000 with an average variable cost of $1.95 and revenue of $3.25 per call. Find the percentage of the capacity that must be placed each year to break even.

8.7 Benjamin used regression analysis to fit quadratic relations to monthly revenue and cost data with the following results:

$$R = -0.007Q^2 + 32Q$$
$$TC = 0.004Q^2 + 2.2Q + 8$$

    a. Plot R and TC. Estimate the quantity Q at which the maximum profit should occur. Estimate the profit at this quantity.

    b. The profit relation Profit $= R - TC$ and calculus can be used to determine the quantity Q_p at which the maximum profit will occur, and the amount of this profit. The equations are:

$$\text{Profit} = aQ^2 + bQ + c$$
$$Q_p = -b/2a$$
$$\text{Maximum profit} = -b^2/4a + c$$

    Use these relations to confirm the graphical estimates you made in (*a*). (Your instructor may ask you to derive the relations above.)

8.8 Brittany is co-oping this semester at Regency Aircraft, which customizes the interiors of private and corporate jets. Her first assignment is to develop the specifications for a new machine to cut, shape, and sew leather or vinyl covers and trims. The first cost is not easy to estimate due to many options, but the annual revenue and M&O costs should net out at $15,000 per year over a 10-year life. Salvage is expected to be 20% of the first cost. Determine what can be paid for the machine now to recover the cost and an MARR of 8% per year under two scenarios:

    I: No outside revenue will be developed.

    II: Outside contracting will occur with estimated revenue of $10,000 the first year, increasing by $5000 per year thereafter.

    Solve using (*a*) tabulated factors, and (*b*) a spreadsheet and the GOAL SEEK tool.

8.9 The National Potato Cooperative purchased a de-skinning machine last year for $150,000. Revenue for the first year was $50,000. Over the total estimated life of 8 years, what must the annual revenue for years 2 through 8 equal to recover the investment, if costs are constant at $42,000 and a return of 10% per year is expected? A salvage value of $20,000 is anticipated.

8.10 ABB purchased fieldbus communication equipment for a project in South Africa for $3.15 million. If net cash flow is estimated at $500,000 per year, and a salvage value of $400,000 is anticipated, determine how many years the equipment must be used just to break even at interest rates ranging from 8% to 15% per year. Solve using (*a*) tabulated factors, and (*b*) a spreadsheet.

Breakeven Analysis Between Alternatives

8.11 A semiautomatic process has fixed costs of $40,000 per year and variable costs of $30 per unit. An automatic process has fixed costs of $88,000 per year and variable costs of $22 per unit. At what production level per year will the two alternatives break even?

8.12 Rent-A-Wreck car rental agency has a contract with PM Warranty, Inc. to do major repairs for $700 per car. The car rental agency estimates that it could repair its own cars for $300 each if it acquires a facility for $300,000 now. A salvage value of $40,000 after 15 years is estimated for the facility. What is the minimum number of cars that must be repaired each year to make the acquisition attractive at an interest rate of 10% per year?

8.13 A rural 2-lane road can be surfaced with concrete for $2.3 million per mile. If signing, mowing, and winter maintenance are not included, the basic maintenance costs for concrete and asphalt roadways are $483 and $774 per mile per year, respectively. If concrete lasts 20 years, what is the maximum amount that should be spent on asphalt that will last only 10 years? Use an interest rate of 8% per year.

8.14 A land development company is considering the purchase of earth-moving equipment. The equipment will have a first cost of $190,000 and a salvage value of $70,000 when the company sells it in 10 years. A service contract for maintenance on the equipment will cost $40,000 per year. The operating cost is expected to be $260 per day. Alternatively, the company can rent the necessary equipment for $1100 per day and hire a driver at $180 per day. If the company's MARR is 10% per year, how many days per year must the company need the equipment in order to justify its purchase?

8.15 The Ascarate Fishing Club, a non-profit organization dedicated to teaching kids how to fish, is considering two options for providing a heavily stocked pond for kids who have never caught a fish before. Option 1 is an above-ground swimming pool made of heavy vinyl plastic that will be assembled and disassembled for each quarterly event. The purchase price will be $400. Leaks from hooks piercing the fabric will be repaired with a vinyl repair kit at a cost of $70 per year, but the pool will have to be replaced when too many repairs have been made.

Option 2 is an in-ground pond that will be excavated by club members at no cost and lined with fabric that costs $1 per square foot. The pond will be 15 feet in diameter and 3 feet deep. Assume 300 ft^2 of liner will be purchased. A chain link fence at $10 per lineal foot will be installed around the pond (100 ft. of fence). Maintenance inside the fence is expected to cost $20 per year. The park where the pond will be constructed has committed the land for only 10 years. At an interest rate of 6% per year, how many years will the above-ground pool have to last for the two options to just break even?

8.16 Microsurfacing is part of a pavement restoration and maintenance program that seals the surface of a street that has minor cracking to prevent water from penetrating into the base material. The annual cost of the equipment (truck, tank, valves, etc.) is $109,000 per year and the material cost is $2.75 per square yard. Alternatively, regular street resurfacing requires equipment that has a first cost of $225,000 with a 15-year life and no salvage value. The variable cost for regular resurfacing is $13 per square yard. At an interest rate of 8% per year, how many square yards per year must be resurfaced for the two methods to break even?

8.17 Two membrane systems are under consideration for treating cooling tower blowdown to reduce its volume. A low-pressure seawater reverse osmosis (SWRO) system will operate at 500 psi and produce 720,000 gallons of permeate per day. It will have a fixed cost of $465 per day and an operating cost of $485 per day. A higher pressure SWRO system operating at 800 psi will produce 950,000 gallons per day at an operating cost of $1280 per day. The fixed cost of the high pressure SWRO system will be only $328 per day because fewer membranes will be required. How many gallons of blowdown water must require treatment each day for the two systems to break even?

8.18 Three methods can be used for producing heat sensors for high-temperature furnaces. Method A will have a fixed cost of $140,000 per year and a production cost of $62 per part. Method B will have a fixed cost of $210,000 per year and a production cost of $28 per part. Method C will require the purchase of equipment costing $500,000. It will have a life of five years and a 25% of first cost salvage value. The production cost will be $53 per part. At an interest rate of 10% per year, determine the breakeven annual production rate between the two lowest cost methods.

8.19 A Yellow Pages directory company must decide whether it should compose the ads for its clients inhouse or outsource them to a production company. To develop the ads inhouse, the company will have to purchase computers, printers, and other peripherals at a cost of $12,000. The equipment will have a useful life of 3 years, after which it will be sold for $2000. The employee who creates the ads will be paid $55,000 per year. In addition, each ad will have an average cost of $5. Alternatively, the company can outsource ad development at a flat fee of $21 per ad. At an interest rate of 10% per year, how many ads must the company sell each year for the alternatives to break even?

8.20 A plant manager has received two estimates from contractors to improve traffic flow and repave the parking areas. Proposal A includes new curbs, grading, and paving at an initial cost of $250,000. The life of the parking lot surface constructed in this manner is expected to be 4 years with an annual cost of $3000 for maintenance and repainting of strips. According to proposal B, the pavement has a higher quality and an expected life of 12 years. The annual maintenance cost will be negligible for the paved parking area, but the markings will have to be repainted every 2 years at a cost of $5000, except in the final year 12 of ownership. If the company's current MARR is 12% per year, how much can it afford to spend on proposal B so the two estimates will break even?

8.21 Alfred Home Construction is considering the purchase of five dumpsters and the transport truck to store and transfer construction debris from building sites. The entire rig is estimated to have an initial cost of $125,000, a life of 8 years, a $5000 salvage value, an operating cost of $40 per day, and an annual maintenance cost of $2000. Alternatively, Alfred can obtain the same services from the city as needed at each construction site for an initial delivery cost of $125 per dumpster per site and a daily charge of $20 per day per dumpster. An estimated 45 construction sites will need debris storage throughout the average year. If the minimum attractive rate of return is 12% per year, how many days per year must the equipment be required to justify its purchase?

8.22 Process X is estimated to have a fixed cost of $40,000 per year and a variable cost of $60 per unit in year 1, decreasing by $5 per unit per year. Process Y will have a fixed cost of $70,000 per year and a variable cost of $10 per unit, increasing by $1 per unit per year. At an interest rate of 12% per year, how many units must be produced *in year 3* for the two processes to break even?

8.23 An effective method to recover water used for regeneration of ion exchange resins is to use a reverse osmosis system in a batch-treatment mode. Such a system involves recirculation of the partially treated water back into the feed tank, causing the water to heat up. The water could be cooled using a single-pass heat exchanger or a closed-loop heat exchange system. The single-pass system would require a small chiller costing $920, plus stainless steel tubing, connectors, valves, etc., costing $360, and it could be used for 3 years. The cost of water, sewer charges, electricity, etc., will be $3.10 per hour. The closed-loop system will cost $3850 to buy, have a useful life of 5 years, and will cost $1.28 per hour of operation. The interest rate is 10% per year and the salvage values are negligible. (*a*) What is the breakeven number of hours per year that the cooling system must be needed to justify purchase of the closed-loop system? (*b*) What is the difference in annual costs if 400 hours per year is the expected usage level?

8.24 An engineering practitioner can lease a fully equipped computer and color printer system for $800 per month or purchase one for $8500 now and pay a $75 per month maintenance fee. If the nominal interest rate is 15% per year, determine the months of use necessary for the two to break even using (*a*) tabulated factors, and (*b*) a single-cell spreadsheet function.

8.25 The Ecology Group wishes to purchase a piece of equipment for metal recycling. Machine 1 costs $123,000, has a life of 10 years, an annual cost of

$5000, and requires one operator at a cost of $24 per hour. It can process 10 tons per hour. Machine 2 costs $70,000, has a life of 6 years, an annual cost of $2500, and requires two operators at a cost of $24 per hour for each operator. It can process 6 tons per hour.

a. Determine the breakeven tonnage of scrap metal at $i = 7\%$ per year and select the better machine for a processing level of 1000 tons per year.

b. Calculate and plot the sensitivity of the breakeven tons per year to $\pm 15\%$ change in the hourly cost of an operator.

8.26 Donny and Barbara want to join a sports and exercise club. The HiPro plan has no upfront charge and the first month is free. It then charges a total of $100 at the end of each subsequent month. Bally charges a membership fee of $100 per person now and $20 per person per month starting the first month. How many months will it take the two plans to reach a breakeven point? Solve by hand or spreadsheet, as requested by your instructor.

8.27 Balboa Industries' Electronics Division is trying to reduce supply chain risk by making more responsible make/buy decisions through improved cost estimation. A high-use component (expected usage is 5000 units per year) can be purchased for $25 per unit with delivery promised within a week. Alternatively, Balboa can make the component inhouse and have it readily available at a cost of $5 per unit, if equipment costing $150,000 is purchased. Labor and other operating costs are estimated to be $35,000 per year over the study period of 5 years. Salvage is estimated at 10% of first cost and $i = 12\%$ per year. Neglecting the element of availability (*a*) determine the breakeven quantity, and (*b*) recommend making or buying at the expected usage level.

8.28 Claris Water Company makes and sells filters for public water drinking fountains. The filter sells for $50 per unit. Recently an inhouse/outsource analysis was completed based on the need for new manufacturing equipment. The equipment first cost of $200,000 and $25,000 annual operation cost comprise the fixed cost, while Claris's variable cost is $20 per filter. The equipment has a 5-year life, no salvage value, and the MARR is 6% per year. The decision to make the filter was based on the breakeven point and the historical sales level of 5000 filters per year.

a. Determine the breakeven point. Should the filters be made inhouse?

b. An engineer at Claris learned that an outsourcing firm offered to make the filters for $30 each, but this offer was rejected by the president as entirely too expensive. Perform the breakeven analysis of the two options and determine if the inhouse decision was correct.

c. Develop and use the profit relations for both options to verify the preceding answers.

Note: Solve all three parts using a single spreadsheet if requested by your instructor.

Sensitivity Analysis

8.29 Josaline, the owner of a construction company, is planning to purchase specialized equipment to complete a contract awarded to her company. The first cost of the equipment is $250,000 with a life of 3 years at which time she will no longer need the equipment. The operating cost is expected to be $75,000 per year. Alternatively, a subcontractor can perform the work for $175,000 per year. Because the equipment is specialized, Josaline is not sure about the salvage value. She estimates a likely salvage of $90,000, but it might have to be scrapped for as little as $10,000 in three years. The MARR is 15% per year.

a. Is her decision to buy the equipment sensitive to the salvage value?

b. Determine the salvage value at which the two alternatives break even.

8.30 A company planning to borrow $10.5 million for a plant expansion is not sure what the interest rate will be when it applies for the loan. The rate could be as low as 10% per year or as high as 12% per year. The company will move forward with the project only if the annual worth of the expansion is below $5.7 million. If the M&O cost is fixed at $3.1 million per year and the salvage will be $2 million if the interest rate is 10% and $2.5 million if it is 12%, is the decision to move forward with the project sensitive to the interest rate? Use a 5-year study period.

8.31 A company that manufactures clear PVC pipe is investigating two production options with the following cash flow estimates.

	Batch	**Continuous**
First cost, $	−80,000	−130,000
Annual cost, $/year	−55,000	−30,000
Salvage value, $	10,000	40,000
Life, years	3 to 10	5

The chief operating officer (COO) has asked you to determine if the batch option would ever have a lower annual worth than the continuous flow system using interest rates over a range of 5% to 15% for the batch option, but only 15% for the continuous flow system. The batch process can be used anywhere from 3 to 10 years. (Note: The continuous flow process was previously determined to have its lowest cost over a 5-year life cycle.)

8.32 Home Automation is considering an investment of $500,000 in a new product line. The company will make the investment only if it will result in a rate of return of 15% per year or higher. If the revenue is expected to be between $138,000 and $165,000 per year for 5 years, use a present worth analysis to determine if the decision to invest is sensitive to the projected range of revenue.

8.33 Amphenol manufactures motor power connectors and is considering upgrading the production equipment to reduce costs over a 6-year planning horizon. The company can invest $80,000 now (year 0), 1 year from now, or 2 years from now. Depending on when the investment is made, the savings will vary. The savings estimates are $25,000, $26,000, and $29,000 per year for investing now, year 1, and year 2, respectively. If the company's MARR is 20% per year, use an FW analysis to determine if the timing of the investment will return at least 20% per year.

8.34 The equivalent annual worth of the process currently used in manufacturing motion controllers is AW = $−62,000 per year. A replacement process is under consideration that will have a first cost of $64,000 and an operating cost of $38,000 per year for the next 3 years. Three different engineers have given their opinion about what the salvage value of the new process will be 3 years from now as follows: $10,000, $13,000, and $18,000. Is the decision to replace the process sensitive to the salvage value estimates at the company's MARR of 15% per year?

8.35 Emerson Electric is considering the purchase of equipment that will allow the company to manufacture a new line of wireless devices for home appliance control. The first cost will be $80,000, and the life is estimated at 6 years with a salvage value of $10,000. Three different salespeople have provided estimates regarding the added revenue the equipment will generate. Salespersons 1, 2, and 3 have made estimates of $10,000, $16,000, and $18,000 per year, respectively. If the company's MARR is 8% per year, use a PW-based relation to determine if these different estimates will change the decision to purchase the equipment.

8.36 MAG Industrial needs 1000 square meters of storage space. Purchasing land for $80,000 and then erecting a temporary metal building at $70 per square meter is one option. The president hopes to sell the land for $100,000 and the building for $20,000 after 3 years. Another option is to lease space for $30 per square meter per year payable at the beginning of each year. The MARR is 20%. Perform a present worth analysis of the building and leasing alternatives to determine the sensitivity of the decision if the construction cost decreases by 10% to $63 per square meter and the lease cost remains at $30 per square meter per year.

8.37 Consider the two air conditioning systems detailed below.

	System 1	**System 2**
First cost, $	−10,000	−17,000
Annual operating cost, $/year	−600	−150
Salvage (disposal) value, $	−100	−300
New compressor and motor cost at midlife, $	−1,750	−3,000
Life, years	8	12

Use AW analysis to determine the sensitivity of the economic decision to MARR values of 4%, 6%, and 8%. Work this problem (a) by hand and (b) by spreadsheet.

8.38 Ned Thompson Labs performs tests on superalloys, titanium, aluminum, and most metals. Tests on metal composites that rely upon scanning electron microscope results can be subcontracted or the labs can purchase new equipment. Evaluate the sensitivity of the economic decision to purchase the equipment over a range of ±20% (in 10% increments) of the estimates for P, AOC, R, n, and MARR (range on MARR is 12% to 18%). Use the AW method and plot the results on a sensitivity graph. For which parameter(s) is the AW most sensitive? least sensitive?

First cost, $P = \$-180,000$
Salvage, $S = \$20,000$
Life, $n = 10$ years
Annual operating cost, AOC $= \$-30,000$ per year
Annual revenue, $R = \$70,000$ per year
MARR $= 15\%$ per year

8.39 Determine if the selection of system 1 or 2 is sensitive to variation in the return required by management. The corporate MARR ranges from 8% to 16% per year on different projects. Use tabulated factors or a spreadsheet, as requested by your instructor.

	System 1	System 2
First cost, $	−50,000	−100,000
AOC, $ per year	−6,000	−1,500
Salvage value, $	30,000	0
Rework at midlife, $	−17,000	−30,000
Life, years	4	12

8.40 Titan manufactures and sells gas-powered electricity generators. It can purchase a new line of fuel injectors from either of two companies. Cost and savings estimates are available, but the savings estimates are unreliable at this time. Use an AW analysis at MARR = 10% per year to determine if the selection between company A and B changes when the estimated savings varies as much as ±40% from the best estimates. Use tabulated factors or a spreadsheet, as requested by your instructor.

	Company A	Company B
First cost, $	−50,000	−37,500
AOC, $ per year	−7,500	−8,000
Savings best estimate, $ per year	15,000	13,000
Salvage, $	5,000	3,700
Life, years	5	5

Alternative Selection Using Multiple Estimates

8.41 A civil engineer involved in construction management must decide between two ways to pump concrete up to the top floors of a seven-story office building under construction. Plan 1 requires the purchase of equipment for $6000 which costs between $0.40 and $0.75 per metric ton to operate, with a most likely cost of $0.50 per metric ton. The asset is able to pump 100 metric tons per day. If purchased, the asset will last for 5 years, have no salvage value, and be used 50 days per year. Plan 2 is an equipment-leasing option and is expected to cost the company $2500 per year for equipment with a low cost estimate of $1800 and a high estimate of $3200 per year. In addition, an extra $5 per hour labor cost will be incurred for operating the leased equipment each 8-hour day. Use $i = 12\%$ per year. (*a*) Which plan should the engineer recommend on the basis of the most likely estimates of costs? (*b*) Will the decision above change if the pessimistic estimates are used?

8.42 When the country's economy is expanding, AB Investment Company is optimistic and expects a MARR of 15% for new investments. However, in a receding economy the expected return is 8%. Normally a 10% return is required. An expanding economy causes the estimates of asset life to go down about 50%, and a receding economy makes the n values increase about 20%. Which plan should be selected if the company president expects the economy to be (*a*) expanding, and (*b*) receding?

	Plan M	Plan Q
Initial investment, $	−200,000	−240,000
Net cash flow, $/year	65,000	71,000
Life, years	10	10

8.43 In evaluating two environmental chambers at Holly Farms, the AW of the 409G model is determined to be $\$-135,143$ with a high degree of certainty. The cost estimates for the D103 model are less certain. The manager requested a "worst case" analysis with three estimates for P and n as shown on the next page. The annual operating cost is fixed at $4000 and the salvage value of the chamber is expected to be 10% of the first cost.

Perform the analysis using (*a*) tabulated factors, and (*b*) a spreadsheet to determine if D103 is favored under any of the first cost and life scenarios indicated. The company's MARR is 10% per year.

	Pessimistic	Most Likely	Optimistic
First cost, $	−500,000	−400,000	−300,000
Life, years	1	3	5

Payback Analysis

8.44 State why payback analysis is best used as a supplemental analysis tool when an economic study is performed.

8.45 The process for producing a fruit-tree pesticide has a first cost of $200,000 with annual costs of $50,000 and revenue of $90,000 per year. What is the payback period at (*a*) $i = 0\%$, and (*b*) $i = 12\%$ per year?

8.46 Two machines can be used to produce an aircraft part from titanium. The costs and other cash flows associated with each alternative are shown. The salvage values are zero regardless of when the machines are replaced. Use the estimates to preliminarily determine which alternative(s) should be selected for further analysis provided they must pay back in 5 years or less. Perform the analysis with (*a*) $i = 0\%$ and (*b*) $i = 10\%$ per year.

	Machine 1	Machine 2
First cost, $	−40,000	−90,000
Net cash flow, $ per year	10,000	15,000
Maximum life, years	10	10
Salvage value, $	0	0

8.47 Laura's grandparents helped her purchase a small self-serve laundry business to make extra money during her five college years. When she completed her electrical engineering degree, she sold the business and her grandparents told her to keep the money as a graduation present. For the net cash flows (NCF) listed, determine:

a. if the total income exceeded the total amount invested in 5 years ($i = 0\%$).
b. the actual rate of return over the 5-year period.

c. how long it took to pay back the $75,000 investment plus a 7% per year return.
d. answers to all three questions above using a spreadsheet.

Year	0	1	2	3	4	5
NCF, $1000 per year	−75	−10.5	18.6	−2	28	105

8.48 A company that manufactures diaphragm seals has identified the cash flows shown below with a certain part of the manufacturing and sales functions. Determine the no-return payback period.

First cost of equipment, $	−130,000
Annual expenses, $/year	−45,000
Annual revenue, $/year	75,000

8.49 In desalting groundwaters that contain a significant amount of sulfates, the concentrate that is generated during the desalting process can sometimes be treated with lime to recover gypsum and other salts. Because the high pH process is tough on equipment, the equipment's useful life is uncertain. A treatment train with an initial cost of $90,000 has an operating cost of $20,000 per month. The revenue from the sale of calcium sulfate is $22,000 per month. Determine how many months the equipment must last to recover the investment at $i = 0.5\%$ per month using (*a*) hand solution, (*b*) a calculator, and (*c*) a single-cell spreadsheet function.

8.50 Ellis Equipment sold a used Massey Ferguson tractor for $55,000 to a South Kansas farmer 10 years ago. (*a*) What is the uniform net cash flow that the farmer had to receive each year to realize payback and a return of 5% per year on his investment over a period of 3 years? 5 years? 8 years? All 10 years? (*b*) If the net cash flow was actually $6000 per year, what is the amount the farmer should have paid for the tractor to realize payback plus the 5% per year return over these 10 years?

8.51 CMS Express has historically owned and maintained its own delivery trucks. Leasing is an option being seriously considered because costs for maintenance, fuel, insurance, and some liability issues will be transferred to United Leasing, the truck leasing company. The study period is no more than 24 months for either alternative. The annual lease cost is paid at the beginning of each year and is not refundable for partially used years.

Purchase: $P = \$30,000$ now; monthly cost = $1200; monthly revenue = $4500

    Lease: $P = \$10,000$ at the beginning of each year (months 0 and 12); monthly cost = $2800; monthly revenue = $4500

a. Use the first cost and net cash flow estimates to determine the payback in months with a nominal 9% per year return for the purchase and lease options.

b. Spreadsheet question: Write the single-cell NPER function, including the optional "type" entry, that will display payback for the lease option, where lease costs are paid at the beginning of the year. Explain how you developed the function.

8.52 Fidelity Life Insurance has a document imaging system that needs replacement. A local salesperson quoted a cost of $10,000 with an estimated salvage of $900 after 5 or more years. If the system is expected to save $1700 per year in clerical time, find the payback time at 8% per year. As a practice, the office manager purchases equipment only when the payback is less than 6 years. Otherwise, he prefers to lease. Should the imaging system be purchased or leased?

The following information is used for Problems 8.53 through 8.56

Darrell, an engineer with TAGHeuer Watches, is considering two alternative processes to water-

proof the new line of scuba-wear watches. Estimates follow.

Process 1: $P = \$-50,000$; $n = 5$ years; NCF = $24,000 per year; no salvage value.

Process 2: $P = \$-120,000$; $n = 10$ years; NCF = $42,000 for year 1, decreasing by $2500 per year thereafter; no salvage value.

8.53 Darrell first decided to use no-return payback to select the process, because his boss told him most investments at TAGHeuer must pay back in 3 to 4 years. Determine which process Darrell will select.

8.54 Next Darrell decided to use the AW method at the corporate MARR of 12% per year that he used previously on another evaluation. Now what process will he select?

8.55 Finally, Darrell decided to calculate the rate of return for the cash flows of each process over its respective life. (*a*) Which process does this analysis indicate is better? (*b*) Explain the fundamental assumptions and errors made when this approach is used. (Hint: Before answering, review Sections 6.3 and 6.4 on ROR.)

8.56 Of the evaluations above, which is the correct method upon which to base the final economic decision? Why is this method the only correct one?

ADDITIONAL PROBLEMS AND FE EXAM REVIEW QUESTIONS

8.57 In linear breakeven analysis, if a company expects to operate at a point *below* the breakeven point, the alternative to select is:
a. the one with the lower fixed cost.
b. the one with the higher fixed cost.
c. the one with the lower variable cost.
d. the one with the higher variable cost.

8.58 A process can be completed using either Alternative X or Y, where Y is an automated version of X. Alternative X has fixed costs of $10,000 per year with a variable cost of $50 per unit. If the

process is automated, the fixed cost for Y will be $5,000 per year and its variable cost will be only $30 per unit. The minimum number of units that must be produced each year for alternative Y to be favored is closest to:
a. Alternative Y will be favored for any level of production
b. 125
c. 375
d. Alternative X will be favored for any level of production

8.59 A company is considering two alternatives for automating a certain process. Alternative A will have fixed costs of $42,000 per year and will require 2 laborers at $48 per day each. Together, these laborers can generate 100 units of product. Alternative B will have fixed costs of $56,000 per year, but with this alternative, the 3 laborers will generate 200 units of product. In determining the breakeven number of units Q, the total cost per year for Alternative B is represented as:

a. $[2(48)/100]Q$
b. $[3(48)/200]Q$
c. $[3(48)/200]Q + 56,000$
d. $[2(48)/100]Q + 42,000$

8.60 The price of a car that you want is $70,000 today. Its price is expected to increase by $3300 per year. You now have $35,000 in an investment which is earning 15% per year. The number of years before you have enough to buy the car without borrowing any money is closest to:

a. 3 years
b. 5 years
c. 7 years
d. 9 years

8.61 In conducting a sensitivity analysis, the only parameter that represents a measure of worth is:

a. the future worth.
b. the breakeven point.
c. a cost index.
d. a sinking fund equation.

8.62 When conducting a sensitivity analysis using multiple estimates, the three estimates are usually:

a. probabilistic, authentic, most likely
b. deterministic, most likely, optimistic
c. pessimistic, strategic, realistic
d. optimistic, pessimistic, most likely

8.63 A process for making a laboratory-grade sodium phosphate will have a first cost of $320,000 with annual costs of $40,000 and revenue of $98,000 per year. At a return requirement of 20% per year, the payback period is closest to:

a. 3 years
b. 5 years
c. 7 years
d. ∞; it will never pay off

8.64 When the variable cost is reduced for linear total cost and revenue lines, the breakeven point decreases. This is an economic advantage because:

a. the revenue per unit will increase.
b. the two lines will now cross at zero.

c. the profit will increase for the same revenue per unit.
d. the total cost line becomes nonlinear.

8.65 The profit relation for the following estimates at a quantity that is 10% above breakeven is:

Fixed cost = $500,000 per year
Cost per unit = $200
Revenue per unit = $250

a. Profit = $200(11,000) - 250(11,000) - 500,000$
b. Profit = $250(11,000) - 500,000 - 200(11,000)$
c. Profit = $250(11,000) - 200(11,000) + 500,000$
d. Profit = $250(10,000) - 200(10,000) - 500,000$

8.66 For these two AW relations, the breakeven point Q_{BE} in miles per year is closest to:

$$AW_1 = -23,000(A/P,10\%,10) + 4000(A/F,10\%,10) - 5000 - 4Q_{BE}$$

$$AW_2 = -8000(A/P,10\%,4) - 2000 - 6Q_{BE}$$

a. 1984
b. 1224
c. 1090
d. 655

8.67 To make an item inhouse, equipment costing $250,000 must be purchased. It will have a life of 4 years, an annual cost of $80,000, and each unit will cost $40 to manufacture. Buying the item externally will cost $100 per unit. At $i = 15\%$ per year, it is cheaper to make the item inhouse if the number per year needed is:

a. above 1047 units
b. above 2793 units
c. equal to 2793 units
d. below 2793 units

8.68 The sensitivity of two parameters (P and n) for one project is evaluated by graphing the AW values versus percentage variation from the most likely estimates. The curve for n has a slope very close to zero, while the P curve has a significant negative slope. One good conclusion from the graph is that:

a. both PW and AW values are more sensitive to variations in P than n.
b. the project should be rejected, since AW values vary with P and n.
c. a better estimate of P needs to be made.
d. the ROR is equally sensitive for both parameters.

The following information is used for Problems 8.69 and 8.70

Four mutually exclusive alternatives are evaluated using three estimates or strategies (pessimistic, most likely, and optimistic) for several parameters. The resulting PW values over the LCM are determined as shown.

	PW Values Over LCM, $			
Strategy	**1**	**2**	**3**	**4**
Pessimistic (P)	4,500	−6,000	3,700	−1,900
Most likely (ML)	6,000	−500	5,000	−100
Optimistic (O)	9,500	2,000	10,000	3,500

8.69 The best alternative to select under the stated condition is:
  a. pessimistic: select alternative 2
  b. optimistic: select alternative 2
  c. pessimistic: select alternative 1
  d. optimistic: select alternative 4

8.70 If no one of the strategies is more likely than any other strategy, the alternative to select is:
  a. 2
  b. 1 and 2 are equally acceptable
  c. 1
  d. 3

Replacement and Retention Decisions

The McGraw-Hill Companies, Inc./ Ken Cavanagh

One of the most commonly performed engineering economy studies is that of replacement or retention of an asset or system that is currently installed. This differs from previous studies where all the alternatives are new. The fundamental question answered by a replacement study about a currently installed asset or system is, *Should it be replaced now or later?* When an asset is currently in use and its function is needed in the future, it will be replaced at some time. So, in reality, a replacement study answers the question of *when,* not *if,* to replace.

A replacement study is usually designed to first make the economic decision to retain or replace *now.* If the decision is to replace, the study is complete. If the decision is to retain, the cost estimates and decision will be revisited in the future, usually on an annual basis, to ensure that the decision to retain is still economically correct. This chapter explains how to perform the initial year and follow-on year replacement studies.

A replacement study is an application of the AW method of comparing unequal-life alternatives, first introduced in Chapter 5. In a replacement study with no specified study period, the AW values are determined by a technique of cost evaluation called the *economic service life (ESL)* analysis. If a study period is specified, the replacement study procedure is different from that used when no study period is set. Both procedures are covered in this chapter.

Purpose: Perform a replacement study between an in-place asset or system and a new one that could replace it.

LEARNING OUTCOMES

1. Understand the fundamentals and terms for a replacement study.

   Basics

2. Determine the economic service life of an asset that minimizes the total AW of costs.

   Economic service life

3. Perform a replacement study between the defender and the best challenger.

   Replacement study

4. Determine the replacement value to make the defender and challenger equally attractive.

   Replacement value

5. Perform a replacement study over a specified number of years.

   Study period specified

6. Use a spreadsheet to determine the ESL, to perform a replacement study, and to calculate the replacement value.

   Spreadsheets

9.1 BASICS OF A REPLACEMENT STUDY

Up to this point, none of the two or more mutually exclusive alternatives compared are currently in place. It is very common to face the situation that the currently used asset (or system or service) could be either replaced with a more economical alternative or retained as is. This is called a *replacement study,* which may be necessary for several reasons—unacceptable performance or reliability, physical deterioration, competitive or technological obsolescence, or updated and new requirements. At some point in time, the function of every currently used asset must be replaced. A replacement study provides an answer to the question: Is replacement with a specified alternative economical at this point?

The in-place asset is referred to as the *defender,* and the replacement alternative is called the *challenger.* The replacement analysis is performed from the nonowner's perspective (viewpoint) of a *consultant* or *outsider.* That is, the analysis assumes that neither of the alternatives is owned currently; selection is between the proposed challenger and the in-place defender.

To conduct the economic analysis, the estimates for the challenger are developed as presented in previous chapters. The challenger first cost is the actual investment needed for acquisition and installation. (A common temptation is to increase the challenger's initial cost by the defender's unrecovered depreciation. This amount is a *sunk cost* related to the defender and must not be borne by the challenging alternative.)

In a replacement study, estimates for the defender's n and P values are obtained as follows:

- The expected life n is the number of years at which the lowest AW of costs occurs. This is called the economic service life or ESL. (Section 9.2 details the technique to find ESL.)
- The defender's "initial investment" P is estimated by the *defender's current market value,* that is, the amount required to acquire the services provided by the in-place asset. If mandatory additional capital investment is necessary for the defender to provide the current services, this amount is also included in the P value. The equivalent annual capital recovery and costs for the defender must be based on the entire amount required *now* to continue the defender's services in the future. This approach is correct because all that matters in an economic analysis is what will happen from now on. Previous costs are sunk costs and are considered *irrelevant* to the replacement study. (A common temptation is to use the defender's current book value for P. Again, this figure is irrelevant to the replacement study as depreciation is for tax purposes, as discussed in Chapter 13.)

An annual worth analysis is most commonly used for the replacement analysis. The length of the replacement study period is either *unlimited* or *specified.* If the *period is not limited,* the assumptions of the AW method discussed in Section 5.1 are made—the service is needed for the indefinite future, and cost estimates are the same for future life cycles changing with the rate of inflation or deflation. If the *period is specified,* these assumptions are unnecessary, since estimates are made for only the fixed time period.

EXAMPLE 9.1   The Arkansas Division of ADM, a large agricultural products corporation, purchased a state-of-the-art ground-leveling system for rice field preparation 3 years
ago for $120,000. When purchased, it had an expected service life of 10 years,
an estimated salvage of $25,000 after 10 years, and AOC of $30,000. Current
account book value is $80,000. The system is deteriorating rapidly; 3 more years
of use and then salvaging it for $10,000 on the international used farm equip-
ment network are now the expectations. The AOC is averaging $30,000.

A substantially improved, laser-guided model is offered today for $100,000
with a trade-in of $70,000 for the current system. The price goes up next week
to $110,000 with a trade-in of $70,000. The ADM division engineer estimates
the laser-guided system to have a useful life of 10 years, a salvage of $20,000,
and an AOC of $20,000. A $70,000 market value appraisal of the current system was made today.

If no further analysis is made on the estimates, state the correct values to
include if the replacement study is performed today.

Solution

Take the nonowner's viewpoint and use the most current estimates.

Defender	Challenger
$P = \$-70,000$	$P = \$-100,000$
$AOC = \$-30,000$	$AOC = \$-20,000$
$S = \$10,000$	$S = \$20,000$
$n = 3$ years	$n = 10$ years

The defender's original cost, AOC, and salvage estimates, as well as its current
book value, are all *irrelevant* to the replacement study. *Only the most current
estimates should be used.* From the outsider's perspective, the services that the
defender can provide could be obtained at a cost equal to the *defender market
value* of $70,000.

9.2 ECONOMIC SERVICE LIFE

Podcast
available

Until now the estimated life n of a project or alternative has been stated. In actuality,
this value is determined prior to an evaluation. An asset should be retained for a time
period that minimizes its cost to the owner. This time is called the economic service
life (ESL) or minimum cost life. *The smallest total AW of costs identifies the ESL value.*
This n value is used in all evaluations for the asset, including a replacement study.

Total AW of costs is the sum of the asset's annual capital recovery and AW
of annual operating costs, that is,

$$\textbf{Total AW} = -\textbf{capital recovery} - \textbf{AW of annual operating costs}$$
$$= -\textbf{CR} - \textbf{AW of AOC} \qquad \textbf{[9.1]}$$

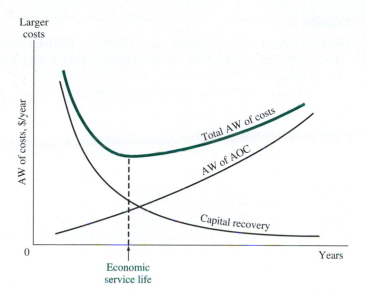

FIGURE 9.1
Annual worth curves
of cost elements
that determine the
economic service life.

These are cost estimates; all, except for salvage value, are negative numbers. Figure 9.1 shows the characteristic concave shape of the total AW curve. The CR component decreases with time and the AOC component increases. For k years of service, the components are calculated using the following formulas:

$$CR_k = -P(A/P,i,k) + S_k(A/F,i,k) \qquad \textbf{[9.2]}$$

$$(\text{AW of AOC})_k = [AOC_1(P/F,i,1) + AOC_2(P/F,i,2) + \dots$$
$$+ AOC_k(P/F,i,k)](A/P,i,k) \qquad \textbf{[9.3]}$$

In Equation [9.2], the salvage S_k after k years of service is the estimated future market value were the asset (defender or challenger) purchased now. The ESL (best n value) is indicated by the smallest total AW in the series calculated using Equation [9.1]. Example 9.2 illustrates ESL calculations. Spreadsheet usage is presented in Section 9.6.

EXAMPLE 9.2

A device that monitors rotational vibration changes in turbines may be purchased for use in southern California wind farms. The first cost is $40,000 with a constant AOC of $15,000 over a maximum service period of 6 years. Use the decreasing future market values and $i = 20\%$ per year to find the best n value for an economic evaluation.

After k years of service	1	2	3	4	5	6
Estimated market value is	$32,000	30,000	24,000	20,000	11,000	0

Solution

Determine the total AW of costs using Equations [9.1] through [9.3] for years 1 through 6. The AW of AOC is constant at $15,000 in Equation [9.3]. For one year of retention, $k = 1$.

$$\text{Total AW}_1 = -40,000(A/P,20\%,1) + 32,000(A/F,20\%,1) - 15,000$$
$$= -16,000 - 15,000$$
$$= \$-31,000$$

For two years of retention, $k = 2$.

$$\text{Total AW}_2 = -40,000(A/P,20\%,2) + 30,000(A/F,20\%,2) - 15,000$$
$$= -12,546 - 15,000$$
$$= \$-27,546$$

Table 9.1 shows the AW values over all possible years of service. The smallest total AW cost value is the ESL, which occurs at $\$-26,726$ for $k = 4$. Use an estimated life of $n = 4$ years in the evaluation.

Comment: The CR component (Table 9.1) does not decrease every year, only through year 4. This is the effect of the changing future market values that are used for the salvage estimates in Equation [9.2].

TABLE 9.1   **Calculation of Total AW of Costs Including Capital Recovery and AOC, Example 9.2**

Years of Retention	1	2	3	4	5	6
Capital recovery, $/year	−16,000	−12,546	−12,395	−11,726	−11,897	−12,028
AW of AOC, $/year	−15,000	−15,000	−15,000	−15,000	−15,000	−15,000
Total AW, $/year	−31,000	−27,546	−27,395	−26,726	−26,897	−27,028

A replacement study may be performed when the n value of the defender, the challenger, or both are specified, that is, fixed. In this case, no ESL analysis is necessary for the alternative(s) that have a specified n value; the AW value over the specified life is the correct one to use in the replacement analysis.

9.3 PERFORMING A REPLACEMENT STUDY

Podcast available

Replacement studies are performed in one of two ways: without a study period specified or with one defined. Figure 9.2 gives an overview of the approach taken for each situation. The procedure discussed in this section applies when no study period (planning horizon) is specified. If a specific number of years is identified for the replacement study, for example, over the next 5 years, with no continuation considered after this time period, the procedure in Section 9.5 is applied.

A replacement study determines when a challenger replaces the in-place defender. The complete study is finished if the challenger (C) is selected to replace

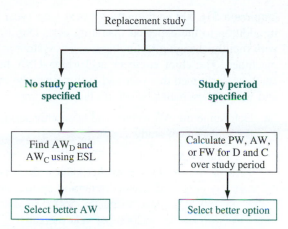

FIGURE 9.2

Overview of replacement study approaches.

the defender (D) now. However, if the defender is retained now, the study may extend over a number of years equal to the life of the defender n_D, after which a challenger replaces the defender. Use the annual worth and life values for C and D determined in the ESL analysis to apply the following replacement study procedure. This assumes the services provided by the defender could be obtained at the AW_D amount.

New replacement study:
1. On the basis of the better AW_C or AW_D value, select the challenger alternative (C) or defender alternative (D). When the challenger is selected, replace the defender now, and expect to keep the challenger for n_C years. This replacement study is complete. If the defender is selected, plan to retain it for up to n_D more years, but next year, perform the following analysis.

One-year-later analysis:
2. Are all estimates still current for both alternatives, especially first cost, market value, and AOC? If no, proceed to step 3. If yes and this is year n_D, replace the defender. If this is not year n_D, retain the defender for another year and repeat this same step. This step may be repeated several times.
3. Whenever the estimates have changed, update them, perform new ESL analyses, and determine new AW_C and AW_D values. Initiate a new replacement study (step 1).

If the defender is selected initially (step 1), estimates may need updating after 1 year of retention (step 2). Possibly there is a new best challenger to compare with D. Either significant changes in defender estimates or availability of a new challenger indicates the need for a new replacement study. In actuality, a replacement study can be performed each year, or at any time, to determine the advisability of replacing or retaining any defender, provided a competitive challenger is available.

EXAMPLE 9.3

Two years ago, Toshiba Electronics made a $15 million investment in new assembly line machinery. It purchased approximately 200 units at $70,000 each and placed them in plants in 10 different countries. The equipment sorts, tests, and performs insertion-order kitting on electronic components in preparation for special-purpose printed circuit boards. A new international industry standard requires a $16,000 additional cost next year (year 1 of retention) on each unit in addition to the expected operating cost. Due to the new standards, coupled with rapidly changing technology, a new system is challenging these 2-year-old machines. The chief engineer at Toshiba USA has asked that a replacement study be performed this year and each year in the future, if need be. At $i = 10\%$ and with the estimates below, do the following:

a. Determine the AW values and economic service lives necessary to perform the replacement study.

> Challenger: First cost: $50,000
> Future market values: decreasing by 20% per year
> Estimated retention period: no more than 5 years
> AOC estimates: $5000 in year 1 with increases of $2000 per year thereafter

> Defender: Current international market value: $15,000
> Future market values: decreasing by 20% per year
> Estimated retention period: no more than 3 more years
> AOC estimates: $4000 next year, increasing by $4000 per year thereafter, plus the extra $16,000 next year

b. Perform the replacement study now.

c. After 1 year, it is time to perform the follow-up analysis. The challenger is making large inroads into the market for electronic components assembly equipment, especially with the new international standards features built in. The expected market value for the defender is still $12,000 this year, but it is expected to drop to virtually nothing in the future—$2000 next year on the worldwide market and zero after that. Also, this prematurely outdated equipment is more costly to keep serviced, so the estimated AOC next year has been increased from $8000 to $12,000 and to $16,000 two years out. Perform the follow-up replacement study analysis.

Solution

a. The results of the ESL analysis, shown in Table 9.2, include all the market values and AOC estimates for the challenger in the top of the table. Note that $P = \$50,000$ is also the market value in year 0. The total AW of costs is shown by year, should the challenger be placed into service for that number of years. As an example, if the challenger is kept for 4 years, AW_4 is

$$\text{Total AW}_4 = -50,000(A/P,10\%,4) + 20,480(A/F,10\%,4)$$
$$-[5000 + 2000(A/G,10\%,4)]$$
$$= \$-19,123$$

The defender costs are analyzed in the same way in Table 9.2 (bottom) up to the maximum retention period of 3 years.

The lowest AW cost (numerically largest) values for the replacement study are

> Challenger: $AW_C = \$-19,123$ for $n_C = 4$ years
> Defender: $AW_D = \$-17,307$ for $n_D = 3$ years

TABLE 9.2 Economic Service Life (ESL) Analysis of Challenger and Defender Costs, Example 9.3

	Challenger		
Challenger Year k	Market Value	AOC	Total AW If Owned k Years
0	$50,000	—	—
1	40,000	$ −5,000	$−20,000
2	32,000	−7,000	−19,524
3	25,600	−9,000	−19,245
4	20,480	−11,000	−19,123 ESL
5	16,384	−13,000	−19,126

	Defender		
Defender Year k	Market Value	AOC	Total AW If Retained k Years
0	$15,000	—	—
1	12,000	$−20,000	$−24,500
2	9,600	−8,000	−18,357
3	7,680	−12,000	−17,307 ESL

b. To perform the replacement study now, apply only the first step of the procedure. Select the defender because it has the better AW of costs ($−17,307), and expect to retain it for 3 more years. Prepare to perform the one-year-later analysis 1 year from now.

c. One year later, the situation has changed significantly for the equipment Toshiba retained last year. Apply the steps for the one-year-later analysis:

2. After 1 year of defender retention, the challenger estimates are still reasonable, but the defender market value and AOC estimates are substantially different. Go to step 3 to perform a new ESL analysis for the defender.

3. The defender estimates in Table 9.2 (bottom) are updated below, and new AW values are calculated. There is now a maximum of 2 more years of retention, 1 year less than the 3 years determined last year.

Year k	Market Value	AOC	Total AW If Retained k More Years
0	$12,000	—	—
1	2,000	$−12,000	$−23,200
2	0	−16,000	−20,819 ESL

The defender ESL is 2 years. The AW and n values for the new replacement study are:

Challenger: unchanged at $AW_C = \$-19{,}123$ for $n_C = 4$ years

Defender: new $AW_D = \$-20{,}819$ for $n_D = 2$ more years

Now select the challenger based on its favorable AW value. Therefore, replace the defender now, not 2 years from now. Expect to keep the challenger for 4 years, or until a better challenger appears on the scene.

In the example above, the challenger's P value is its estimated first cost and the defender's P value is its current market value. This approach is called the *conventional* or *opportunity-cost approach* to a replacement study. *This is the correct approach.* Another method is entitled the cash-flow approach, in which the defender's market value is subtracted from the challenger's first cost and the defender's first cost is set equal to zero. The same economic decision results using either approach, but a falsely low challenger capital recovery (CR) amount is calculated, since the challenger's P estimate is lowered for the evaluation. This, plus the fact that the equal service assumption is violated when defender and challenger lives are unequal, makes it important that the opportunity-cost approach illustrated in Example 9.3 be taken in all replacement studies. (Only if the lives of the defender and challenger are known to be equal should either approach be considered when performing the replacement study.)

9.4 DEFENDER REPLACEMENT VALUE

Oftentimes it is helpful to know the minimum defender market value that, if exceeded, will make the challenger the better alternative. This defender value, called its *replacement value (RV)*, yields a breakeven value between challenger and defender. RV is found by setting $AW_C = AW_D$ with the defender's first cost P replaced by the unknown RV. In Example 9.3b, $AW_C = \$-19{,}123$, which is larger than $AW_D = \$-17{,}307$. Using the estimates in Table 9.2, the alternatives are equally attractive at $RV = \$22{,}341$, determined from the breakeven relation.

$$-19{,}123 = -RV(A/P,10\%,3) + 0.8^3 RV(A/F,10\%,3) - [20{,}000(P/F,10\%,1)$$
$$+ 8{,}000(P/F,10\%,2) + 12{,}000(P/F,10\%,3)](A/P,10\%,3)$$
$$RV = \$22{,}341$$

Any trade-in offer (market value) above this amount is an economic indication to replace the defender now. The current market value was estimated at \$15,000, thus the defender's selection in Example 9.3b.

9.5 REPLACEMENT STUDY OVER A SPECIFIED STUDY PERIOD

The right branch of Figure 9.2 applies when the time period for the replacement study is limited to a specified study period or planning horizon, for example, 3 years. In this case, the only relevant cash flows are those that occur within the 3-year period. Situations such as this often arise because of international competition and swift obsolescence of in-place technologies. Skepticism and uncertainty about the future are often reflected in management's desire to impose *abbreviated study periods* upon all economic evaluations, knowing that it may be necessary to consider yet another replacement in the near future. This approach, though reasonable from management's perspective, usually forces recovery of the initial investment and the required MARR over a shorter period of time than the ESL of the asset. In fixed study period analyses, the AW, PW, or FW is determined based on the estimates that apply only from the present time through the end of the study period.

For the data shown in Table 9.3 (partially developed from Table 9.2), determine which alternative is better at $i = 10\%$ per year, if the study period is (*a*) 1 year and (*b*) 3 years.

EXAMPLE 9.4

Solution

a. Use AW relations for a 1-year study period.

$$AW_C = -50{,}000(A/P{,}10\%{,}1) + 40{,}000(A/F{,}10\%{,}1) - 5000$$
$$= \$-20{,}000$$
$$AW_D = -15{,}000(A/P{,}10\%{,}1) + 12{,}000(A/F{,}10\%{,}1) - 20{,}000$$
$$= \$-24{,}500$$

Select the challenger.

TABLE 9.3 Challenger and Defender Estimates for Replacement Study, Example 9.4.

	Challenger	
Challenger Year k	**Market Value**	**AOC**
0	$50,000	—
1	40,000	$ −5,000
2	32,000	−7,000
3	25,600	−9,000
4	20,480	−11,000
5	16,384	−13,000

	Defender	
Defender Year k	**Market Value**	**AOC**
0	$15,000	—
1	12,000	$−20,000
2	9,600	−8,000
3	7,680	−12,000

b. For a 3-year study period, the AW equations are

$$AW_C = -50{,}000(A/P,10\%,3) + 25{,}600(A/F,10\%,3)$$
$$- [5000 + 2000(A/G,10\%,3)]$$
$$= \$-19{,}245$$
$$AW_D = -15{,}000(A/P,10\%,3) + 7680(A/F,10\%,3)$$
$$- [20{,}000(P/F,10\%,1) + 8000(P/F,10\%,2)$$
$$+ 12{,}000(P/F,10\%,3)](A/P,10\%,3)$$
$$= \$-17{,}307$$

Select the defender.

If there are several options for the number of years that the defender may be retained before replacement with the challenger, the first step is to develop the succession options and their AW values. For example, if the study period is 5 years, and the defender will remain in service 1 year, or 2 years, or 3 years, cost estimates must be made to determine AW values for each defender retention period. In this case, there are four options; call them W, X, Y, and Z.

Option	Defender Retained	Challenger Serves
W	3 years	2 years
X	2	3
Y	1	4
Z	0	5

The respective AW values for defender retention and challenger service define the cash flows for each option. Example 9.5 illustrates the procedure.

EXAMPLE 9.5   Amoco Canada has oil field equipment placed into service 5 years ago for which a replacement study has been requested. Due to its special purpose, it has been decided that the current equipment will have to serve for either 2, 3, or 4 more years before replacement. The equipment has a current market value of $100,000, which is expected to decrease by $25,000 per year. The AOC is constant now, and is expected to remain so, at $25,000 per year. The replacement challenger is a fixed-price contract to provide the same services at $60,000 per year for a minimum of 2 years and a maximum of 5 years. Use MARR of 12% per year to perform a replacement study over a 6-year period to determine when to sell the current equipment and purchase the contract services.

Solution

Since the defender will be retained for 2, 3, or 4 years, there are three viable options (X, Y, and Z).

Option	Defender Retained	Challenger Serves
X	2 years	4 years
Y	3	3
Z	4	2

The defender annual worth values are identified with subscripts D2, D3, and D4 for the number of years retained.

$$AW_{D2} = -100,000(A/P,12\%,2) + 50,000(A/F,12\%,2) - 25,000$$
$$= \$-60,585$$
$$AW_{D3} = -100,000(A/P,12\%,3) + 25,000(A/F,12\%,3) - 25,000$$
$$= \$-59,226$$
$$AW_{D4} = -100,000(A/P,12\%,4) - 25,000 = \$-57,923$$

For all options, the challenger has an annual worth of

$$AW_C = \$-60,000$$

Table 9.4 presents the cash flows and PW values for each option over the 6-year study period. A sample PW computation for option Y is

$$PW_Y = -59,226(P/A,12\%,3) - 60,000(F/A,12\%,3)(P/F,12\%,6)$$
$$= \$-244,817$$

Option Z has the lowest cost PW value ($-240,369). Keep the defender all 4 years, then replace it. Obviously, the same answer will result if the annual worth, or future worth, of each option is calculated at the MARR.

TABLE 9.4 **Equivalent Cash Flows and PW Values for a 6-Year Study Period Replacement Analysis, Example 9.5**

Option	Time in Service, Years Defender	Challenger	AW Cash Flows for Each Option, $/Year 1	2	3	4	5	6	Option PW, $
X	2	4	−60,585	−60,585	−60,000	−60,000	−60,000	−60,000	−247,666
Y	3	3	−59,226	−59,226	−59,226	−60,000	−60,000	−60,000	−244,817
Z	4	2	−57,923	−57,923	−57,923	−57,923	−60,000	−60,000	−240,369

Comment: If the study period is long enough, it is possible that the ESL of the challenger should be determined and its AW value used in developing the options and cash flow series. An option may include more than one life cycle of the challenger for its ESL period. Partial life cycles of the challenger can be included. Regardless, any years beyond the study period must be disregarded for the replacement study, or treated explicitly, in order to ensure that equal-service comparison is maintained, especially if PW is used to select the best option.

9.6 USING SPREADSHEETS FOR A REPLACEMENT STUDY

This section includes two examples. The first illustrates the use of a spreadsheet to determine the ESL of an asset. The second demonstrates a one-worksheet replacement study, including ESL determination followed by a breakeven analysis to find the defender's replacement value using the GOAL SEEK tool.

Using the PMT function each year k to find the components in Equation [9.1]—CR and AW of AOC—and adding the two offers a rapid way to determine the ESL. The function formats for each component follow.

For CR component: $= -\text{PMT}(i\%,k,P,S)$

For AW of AOC component: $= -\text{PMT}(i\%,k,\text{NPV}(i\%,AOC_1{:}AOC_k))$

The minus sign retains the sign sense of the cash flow values. For the AOC component, the NPV function is embedded in the PMT function to calculate the AW value of all AOC estimates from years 1 through k in one operation. Example 9.6 demonstrates this technique.

A calculator is as helpful as a spreadsheet in determining the ESL when the predicted market value MV is constant and the AOC is a uniform series. For example, to determine the capital recovery for k years, the PMT(i,n,P,F) function for a calculator is PMT($i,k,P,$MV). However, when the market value and AOC series are more complex, a spreadsheet is much faster and less error-prone.

EXAMPLE 9.6   The Navarro County District has purchased new $850,000 rugged trenching equipment for preparing utility cuts in very rocky areas. Because of physical deterioration and heavy wear and tear, market value will decrease 30% per year annually and AOC will increase 30% annually from the expected AOC of $13,000 in year 1. Capital equipment policy dictates a 5-year retention period. Benjamin, a young engineer with the county, wants to confirm that 5 years is the best life estimate. At $i = 5\%$ per year, help Benjamin perform the spreadsheet analysis.

Solution

Equations [9.2] and [9.3] are determined using the PMT functions. Figure 9.3, columns B and C, detail the annual S and AOC values over a 13-year period. The right side (columns D and E) shows the CR and AW of AOC values for 1 through k years of ownership.

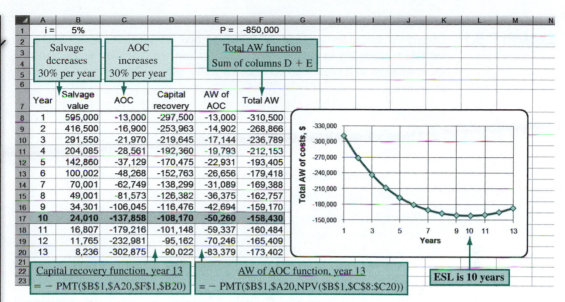

FIGURE 9.3 Determination of ESL using PMT functions, Example 9.6.

Once the functions are set using cell reference formatting, they can be dragged down through the years. (As an example, when $k = 2$, the function $= -PMT(5\%,2,-850000,416500)$ displays CR $= \$-253{,}963$; and $= -PMT(5\%,2,NPV(5\%,2,C8:C9))$ displays AW of AOC as $\$-14{,}902$. The total AW is $\$-268{,}866$.) Cell tags for columns D and E detail the functions for year 13 using cell reference formatting.

Care is needed when inserting minus signs on estimates and functions in order to ensure correct answers. For example, the first cost P has a minus sign, so the (positive) salvage value is calculated using a plus sign. Likewise, the PMT functions are preceded by a minus to ensure that the answer remains a negative (cost) amount.

The lowest total AW value in the spreadsheet and the chart indicate an ESL of 10 years. The differences in total AW are significantly lower for 10 than for 5 years ($\$-158{,}430$ vs. $\$-193{,}405$). Therefore, the 5-year retention period dictated by county policy will have a higher equivalent annual cost based on the estimates.

It is quite simple to use one spreadsheet to find the ESL values, perform the replacement study, and, if desired, find the RV value for the defender. The PMT functions determine the total AW values in the same way as the previous example. The resulting ESL and AW values are used to make the replace or retain decision. Then, the equivalent of setting up the RV relation of $AW_C = AW_D$ is accomplished by applying the GOAL SEEK tool. An illustration of this efficient use of a spreadsheet follows.

EXAMPLE 9.7 Utilize the estimates in Example 9.3*a* and one spreadsheet (*a*) to determine the ESL values, (*b*) to decide to replace or retain the defender, and (*c*) to find the minimum defender market value to make the challenger more attractive economically.

Solution

a. Figure 9.4 includes all estimates and functions that determine the total AW and ESL values.

Challenger: minimum total AW = $-19,123; ESL = 4 years

Defender: minimum total AW = $-17,307; ESL = 3 years

Remember to use minus signs correctly so that costs and salvage values have the correct sign sense.

b. Retaining the defender is economically favorable since its total AW is smaller.

c. Use GOAL SEEK to set the defender total AW for 3 years equal to $-19,123, the challenger total AW at its ESL of 4 years. This is normally accomplished on the same worksheet. However, for clarity, the defender portion from Figure 9.4 is duplicated in Figure 9.5 with the

Challenger P = -50,000

Salvage decreases 20% per year | AOC increases $2000 per year

Year	Salvage value	AOC	Capital recovery	AW of AOC	Total AW
1	40,000	-5,000	-15,000	-5,000	-20,000
2	32,000	-7,000	-13,571	-5,952	-19,524
3	25,600	-9,000	-12,372	-6,873	-19,245
4	**20,480**	**-11,000**	**-11,361**	**-7,762**	**-19,123** ← ESL is 4 years
5	16,384	-13,000	-10,506	-8,620	-19,126

CR function, year 4
= − PMT(10%,$A10,$F$1,B10)

AW of AOC function, year 4
= − PMT(10%,$A10,NPV(10%,C$7:C10))

Defender P = -15,000

Salvage decreases 20% per year | AOC increases $4000 per year

Year	Salvage value	AOC	Capital recovery	AW of AOC	Total AW
1	12,000	-20,000	-4,500	-20,000	-24,500
2	9,600	-8,000	-4,071	-14,286	-18,357
3	**7,680**	**-12,000**	**-3,711**	**-13,595**	**-17,307** ← ESL is 3 years

FIGURE 9.4 Spreadsheet determination of minimum total AW of costs and ESL values, Example 9.7.

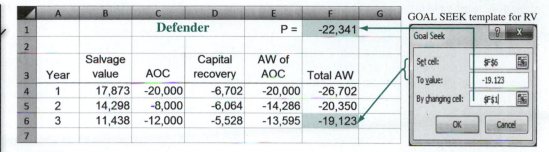

FIGURE 9.5   Use of GOAL SEEK to determine defender replacement value on the spreadsheet in Figure 9.4, Example 9.7.

GOAL SEEK template added. In short, this forces the defender market value/first cost higher (currently $-15,000$) so its total AW goes from $-17,307$ to the challenger's AW of $-19,123$. This is the point of breakeven between the two alternatives. Initiation of GOAL SEEK displays the RV of $-22,341$ in cell F1 with updated salvage and AW values as shown in Figure 9.5. As before, a defender trade-in offer greater than RV indicates that the challenger is better.

SUMMARY

It is important in a replacement study to compare the best challenger with the defender. *Best (economic) challenger is described as the one with the lowest AW of costs for some period of years.* However, if the expected remaining life of the defender and the estimated life of the challenger are specified, the AW values over these years must be used in the replacement study.

The economic service life (ESL) analysis is designed to determine the best challenger's years of service and the resulting lowest total AW of costs. The resulting n_C and AW_C values are used in the replacement study procedure. The same analysis can be performed for the ESL of the defender.

Replacement studies in which no study period (planning horizon) is specified utilize the annual worth method of comparing two unequal-life alternatives. The better AW value determines how long the defender is retained before replacement.

When a study period is specified, it is vital that market value and cost estimates for the defender be as accurate as possible. All the viable time options for using the defender and challenger are enumerated, and their AW equivalent cash flows determined. For each option, the PW, AW, or FW value is used to select the best option. This option determines how long the defender is retained before replacement.

PROBLEMS

Foundations of Replacement

9.1 Briefly explain what is meant by the defender/challenger concept.

9.2 List three reasons why a replacement study might be needed.

9.3 In conducting a replacement study of assets with different lives, can the annual worth values over the asset's own life cycle be used in the comparison, if the study period is (*a*) unlimited, (*b*) limited and the study period *is not* an even multiple of asset lives, and (c) limited wherein the period *is a* multiple of asset lives? Explain your answers.

9.4 An entrepreneurial civil engineer who owns his own design/build company purchased a small crane 2 years ago at a cost of $71,000. At that time, it was expected to be used for 10 years and then traded in for its salvage value of $10,000. Due to increased construction activities, the company would prefer to trade for a new, larger crane now which will cost $93,000. The company estimates that the old crane can be used, if necessary, for another 4 years, at which time it will have a $25,000 estimated market value. Its current market value is estimated to be $39,000, and if it is used for another 4 years, it will have M&O costs of $17,000 per year. Determine the values of *P, n, S,* and AOC that should be used for the existing crane in a replacement analysis performed today.

9.5 A mechanical engineer who designs and sells equipment that automates manual labor processes is offering a machine/robot combination that will significantly reduce labor costs associated with manufacturing garage-door opener transmitters. The equipment has a first cost of $170,000, an estimated annual operating cost of $54,000, a maximum useful life of 5 years, and a $20,000 salvage value anytime it is replaced. The existing equipment was purchased 12 years ago for $65,000 and has an annual operating cost of $78,000. At most, the currently owned equipment can be used two more years, at which time it will be auctioned off for an expected amount of $6000, less 33% paid to the company handling the auction. The same scenario will occur if the currently-owned equipment is replaced now. Determine the defender and challenger estimates of *P, n, S,* and AOC in conducting a replacement analysis today at an interest rate of 20% per year.

9.6 A machine tool purchased two years ago for $40,000 has a market value that can be described by the relation $40,000 - 3000k$, where k is the number of years from time of purchase. Experience with this type of asset has shown that its annual operating cost is described by the relation $30,000 + 1000k$. The asset's salvage value was originally estimated to be $10,000 after a predicted 10-year useful life. Determine the current estimates for *P, S,* and AOC for a replacement study, assuming it will be kept only

one more year, which will be the third year of ownership.

Economic Service Life

9.7 The AW values for retaining a presently owned machine for additional years are shown in the table. (Note that the values are the AW amount per *each of the n years* that the asset is kept.) A challenger has an economic service life of 7 years with an $AW_C = \$-86,000$ per year. Assuming future costs remain as estimated for the replacement study, what is the economic service life of the defender, if the company's MARR is 12% per year? Assume used machines like the one presently owned will always be available.

Retention Period, Years	AW Value, $ per Year
1	−92,000
2	−86,000
3	−85,000
4	−89,000
5	−95,000

9.8 To improve package tracking at a UPS transfer facility, conveyor equipment was upgraded with RFID sensors at a cost of $345,000. The operating cost is expected to be $148,000 per year for the first 3 years and $210,000 for the next 3 years. The salvage value of the equipment is expected to be $140,000 for the first 3 years, but due to obsolescence, it won't have a significant value after that. At an interest rate of 10% per year, determine the economic service life and equivalent annual worth of the equipment. Use tabulated factors or a spreadsheet, as requested by your instructor.

9.9 From the data shown, determine the ESL of the defender and challenger.

Years Retained	AW of Defender, $	AW of Challenger, $
1	−145,000	−136,000
2	−96,429	−126,000
3	−63,317	−92,000
4	−39,321	−53,000
5	−49,570	−38,000

9.10 From the data shown, determine the economic service life of the asset.

Years Retained	AW of First Cost, $	AW of Operating Cost, $	AW of Salvage Value, $
1	−165,000	−36,000	99,000
2	−86,429	−36,000	38,095
3	−60,317	−42,000	18,127
4	−47,321	−43,000	6,464
5	−39,570	−48,000	3,276

9.11 In trying to determine the economic service life of a new piece of equipment, an engineer made the calculations shown below. She forgot to enter the annual worth of the salvage value for two years of retention. Determine the following to make the ESL equal two years: (*a*) minimum AW of the salvage value, and (*b*) estimated salvage value for year two if $i = 10\%$ per year.

Years Retained	AW of First Cost, $	AW of Operating Cost, $	AW of Salvage Value, $
1	−88,000	−45,000	50,000
2	−46,095	−46,000	?
3	−32,169	−51,000	6,042
4	−25,238	−59,000	3,232
5	−21,104	−70,000	1,638

9.12 A general manager wants to know the economic service life of currently owned machines. The market value of the machines is $30,000, but this value is expected to decrease as shown in the table below. The maintenance and operating cost associated with each additional year of retention is also shown. Use the company's MARR of 15% per year to determine the ESL.

Year	Market Value, $	M&O cost, $
0	30,000	—
1	25,000	−49,000
2	20,000	−51,000
3	15,000	−53,000
4	10,000	−55,000

9.13 A construction company bought a 180,000 metric ton earth sifter at a cost of $65,000. The company expects to keep the equipment a maximum of 7 years. The operating cost is expected to follow the series described by $40,000 + 10,000k$, where k is the number of years since it was purchased ($k = 1,\ldots, 7$). The salvage value is estimated to be $30,000 for years 1 and 2 and $20,000 for years 3 through 7. At $i = 10\%$ per year, determine the economic service life and equivalent annual worth of the sifter using (*a*) tabulated factors, and (*b*) a spreadsheet.

9.14 Cetec Aviation Services has cost estimates associated with operating and maintaining the currently owned filter analysis system as shown below. It is considering the acquisition of a replacement system that can identify residual particles in industrial filters, report its findings, and archive images and information for future retrieval. Determine the cost of keeping the current system one more year at an interest rate of 10% per year.

Year	Market value, $	Operating cost, $
0	30,000	—
1	25,000	−15,000
2	14,000	−15,000
3	10,000	−15,000

Replacement Study

9.15 An engineer with Haliburton calculated the AW values shown for retaining a presently owned machine additional years. A challenger has an economic service life of 7 years with AW = $−86,000 per year. Assuming all future costs remain as estimated for the analysis, (*a*) when should the company replace the defender and with what machine, and (*b*) when should the company purchase the challenger? The MARR is 12% per year. Assume used machines like the one presently owned will always be available.

Retention Period, Years	Defender AW, $ per Year
1	−92,000
2	−81,000
3	−85,000
4	−89,000
5	−95,000

9.16 A presently owned machine can last for three more years, if properly maintained at a cost of $15,000 per year. Its operating cost will be $31,000 per year. After three years, it can be sold for an estimated $9000. A replacement costs $80,000 with a $10,000 salvage value after three years and an operating cost of $19,000 per year. Different vendors have offered $10,000 and $20,000, respectively, for the current system in trade for the replacement. At 12% per year interest, perform a replacement study for the two trade-in offers.

9.17 A piece of imaging equipment was purchased two years ago for $50,000 with an expected useful life of 5 years and a $5000 salvage value. Since its installation performance was poor, it was upgraded for $20,000 one year ago. Increased demand now requires another upgrade for an additional $22,000 so that it can be used for 3 more years. Its new annual operating cost will be $27,000 with a $12,000 salvage after the 3 years. Alternatively, it can be replaced with new equipment costing $65,000, operating costs of $14,000 per year and an expected salvage of $23,000 after 3 years. If replaced now, the existing equipment will be sold for $7000. Determine whether the company should keep or replace the defender at a MARR of 10% per year.

9.18 The plant manager asked you to do a cost analysis to determine when currently owned equipment should be replaced. The manager stated that under no circumstances will the existing equipment be retained longer than two more years It can be replaced any year with an outside contractor at a cost of $97,000 per year. The market value of the currently owned equipment is estimated to be $37,000 now, $30,000 in one year, and $19,000 two years from now. The operating cost is $85,000 per year. Using an interest rate of 10% per year, determine when the defending equipment should be retired.

9.19 A small company that manufactures vibration isolation platforms is trying to decide whether it should immediately upgrade the current assembly system D, which is rather labor-intensive, with the more highly automated system C one year from now. Some components of the current system can be sold now for $9000, but they will be worthless hereafter. The operating cost of the existing system is $192,000 per year. System C will cost $320,000 with a $50,000 salvage value after four years. Its operating cost will be $68,000 per year. If you are told to do a replacement analysis using

an interest rate of 10% per year, which system do you recommend?

9.20 A critical machine in the Freeport copper refining operation was purchased 7 years ago for $160,000. Last year a replacement study was performed with the decision to retain for 3 more years. The situation has changed. The equipment is estimated to have a value of $8000 if "scavenged" for parts now or anytime in the future. If kept in service, it can be minimally upgraded at a cost of $43,000 to make it usable for up to 2 more years. Its operating cost is estimated at $22,000 the first year and $25,000 the second year. Alternatively, the company can purchase a new system that will have an equivalent annual worth of $54,063 per year over its ESL. The company uses a MARR of 10% per year. Use annual worth analysis to determine when the company should replace the machine.

9.21 A biotech company planning a plant expansion is trying to determine whether it should upgrade the existing controlled-environment rooms or purchase new ones. The presently owned rooms were purchased 4 years ago for $250,000. They have a current "quick sale" value of $30,000. However, for an investment of $100,000 now, they can be adequate for another 4 years, after which they could be sold for an estimated $40,000. Alternatively, new controlled-environment rooms cost $300,000, have an expected 10-year economic life, and a $50,000 salvage value after that time. Determine whether the company should upgrade the existing controlled-environment rooms or purchase new ones. Use a MARR of 12% per year and assume that used controlled-environment rooms will always be available.

Defender Replacement Value

9.22 From the data shown below, determine the trade-in value of machine X that will render its AW the same as that of machine Y. Use an interest rate of 8% per year.

	Machine X	Machine Y
Market value, $	?	−80,000
Annual cost, $ per year	−60,000	−40,000 year 1, increasing by 2000 per year thereafter
Salvage value, $	15,000	20,000
Life, years	3	5

9.23 A company that makes micro motion compact coriolis meters purchased a new packaging system for $600,000. The estimated salvage value was $28,000 after 10 years. Currently the expected remaining life is 7 years with an AOC of $27,000 per year and an estimated salvage value of $40,000. The company is considering early replacement of the system with one that costs $370,000 and has a 12-year economic service life, a $22,000 salvage value, and an estimated AOC of $50,000 per year. If the MARR for the corporation is 12% per year, find the minimum trade-in value necessary now to make the replacement economically advantageous.

9.24 Hydrochloric acid, which fumes at room temperatures, creates a very corrosive work environment. A mixing machine, working in this environment, is deteriorating fast and can be used for only one more year, at which time it will be scrapped. It was purchased 3 years ago for $88,000 and its operating cost for the next year is expected to be $63,000. A more corrosion-resistant challenger will cost $226,000 with an operating cost of $48,000 per year. It is expected to have a $60,000 salvage value after its 10-year ESL. At an interest rate of 15% per year, what minimum trade-in value will make the challenger economically attractive?

9.25 Machine A was purchased 5 years ago for $90,000. Its operating cost is higher than expected, so it will be used for only 4 more years. Its operating cost this year will be $40,000, increasing by $2000 per year through the end of its useful life. The challenger, machine B, will cost $150,000 with a $50,000 salvage value after its 10-year ESL. Its operating cost is expected to be $10,000 for year 1, increasing by $500 per year thereafter. What is the market value for machine A that would make the two machines equally attractive at an interest rate of 12% per year?

Replacement Study Over a Study Period

9.26 The market values and M&O costs associated with a presently owned machine and a possible replacement are shown below. The plant manager has told you that he is interested only in what happens over the next three years and that if the defender is to be replaced, it must be replaced now or kept for the entire 3-year study period. Using an interest rate of 10% per year, determine whether or not the defender should be replaced.

	Defender		Challenger	
Year	Market value, $	M&O cost, $	Market value, $	M&O cost, $
0	40,000		80,000	
1	32,000	−55,000	65,000	−37,000
2	23,000	−55,000	39,000	−37,000
3	11,000	−55,000	20,000	−37,000
4			19,000	−38,000
5			11,000	−39,000

9.27 An engineer at a fiber optic manufacturing company is considering two robots to reduce costs in a production line. In-place robot X has a current market value of $82,000, annual maintenance and operation (M&O) costs of $30,000, and salvage values of $50,000, $42,000, and $35,000 if retained 1, 2, and 3 more years, respectively. The challenging robot Y has a first cost of $97,000, annual M&O costs of $27,000 whenever it is purchased, and salvage values of $66,000, $51,000, and $42,000 after 1, 2, and 3 years, respectively. What is the best economic plan if a 2-year study period is used at an interest rate of 12% per year?

9.28 A machine purchased 3 years ago for $140,000 is now too slow to satisfy increased demand. The machine can be upgraded now for $70,000 or sold to a smaller company for $40,000. The current machine will have an annual operating cost of $85,000 per year and a $30,000 salvage value in 3 years. If upgraded, the presently owned machine will definitely be retained for 3 more years. The replacement, which will serve the company now and for at least 8 years, will cost $220,000. Its salvage value will be $50,000 for years 1 through 5; $20,000 after 6 years; and $10,000 thereafter. It will have an estimated operating cost of $65,000 per year. The company asks you to perform an economic analysis at 15% per year using a 3-year planning horizon. Should the company replace the presently owned machine now, or do it 3 years from now? What are the AW values?

9.29 Two processes can be used for producing a polymer that reduces friction loss in engines. Process K will use a presently owned machine that has a current market value of $160,000, an operating cost of $7000 per month, and a salvage value of $50,000 after 1 year and $40,000 after its maximum 2-year life. Used machines of this type can be purchased and the same estimates can be used for a period of 1 or 2 years. Process L will utilize a new machine that has a first cost of $210,000, an operating cost of $5000 per month, and salvage values of $100,000 after 1 year, $70,000 after 2 years, $45,000 after 3 years, and $26,000 after its maximum 4-year life. You have been asked to determine which process is better using a study period of (a) 1 year, (b) 2 years, and (c) 3 years. The company's MARR is 12% per year compounded monthly.

9.30 Excelon is looking for cost-cutting measures. One of the engineers determined that the equivalent annual worth of an existing machine over its remaining useful life of 1 or 2 years will be $−70,000 per year. The engineer also determined that used machines like the one currently in use are no longer available. However, the machine can be replaced with one that is more advanced that will have an AW of −$80,000 if it is kept for 2 years or less, −$75,000 if it is kept between 3 and 4 years, and −$65,000 if it is kept for 5 to 10 years. If the company uses a 3-year planning period and an interest rate of 15% per year, when should the company replace the machine – now or in 2 years – and at what AW for the next 3 years?

ADDITIONAL PROBLEMS AND FE EXAM REVIEW QUESTIONS

9.31 In a replacement study, the correct value for the first cost of the challenger is:
a. the cost when it is purchased.
b. the first cost minus the trade-in value of the defender.
c. its first cost plus the trade-in value of the defender.
d. the book value of the defender.

9.32 A replacement analysis is most objectively conducted from the viewpoint of:
a. an outsider.
b. a consultant.
c. a non-owner.
d. any of the above.

9.33 The economic service life of an asset is:
a. the length of time required to recover the first cost of the asset.
b. the time when the operating cost is at a minimum.
c. the time when the salvage value goes below 25% of the first cost.
d. the time when the AW of the asset is at a minimum.

9.34 In a one-year-later replacement analysis, if all estimates are still current and the year is n_D, the action that should be taken is:
a. keep the defender one more year.
b. replace the defender with the challenger.
c. look for a new challenger and calculate its AW.
d. keep the defender until its market value is equal to the estimated salvage value of the challenger.

9.35 For the data shown, the economic service life of the challenger is:
a. 2 years
b. 3 years
c. 4 years
d. 5 years

Years Retained	AW of Defender, $	AW of Challenger, $
1	−145,000	−136,000
2	−96,429	−126,000
3	−63,317	−92,000
4	−39,321	−53,000
5	−49,570	−38,000

9.36 A milling machine with enhanced CNC controls that allow for high-speed machining of free-form parts was purchased two years ago for $195,000. The company wants to purchase a recently available faster model with 8-axis control for $240,000. The presently owned machine can be sold today for $105,000. Its operating costs over the past 2 years have been $30,000 per year. The value that should be used as P for the presently owned machine is:
a. $240,000
b. $195,000
c. $105,000
d. $30,000

9.37 An industrial engineer with an MBA degree is trying to get the company's stock to rise by cutting costs. He determined that the equivalent annual worth of an existing machine over its remaining useful life of 1 or 2 years will be −$74,000. The IE also determined that the machine can be replaced with a more advanced model that will have AW = −$84,000 if it is kept for 2 years or less, −$73,000 if it is kept between 3 and 4 years, and −$65,000 if it is kept for 5 to 10 years. The company uses a 3-year planning period. At an interest rate of 15% per year, the IE should recommend that the existing machine be replaced:

a. now.
b. one year from now.
c. two years from now.
d. it should not be replaced.

9.38 At an interest rate of 10% per year, the economic service life of an asset that has a current market value of $15,000 and the expected cash flows shown is:

a. 1 year
b. 2 years
c. 3 years
d. 4 years

Year	Salvage Value at End of Year, $	Operating Cost, $
1	10,000	−50,000
2	8,000	−53,000
3	5,000	−60,000
4	0	−68,000

9.39 In trying to decide whether or not to replace a sorting/baling machine in a solid waste recycling operation, an engineer calculated the annual worth values for the in-place machine and a challenger. On the basis of these costs, the defender should be replaced:

a. now.
b. 1 year from now.
c. 2 years from now.
d. 3 years from now.

Year	AW of Defender, $ per year	AW of Challenger, $ per year
1	−24,000	−31,000
2	−25,500	−28,000
3	−26,900	−25,000
4	−27,000	−25,900
5	−28,000	−27,500

Effects of Inflation

Digital Vision/Getty Images

This chapter concentrates upon understanding and calculating the effects of inflation in time value of money computations. Inflation is a reality that we deal with nearly every day in our professional and personal lives.

The annual inflation rate is closely watched and historically analyzed by government units, businesses, and industrial corporations. An engineering economy study can have different outcomes in an environment in which inflation is a serious concern compared to one in which it is of minor consideration. The inflation rate is sensitive to real, as well as perceived, factors of the economy. Factors such as the cost of energy, interest rates, availability and cost of skilled people, scarcity of materials, political stability, and other less tangible factors have short-term and long-term impacts on the inflation rate. In some industries, it is vital that the effects of inflation be integrated into an economic analysis. The basic techniques to do so are covered here.

LEARNING OUTCOMES

Impact of inflation	1. Determine the difference inflation makes between money now and money in the future.
PW with inflation	2. Calculate present worth with an adjustment for inflation.
FW with inflation	3. Determine the real interest rate and the inflation-adjusted MARR, and calculate a future worth with an adjustment for inflation.
AW with inflation	4. Calculate an annual amount in future dollars that is equivalent to a specified present or future sum.
Spreadsheets	5. Use a spreadsheet to perform inflation-adjusted equivalence calculations.

10.1 UNDERSTANDING THE IMPACT OF INFLATION

Podcast available

All of us are very aware that $20 now does not purchase the same amount as $20 did in 2005 or 2010 and significantly less than in 2000. This is primarily because of inflation. *Inflation is the increase in the amount of money necessary to purchase the same amount of a product or service before the inflated price was present.* It occurs because the value of the currency has decreased, so it takes more of the currency to obtain the same amount of goods or services. Associated with inflation is an increase in the money supply, that is, the government prints more dollars, while the supply of goods and services does not increase.

To consider inflation when making comparisons between monetary amounts that occur at different time periods, the *different-value* dollars must be converted to *constant-value* dollars so that they represent the same purchasing power. *Purchasing power measures the value of currency in terms of the quantity and quality of the goods or services that one unit of money will buy.* Over time, inflation decreases purchasing power because less goods or services can be purchased for the same amount of money.

There are two ways to make meaningful economic calculations when the currency is changing in value, that is, when inflation is considered:

- Convert the amounts that occur in different time periods into equivalent amounts that force the currencies to have the same value. This is accomplished *before* any time value of money calculation is made.
- Change the interest rate used in the economic evaluation to account for the changing currencies (inflation) *plus* the time value of money.

All the computations that follow work for any country's currency; dollars are used here.

The first method above is referred to as making calculations in constant-value (CV) dollars. Money in one time period is brought to the same value as money in a later period as follows:

$$\text{Dollars in period } t_1 = \frac{\text{Dollars in period } t_2}{(1 + \text{Inflation rate between } t_1 \text{ and } t_2)} \qquad \textbf{[10.1]}$$

where:

Dollars in period t_1 = Constant-value (CV) dollars, also called *today's dollars*

Dollars in period t_2 = Future dollars, also called *inflated* or *then-current dollars.*

If f represents the inflation rate *per period* (year) and n is the number of periods (years) between t_1 and t_2, Equation [10.1] may be used to express future dollars in CV dollar terms, or vice versa.

$$\textbf{CV dollars} = \frac{\textbf{Future dollars}}{\mathbf{(1 + f)^n}} \qquad \textbf{[10.2]}$$

$$\textbf{Future dollars} = \textbf{CV dollars } \mathbf{(1 + f)^n} \qquad \textbf{[10.3]}$$

Equations [10.2] and [10.3] are used to find the average inflation rate over a given period of time by substituting the actual monetary amounts and solving for f.

(Note that the average inflation rate cannot be determined by taking an arithmetic average.) As an example, use the price of 87 octane (regular unleaded) gasoline. From 1986 to 2006 and on to 2012, the average pump price in the United States went from \$0.92 to \$2.86 to \$3.64 per gallon. Solving Equation [10.3] for f from 1986 to 2006 yields an average price increase of 5.83% per year over the 20 years.

$$2.86 = 0.92(1 + f)^{20}$$
$$(1 + f) = 3.1087^{0.05}$$
$$f = 5.83\% \text{ per year}$$

A similar computation over the 6 years from 2006 to 2012 results in $f = 4.10\%$ per year.

Assume, for a minute, that gas price increases continue at an average of 4.10% per year from 2012 on. Predicted average prices are:

$$\text{In 2014: } 3.64(1.0410)^2 = \$3.94 \text{ per gallon}$$
$$\text{In 2015: } 3.64(1.0410)^3 = \$4.11 \text{ per gallon}$$
$$\text{In 2016: } 3.64(1.0410)^4 = \$4.27 \text{ per gallon}$$

A 4.10% annual increase causes a 17.3% increase from 2012 to 2016. In some areas of the world, hyperinflation may average 40 to 50% per year, with some spikes as high as 1200% for short periods of time. Hyperinflation is discussed further in Section 10.3.

The *Consumer Price Index (CPI)* measures the average change over time in the price level of consumer goods and services purchased by households. In the United States, the CPI is published monthly by the federal government's Bureau of Labor Statistics. Typically, the CPI is based on a *"market basket" of goods and services* such as food, housing, apparel, utilities, and education; items that do not tend to have large price spikes (increases or decreases) over short periods of time. There is considerable criticism of the bases used to calculate the CPI, because it does not consider quality differences, and some items that were not available historically and have had significant price decreases, such as computers and other electronics, are not included in the market basket. Additionally, it does not include items such as taxes, and the cost of crime. However, it is a very widely quoted index, used as the basis for other rates, one being the interest rate for all credit cards issued by banks and other financial institutions.

Placed into an industrial or business context, even at a reasonably low inflation rate averaging, say, 4% per year, equipment or services with a first cost of \$209,000 will increase by 48% to \$309,000 over a 10-year span. This is before any rate of return requirement is placed upon the equipment's revenue-generating ability. Clearly, inflation must be taken into account.

There are actually three different inflation-related rates that are important: the real interest rate (i), the market interest rate (i_f), and the inflation rate (f). Only the first two are interest rates.

Real or inflation-free interest rate i. This is the rate at which interest is earned when the effects of changes in the value of currency (inflation) have been removed. Thus, the real interest rate presents an actual gain in purchasing power. (The equation used to calculate i, with the influence of inflation removed, is derived later in Section 10.3.) The real rate of return that generally applies for individuals is approximately 3.5% per year. This is the "safe investment" rate. The required real rate for corporations (and many individuals) is set above this safe rate when a MARR is established without an adjustment for inflation.

Inflation-adjusted interest rate i_f.   As its name implies, this is the interest rate that has been adjusted to take inflation into account. The *market interest rate,* which is the one we hear everyday, is an inflation-adjusted rate. This rate is a combination of the real interest rate i and the inflation rate f, and, therefore, it changes as the inflation rate changes.

Inflation rate f.   As described above, this is a measure of the rate of change in the value of the currency.

A company's MARR adjusted for inflation is referred to as the inflation-adjusted MARR. The determination of this value is discussed in Section 10.3.

Deflation is the opposite of inflation in that when deflation is present, the purchasing power of the monetary unit is greater in the future than at present. That is, it will take fewer dollars in the future to buy the same amount of goods or services as it does today. In deflationary economic conditions, the market interest rate is always less than the real interest rate.

Temporary price deflation may occur in specific sectors of the economy due to the introduction of improved products, cheaper technology, or imported materials or products that force current prices down. In normal situations, prices equalize at a competitive level after a short time. However, deflation over a short time in a specific sector of an economy can be orchestrated through *dumping*. An example of dumping may be the importation of materials, such as steel, cement, or cars, into one country from international competitors at very low prices compared to current market prices in the targeted country. The prices will go down for the consumer, thus forcing domestic manufacturers to reduce their prices in order to compete for business. If domestic manufacturers are not in good financial condition, they may fail, and the imported items replace the domestic supply. Prices may then return to normal levels and, in fact, become inflated over time, if competition has been significantly reduced.

On the surface, having a moderate rate of deflation sounds good when inflation has been present in the economy over long periods. However, if deflation occurs at a more general level, say nationally, it is likely to be accompanied by the lack of money for new capital. Another result is that individuals and families have less money to spend due to fewer jobs, less credit, and fewer loans available; an overall "tighter" money situation prevails. As money gets tighter, less is available to be committed to industrial growth and capital investment. In the extreme case, this can evolve over time into a deflationary spiral that disrupts the entire economy. This has happened on occasion, notably in the United States during the Great Depression of the 1930s.

Engineering economy computations that consider deflation use the same relations as those for inflation. Equations [10.2] and [10.3] are used, except the deflation rate is a $-f$ value. For example, if deflation is estimated to be 2% per year, an asset that costs $10,000 today would have a first cost 5 years from now determined by Equation (10.3).

$$10{,}000(1 - f)^n = 10{,}000(0.98)^5 = 10{,}000(0.9039) = \$9039$$

10.2 PW CALCULATIONS ADJUSTED FOR INFLATION

Present worth calculations may be performed using either of the two methods described at the beginning of Section 10.1—constant-value (CV) dollars and the real interest rate i, or future dollars and the inflation-adjusted rate i_f. By way of introduction to these two methods, consider an asset that may be purchased now or in any of the next four years at the future dollar equivalent. Table 10.1 (columns 2 and 3) indicates a first cost of $5000 now, increasing each future year at 4% per year inflation. The estimated cost four years from now is $5849. However, the CV dollar cost is always $5000 (column 4).

By the first method, use CV dollars and a real interest rate i to find the PW now of the first cost for any year in the future. By this method, all effects of inflation are removed before the real interest rate is applied. If the real interest rate is 10% per year, the equivalent cost now for each future year t is PW $= 5000(P/F,10\%,t)$, as calculated in column 5. In CV dollar terms, $3415 now will buy an asset worth $5000 four years hence. Similarly, $3757 now will buy the same asset if it is worth $5000 three years hence.

Figure 10.1 shows the differences over a 4-year period of the constant-value amount of $5000, the future-dollar costs at 4% inflation, and the present worth at 10% real interest with inflation considered. The effect of compounded inflation and interest rates increases rapidly, as indicated by the shaded area.

The second method that utilizes future-dollar estimates and adjusts the interest rate for inflation is more commonly used to calculate PW values. Consider the P/F formula, where i is the real interest rate.

$$P = F\frac{1}{(1 + i)^n}$$

TABLE 10.1 **Inflation Calculations and Present Worth Calculation Using Constant-Value Dollars (f = 4%, i = 10%)**

Year t (1)	Cost Increase Due to 4% Inflation (2)	Cost in Future Dollars (3)	Future Cost in Constant-Value Dollars (4) = (3)$(P/F,4\%,t)$	Present Worth at Real i = 10% (5) = (4)$(P/F,10\%,t)$
0		$5000	$5000	$5000
1	$5000(0.04) = \$200	5200	5000	4545
2	5200(0.04) = 208	5408	5000	4132
3	5408(0.04) = 216	5624	5000	3757
4	5624(0.04) = 225	5849	5000	3415

FIGURE 10.1

Comparison of constant-value dollars, future dollars, and their present worth values.

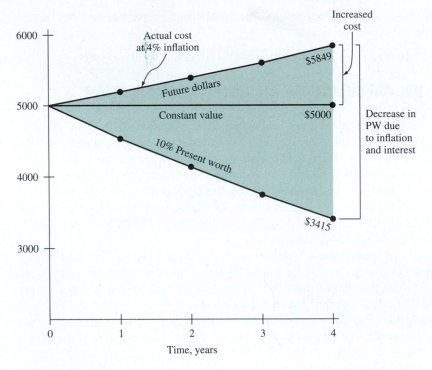

The *F*, which is a future-dollar amount with inflation built in, can be converted into CV dollars by using Equation [10.2].

$$P = \frac{F}{(1+f)^n} \frac{1}{(1+i)^n}$$

$$= F\frac{1}{(1+i+f+if)^n} \qquad [10.4]$$

If the term $i + f + if$ is defined as i_f, the equation becomes

$$P = F\frac{1}{(1+i_f)^n} = F(P/F,i_f,n) \qquad [10.5]$$

The *inflation-adjusted interest rate* i_f is defined as

$$i_f = i + f + if \qquad [10.6]$$

For a real interest rate of $i = 10\%$ per year and an inflation rate of $f = 4\%$ per year, Equation [10.6] yields an inflated interest rate of 14.4%.

$$i_f = 0.10 + 0.04 + 0.10(0.04) = 0.144$$

Table 10.2 illustrates the use of $i_f = 14.4\%$ in PW calculations for $5000 now, which inflates to $5849 in future dollars 4 years hence. As shown in column 5, the present worth for each year is the same as column 5 of Table 10.1.

TABLE 10.2 **Present Worth Calculation Using an Inflation-Adjusted Interest Rate**

Year n (1)	Cost Increase at $f = 4\%$ (2)	Cost in Future Dollars (3)	$(P/F,14.4\%,n)$ (4)	PW (5) = (3)(4)
0	—	$5000	1	$5000
1	$200	5200	0.8741	4545
2	208	5408	0.7641	4132
3	216	5624	0.6679	3757
4	225	5849	0.5838	3415

The present worth of any series of cash flows—equal, arithmetic gradient, or geometric gradient—can be found similarly. That is, either i or i_f is introduced into the P/A, P/G, or P_g factors, depending upon whether the cash flow is expressed in CV dollars or future dollars.

The following two examples and the spreadsheet Example 10.6 at the end of the chapter illustrate how inflation can be removed from or included in PW calculations. When worked correctly, the PW method will result in the same numerical values for both constant-value dollar calculations using a real interest rate i, and future-dollar calculations using an inflation-adjusted rate i_f.

EXAMPLE 10.1

Joey is one of several winners who shared a lottery ticket. There are three plans offered to receive the after-tax proceeds.

Plan 1: $100,000 now

Plan 2: $15,000 per year for 8 years beginning 1 year from now. Total is $120,000.

Plan 3: $45,000 now, another $45,000 four years from now, and a final $45,000 eight years from now. Total is $135,000.

Joey, a quite conservative person financially, plans to invest all of the proceeds as he receives them. He expects to make a real return of 6% per year. Use the 8-year time frame and an average inflation of 4% per year to determine which plan provides the best deal.

Solution

Select the plan with the largest PW now with inflation considered. Either of the two methods discussed earlier may be used. In the first method, all amounts are converted to CV dollars at 4% inflation, and the PW value is determined at the real rate of 6% per year. For the second method, find the PW of the quoted future amounts at the inflation-adjusted rate of 10.24% determined by Equation [10.6].

$$i_f = 0.06 + 0.04 + (0.06)(0.04) = 0.1024$$

Using the i_f method to obtain PW values for each plan,

$PW_1 = \$100,000$ since all proceeds are taken immediately

$PW_2 = 15,000(P/A,10.24\%,8) = 15,000(5.2886) = \$79,329$

$PW_3 = 45,000[1 + (P/F,10.24\%,4) + (P/F,10.24\%,8)]$
$\quad\quad = 45,000(2.1355) = \$96,099$

It is economically best to take the money now (plan 1).

Comment: This result supports the principle that it is better to take the money as early as it is available because it is usually possible to make a better return investing it yourself. In this case, plan 1 is marginally the best only if all proceeds are invested and they earn at a real rate of 6% per year. Plan 3 is a close contender; if i_f goes down to 8.81% per year, plans 1 and 3 are economically equivalent with PW values of $100,000.

EXAMPLE 10.2   A self-employed chemical engineer is on contract with Dow Chemical, currently working in a relatively high-inflation country. She wishes to calculate a project's PW with estimated costs of $35,000 now and $7000 per year for 5 years beginning 1 year from now with increases of 12% per year thereafter for the next 8 years. Use a real interest rate of 15% per year to make the calculations (*a*) without an adjustment for inflation and (*b*) considering inflation at a rate of 11% per year.

Solution

a. Figure 10.2 presents the cash flows. The PW without an adjustment for inflation is found using $i = 15\%$ and $g = 12\%$ in Equation [2.7] for the geometric series.

$$PW = -35,000 - 7000(P/A,15\%,4)$$

$$-\left\{ \frac{7000\left[1 - \left(\dfrac{1.12}{1.15}\right)^9\right]}{0.15 - 0.12} \right\}(P/F,15\%,4)$$

$$= -35,000 - 19,985 - 28,247$$

$$= \$-83,232$$

In the P/A factor, $n = 4$ because the $7000 cost in year 5 is the A_1 term in Equation [2.7].

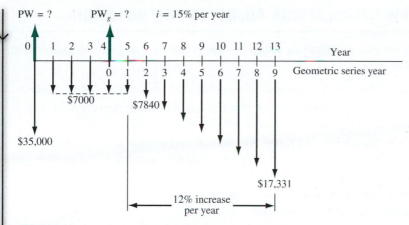

FIGURE 10.2   Cash flow diagram, Example 10.2.

b. To adjust for inflation, calculate the inflated interest rate by Equation [10.6].

$$i_f = 0.15 + 0.11 + (0.15)(0.11) = 0.2765$$

$$PW = -35,000 - 7000(P/A,27.65\%,4)$$

$$-\left\{ \frac{7000\left[1 - \left(\dfrac{1.12}{1.2765}\right)^9\right]}{0.2765 - 0.12} \right\}(P/F,27.65\%,4)$$

$$= -35,000 - 7000(2.2545) - 30,945(0.3766)$$

$$= \$-62,436$$

This result demonstrates that in a high-inflation economy, when negotiating the amount of the payments to repay a loan, it is economically advantageous for the borrower to use future (inflated) dollars whenever possible to make the payments. The present value of future inflated dollars is significantly less when the inflation adjustment is included. And the higher the inflation rate, the larger the discounting because the P/F and P/A factors decrease in size.

The last example seems to add credence to the "buy now, pay later" philosophy of financial management. However, at some point, the debt-ridden company or individual will have to pay off the debts and the accrued interest with the inflated dollars. If cash is not readily available, the debts cannot be repaid. This can happen, for example, when a company unsuccessfully launches a new product, when there is a serious downturn in the economy, or when an individual loses a salary. In the longer term, this "buy now, pay later" approach must be tempered with sound financial practices now, and in the future.

10.3 FW CALCULATIONS ADJUSTED FOR INFLATION

Podcast available

In future worth calculations, a future amount F can have any one of four different interpretations:

Case 1. The *actual amount* of money that will be accumulated at time n.

Case 2. The *purchasing power* of the actual amount accumulated at time n, but stated in today's (constant-value) dollars.

Case 3. The number of *future dollars required* at time n to maintain the same purchasing power as a given amount today; that is, inflation is considered, but interest is not.

Case 4. The number of dollars required at time n to *maintain purchasing power and earn a stated real interest rate.*

Depending upon which interpretation is intended, the F value is calculated differently, as described below. Each case is illustrated.

Case 1: Actual Amount Accumulated

It should be clear that F, the actual amount of money accumulated, is obtained using the inflation-adjusted (market) interest rate.

$$F = P(1 + i_f)^n = P(F/P,i_f,n) \qquad [10.7]$$

For example, when a market rate of 10% is quoted, the inflation rate is included. Over a 7-year period, $1000 will accumulate to

$$F = 1000(F/P,10\%,7) = \$1948$$

Case 2: Constant-Value with Purchasing Power

The purchasing power of future dollars is determined by first using the market rate i_f to calculate F and then deflating the future amount through division by $(1 + f)^n$.

$$F = \frac{P(1 + i_f)^n}{(1 + f)^n} = \frac{P(F/P,i_f,n)}{(1 + f)^n} \qquad [10.8]$$

This relation, in effect, recognizes the fact that inflated prices mean $1 in the future purchases less than $1 now. The percentage loss in purchasing power is a measure of how much less. As an illustration, consider the same $1000 now, a 10% per year market rate, and an inflation rate of 4% per year. In 7 years, the purchasing power has risen, but only to $1481.

$$F = \frac{1000(F/P,10\%,7)}{(1.04)^7} = \frac{\$1948}{1.3159} = \$1481$$

This is $467 (or 24%) less than the $1948 actually accumulated at 10% (case 1). Therefore, we conclude that 4% inflation over 7 years reduces the purchasing power of money by 24%.

Also for case 2, the future amount of money accumulated with today's buying power could equivalently be determined by calculating the real interest rate and using it in the F/P factor to compensate for the decreased purchasing power of the dollar. This *real interest rate* is the i in Equation [10.6].

$$i_f = i + f + if$$
$$= i(1 + f) + f$$
$$i = \frac{i_f - f}{1 + f} \qquad \text{[10.9]}$$

The real interest rate i represents the rate at which today's dollars expand with their *same purchasing power* into equivalent future dollars. An inflation rate larger than the market interest rate leads to a negative real interest rate. The use of this interest rate is appropriate for calculating the future worth of an investment (such as a savings account or money market fund) when the effect of inflation must be removed. For the example of $1000 in today's dollars from Equation [10.9]

$$i = \frac{0.10 - 0.04}{1 + 0.04} = 0.0577, \text{ or } 5.77\%$$
$$F = 1000(F/P,5.77\%,7) = \$1481$$

The market interest rate of 10% per year has been reduced to a real rate that is less than 6% per year because of the erosive effects of inflation.

Case 3: Future Amount Required, No Interest

This case recognizes the fact that prices increase when inflation is present. Simply put, future dollars are worth less, so more are needed. No interest rate is considered at all in this case. This is the situation if someone asks "How much will a car cost in 5 years if its current cost is $25,000 and its price will increase by 3% per year?" (The answer is $28,982.) No interest rate, only inflation, is involved. To find the future cost, substitute f for the interest rate in the F/P factor.

$$F = P(1 + f)^n = P(F/P,f,n) \qquad \text{[10.10]}$$

Reconsider the $1000 used previously. If it is escalating at exactly the inflation rate of 4% per year, the amount 7 years from now will be

$$F = 1000(F/P,4\%,7) = \$1316$$

Case 4: Inflation and Real Interest

This is the case applied when a MARR is established. Maintaining purchasing power and earning interest must account for both increasing prices (case 3) and the time value of money. If the growth of capital is to keep up, funds must grow at a rate equal to or above the real interest rate i plus the inflation rate f. Thus, to make a *real rate of return* of 5.77% when the inflation rate is 4%, i_f is the market (inflation-adjusted) rate that must be used. For the same $1000 amount,

$$i_f = 0.0577 + 0.04 + 0.0577(0.04) = 0.10$$
$$F = 1000(F/P,10\%,7) = \$1948$$

This calculation shows that $1948 seven years in the future will be equivalent to $1000 now with a real return of $i = 5.77\%$ per year and inflation of $f = 4\%$ per year. The calculation is exactly the same as that for case 1.

Table 10.3 summarizes which rate is used in the equivalence formulas for the different interpretations of F. The calculations made in this section reveal that:

- Case 1: $1000 now at a market rate of 10% per year accumulates to $1948 in 7 years.
- Case 2: The $1948 has the purchasing power of $1481 of today's dollars if $f = 4\%$ per year.
- Case 3: An item with a cost of $1000 now will cost $1316 in 7 years at an inflation rate of 4% per year.
- Case 4: It takes $1948 of future dollars to be equivalent to the $1000 now at a real interest rate of 5.77% with inflation considered at 4%.

Most corporations evaluate alternatives at a MARR large enough to cover inflation plus some return greater than their cost of capital, and significantly higher than the safe investment return of approximately 3.5% mentioned earlier. Therefore, for case 4, the resulting MARR will normally be higher than the market rate i_f. Define the symbol MARR_f as the inflation-adjusted MARR, which is calculated in a fashion similar to i_f.

$$\text{MARR}_f = i + f + i(f) \qquad \text{[10.11]}$$

TABLE 10.3 Calculation Methods for Various Future Worth Interpretations

Future Worth Desired	Method of Calculation	Example for $P = \$1000$, $n = 7$, $i_f = 10\%$, $f = 4\%$
Case 1: Actual dollars accumulated	Use stated market rate i_f in equivalence formulas	$F = 1000(F/P,10\%,7)$ $= \$1948$
Case 2: Purchasing power of accumulated dollars in terms of today's dollars	Use market rate i_f in equivalence and divide by $(1 + f)^n$ or Use real i	$F = \dfrac{1000(F/P,10\%,7)}{(1.04)^7}$ or $F = 1000(F/P,5.77\%,7)$ $= \$1481$
Case 3: Dollars required for same purchasing power	Use f in place of i in equivalence formulas	$F = 1000(F/P,4\%,7)$ $= \$1316$
Case 4: Future dollars to maintain purchasing power and to earn a return	Calculate i_f and use in equivalence formulas	$F = 1000(F/P,10\%,7)$ $= \$1948$

The real rate of return i used here is the required rate for the corporation relative to its cost of capital. (Cost of capital was introduced in Section 1.3 and is detailed in Section 13.5.) Now the future worth F, or FW, is calculated as

$$F = P(1 + MARR_f)^n = P(F/P, MARR_f, n) \qquad [10.12]$$

For example, if a company has a cost of capital of 10% per year and requires that a project return 3% per year, the real return is $i = 13\%$. The inflation-adjusted MARR is calculated by including the inflation rate of, say, 4% per year. Then, the project PW, AW, or FW will be determined at the rate

$$MARR_f = 0.13 + 0.04 + 0.13(0.04) = 17.52\%$$

Abbott Mining Systems wants to determine whether it should buy now or buy later for upgrading a piece of equipment used in deep mining operations. If the company selects plan N, the equipment will be purchased now for $200,000. However, if the company selects plan L, the purchase will be deferred for 3 years when the cost is expected to rise rapidly to $340,000. Abbott is ambitious; it expects a real MARR of 12% per year. The inflation rate in the country has averaged 6.75% per year. From only an economic perspective, determine whether the company should purchase now or later (*a*) when inflation is not considered and (*b*) when inflation is considered.

EXAMPLE 10.3

Solution

a. *Inflation not considered:* This is case 2 with a real rate, or MARR, of $i = 12\%$ per year. The cost of plan L is $340,000 three years hence. Calculate the FW value for plan N three years from now.

$$FW_N = -200,000(F/P,12\%,3) = \$-280,986$$
$$FW_L = \$-340,000$$

Purchase the upgrade now.

b. *Inflation considered:* This is case 4; there is a real rate (12%), and inflation is 6.75%. First, compute the inflation-adjusted MARR by Equation [10.11].

$$MARR_f = 0.12 + 0.0675 + 0.12(0.0675) = 0.1956$$

Compute the FW value for plan N in future dollars.

$$FW_N = -200,000(F/P,19.56\%,3) = \$-341,812$$
$$FW_L = \$-340,000$$

Purchasing later is now selected, because it requires fewer equivalent future dollars. The inflation rate of 6.75% per year has raised the equivalent future worth of costs by 21.6% to $341,812. This is the same as an increase of 6.75% per year, compounded over 3 years, or $(1.0675)^3 - 1 = 21.6\%$.

Most countries have inflation rates in the range of 2% to 5% per year, but *hyperinflation* is a problem in countries where political instability, overspending by the government, weak international trade balances, etc., are present. Hyperinflation rates may be very high—10% to 100% *per month*. In these cases, the government may take drastic actions to decrease inflation: redefine the currency in terms of the currency of another country, control banks and corporations, or control the flow of capital into and out of the country.

In a hyperinflated environment, people usually spend all their money immediately since the cost will be so much higher the next month, week, or day. To appreciate the disastrous effect of hyperinflation on a company's ability to keep up, rework Example 10.3*b* using an inflation rate of 10% per month, that is, a nominal 120% per year (not considering the compounding of inflation). The FW_N amount skyrockets and plan L is a clear choice. Of course, in such an environment the $340,000 purchase price for plan L 3 years hence would obviously not be guaranteed, so the entire economic analysis is unreliable. Good economic decisions in a hyperinflated economy are very difficult to make, since the estimated future values are totally unreliable and the future availability of capital is uncertain.

10.4   AW CALCULATIONS ADJUSTED FOR INFLATION

Podcast
available

It is particularly important in capital recovery calculations used for AW analysis to include inflation because current capital dollars must be recovered with future inflated dollars. Since future dollars have less purchasing power than today's dollars, it is obvious that more dollars will be required to recover the present investment. This suggests the use of the inflated interest rate in the A/P formula. For example, if $1000 is invested today at a real interest rate of 5.77% per year when the inflation rate is 4% per year, the equivalent amount that must be recovered each year for 5 years in future dollars is determined at $i_f = 10\%$ per year.

$$A = 1000(A/P,10\%,5) = \$263.80$$

On the other hand, the decreased value of dollars through time means that investors can spend fewer present (higher-value) dollars to accumulate a specified amount of future (inflated) dollars. This suggests the use of an inflation-adjusted interest rate in the A/F formula to obtain a lower A value. The annual equivalent (with adjustment for inflation) of $F = \$1000$ five years from now in future dollars is

$$A = 1000(A/F,10\%,5) = \$163.80$$

This result is discussed further in the next example.

For comparison, the equivalent annual amount to accumulate $F = \$1000$ at a real $i = 5.77\%$, (without adjustment for inflation) is $1000(A/F,5.77\%,5) = \$178.21$. Thus, when F is fixed, uniformly distributed future *costs* (not income) should be spread over as long a time period as possible so that the leveraging effect of inflation will reduce the payment ($163.80 versus $178.21 without an inflation adjustment).

a. What annual deposit is required for 5 years to accumulate an amount of money with the same purchasing power as $821.93 today, if the market interest rate is 10% per year and inflation is 4% per year?

b. What is the real interest rate?

EXAMPLE 10.4

Solution

a. First, find the actual number of future (inflated) dollars required 5 years from now. This is case 3, as described earlier.

$$F = \text{(present purchasing power)}(1 + f)^5 = 821.93(1.04)^5 = \$1000$$

The actual amount of the annual deposit is calculated using the market (inflated) interest rate of 10%. This is case 4 using A instead of P.

$$A = 1000(A/F,10\%,5) = \$163.80$$

b. Using Equation [10.9], i is calculated.

$$i = (0.10 - 0.04)/(1.04)$$
$$= 0.0577 \text{ or } 5.77\% \text{ per year}$$

Comment: To put these calculations into perspective, consider the following: If the *inflation rate is* $f = 0\%$ when the real interest rate is 5.77%, the future amount of money with the same purchasing power as $821.93 today is obviously $821.93. Then the annual amount required to accumulate this future amount in 5 years is $A = 821.93(A/F,5.77\%,5) = \$146.48$. This is $31.73 lower than the $178.21 calculated above to accumulate $1000 based on $f = 4\%$. This difference is due to the fact that during inflationary periods, dollars deposited at the beginning of a period have more purchasing power than the dollars returned at the end of the period. To make up the purchasing power difference, more lower-value dollars are required. That is, to maintain equivalent purchasing power at $f = 4\%$ per year, an extra $31.73 per year is required.

This logic explains why, in times of increasing inflation, lenders of money (credit card companies, mortgage companies, and banks) tend to further increase their market interest rates. People tend to pay off less of their incurred debt at each payment because they use any excess money to purchase other items before the price is further inflated. Also, the lending institutions must have more dollars in the future to cover the expected higher costs of lending money. All this is due to the spiraling effect of increasing inflation. Breaking this cycle is difficult for the individual and family to do, and much more difficult to alter at a national level.

10.5　USING SPREADSHEETS TO ADJUST FOR INFLATION

When using a spreadsheet, adjust for inflation in PW, AW, and FW calculations by entering the correct percentage rate into the appropriate function. For example, suppose PW is required for a receipt of $A = \$15,000$ per year for 8 years at a real rate of $i = 5\%$ per year. To adjust for an expected inflation rate $f = 3\%$ per year, first determine the inflation-adjusted rate i_f by using Equation [10.6] or by entering its equivalent relation in decimal form = 0.05 + 0.03 + 0.05*0.03 to obtain 0.0815 or 8.15%.

The single-cell PV functions used to determine the PW value with and without inflation considered follow. The minus keeps the sign of the answer positive.

With inflation: = −PV(8.15%,8,15000)   PW = $85,711

Without inflation: = −PV(5%,8,15000)   PW = $96,948

The next example uses different spreadsheet functions when inflation is considered and the timing of the cash flows varies between alternatives.

EXAMPLE 10.5   This is a reconsideration of the three plans presented in Example 10.1. To recap, the plans are:

Plan 1: Receive $100,000 now.

Plan 2: Receive $15,000 per year for 8 years beginning 1 year from now.

Plan 3: Receive 45,000 now, another $45,000 four years from now, and a final $45,000 eight years from now.

If $i = 6\%$ per year and $f = 4\%$ per year, (a) find the best plan based on PW, (b) determine the worth in future dollars of each plan 8 years from now in inflated dollars, and (c) determine the future value of the plans in terms of today's purchasing power.

Solution

a. Figure 10.3 shows the computation for $i_f = 10.24\%$, which is necessary for each plan's PW value. For plan 1, $PW_1 = \$100,000$. The value $PW_2 = \$79,329$

	A	B	C	D	E	F	G	H
1	Inflated-adjusted rate =	10.24%			= 6% + 4% + 6%*4%			
2								
3	**Year**	**Plan 1**	**Plan 2**	**Plan 3**				
4	0	100,000		45,000				
5	1		15,000	0				
6	2		15,000	0				
7	3		15,000	0				
8	4		15,000	45,000				
9	5		15,000	0				
10	6		15,000	0				
11	7		15,000	0				
12	8		15,000	45,000				
13	**a. PW with inflation**	$100,000	$ 79,329	$ 96,099		a. PW with inflation		
14						Use NPV functions at **10.24%**		
15	**b. FW with inflation**	$218,129	$173,041	$209,619		= − FV(B1,8,,D13)		
16								
17	**c. FW without inflation**	$159,385	$126,439	$153,167		c. Today's purchasing power		
18						Use FV functions at **6%**		
19						= − FV(6%,8,,D13)		
20								
21								

FIGURE 10.3   PW and FW calculations with inflation considered and with today's purchasing power maintained, Example 10.5.

is found using either the NPV or PV function, since the series is uniform. However, for PW_3 the cash flows must be entered with intervening zeros and the NPV function applied. Plan 1 is best with the largest PW value.
  b. Row 15 of Figure 10.3 shows the equivalent future worth of each plan 8 years from now at 10.24% per year using the FV function.
  c. To determine FW while retaining today's purchasing power (case 2), use the real rate of 6% in the FV function. Row 17 shows the significant decrease from the previous FW values with inflation considered. The impact of inflation is large. For plan 3, as an example, the winner could have the equivalent of over $209,000 in 8 years. However, this will purchase only approximately $153,000 worth of goods in terms of today's dollars.

In all cases, plan 1 is the best economically with the largest PW and FW values. Plan 3 runs a close second.

EXAMPLE 10.6

The president of Harling+Down, a company specializing in health care project consulting, just announced today's purchase of a competing corporation for a total of $10 billion. During a brief conversation after the announcement, Dave and Carol, who both work at the headquarters, agreed that a fair estimate for increased net revenue resulting from the acquisition is $2.8 billion per year. Using a study period of 6 years, Dave, who works in engineering, returned to his office and used a spreadsheet's IRR function to determine a rate of return ($i*$) of 17.2%, which exceeds the corporation's expected rate of return of 14% per year on acquisitions. Also, Dave calculated the PW at 14% to obtain $888. Figure 10.4 (left side) shows the results of his analysis (in $ million units).
  At the same time, Carol, who works in finance, completed a similar analysis; however, she took the expected national inflation estimate of 4% into account in her predictions. Thus, the estimates are in constant-value (CV) dollars. The right side of Figure 10.4 details her analysis and the resulting rate of return of 12.7%, which shows the purchase not to be economically justified.
  When Dave and Carol met this afternoon with the president, they disagreed concerning the economic viability of the purchase. Determine which of the two individuals is correct. How should the MARR of 14% and calculated PW values be interpreted?

Solution

Based on Equation [10.6] and a 4%-per-year inflation rate, Carol's expected return of 12.7% is the same as Dave's rate of 17.2%.

$$0.127 + 0.04 + 0.127(0.04) = 0.172 \qquad (17.2\%)$$

They are both correct; Carol's real rate is $i* = 12.7\%$ and Dave's inflation-adjusted rate is $i_f* = 17.2\%$ at an inflation rate of $f = 4\%$. This comparison shows that, when correctly considered, CV-dollar estimates and future-dollar estimates result in the same conclusion.

The dilemma in this case is the MARR of 14%; it is not clear whether it is inflation-adjusted or not. If it is inflation-adjusted, then $MARR_f = 14\%$ and the purchase is viable, since $17.2\% > 14\%$. But, if the MARR is not inflation-adjusted, the 14% real MARR becomes an inflation-adjusted $MARR_f$ of 18.56%. Now, the purchase is not economically justified since $17.2\% < 18.56\%$. Similarly, the 14% real MARR must be compared to Carol's result of 12.7%; the acquisition is not economically justified.

Figure 10.4, row 14, indicates that PW = $888 at both $MARR_f = 14\%$ and at the real MARR = 9.6%. When correctly performed, the PW values will always be the same for cash flows and MARR values that are in future-dollars (left side) and in CV-dollars (right side).

	A	B	C	D	E	F	G	H
1	Dave's estimates			Carol's estimates				
2		Future-dollar			CV-dollar			
3	Year	cash flows, $		Year	cash flows, $			
4	0	-10,000		0	-10,000		Inflation-removed	
5	1	2800		1	2692		estimate, year 1	
6	2	2800		2	2589		= 2800/1.04	
7	3	2800		3	2489			
8	4	2800		4	2393			
9	5	2800		5	2301			
10	6	2800		6	2213			
11	i*	17.2%		i*	12.7%		MARR with	
12							inflation removed	
13	$MARR_f$	14.0%		MARR	9.6%		$= \dfrac{(0.14 - 0.04)}{(1 + 0.04)}$	
14	PW @ $MARR_f$	$888		PW @ MARR	$888			

FIGURE 10.4 ROR and PW values for future-dollar and constant-value dollar estimates, Example 10.6.

SUMMARY

Inflation, treated computationally as an interest rate, makes the cost of the same product or service increase over time due to the decreased value of money. The chapter covers two ways to consider inflation in engineering economy computations: (1) in terms of today's (constant-value) dollars, and (2) in terms of future dollars. Some important relations are:

Inflated interest rate: $i_f = i + f + if$

Real interest rate: $i = (i_f - f)/(1 + f)$

PW of a future amount with inflation considered: $P = F(P/F, i_f, n)$

Future worth of a present amount in constant-value dollars with the same purchasing power: $F = P(F/P, i, n)$

Future amount to cover a current amount with no interest: $F = P(F/P, f, n)$

Future amount to cover a current amount with interest: $F = P(F/P, i_f, n)$

Annual equivalent of a future dollar amount: $A = F(A/F, i_f, n)$

Annual equivalent of a present amount in future dollars: $A = P(A/P, i_f, n)$

PROBLEMS

Understanding Inflation

10.1 How do you convert inflated dollars into constant-value dollars?

10.2 Find the inflation rate necessary for something to cost exactly twice as much as it did 10 years earlier.

10.3 In an inflationary period for a dollar-based currency, what is the difference between (*a*) inflated dollars and "then-current" dollars, and (*b*) "then-current" (future) dollars and constant-value dollars?

10.4 For many years, college cost increases have been about twice the inflation rate, averaging 5% to 8% per year. According to the College Board's *Trends in College Pricing*, the 2011–2012 average annual total costs (including tuition, fees, room and board) were $17,131 for students attending four-year public colleges and universities in-state, $29,657 out-of-state, and $38,589 at four-year private colleges and universities. Use a 7% per year inflation rate. (*a*) Determine how much a sophomore high-school student can expect to spend on in-state tuition, fees, room and board when he or she starts at a four-year public college in 3 years. (*b*) Determine the *total annual cost* if textbooks, supplies, etc. currently average $4000 per year and they also increase at 7% per year.

10.5 The Pell Grant program of the federal government provides financial aid to needy college students. A supplemental grant of $690 per recipient in the 2012–2013 school year (toward the maximum total of $5550) will increase annually until 2017 to account for inflation. Beginning in 2018, grants will no longer increase with inflation. If the maximum total Pell Grant award increases to $5645 in the 2013–2014 school year, what inflation rate per year was used in the calculation?

10.6 For 2012, the USDA determined that "The Twelve Days of Christmas" cost index was $101,120, based on prices of the 364 items listed in the classic carol. If the "True Cost of Christmas" index rose by 4.4%, (*a*) what was the value of the index in 2011, and (*b*) if seven swans-a-swimming increased by $700 to $6300, what was their percentage increase?

10.7 Midstate Independent School District signed an agreement with a local law firm that increased the firm's hourly billing rate of $185 per hour, established in 2008, to $225 per hour beginning in April of 2013. What was the percent increase *per year* in the hourly billing rate?

10.8 Emissions of heat-trapping carbon dioxide (CO_2) reached an all-time high of 31.6 gigatons (giga $= 10^9$) in 2011, a 3.2% increase over 2010. The International Energy Agency said this further reduces the chance that the world can avoid a dangerous rise in global average temperature by 2020. (*a*) If the discharge increase continues at the same 3.2% rate per year for the next 9 years, how many gigatons will be released in 2020? (*b*) What will be the total percent increase between 2011 and 2020?

10.9 When the inflation rate is 4% per year, how many inflated dollars will be required 20 years from now to buy the same things that $10,000 buys now?

10.10 According to data from the National Association of Colleges and Employers (NACE), engineering graduates' average salaries of $61,872 were among the highest of the class of 2011. What will the average starting amount be in 2017, if salaries increase at a rate of (*a*) 2% per year, and (*b*) twice the assumed inflation rate of 3.5% per year?

10.11 The inflation rate over a 10-year period for an item that now costs $1000 is shown in the following table. (*a*) What will be the cost at the end of year 10? (*b*) Do you get the same cost using an average inflation rate of 5% per year through the 10-year period? Why?

Year	Inflation Rate
1	10%
2	0%
3	10%
4	0%
5	10%
6	0%
7	10%
8	0%
9	10%
10	0%

10.12 An engineer who is now 65 years old began planning for retirement 40 years ago. At that time, he thought that if he had $1 million when he retired, he would have more than enough money to live his remaining life in luxury. Use an average inflation rate of 4% per year over the 40-year time period. (*a*) What is the constant-value dollar amount of his $1 million now at age 65? Use the day he started 40 years ago as the base year. (*b*) How many then-current (future) dollars should he have accumulated over the 40 years to have a constant-value purchasing power equal to $1 million at retirement age?

10.13 In 2012, every dollar (100¢) spent on grocery items was broken down as follows:

> Labor = 38.5¢
> Farm value of food products = 19.5¢
> Advertising and packaging = 12¢
> Transportation = 7.5¢
> Rent and insurance = 7¢
> Taxes = 6¢
> Depreciation and repairs = 5¢
> Profit = 4.5¢

Assume the transportation cost increases by 10% per year and the labor cost increases by 4% per year for five years, with all other costs remaining constant. Determine the percent profit in the years 2012 and 2017, assuming the profit amount remains at 4.5¢.

Present Worth Calculations with Inflation

10.14 For a young, growing company that has a required ROR of a real 25% per year, what is the inflation-adjusted ROR for an inflation rate of 5% per year?

10.15 The CEO of a high-tech incubator company wants to entice venture capitalists by promising a growth rate of 40% per year for at least 3 years. Therefore, the company's MARR was set at 40%. If the company did get a 40% ROR over that time period, but it didn't account for the 8% per year inflation rate that occurred during that time, what was the real growth rate of the company?

10.16 For a nominal inflation-adjusted interest rate of 24% per year compounded monthly, calculate the real interest rate per month when the inflation rate is 0.5% per month.

10.17 Find the present worth of the following estimated cash flows. As indicated, some are expressed in then-current (future) terms and others in today's dollars. Use a real interest rate of 10% per year and an inflation rate of 6% per year.

Year	Cash Flow, $	Expressed as
0	16,000	Today's
3	40,000	Then-current
4	12,000	Then-current
7	26,000	Today's

10.18 The company you work for is considering a new product line that is projected to have the net cash flows below (in $1000). The values are in future dollars which have been inflated by 5% per year. The plant manager isn't sure about how the present worth of the cash flows should be calculated, so he asked you to do it two ways over the 4-year planning horizon: (1) using an inflation-adjusted rate of 20% per year, and (2) converting all of the cash flows to current-value dollars and using a real interest rate. You said both ways will provide the same answer, but he asked you to show him the calculations. Prepare the present worth computations by methods (1) and (2).

Year	Cash Flow, $1000
0	−10,000
1	2,000
2	5,000
3	5,000
4	5,000

10.19 Find the present worth of a piece of equipment that has a first cost of $150,000, an annual operating cost of $60,000, and a salvage value of 20% of the first cost after 5 years. Assume that the real interest rate is 10% per year, that the inflation rate is 7% per year, and that inflation is to be accounted for. Also, assume that all costs are future dollar estimates.

10.20 A very generous grandfather is planning to leave his only granddaughter well off when she reaches the age of 25. He plans to deposit a lump sum now, which is her 2nd birthday, such that she will have enough money to live comfortably without working for a salary. He wants her to receive an amount that will have the same purchasing power as $2 million today. If he can invest the money and earn an average market interest rate of 8% per year while the inflation rate averages 4% per year, how much must he deposit now?

10.21 The president of a medium-sized oil company wants to buy a private plane to reduce the total travel time between cities where refineries are located. The company can buy a used Lear jet now or wait for a new very light jet (VLJ) that will be available 3 years from now. The cost of the VLJ will be $1.5 million, payable when the plane is delivered in 3 years. The president has asked you to determine the present worth of the plane so that he can decide whether to buy the used Lear now or wait for the VLJ. If the company uses a market MARR of 18.45% per year and inflation is projected to be 3% per year, what is the present worth of the VLJ with inflation considered?

10.22 A salesman from vendor A, who is trying to get his foot in the door at Filerbee, Inc., offered water desalting equipment for $2.1 million. This is $400,000 more than the price that a saleswoman from vendor B offered, if purchased and paid for now. However, as a special offer, vendor A said Filerbee won't have to pay until the warranty runs out. If the equipment has a 2-year warranty, determine which offer is better. The company's real MARR is 12% per year and the inflation rate is 4% per year.

10.23 How much can the manufacturer of superconducting magnetic energy storage systems afford to spend now on new equipment in lieu of spending $75,000 four years from now? The company's real MARR is 12% per year and the inflation rate is 3% per year.

Future Worth Calculations with Inflation

10.24 Explain the difference between a future worth amount calculated using case 1 and case 2 if the inflation-adjusted interest rate i_f and inflation rate f are the same for both computations.

10.25 An engineer planning for her son's college education made deposits into a separate brokerage account every time she earned extra money from side consulting jobs. The amounts and timing of the deposits are shown below.

Year	Amount, $
0	5,000
3	8,000
4	9,000
7	15,000
11	16,000
17	20,000

If the account increased at a market rate of 15% per year and inflation averaged 3% per year over the entire deposit period, determine the purchasing power in terms of year-zero dollars immediately after the last deposit in year 17.

10.26 A plant manager is not sure whether he will get the approval to buy new equipment for automating an engine assembly line now or at some future time within the next three years. In order to have the money whenever he is given the go-ahead, he has asked you to tell him what the equipment is likely to cost in each of the next three years. The cost of the equipment today is $300,000. How much will it cost at the end of years 1, 2, and 3, if the cost increases only by the inflation rate of 4% per year? Use a market interest rate of 15% per year.

10.27 With hopes to retire at a decent age and move to Hawaii, an engineer plans to turn her investment account over to a professional management firm that promises to make a real return of 10% per year when the inflation rate is 4% per year. The account currently is valued at $422,000 and she wants to retire in 15 years. (*a*) How much in future-dollar terms will have to be in the account to realize the promised 10% per year return? (*b*) Is this an example of case 1 or case 2 as described in the text?

10.28 An engineer deposits $10,000 into an account when the market interest rate is 10% per year and the inflation rate is 5% per year. The account is left undisturbed for 5 years. (*a*) How much money will be in the account? (*b*) What will be the purchasing power in terms of today's dollars? (*c*) What is the real rate of return earned?

10.29 A Division of Dow Chemical wants to set aside money now so that it can purchase new air fin coolers three years from now. The total cost now is $45,000. The price of the coolers is expected to increase only by the inflation rate of 3.7% per year for each of the next three years. Dow earns interest at a market rate of 8% per year on its investments. (*a*) What is the coolers' cost expected to be three years from now? (*b*) How much will the company have to set aside now to buy the coolers in 3 years? (*c*) If the $45,000 is set aside now, how much money will be available in 3 years to pay for the coolers?

10.30 The cost of constructing a roundabout in a low-traffic residential neighborhood five years ago was $625,000. A civil engineer designing another one that is almost exactly the same estimates the cost today will be $740,000. (*a*) If the cost had increased only by the inflation rate over the five years, what was the actual inflation rate per year? (*b*) What is the interest rate involved in this computation?

10.31 If you make an investment in an insurance policy that is guaranteed to pay you $1.8 million 20 years from now, provided you live that long, what will be the buying power of the money with respect to today's dollars? The market interest rate is 8% per year and the inflation rate stays at 3.8% per year.

10.32 Colgate-Palmolive can purchase a piece of equipment now for $80,000 or buy it 3 years from now for an estimated $128,000. The MARR requirement is a real return of 15% per year. If an inflation rate of 4% per year must be accounted for, should the company buy the machine now or later?

10.33 An offshore services company is considering the purchase of equipment that has a cost today of $96,000. In a period of 5% per year inflation, how much will the equipment cost 3 years from now in terms of *constant-value dollars*? The manufacturer plans to raise the price exactly in accordance with the inflation rate.

10.34 In a period of 4% per year inflation, how much will a machine cost 3 years from now in terms of *constant-value dollars,* if the cost today is $40,000 and the price increases such that the manufacturer will make a real rate of return of 5% per year over the 3 years?

10.35 A pulp and paper company is planning to set aside $150,000 now for possibly replacing its large synchronous refiner motors. If the replacement isn't needed for 5 years, how much will the company have in the account provided it earns a market rate of 10% per year and the inflation rate is 4% per year?

10.36 Well-managed companies set aside money to pay for emergencies that inevitably arise in the course of doing business. If a commercial solid waste recycling and disposal company puts 0.5% of its after-tax income into such an account, how much will the company have after 7 years, provided the company's after-tax income averages $15.2 million per year? The inflation and market rates are 5% per year and 9% per year, respectively.

10.37 A small mechanical consulting company is examining its future cash flow requirements. The

president expects to replace office machines and IT equipment at various times over a 6-year planning period. Specifically, the company expects to spend $6000 two years from now, $9000 three years from now, and $5000 six years from now. What is the purchasing power (with respect to today's dollars) of each expenditure in its respective year, if the inflation rate is 4% per year?

10.38 The strategic plan of a solar energy company that manufactures high-efficiency solar cells includes an expansion of its physical plant in 4 years. The engineer in charge of planning estimates the expenditure required now to be $8 million, but in 4 years, the cost will be higher by an amount equal to the inflation rate. If the company sets aside $7,000,000 now into an account that earns interest at an advertised 7% per year, what can the annual inflation rate be to have exactly the right amount of money for the expansion? Solve (*a*) by hand, and (*b*) using a spreadsheet.

Capital Recovery with Inflation

10.39 An entrepreneur engaged in wildcat oil well drilling is seeking investors who will put up $500,000 for an opportunity to reap high returns, if the venture is successful. The prospectus states that a real return of at least 22% per year for 5 years is likely, but not promised. How much will the investors have to receive each year to recover their money and the 22% return, if an inflation rate of 5% per year is to be included in the calculation?

10.40 Veri-Trol, Inc. manufactures in-situ calibration verification systems that confirm flow measurement accuracies without removing the meters. The company is considering modifying the main assembly line with one of the enhancements shown below. If the company's real MARR is 15% per year, which process has the lower annual cost? Include an inflation rate of 5% per year in the analysis?

	Process X	Process Y
First cost, $	−65,000	−90,000
Operating cost, $ per year	−40,000	−34,000
Salvage value, $	0	10,000
Life, years	5	5

10.41 Aquatech Microsystems spent $183,000 for a communications protocol to achieve interoperability among its utility systems. If the company uses a real interest rate of 15% per year on such investments and a recovery period of 5 years, what is the equivalent annual worth of the expenditure in then-current dollars at an inflation rate of 6% per year?

10.42 A European-based cattle genetics engineering research lab is planning for a major expenditure on research equipment. The lab needs $5 million of today's dollars so it can make the acquisition 4 years from now. The inflation rate is steady at 5% per year. (*a*) How many future dollars will be needed when the equipment is purchased, if purchasing power is maintained? (*b*) What is the required amount of the annual deposit into a fund that earns the market rate of 10% per year to ensure that the amount calculated in part (*a*) is accumulated?

10.43 The costs associated with a small X-ray inspection system are $40,000 now and $24,000 per year, with a $6000 salvage value after 3 years. Determine the equivalent annual cost of the system if the real interest rate is 10% per year and the inflation rate is 4% per year.

10.44 Maintenance costs for pollution control equipment on a pulverized coal cyclone furnace are expected to be $80,000 now and another $90,000 three years from now. The CFO of Monongahela Power wants to know the equivalent annual cost of the equipment in years 1 through 5. If the company uses a real interest rate of 12% per year and the inflation rate averages 4% per year, what is the equivalent annual cost of the equipment?

10.45 In wisely planning for your retirement, you invest $12,000 per year for 20 years into a 401(k) account. How much can you withdraw each year for 10 years, starting one year after your last deposit, if you obtain a real return of 10% per year? Assume the inflation rate averages 2.8% per year.

ADDITIONAL PROBLEMS AND FE EXAM REVIEW QUESTIONS

10.46 When all future cash flows are expressed in constant-value dollars, the rate that should be used to find the present worth is the:
 a. real MARR.
 b. inflation rate.
 c. inflated interest rate.
 d. inflated MARR.

10.47 In order to convert inflated dollars into constant-value dollars, it is necessary to:
 a. divide by $(1 + i_f)^n$
 b. divide by $(1 + f)^n$
 c. divide by $(1 + i)^n$
 d. multiply by $(1 + f)^n$

10.48 For a real interest rate of 12% per year and an inflation rate of 7% per year, the market interest rate per year is closest to:
 a. 4.7%
 b. 7%
 c. 12%
 d. 19.8%

10.49 If the market interest rate is less than the real interest rate, then:
 a. the inflated interest rate is higher than the real interest rate.
 b. the real interest rate is zero.

 c. a deflationary condition exists.
 d. an inflationary condition exists.

10.50 If the market interest rate is 16% per year when the inflation rate is 9% per year, the real interest rate is closest to:
 a. 6.4%
 b. 7.3%
 c. 9.4%
 d. 16.1%

10.51 The cost of an F-150 pick-up truck was $29,350 three years ago. If the cost increased only by the inflation rate and the price today is $33,015, the inflation rate was closest to:
 a. 3%
 b. 4%
 c. 5%
 d. 6%

10.52 If the market interest rate is 12% per year and the inflation rate is 5% per year, the number of future dollars in year 7 that will be equivalent to $2000 now can be determined by the relation:
 a. $F = 2000(1 + 0.176)^7$
 b. $F = 2000/(1 + 0.176)^7$
 c. $F = 2000(1 + 0.120)^7$
 d. $F = 2000(1 + 0.198)^7$

10.53 The number of dollars that have been accumulated now from an investment of $1000 twenty-five years ago if the market interest rate was 5% per year and the inflation rate averaged 2% per year is closest to:

a. $1640

b. $3385

c. $5430

d. $5556

10.54 If you are promised $50,000 six years from now, the present worth at a real rate of return of 4% per year and an inflation rate of 3% per year is closest to:

a. $27,600

b. $29,800

c. $33,100

d. $37,200

10.55 For a real interest rate of 1% per month and an inflation rate of 1% per month, the *nominal inflated* interest rate *per year* is closest to:

a. 1%

b. 2.0%

c. 24.1%

d. 25.4%

10.56 Provided the inflation rate is f percent per year, to determine the purchasing power of $10,000 ten years from now, the $10,000 must be:

a. divided by $(1 + f)^{10}$

b. multiplied by $(1 + f)^{10}$

c. divided by $(1 + 0.10)^{f}$

d. divided by $(1 + f)$

Estimating Costs

C. Borland/PhotoLink/Getty Images

U p to this point, cost and revenue cash flow values have been stated or assumed as known. In reality, they are not; they must be estimated. This chapter explains what cost estimation involves, and applies cost estimation techniques. *Cost estimation* is important in all aspects of a project, but especially in the stages of project conception, preliminary design, detailed design, and economic analysis. In engineering practice, the estimation of costs receives much more attention than revenue estimation; costs are the topic of this chapter.

Unlike direct costs for labor and materials, indirect costs are not easily traced to a specific department, machine, system, or processing line. Therefore, *allocation of indirect costs* for functions such as IT, utilities, safety, management, purchasing, and quality is made using some rational basis. Both the traditional method of allocation and the Activity-Based Costing (ABC) method are summarized in this chapter.

Purpose: Learn to make cost estimates using basic models and to allocate indirect costs.

LEARNING OUTCOMES

1. Describe different approaches to cost estimation.

   Approaches

2. Use a unit cost factor to estimate preliminary cost.

   Unit method

3. Use a cost index to estimate present cost based on historic data.

   Cost indexes

4. Estimate the cost of a component, system, or plant by using a cost-capacity equation.

   Cost-capacity equations

5. Estimate total plant cost using the factor method.

   Factor method

6. Estimate time and cost using the learning curve relationship.

   Learning curve

7. Allocate indirect cost using traditional indirect cost rates and the Activity-Based Costing (ABC) method.

   Indirect costs

11.1 HOW COST ESTIMATES ARE MADE

Podcast available

Cost estimation is a major activity performed in the initial stages of virtually every effort in industry, business, and government. In general, most cost estimates are developed for either a *project* or a *system;* however, combinations of these are very common. A *project* usually involves physical items, such as a building, bridge, manufacturing plant, or offshore drilling platform, to name just a few. A *system* is usually an operational design that involves processes, software, and other non-physical items. Examples might be a purchase order system, a software package to design highways, or an Internet-based remote-control system. The cost estimates are usually made for the initial development of the project or system, with the life-cycle costs of maintenance and upgrade estimated as a percentage of first cost. Much of the discussion that follows concentrates on physical-based projects. How-ever, the logic is widely applicable to cost estimation in all areas.

Costs are comprised of *direct costs* (largely humans, machines, and materials) and *indirect costs* (mostly support functions, utilities, management, taxes, etc.). Normally direct costs are estimated with some detail, then the indirect costs are added using standard rates and factors. Direct costs in many industries have become a relatively small percentage of overall product cost, while indirect costs have become much larger. Accordingly, many industrial settings require some estimat-ing for indirect costs as well. Indirect cost allocation is discussed in Section 11.7. Primarily, direct costs are discussed here.

Because cost estimation is a complex activity, the questions below form a structure for the sections that follow.

- What cost components must be estimated?
- What approach to cost estimation will be applied?
- How accurate should the estimates be?
- What estimation techniques will be utilized?

Costs to Estimate

If a project revolves around a single piece of equipment, for example, a multistory building, the *cost components* will be significantly simpler and fewer than the com-ponents for a complete system, such as the design, manufacturing, and testing of a new commercial aircraft. Therefore, it is important to know up front how much the cost estimation task will involve. Examples of direct cost components are the first cost P and the annual operating cost (AOC), also called the M&O costs (mainte-nance and operating). Each component will have several *cost elements,* some that are directly estimated, others that require examination of records of similar projects, and still others that must be modeled using an estimation technique. Following are sample elements of the first cost and AOC components.

First cost component P:
Elements: Equipment cost
Delivery charges
Installation cost

Insurance coverage

Initial training of personnel for equipment use

The term delivered-equipment cost is the sum of the first two elements; the term installed-equipment cost includes the third element of installation cost.

AOC component, a part of the equivalent annual cost A:

Elements: Direct labor cost for operating personnel

Direct materials

Maintenance (daily, periodic, repairs, etc.)

Rework and rebuild

Some of these elements, such as equipment cost, can be determined with high accuracy; others, such as maintenance costs, may be harder to estimate. However, a wide variety of data are available, such as McGraw-Hill Construction, R. S. Means cost books, Marshall & Swift, Bureau of Labor Statistics, NASA Cost Estimating website, and many others. When costs for an entire system must be estimated, the number of cost components and elements is likely to be in the hundreds. It is then necessary to prioritize the estimation tasks.

For familiar projects (houses, office buildings, highways, and some chemical plants) there are standard cost estimation packages available in paper or software form. For example, state highway departments utilize software packages that prompt for the correct cost components (bridges, pavement, cut-and-fill profiles, etc.) and estimate costs with time-proven, built-in relations. Once these components are estimated, exceptions for the specific project are added. One such exception is location cost adjustments. (R. S. Means has city index values for over 700 cities in the United States and Canada.)

Cost Estimation Approach

Traditionally in industry, business, and the public sector, a "bottom-up" approach to cost estimation is applied. For a simple rendition of this approach, see Figure 11.1 (left). The progression is as follows: cost components and their elements are identified, cost elements are estimated, and estimates are summed to obtain total direct cost. The price is then determined by adding indirect costs and the profit margin, which is usually a percentage of the total cost. This approach works well when competition is not the dominant factor in pricing the product or service.

The bottom-up approach treats the required price as an output variable and the cost estimates as input variables.

Figure 11.1 (right) shows a simplistic progression for the design-to-cost, or top-down, approach. The competitive price establishes the target cost.

The design-to-cost, or top-down, approach treats the competitive price as an input variable and the cost estimates as output variables.

This approach places greater emphasis on the accuracy of the price estimation activity. The target cost must be realistic, or it becomes very difficult to make sensible, realizable cost estimates for the various components.

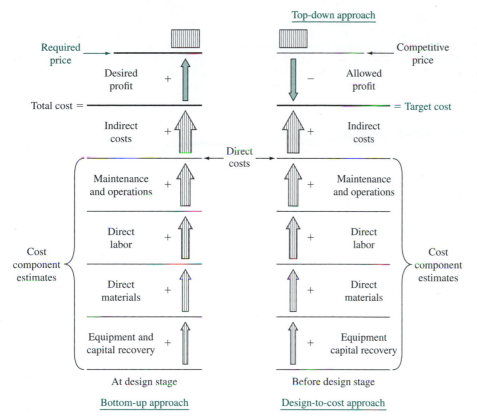

FIGURE 11.1
Simplified cost estimation processes for bottom-up and top-down approaches.

The design-to-cost approach is best applied in the early stages of a product design. This approach is useful in encouraging innovation, new design, process improvement, and efficiency. These are some of the essentials of *value engineering*.

Usually, the resulting approach is some combination of these two cost estimation philosophies. However, it is helpful to understand up front what approach is to be emphasized.

Accuracy of the Estimates

No cost estimates are expected to be exact; however, they are expected to be reasonable and accurate enough to support economic scrutiny. The accuracy required increases as the project progresses from preliminary design to detailed design and on to economic evaluation. Cost estimates made before and during the preliminary design stage are expected to be good "first-cut" estimates that serve as input to the project budget.

When utilized at early and conceptual design stages, estimates are referred to as *order-of-magnitude* estimates and generally range within ±20% of actual cost. At the detailed design stage, cost estimates are expected to be accurate enough to support economic evaluation for a go–no go decision. Every project setting has its own characteristics, but a range of ±5% of actual costs is expected

FIGURE 11.2

Characteristic curve
of estimate's
accuracy versus time
spent to estimate
construction cost of
a building.

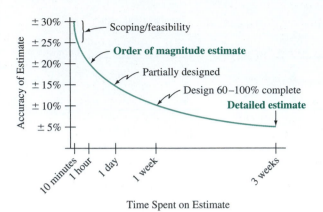

Time Spent on Estimate

at the detailed design stage. Figure 11.2 shows the general range of estimate accuracy for the construction cost of a building versus time spent in preparing the estimate. Obviously, the desire for better accuracy has to be balanced against the cost of obtaining it.

Cost Estimation Techniques

Methods such as expert opinion and comparison with comparable installations serve as excellent estimators. The use of the *unit method* and *cost indexes* base the present estimate on past cost experiences, with inflation considered. Models such as *cost-capacity equations,* the *factor method,* and the *learning curve* are discussed below. They are simple mathematical techniques applied at the preliminary design stage. They are called cost-estimating relationships (CER). There are many additional methods discussed in the handbooks and publications of different industries.

Most cost estimates made in a professional setting are accomplished using *software packages.* These are usually linked to a database that contains cost indexes and rates for the locations and product or process types being studied.

11.2 UNIT METHOD

Podcast
available

The *unit method* is a popular preliminary estimation technique applicable to virtually all professions. The total estimated cost C_T is obtained by multiplying the number of units N by a per unit cost factor U.

$$C_T = U \times N \qquad [11.1]$$

Unit cost factors must be updated frequently to remain current with changing costs, areas, and inflation. Some sample unit cost factors (and values) are

Total average cost of operating an automobile (55¢ per mile)

Cost to construct a parking garage ($12,500 per space)

Cost of constructing interstate highway with interchanges ($7.3 million
  per mile)

Cost of house construction per livable area ($225 per square foot)

Applications of the unit method to estimate costs are easily found. If house construction costs average $225 per square foot, a preliminary cost estimate for an 1800-square-foot house, using Equation [11.1], is $405,000. Similarly, a 200-mile trip should cost about $110 for the car only at 55¢ per mile.

When there are several components to a project or system, the unit cost factors for each component are multiplied by the amount of resources needed and the results are summed to obtain the total cost C_T. This is illustrated in the next example.

Justin, an ME with Dynamic Castings, has been asked to make a preliminary estimate of the total cost to manufacture 1500 sections of high pressure gas pipe using an advanced centrifugal casting method. Since a ±20% estimate is acceptable at this preliminary stage, a unit method estimate is sufficient. Use the following resource and unit cost factor estimates to help Justin.

EXAMPLE 11.1

    Materials: 3000 tons at $45.90 per ton

    Machinery and tooling: 1500 hours at $120 per hour

    Direct labor in-plant:

        Casting and treating: 3000 hours at $55 per hour

        Finishing and shipping: 1200 hours at $45 per hour

    Indirect labor: 400 hours at $100 per hour

Solution

Apply Equation [11.1] to each of the five areas and sum the results to obtain the total cost estimate of $576,700. Table 11.1 provides the details.

TABLE 11.1   **Total Cost Estimate Using Unit Cost Factors for Several Resource Areas, Example 11.1**

Resource	Amount, N	Unit Cost Factor, U	Cost Estimate, $U \times N$
Materials	3000 tons	$ 45.90 per ton	$137,700
Machinery, tooling	1500 hours	$120 per hour	180,000
Labor, casting	3000 hours	$ 55 per hour	165,000
Labor, finishing	1200 hours	$ 45 per hour	54,000
Labor, indirect	400 hours	$100 per hour	40,000
Total cost estimate			**$576,700**

11.3 COST INDEXES

A *cost index* is a ratio of the cost of something today to its cost sometime in the past. As such, the index is a dimensionless number that shows the relative cost change over time. One such index that most people are familiar with is the Consumer Price Index (CPI), which shows the relationship between present and past costs for many of the things that "typical" consumers must buy. This index, discussed previously in Section 10.1, includes such items as rent, food, transportation, and certain services. Other indexes track the costs of equipment, and goods and services that are more pertinent to the engineering disciplines. Table 11.2 is a listing of some common indexes.

TABLE 11.2 Types and Sources of Various Cost Indexes

Type of Index	Source
Overall prices	
Consumer (CPI)	Bureau of Labor Statistics
Producer (wholesale)	U.S. Department of Labor
Construction	
Chemical plant overall	*Chemical Engineering*
Equipment, machinery, and supports	
Construction labor	
Buildings	
Engineering News Record overall	*Engineering News Record (ENR)*
Construction	(published by McGraw-Hill
Building	Construction)
Common labor	
Skilled labor	
Materials	
EPA treatment plant indexes	Environmental Protection Agency; *ENR*
Large-city advanced treatment (LCAT)	
Small-city conventional treatment (SCCT)	
Federal highway	
Contractor cost	
Equipment	
Marshall and Swift (M&S) overall	*Chemical Engineering*
M&S specific industries	
Labor	
Output per man-hour by industry	U.S. Department of Labor

The general equation for updating costs through the use of a cost index over a period from time $t = 0$ (base) to another time t is

$$C_t = C_0\left(\frac{I_t}{I_0}\right)$$

[11.2]

where C_t = estimated cost at present time t
       C_0 = cost at previous time t_0
       I_t = index value at time t
       I_0 = index value at base time t_0

Generally, the indexes for equipment and materials are made up of a mix of components that are assigned certain weights, with the components sometimes further subdivided into more basic items. For example, the equipment, machinery, and support component of the chemical plant cost index is subdivided into process machinery, pipes, valves and fittings, pumps and compressors, and so forth. These subcomponents, in turn, are built up from even more basic items such as pressure pipe, black pipe, and galvanized pipe. Table 11.3 presents several years' values of the *Chemical Engineering* plant cost index (PCI), the *Engineering News Record (ENR)* construction cost index, and the Marshall and Swift (M&S) equipment cost index. The base period of 1957 to 1959 is assigned a value of 100 for the PCI, 1913 = 100 for the *ENR* index, and 1926 = 100 for the M&S equipment cost index.

Current and past values of several of the indexes may be obtained online (for a fee). For example, the PCI is available at www.che.com/pci. The *ENR* construction cost index is found at www.construction.com. This latter site offers a comprehensive series of construction-related resources to include materials, labor, building and overall construction indexes. A couple of websites used by many

TABLE 11.3   **Sample Values for Selected Engineering Indexes**

Year	Chem. Eng. Plant Cost Index	*ENR* Construction Cost Index	M&S Equipment Cost Index
2002	395.6	6538	1104.2
2003	402.0	6695	1123.6
2004	444.2	7115	1178.5
2005	468.2	7446	1244.5
2006	499.6	7751	1302.3
2007	525.4	7967	1373.3
2008	575.4	8310	1449.3
2009	521.9	8570	1468.6
2010	550.9	8802	1457.4
2011	585.7	9088	1490.2
2012 (estimated)	593.1	9324	1536.5

engineering professionals in the form of a "technical chat room" for all types of topics, including estimation, are at www.eng-tips.com and www.cyburbia.com.

EXAMPLE 11.2   In evaluating the feasibility of a major construction project, an engineer is interested in estimating the cost of skilled labor for the job. The engineer finds that a project of similar complexity and magnitude was completed 5 years ago at a skilled labor cost of $360,000. The *ENR* skilled labor index was 4038 then and is now 4681. What is the estimated skilled labor cost?

Solution

The base time t_0 is 5 years ago. Using Equation [11.2], the present cost estimate is

$$C_t = 360,000\left(\frac{4681}{4038}\right)$$
$$= \$417,325$$

In the manufacturing and service industries, tabulated cost indexes may be difficult to find. The cost index will vary, perhaps with the region of the country, the type of product or service, and many other factors. The development of the cost index requires the actual cost at different times for a prescribed quantity and quality of the item. The *base period* is a selected time when the index is defined with a basis value of 100 (or 1). The index each year (period) is determined as the cost divided by the base-year cost and multiplied by 100. Future index values may be forecast using simple extrapolation or more refined mathematical techniques. Cost indexes are sensitive over time to technological change, which causes "index creep." Updating the definition and bases of an index is necessary when identifiable changes occur.

EXAMPLE 11.3   An engineer with Hughes Industries is in the process of estimating costs for a plant expansion. Two important items used in the process are a subcontracted 4 gigahertz microchip and a preprocessed platinum alloy. Spot checks on the contracted prices through the Purchasing Department at 6-month intervals show the following historical costs. Make January 2013 the base period, and determine the cost indexes using a basis of 100.

Year	2011		2012		2013		2014
Month	Jan	Jul	Jan	Jul	Jan	Jul	Jan
Chip, $/unit	57.00	56.90	56.90	56.70	56.60	56.40	56.25
Platinum alloy, $/ounce	446.00	450.00	455.00	575.00	610.00	625.00	705.00

Solution

For each item, the index (I_t/I_0) is calculated with January 2013 cost used for the I_0 value. As indicated by the cost indexes shown, the index for the chip is stable, while the platinum alloy index is rising rapidly.

Year	2011		2012		2013		2014
Month	Jan	Jul	Jan	Jul	Jan	Jul	Jan
Chip cost index	100.71	100.53	100.53	100.17	100.00	99.65	99.38
Platinum alloy cost index	73.11	73.77	74.59	94.26	100.00	102.46	115.57

11.4 COST-ESTIMATING RELATIONSHIPS: COST-CAPACITY EQUATIONS

Podcast available

Design variables (speed, weight, thrust, physical size, etc.) for plants, equipment, and construction are determined in the early design stages. Cost-estimating relationships (CER) use these design variables to predict costs. Thus, a CER is generically different from the cost index method, because the index is based on the cost history of a defined quantity and quality of a variable.

One of the most widely used CER models is a *cost-capacity equation*. As the name implies, an equation relates the cost of a component, system, or plant to its capacity. This is also known as the *power law and sizing model*. Since many cost-capacity equations plot as a straight line on log-log paper, a common form is

$$C_2 = C_1 \left(\frac{Q_2}{Q_1} \right)^x \qquad [11.3]$$

where C_1 = cost at capacity Q_1
 C_2 = cost at capacity Q_2
 x = correlating exponent

The value of the exponent for various components, systems, or entire plants can be obtained from a number of sources, including *Plant Design and Economics for Chemical Engineers, Chemical Engineers' Handbook,* technical journals, the U.S. Environmental Protection Agency, professional and trade organizations, consulting firms, handbooks, and equipment companies. Table 11.4 is a sample listing of typical values of the exponent for various units. When an exponent value for a particular unit is not known, it is common practice to use the average value of 0.6. In fact, in the chemical processing industry, Equation [11.3] is referred to as the six-tenth model. Commonly, $0 < x \le 1$. For values $x < 1$, the economies of scale are taken advantage of; if $x = 1$, a linear relationship is present. When

TABLE 11.4   Sample Exponent Values for Cost-Capacity Equations

Component/System/Plant	Size Range	Exponent
Activated sludge plant	1–100 MGD	0.84
Aerobic digester	0.2–40 MGD	0.14
Blower	1000–7000 ft/min	0.46
Centrifuge	40–60 in	0.71
Chlorine plant	3000–350,000 tons/year	0.44
Clarifier	0.1–100 MGD	0.98
Compressor, reciprocating (air service)	5–300 hp	0.90
Compressor	200–2100 hp	0.32
Cyclone separator	20–8000 ft^3/min	0.64
Dryer	15–400 ft^2	0.71
Filter, sand	0.5–200 MGD	0.82
Heat exchanger	500–3000 ft^2	0.55
Hydrogen plant	500–20,000 scfd	0.56
Laboratory	0.05–50 MGD	1.02
Lagoon, aerated	0.05–20 MGD	1.13
Pump, centrifugal	10–200 hp	0.69
Reactor	50–4000 gal	0.74
Sludge drying beds	0.04–5 MGD	1.35
Stabilization pond	0.01–0.2 MGD	0.14
Tank, stainless	100–2000 gal	0.67

NOTE: MGD = million gallons per day; hp = horsepower; scfd = standard cubic feet per day.

$x > 1$, there are diseconomies of scale in that a larger size is expected to be more costly than that of a purely linear relation.

It is especially powerful to combine the time adjustment of the cost index (I_t/I_0) from Equation [11.2] with a cost-capacity equation to estimate costs that change over time. If the index is embedded into the cost-capacity computation in Equation [11.3], the cost at time t and capacity level 2 may be written as the product of two independent terms.

$$C_2 = C_1 \left(\frac{Q_2}{Q_1}\right)^x \left(\frac{I_t}{I_0}\right) \qquad [11.4]$$

EXAMPLE 11.4

The total design and construction cost for a digester to handle a flow rate of 0.5 million gallons per day (MGD) was $1.7 million in 2006. Estimate the cost today for a flow rate of 2.0 MGD. The exponent from Table 11.3 for the MGD range of 0.2 to 40 is 0.14. The cost index in 2006 of 131 has been updated to 225 for this year.

Solution

Equation [11.3] estimates the cost of the larger system in 2006, but it must be updated by the cost index to today's dollars. Equation [11.4] performs both operations at once. The estimated cost in current-value dollars is

$$C_2 = 1,700,000 \left(\frac{2.0}{0.5}\right)^{0.14} \left(\frac{225}{131}\right)$$

$$= 1,700,000(1.214)(1.718) = \$3,546,178$$

11.5 COST-ESTIMATING RELATIONSHIPS: FACTOR METHOD

Another widely used CER model for preliminary cost estimates of process plants is the *factor method*. While the methods discussed previously can be used to estimate the costs of major items of equipment, processes, and the total plant costs, the factor method was developed specifically for total plant costs. The method is based on the premise that fairly reliable total plant costs can be obtained by multiplying the cost of the major equipment by certain factors. These factors are commonly referred to as Lang factors after Hans J. Lang, who first proposed the method.

In its simplest form, the factor method relation has the same form as the unit method.

$$C_T = h \times C_E \qquad\qquad [11.5]$$

where C_T = total plant cost

h = overall cost factor or sum of individual cost factors

C_E = total cost of major equipment

The h may be one overall cost factor or, more realistically, the sum of individual cost components such as construction, maintenance, direct labor, materials, and indirect cost elements.

In his original work, Lang showed that direct cost factors and indirect cost factors can be combined into one overall factor for some types of plants as follows: solid process plants, 3.10; solid-fluid process plants, 3.63; and fluid process plants, 4.74. These factors reveal that the total installed-plant cost is many times the first cost of the major equipment. Often an estimator will simply use $h = 4$ in Equation [11.5] for overall cost estimating because accurate equipment costs are easy to obtain.

An engineer with Phillips Petroleum has learned that an expansion of the solid-fluid process plant is expected to have a delivered equipment cost of $1.55 million. If a specific overall cost factor for this type of plant is not known now, make a preliminary estimate of the plant's total cost.

EXAMPLE 11.5

Solution

At this early stage, the total plant cost is estimated by Equation [11.5] using $h = 4$.

$$C_T = 4.0(1,550,000) = \$6.2 \text{ million}$$

Subsequent refinements of the factor method have led to the development of separate factors for direct and indirect cost components. Direct costs (as discussed in Section 11.1) are specifically identifiable with a product, function, or process. Indirect costs are not directly attributable to a single product but are shared by several. Examples of indirect costs are administration, computer services, quality, safety, taxes, security, legal, and other support functions. For indirect costs, some of the factors apply to *equipment costs only,* while others apply to the total direct cost. In the former case, the simplest procedure is to *add the direct and indirect cost factors before multiplying by the equipment cost C_E.* In Equation [11.5], the overall cost factor h is now

$$h = 1 + \sum_{i=1}^{n} f_i \qquad\qquad [11.6]$$

where f_i = factor for each cost component, including indirect cost
       i = 1 to n components

If the *indirect* cost factor is applied to the *total direct cost,* only the direct cost factors are added to obtain h. Therefore, Equation [11.5] is rewritten.

$$C_T = \left[C_E\left(1 + \sum_{i=1}^{n} f_i\right)\right](1 + f_{IC}) \qquad\qquad [11.7]$$

where f_{IC} = indirect cost factor
       f_i = factors for direct cost components only

EXAMPLE 11.6   A small activated sludge wastewater treatment plant is expected to have a delivered-equipment first cost of \$273,000. The cost factor for the installation of piping, concrete, steel, insulation, supports, etc., is 0.49. The construction factor is 0.53, and the indirect cost factor is 0.21. Determine the total plant cost if the indirect cost is applied to (*a*) the cost of the delivered equipment and (*b*) the total direct cost.

Solution

a. Total equipment cost is \$273,000. Since both the direct and indirect cost factors are applied to only the equipment cost, the overall cost factor from Equation [11.6] is

$$h = 1 + 0.49 + 0.53 + 0.21 = 2.23$$

The total plant cost estimate is

$$C_T = 2.23(273,000) = \$608,790$$

b. Now the total direct cost is calculated first, and Equation [11.7] is used to estimate the total plant cost.

$$h = 1 + \sum_{i=1}^{n} f_i = 1 + 0.49 + 0.53 = 2.02$$

$$C_T = [273,000(2.02)](1.21) = \$667,267$$

Comment: Note the smaller estimated plant cost when the indirect cost is applied to the equipment cost only in part (*a*). This illustrates the importance of determining exactly what the factors apply to before they are used.

11.6 COST-ESTIMATING RELATIONSHIPS: LEARNING CURVE

Observation of completion times of repetitive operations in the aerospace industry several decades ago led to the conclusion that efficiency and improved performance do occur with more units. This fact is used to estimate the time to completion and cost of future units. The resulting CER is the *learning curve,* primarily used to predict the time to complete a specific repeated unit. The model incorporates a constant decrease in completion time every time the production is *doubled.* As an illustration, assume that Shell's Division of Offshore Rig Operations requires 32 fully loaded laptops of the same, ruggedized configuration. The time to assemble and test the first unit was 60 minutes. If learning generates a reduction of 10% in completion time, 90% is the learning rate. Therefore, each time 2X units are finished, the estimated completion times are 90% of the previous doubling. Here, the second unit takes 60(0.90) = 54 minutes, the fourth takes 48.6 minutes, and so on.

The constant reduction of estimated time (and cost) for doubled production is expressed as an *exponential* model.

$$T_N = T_1 N^s \qquad [11.8]$$

where: N = unit number
T_1 = time or cost for the first unit
T_N = time or cost for the Nth unit
s = learning curve slope parameter, decimal

This equation estimates the time to complete a specific unit (1st or 2nd, . . . , or Nth), not total time or average time per unit. The slope s is a negative number, since it is defined as

$$s = \frac{\log \text{ (learning rate)}}{\log 2} \qquad [11.9]$$

Unit, N	Time, minutes
1	60.0
2	54.0
4	48.6
8	43.7
16	39.4
32	35.4

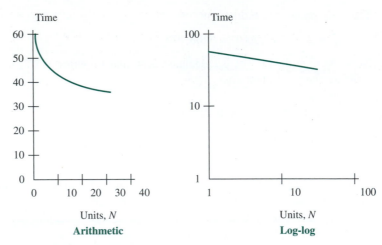

FIGURE 11.3 Plots of time estimates for a learning rate of 90% for 32 units with T_1 = 60 minutes.

Equation [11.8] plots as an exponentially decreasing curve on xy coordinates. Take the logarithm and the plot is a straight line on arithmetic paper, or plot the original data on log-log paper.

$$\log T_N = \log T_1 + s \log N \qquad [11.10]$$

Figure 11.3 shows the per unit completion time estimates (numerical in the table and graphical on arithmetic and log-log scales) for the 32 laptops described previously. As production doubles, the time to produce each unit is 10% less than that of the previous time. From Equation [11.9], this learning rate of 90% results in a slope s of -0.152.

$$s = \frac{\log 0.90}{\log 2} = \frac{-0.04576}{0.30103} = -0.152$$

Now cost estimation is possible. To estimate the cost C_N for the Nth unit, or the total cost C_T for all N units, multiply the cost per unit c by the appropriate time estimate.

$$C_N = (\text{cost per unit}) \times (\text{time for } N\text{th unit}) = c \times T_N \qquad [11.11]$$

$$C_T = (\text{cost per unit})(\text{total time for all } N \text{ units}) = c(T_1 + T_2 + \cdots + T_N) \; [11.12]$$

The learning curve offers good time estimates for larger-scale projects where the batch size is relatively small, say, 5 to 100 or 200. When the production is a routine process and repeated a large number of times, continued learning is clearly not present, and the model may underestimate costs.

The learning curve has several other names, depending upon the application area, but the relationship is the same as Equation [11.8]. Some of these names include *manufacturing progress function* (commonly used in production and manufacturing sectors), *experience curve* (often used in process industries), and *improvement curve.*

FEMA (Federal Emergency Management Agency) has ordered 25 specialized test units capable of field checking 15 separate elements in potable water in emergency situations. Thompson Water Works, Inc., the contractor, took 200 hours to build the first unit. If direct and indirect labor costs average $50 per hour, and an 80% learning rate is assumed, estimate (a) the time needed to complete units 5 and 25, and (b) the total labor cost for the 25 units.

EXAMPLE 11.7

Solution

a. The 80% learning rate and Equation [11.9] are used to find the learning curve slope $s = \log 0.80/\log 2 = -0.322$. Equation [11.8] estimates the time for a specific unit.

$$T_5 = 200(5^{-0.322}) = 119.1 \text{ hours}$$
$$T_{25} = 200(25^{-0.322}) = 70.9 \text{ hours}$$

b. In order to apply Equation [11.12] for the total cost estimate, the individual time estimates for all 25 units must be made then multiplied by the $50 per hour labor cost. If T_1 through T_{25} are added, the total is 2461.4 hours. The total labor cost is estimated at $(50)2461.4 = \$123,070$.

11.7 INDIRECT COST ESTIMATION AND ALLOCATION

It is vitally important to estimate both direct *and* indirect costs for a project, process, system, or product prior to its economic evaluation. Previous sections have dealt with direct costs; this section discusses indirect costs, which are also referred to as *overhead*. These costs include support and infrastructure expenses such as maintenance, human resources, quality, safety, supervision and administration, planning and scheduling, taxes, legal, payroll, accounting, utilities, and a host of other costs. Indirect costs are commonly too difficult and cost-prohibitive to track in detail; some allocation method must be developed and applied. At the end of each fiscal period (quarter or year) or when a project is complete, the cost accounting system uses this method to charge overhead to the appropriate cost center. However, when preliminary and detailed cost estimates are developed, a method to allocate expected indirect costs is necessary. Some common methods are:

- Make indirect costs an implicit element in the estimate. The factor method, discussed earlier, does this by including an indirect cost factor. Process industries, especially chemical and petroleum related, do this using the Lang factors (see Section 11.5).
- Apply established indirect cost rates that are calculated using some reasonable basis. This traditional method is presented next.
- Utilize Activity-Based Costing (ABC), which is especially applicable to high indirect cost industries. ABC is summarized later in this section.

TABLE 11.5 Indirect Cost Allocation Bases and Sample Costs

Allocation Basis	Indirect Cost Category
Direct labor hours	Machine shop, human resources, supervision
Direct labor cost	Machine shop, supervision, accounting
Machine hours	Utilities, IT network servers
Cost of materials	Purchasing, receiving, inspection
Space occupied	Taxes, utilities, building maintenance
Amount consumed	Utilities, food services
Number of items	Purchasing, receiving, inspection
Number of accesses	Software
Number of inspections	Quality assurance

11.7.1 Indirect Cost Rates and Allocation

Using the traditional method, indirect costs are estimated using an *indirect cost rate* that is developed using some *basis*. Table 11.5 includes possible bases and sample cost categories that may be allocated using each basis. The indirect cost rate is calculated using the relation

$$\text{Indirect cost rate} = \frac{\text{estimated total indirect costs}}{\text{estimated basis level}} \qquad [11.13]$$

EXAMPLE 11.8

A make/buy (inhouse/outsource) decision is necessary for several steel components in the new, heavy duty transmission gearbox on off-road recreation vehicles. For the Make alternative, accounting provided the project engineer, Geraldine, the following standard rates for her use in estimating indirect costs:

Machine	Indirect Cost Rate
Vertical milling cutter	$ 1.00 per labor dollar
Turret lathe	$25.00 per labor hour
Numerical control drill press	$ 0.20 per material dollar

Geraldine had intended to apply a single (blanket) rate per direct labor hour until accounting showed her the wide-ranging rates above and the bases and usage data in Table 11.6. Determine how the indirect cost rates were derived.

TABLE 11.6 Indirect Cost Bases and Activity, Example 11.8

Machine	Indirect Cost Basis	Expected Annual Activity	Budgeted Indirect to Be Charged for the Year
Vertical milling cutter	Direct labor costs	$100,000	$100,000
Turret lathe	Direct labor hours	2000 hours	$ 50,000
Numerical control drill press	Materials cost	$250,000	$ 50,000

Solution

The rates would be determined at the beginning of the year using the total indirect costs budgeted for each machine. Equation [11.13] is used to determine the rates provided to Geraldine.

Vertical milling cutter: Rate = 100,000/100,000 = $1.00 per direct labor dollar

Turret lathe: Rate = 50,000/2000 = $25 per direct labor hour

Numerical control drill press: Rate = 50,000/250,000 = $0.20 per material dollar

When the same basis applies to allocate indirect costs to multiple cost centers, a *blanket rate* may be used. For example, if materials cost is a reasonable basis for four separate projects,

$$\text{Blanket indirect cost rate} = \frac{\text{total expected indirect costs}}{\text{total estimated materials cost}}$$

Blanket rates are easy to calculate and apply, but they do not account for the different accomplishments and functions of assets or people in the same department. Consider a chemical processing line in which automated equipment is mixed with nonautomated (lower value added) methods. A blanket rate will over-accumulate indirect costs for the lower-value contribution. The correct approach is to apply different rates for the different machines and methods. When the rate is sensitive to the value added, it is called the *productive hour rate method*. Realization that more than one basis should be normally used to allocate indirect costs has led to the ABC method, discussed below.

The sum of direct and indirect costs is called *cost of goods sold* or *factory cost*. These are presented in financial statements as shown in Appendix B. When indirect costs are estimated, they are included in the economic evaluation along with direct costs. The next example illustrates this for a make/buy evaluation.

EXAMPLE 11.9 For several years the Cuisinart Corporation has outsourced the production of the stainless-steel steam nozzle assembly of its programmable espresso maker line at an annual cost of $1.5 million. The suggestion to make the component inhouse has been made. For the three departments involved, the annual indirect cost rates and hours, plus estimated direct material and direct labor costs are found in Table 11.7. The allocated hours column is the time necessary to produce the assemblies for a year.

Equipment must be purchased with the following estimates: first cost of $2 million, salvage value of $50,000, and life of 10 years. Perform a make/buy analysis at market MARR = 15% per year.

TABLE 11.7 Production Cost Estimates for Example 11.9

Department	Indirect Costs Basis, Hours	Indirect Costs Rate Per Hour	Indirect Costs Allocated Hours	Direct Material Cost	Direct Labor Cost
A	Labor	$10	25,000	$200,000	$200,000
B	Machine	5	25,000	50,000	200,000
C	Labor	15	10,000	50,000	100,000
				$300,000	$500,000

Solution

For making the components inhouse, the AOC is comprised of direct labor, direct material, and indirect costs. Use the data of Table 11.7 to distribute indirect costs.

Department A: 25,000(10) = $250,000
Department B: 25,000(5) = 125,000
Department C: 10,000(15) = 150,000
                            ———————
                            $525,000

AOC = direct labor + direct material + indirect costs
    = 500,000 + 300,000 + 525,000 = $1,325,000

The Make alternative equivalent annual worth is the total of capital recovery and AOC.

$$AW_{Make} = -P(A/P,i,n) + S(A/F,i,n) - AOC$$
$$= -2,000,000(A/P,15\%,10) + 50,000(A/F,15\%,10)$$
$$- 1,325,000$$
$$= \$-1,721,037$$

Currently, the assemblies are purchased at $AW_{Buy} = \$-1,500,000$. It is cheaper to outsource, because the AW of costs is less.

11.7.2 Activity-Based Cost Allocation

As automation, software, and manufacturing technologies have advanced, the number of direct labor hours necessary to manufacture a product has greatly decreased. Where once as much as 35% to 45% of the final cost was labor, now it is commonly 5% to 15%. However, the indirect cost may represent as much as 35% to 45% of the total cost. The use of traditional bases such as direct labor hours to determine indirect cost rates is not accurate enough for automated and technologically advanced environments. As a result, a product or service that by traditional overhead allocation methods appears to contribute to bottom-line profit may actually be a loser when indirect costs are allocated more realistically. This has led to the development of distribution methods that supplement (or replace) traditional ones, and allocation bases different from those previously used.

The best method for high indirect cost industries is *Activity-Based Costing (ABC)*. Its approach is to identify *activities* and *cost drivers*. Implementing ABC includes several steps.

1. Identify each activity and its total cost.
2. Identify the cost drivers and their usage volumes.
3. Calculate the indirect cost rate for each activity.

$$\text{ABC indirect cost rate} = \frac{\textbf{total cost of activity}}{\textbf{total volume of cost driver}} \qquad [11.14]$$

This formula has the same format as Equation [11.13] for the traditional method.

4. Use the rate to allocate indirect costs to cost centers (products, departments, etc.)

Activities are usually support departments or functions—purchasing, quality, engineering, IT. Cost drivers are usually expressed in volumes—number of purchase orders, number of construction approvals, number of machine setups, or cost of engineering changes. As an illustration, suppose a company produces two models of laparoscopic devices (cost centers) for carpal tunnel syndrome operations and has three primary support departments (activities)—purchasing, quality, and personnel. Purchasing is an indirect cost activity with a cost driver of number of purchase orders issued. The ABC allocation rate, in $ per purchase order, is used to distribute budgeted indirect costs to the two cost centers.

EXAMPLE 11.10

A U.S.-based multinational pharmaceutical firm with four plants in Europe uses traditional methods to distribute the annual business travel allocation on the basis of workforce size. Last year $500,000 in travel expenses were distributed, according to Equation [11.13], at a rate of 500,000/29,100 = $17.18 per employee.

City	Employees	Allocation
Paris	12,500	$214,777
Florence	8,600	147,766
Hamburg	4,200	72,165
Athens	3,800	65,292
	29,100	$500,000

A switch to ABC allocates the $500,000 in travel expenses on the basis of the number of travel vouchers, categorized by travelers working on each product line. In ABC terminology, travel is the activity and a travel voucher is the cost driver. Table 11.8 details the distribution of 500 vouchers to each product line by plant. Not all products are produced at each plant.

Use the ABC method to allocate travel expenses to each product line and each plant. Compare plant-by-plant allocations based on workforce size (traditional) and number of travel vouchers (ABC).

TABLE 11.8   **Travel Vouchers Submitted by Five Product Lines at Four Plants**

Plant	Product line 1	2	3	4	5	Total
Paris	50	25				75
Florence	80		30		30	140
Hamburg	100	25		20		145
Athens					140	140
Totals	**230**	**50**	**30**	**20**	**170**	**500**

Solution

The ABC allocation is to products, not to plants. Plant allocation is determined after allocation to products, based on where a product line is manufactured. Equation [11.14] determines the ABC allocation rate per voucher.

ABC allocation rate = 500,000/500 = $1000 per voucher

With the round-number rate of $1000, allocations to product-plant combinations are the same as the values in Table 11.8 times $1000. For example, total allocation to Product 1 is $230,000 and total allocation to the Paris plant is $75,000.

Products 1 and 5 dominate the ABC allocation. Comparison of by-plant allocations for ABC with the respective traditional method indicates a substantial difference in the amounts for all plants, except Florence.

Plant	ABC Allocation	Traditional Allocation
Paris	$ 75,000	$214,777
Florence	140,000	147,766
Hamburg	145,000	72,165
Athens	140,000	65,292

This comparison supports the premise that product lines, not plants, drive travel expenses and that travel vouchers are a good cost driver for ABC allocation.

The traditional method is better and easier to use for preliminary and detailed cost estimates, whereas ABC provides more detailed information once the project or fiscal period is completed. Some proponents of ABC recommend discarding the traditional method completely. However, from the viewpoints of cost estimation on one hand, and cost tracking and control on the other, the two methods work well together. The traditional method is good for estimation and allocation, while the ABC method tracks costs more closely. ABC is more costly to implement and operate, but it assists in understanding the impact of management decisions and in controlling selected indirect costs.

SUMMARY

Cost estimates are not expected to be exact, but they should be accurate enough to support a thorough economic analysis using an engineering economy approach. There are bottom-up and top-down approaches; each treats price and cost estimates differently.

Costs can be estimated in the design phase using the unit method or updated via a cost index, which is a ratio of costs for the same item at two different times.

Cost estimating may also be accomplished with a variety of models called Cost-Estimating Relationships. Three of them are

Cost-capacity equation. Good for estimating costs from design variables for equipment, materials, and construction.

Factor method. Good for estimating total plant cost.

Learning curve. Estimates time and cost for specific units produced. Good for manufacturing industries.

Traditional indirect cost rates are determined for a machine, department, product line, etc. Bases such as direct labor cost or direct material cost are used. For high automation and information technology environments, the Activity-Based Costing (ABC) method is an excellent alternative.

PROBLEMS

Cost Estimating Approaches and Unit Costs

11.1 For cost estimation purposes, there are several elements of the first cost component P. List three of them.

11.2 When competition is the dominant factor in pricing a product or service, what is the best cost estimation approach to use?

11.3 Identify the following as elements of *first cost* or *annual operating cost*: (*a*) direct materials, (*b*) delivery charges, (*c*) periodic repairs, (*d*) rework and rebuild, (*e*) installation cost, (*f*) equipment cost.

11.4 If the preliminary cost for a concrete block, reinforced steel frame aircraft hangar is estimated to be $98.23 per square foot, what is the estimated cost of a 50,000 ft^2 hangar?

11.5 Low, medium, and high cost ranges for a 30,000 ft^2 apartment building in ZIP code 79912 were estimated to be $3,111,750, $3,457,500, and $4,321,875, respectively. (*a*) What is the cost per square foot for the *medium* cost estimate? (*b*) What is the percent increase between *low* and *high* cost estimates per ft^2?

11.6 Estimate the total cost of a house (from purchasing a lot to furnishing the house) using the following estimates:

Cost Data	House Data
Purchasing a lot = $2.50/ft^2	Lot size = 100 × 150 ft
Construction = $125/ft^2 livable space	House size = 50 × 46, with 75% livable
Furnishings = $3000 per room	Number of rooms = 6

11.7 The Department of Defense uses area cost factors (ACFs) to compensate for differences in construction costs in different parts of the country and world. The ACF for Andros Island in the Bahamas is 1.70 while the ACF for Rapid City, South Dakota, is 0.93. If a cold storage processing warehouse costs $1,350,000 in Rapid City, what is the estimated cost of the same facility on Andros Island?

11.8 Preliminary cost estimates for jails can be made using costs based on either unit area (square feet) or unit volume (cubic feet). If the unit area cost is $185 per square foot and the average height of the ceilings is 10 feet, what is the unit volume cost?

11.9 The unit area and unit volume total project costs for a library are $114 per square foot and $7.55 per cubic foot, respectively. On the basis of these numbers, what is the average height of a library room?

Cost Indexes

11.10 Equipment necessary to manufacture adjustable ball-lock pins for pulling two components together had a cost of $185,000 when the M&S Equipment Cost Index was 1375.4. What is the cost estimate for a similar system when the index value is 1634.9?

11.11 In 1999, the Office of the Under Secretary of Defense projected that the ENR Building Cost Index (BCI) would increase from its 1999 value of 3423 to 4098 in 2012. If the actual value in 2012 was 5167, what was the difference between the projected and actual BCI *annual inflation rates* from 1999 to 2012?

11.12 From historical data, you discover that the national average construction cost of middle schools is $10,500 per student. If the city index for Oklahoma City is 76.9 and for Los Angeles it is 108.5, estimate the total construction cost for a middle school of 800 students in each city.

11.13 A consulting engineering firm is preparing a preliminary cost estimate for a design/construct project of a brackish groundwater desalting plant. The firm, which completed a similar project in 2003 with a construction cost of $30 million, wants to use the *ENR* construction cost index to update the cost. If the index value in 2012 was 9290, estimate the cost of construction for a similar-size plant in 2012.

11.14 If the editors at *ENR* decide to redo the construction cost index so that the year 2003 has a base value of 100 instead of 6695, determine what the value was for 2011.

11.15 An engineer who owns a construction company that specializes in large commercial projects noticed that material costs increased at a rate of 1% *per month* over the past 12 months. If a material cost index were created for that year with the value of the index set at 100 at the beginning of the year, what is the value of the index at the end of the year? Express your answer to two decimal places.

11.16 Omega electropneumatic general-purpose pressure transducers convert supply pressure to regulated output pressure in direct proportion to an electrical input signal. If the cost of a certain model was $328 in 2011 when the M&S equipment index was at 1490.2, what was the cost in 2002 when the M&S index value was 1104.2? Assume the cost changed in proportion to the index.

11.17 The *ENR* construction cost index for New York City had a value of 12,381.40 when the values in Pittsburgh and Atlanta were 7341.32 and 4874.06, respectively. If a general contractor in Atlanta won construction jobs totaling $54.3 million, determine their equivalent total value in New York City.

11.18 The *ENR* materials cost index (MCI) had a value of 2583.52 when the cost for cement was $95.90 per ton. If cement increased in price exactly in accordance with the MCI, what was the cost of cement per ton in 1913 when the MCI index value was 100?

11.19 The *ENR* 20-city construction common labor index had a value of 16,520.53 when the wage rate was $31.39 per hour. Assuming that labor rates increased in proportion to the common labor index, what was the common labor wage rate per hour in 1913 when the common labor index had a value of 100?

11.20 The cost of lumber per million board feet (MBF) was $464.49 when the value of the *ENR* materials cost index (MCI) was 2583.52. If the cost of lumber increased in proportion to the MCI, what was the value of the index when the cost of lumber was $400 per MBF?

Cost-Estimating Relationships

11.21 A high pressure (1000 psi) stainless steel pump with a variable frequency drive is installed in a seawater reverse osmosis pilot plant that is recovering water from membrane concentrate at a rate of 4 gallons per minute (gpm). The cost of the pump was $13,000. Because of favorable results from the pilot study, the water utility wants to go with a full-scale system that will produce 500 gpm. Estimate the cost of the larger pump, if the exponent in the cost-capacity equation has a value of 0.37.

11.22 The cost of a high-quality 250-horsepower compressor was $9500 when recently purchased. What would a 450-horsepower compressor be expected to cost if the exponent in the cost-capacity equation has a value of 0.32?

11.23 The variable frequency drive (VFD) for a 200 hp motor costs $20,000. How much will the VFD for a 100 hp motor cost if the exponent in the cost-capacity equation is 0.61?

11.24 The cost of a 68 m^2 falling-film evaporator was 1.65 times the cost of the 30 m^2 unit. What

exponent value in the cost-capacity equation yielded this result?

11.25 Instead of using a cost-capacity equation for relating project size and construction cost, the Department of Defense uses Size Adjustment Factors (SAFs) that are based on Size Relationship Ratios (SRRs). For example, a SRR of 0.50 has a SAF of 1.04. Determine the value of the exponent in the cost-capacity relation for C_2 to equal $1.04C_1$ when the quantity Q_2 is 50% of Q_1.

11.26 Reinforced concrete pipe (RCP) that is 12 inches in diameter has a cost of $12.54 per linear foot in Dallas, Texas. The cost for 24-inch RCP is $27.23 per foot. If the cross-sectional area of the pipe is considered the "capacity," determine the value of the exponent (to two decimal places) in the cost-capacity equation that exactly relates the cost of the two pipe sizes.

11.27 A 100,000 barrel per day (BPD) fractionation tower cost $1.2 million when the *Chemical Engineering* plant cost index value was 394.3. Estimate the cost of a 300,000 BPD plant when the index value is 575.8, provided the exponent in the cost-capacity equation is 0.67.

11.28 A mini wind tunnel for calibrating vane or hotwire anemometers cost $3750 when the M&S equipment index value was 1104.2. If the index value is now 1520.6, estimate the cost of a tunnel twice as large. The cost-capacity equation exponent is 0.89.

11.29 In 2011, an engineer estimated the cost for laser-based equipment to be $400,000. The engineer used the M&S equipment index values of 1178.5 for 2004 and 1490.2 for 2011 with the cost-capacity equation and an exponent value of 0.61. The original equipment had only one-fourth the capacity of the new equipment. (*a*) What was the cost of the original equipment in 2004? (*b*) If the index continues to increase at the same annual rate as it did from 2004 to 2011 for the next 5 years, estimate the cost of the same equipment in 2016.

11.30 The equipment cost for removing arsenic from a well that delivers 800 gallons per minute is $1.8 million. If the overall cost factor for this type system is 2.25, what is the total plant cost expected to be?

11.31 A closed-loop filtration system for waterjet cutting industries eliminates the cost of makeup water treatments (water softeners, reverse osmosis, etc.) while maximizing orifice life and machine performance. If the equipment cost is $225,000 and the total plant cost is $1.32 million, what is the overall cost factor for the system?

11.32 The equipment cost for a laboratory that plans to specialize in analyzing for endocrine disruptors, pharmaceuticals, and personal care products is $920,000. If the direct cost factor is 1.54 and the indirect cost factor is 0.49 (applied to equipment only), determine the expected cost of the laboratory.

11.33 The Pavonka family is contracted to frame 32 wooden houses of the same design in a new subdivision of 300 homes. The fourth unit took 400 hours and the carpenter knows they cut about 10% off the time for each replication of the same design. Use the learning curve assumption of a constant decrease with each doubling of production to estimate past (units 1 and 2) and future completion times.

11.34 The Department of the Navy estimates that the learning rate for shipbuilding is 80 to 85%. Assuming the 85% rate applies to nuclear submarines, how long should it take to build the 12th vessel, if the first one took 56 months to complete?

11.35 An engineer wants to predict the completion time per unit for KBR Construction to build 175 field-ready office/storage units in their plant. The Red Cross uses the units in international hunger and medical relief efforts. Use a spreadsheet to plot the learning curves on arithmetic and log-log scales for a 90% learning rate if the first-unit time to completion is 90 hours.

11.36 An engineer has been asked to estimate the cost per unit for KBR Construction to build 175 field-ready office/storage units in their plant for use by the Red Cross. Use a spreadsheet to plot the learning curves on arithmetic and log-log scales for a 95% learning rate applied to costs if the first unit costs $3000.

Indirect Costs

11.37 SIS Technologies currently allocates insurance costs on the basis of direct labor hours. If the direct labor hours for departments A, B, and C are expected to be 3000, 9000, and 5000, respectively, this year, determine the allocation to each department based on an indirect cost budget of $34,000.

11.38 A city utility currently allocates costs associated with the maintenance shop and warehouse to pumping stations based on the number of the pumps the stations contain. However, a suggestion has been made to change the allocation basis to the number of trips service personnel make to each station because some stations have old pumps that require more maintenance. Information about the stations is shown in the table below. If the indirect cost budget is $1000 per pump, reallocate the budget to each station based on the number of service trips.

Station ID	No. pumps	Service trips/year
Stanton	5	190
Main	7	55
7th St	3	38
Northeast	4	104

11.39 The director of public works needs to distribute the indirect cost allocation of $1.2 million to the three branches around the city. She will use the information from last year to determine the rates for this year.

Branch	Miles Driven	Direct Labor Hours	Basis	Allocation Last Year
North	350,000	40,000	Miles	$300,000
South	200,000	20,000	Labor	$200,000
Midtown	500,000	64,000	Labor	$450,000

a. Determine this year's indirect cost rates for each branch.

b. Use the rates from last year and records for this year to distribute the allocation for this year. How much of the $1.2 million is actually distributed?

	Records for This Year	
Branch	Miles Driven	Direct Labor Hours
North	275,000	38,000
South	247,000	31,000
Midtown	395,000	55,500

11.40 The Mechanical Components Division manager asks you to recommend a make/buy decision on a major automotive subassembly that is currently outsourced for a total of $4.5 million this year. This cost is expected to continue rising at a rate of $200,000 per year. Your manager asks that both direct and indirect costs be included when in-house manufacturing (the make alternative) is evaluated. New equipment will cost $3 million, have a salvage of $0.5 million and a life of 6 years. Estimates of materials, labor costs, and other direct costs are $1.5 million per year. Typical indirect rates, bases, and expected usage are shown below. Perform the AW evaluation at MARR = 12% per year over a 6-year study period. Perform the analysis using (*a*) hand solution, and (*b*) a spreadsheet.

Department	Basis	Rate	Expected usage
M	Direct labor in $	$2.40 per $	$450,000
P	Materials in $	$0.50 per $	$850,000
Q	Number of inspections	$25 per inspection	4,800

The following information is used in Problems 11.41 through 11.43.

Blue Sky Airways traditionally distributes the indirect cost of lost baggage to its three major hubs using a basis of annual number of flights in and out of each airport. Last year $667,500 was distributed as follows:

Hub Airport	Flights	Rate	Allocation
DFW	110,000	$3/flight	$330,000
YYZ	62,500	$3/flight	187,500
MEX	75,000	$2/flight	150,000

The head of Baggage Management suggests that an allocation on the basis of baggage traffic, not flights, will be better at representing the distribution of lost bag costs. Total number of bags handled during the year are: 2,490,000 at DFW; 1,582,400 at YYZ, and 763,500 at MEX.

11.41 What is the activity and the cost driver for the baggage traffic basis?

11.42 Using the baggage traffic basis, determine the allocation rate using last year's allocation of $667,500 and distribute this amount to the hubs.

11.43 What are the percentage changes in allocation at each hub using the two different bases?

ADDITIONAL PROBLEMS AND FE EXAM REVIEW QUESTIONS

11.44 An order-of-magnitude cost estimate should be accurate to within the following percentage of the actual cost:
a. 5%
b. 10%
c. 15%
d. 20%

11.45 The total mechanical and electrical unit costs for a library are listed as $40 per square foot, while total project cost is estimated at $119 per square foot. Based on these values, the percentage of total costs represented by mechanical and electrical costs is closest to:
a. 21%
b. 29%
c. 34%
d. 38%

11.46 In the bottom-up approach to cost estimating, the:
a. required price is an input variable.
b. cost estimates are an output variable.
c. required price is an output variable.
d. both (*a*) and (*b*) are correct.

11.47 The cost of constructing an 8000 ft² warehouse was $400,000. The *ENR* construction cost index was 6770 at that time. The cost of constructing a similar building using an *ENR* index of 9088 is closest to:
a. less than $450,000
b. $537,000
c. $672,000
d. more than $700,000

11.48 An assembly line robot arm had a first cost of $25,000 when the M&S equipment cost index was 1201.9. If the robot cost increased exactly in proportion to the index, the value of the index if the robot arm now costs $29,860 is closest to:
a. less than 1000
b. 1006.3
c. 1245.8
d. more than 1250

11.49 The cost for a 200-horsepower (hp) pump, with controller, is $22,000. The cost estimate of a similar pump of 300 hp capacity, provided the exponent in the cost-capacity equation is 0.64, is closest to:
a. $12,240
b. $28,520
c. $33,780
d. $39,550

11.50 If the cost of an 8000-operation-per-second assembly line robot is $80,000 and the cost for one with twice the capacity is $120,000, the value of the exponent in the cost-capacity equation is closest to:
 a. 0.51
 b. 0.58
 c. 0.62
 d. 0.69

11.51 The delivered equipment cost for setting up a production and assembly line for high-sensitivity, gas-damped accelerometers is $650,000. If the direct cost and indirect cost factors are 1.36 and 0.31, respectively, and both factors apply to delivered equipment cost, the total plant cost estimate is approximately:
 a. $2,034,500
 b. $1,735,500
 c. $1,384,500
 d. $1,183,000

11.52 The equipment for applying specialty coatings that provide a high angle of skid for the paperboard and corrugated box industries has a delivered cost of $390,000. If the *overall* cost factor for the complete system is 2.45, the total plant cost is approximately:
 a. $955,500
 b. $1,054,400
 c. $1,154,400
 d. $1,544,400

11.53 If the processing cost decreases by a constant 6% every time the output doubles, the slope parameter of the learning curve is closest to:
 a. −0.009
 b. −0.059
 c. −0.089
 d. −1.699

11.54 The total plant cost for manufacturing CO_2 warning devices that have a compact gas detector and a signaling unit is $1,154,400. If the overall cost factor for the plant was 2.61, the delivered equipment cost was closest to:
 a. $442,300
 b. $501,200
 c. $593,700
 d. $613,900

11.55 The cost for implementing a manufacturing process that has a capacity of 6000 units per day was $550,000. If the cost for a plant with the capacity of 100,000 units per day is $3 million, the value of the exponent in the cost-capacity equation is closest to:
 a. 0.26
 b. 0.39
 c. 0.45
 d. 0.60

11.56 The first golf-cart chassis off a new fabrication line took 24.5 seconds to paint. If during the first 100 units a learning rate of 95% is assumed, the paint time for unit 10 is closest to:
 a. 20.7 seconds
 b. 0.46 minutes
 c. 2.75 seconds
 d. 15.5 seconds

11.57 A police department wants to allocate the indirect cost of speed monitoring to the three toll roads around the city. An allocation basis that may *not* be reasonable is:
 a. miles of toll road monitored.
 b. average number of cars patrolling per hour.
 c. amount of car traffic per section of toll road.
 d. cost to operate a patrol car.

11.58 The IT department allocates indirect costs to user departments on the basis of CPU time at the rate of $2000 per second. For the first quarter, the two heaviest use departments logged 900 and 1300 seconds, respectively. If the IT indirect budget for the year is $8.0 million, the percentage of this year's allocation consumed by these departments is closest to:
 a.   32%
 b. 22.5%
 c.   55%
 d. not enough information to determine

11.59 If engineering change order is the activity for an application of the ABC method of overhead allocation, the most reasonable cost driver(s) may be:

 1 – number of changes processed
 2 – size of the work force
 3 – management cost to process the change order

 a. 1
 b. 2
 c. 3
 d. 1 and 3

Depreciation Methods

The capital investments of a corporation in tangible assets—equipment, computers, vehicles, buildings, and machinery—are commonly recovered on the books of the corporation through *depreciation*. Although the depreciation amount is not an actual cash flow, the process of depreciating an asset, also referred to as *capital recovery* or *amortizing,* accounts for the decrease in an asset's value because of age, wear, and obsolescence.

Why is depreciation important to engineering economy? Depreciation is a *tax-allowed deduction* included in tax calculations in virtually all industrialized countries. Depreciation lowers income taxes via the relation

$$\text{Taxes} = (\text{income} - \text{expenses} - \text{depreciation})(\text{tax rate})$$

Income taxes are discussed further in Chapter 13.

This chapter includes an introduction to two methods of *depletion,* which are used to recover capital investments in deposits of natural resources such as minerals, ores, and timber.

Important note: **To consider depreciation and after-tax analysis early in a course, cover this chapter and the next one (After-Tax Economic Analysis) after Chapter 5 (AW), Chapter 7 (B/C), or Chapter 9 (Replacement Analysis).**

Purpose: Use a specific method to reduce the book value of capital invested in an asset or natural resource.

LEARNING OUTCOMES

1. Understand and use the basic terminology of depreciation.

   | Depreciation terms |

2. Apply the straight line method.

   | Straight line |

3. Apply the declining balance method.

   | Declining balance |

4. Apply the Modified Accelerated Cost Recovery System (MACRS).

   | MACRS |

5. Understand the basis and application of the tax depreciation system used in Canada.

   | CCA system in Canada |

6. Switch from one classical method to another; explain how MACRS uses switching.

   | Switching |

7. Utilize percentage depletion and cost depletion methods for natural resource investments.

   | Depletion |

8. Use various spreadsheet functions to determine depreciation and book value schedules.

   | Spreadsheets |

12.1 DEPRECIATION TERMINOLOGY

Primary terms used in depreciation are defined here.

Depreciation is the reduction in value of an asset. The method used to depreciate an asset is a way to account for the decreasing value of the asset to the owner *and* to represent the diminishing value (amount) of the capital funds invested in it. The annual depreciation amount D_t does not represent an actual cash flow, nor does it necessarily reflect the actual usage pattern.

Book depreciation and **tax depreciation** are terms used to describe the purpose for reducing asset value. Depreciation may be performed for two reasons:

1. Use by a corporation or business for internal financial accounting. This is book depreciation.
2. Use in tax calculations per government regulations. This is tax depreciation.

The methods applied for these two purposes may or may not be the same. *Book depreciation* uses methods to indicate the reduced investment in an asset throughout its expected useful life. The amount of *tax depreciation* is important in an after-tax engineering economy study because the annual tax depreciation is usually tax deductible; that is, it is subtracted from income when calculating the amount of income taxes.

Tax depreciation may be calculated and referred to differently in countries outside the United States. For example, in Canada the equivalent is CCA (capital cost allowance), which is calculated based on the undepreciated value of all corporate properties that form a particular class of assets, whereas in the United States, depreciation may be determined for each asset separately.

First cost *P* or basis *B* is the delivered and installed cost of the asset including purchase price, installation fees, and any other depreciable direct costs.

Book value BV represents the remaining, undepreciated investment after the total amount of depreciation charges to date have been removed. The book value is determined at the end of each year, which is consistent with the end-of-year convention.

Recovery period *n* is the depreciable life in years. Often there are different *n* values for book and tax depreciation. Both of these values may be different from the asset's estimated productive life.

Market value MV is the estimated amount realizable if the asset were sold on the open market. Because of the structure of depreciation laws, the book value and market value may be substantially different. For example, a commercial building tends to increase in market value, but the book value will decrease with time. However, IT equipment usually has a market value much lower than its book value due to rapidly changing technology.

Salvage value S is the estimated trade-in or market value at the end of the asset's depreciable life. The salvage value, expressed as a dollar amount or a percentage of first cost, may be positive, zero, or negative (due to carry-away costs).

Depreciation rate or **recovery rate** d_t is the fraction of the first cost removed by depreciation each year.

Personal property, one of the two types of property for which depreciation is allowed, is the income-producing, tangible possessions of a corporation. Examples are vehicles, manufacturing equipment, computer equipment, chemical processing equipment, and construction assets.

Real property includes real estate and all improvements—office buildings, factories, warehouses, apartments, and other structures. *Land itself is considered real property, but it is not depreciable.*

Half-year convention assumes that assets are placed in service or disposed of in midyear, regardless of when these events actually occur. This convention is utilized in U.S. required tax depreciation methods.

As mentioned before, there are several methods for depreciating assets. The straight line (SL) model is well-known, historically and internationally. Accelerated models, such as the declining balance (DB) model, decrease the book value to zero (or to the salvage value) more rapidly than the straight line method, as shown by the general book value curves in Figure 12.1. Accelerated methods defer some of the income tax burden to later in the asset's life; they do not reduce the total tax burden. In the 1980s the U.S. government standardized accelerated methods for *federal tax depreciation* purposes. All classical methods (straight line, declining balance,

FIGURE 12.1
General shape of book value curves for different depreciation models.

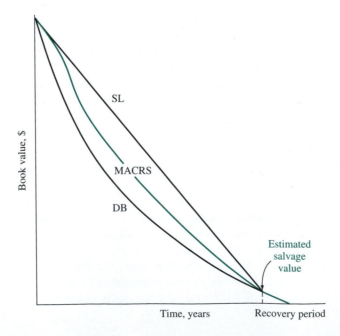

and sum-of-year digits depreciation) were disallowed as acceptable procedures for income tax purposes and replaced by the Accelerated Cost Recovery System (ACRS). In a second round of standardization, MACRS (Modified ACRS) was made the required tax depreciation method. To this date, the following is the law in the United States: *tax depreciation must be calculated using MACRS; book depreciation may be calculated using any classical method or MACRS.* This chapter discusses the straight line, declining balance, and MACRS methods.

There are many exceptions to the standardized depreciation methods. One of interest for an after-tax economic analysis is the *Section 179 Deduction*. This is an economic incentive for businesses, especially small and medium-sized ones, to invest their capital in new and improved equipment through purchase or leasing. Section 179 deduction allows full depreciation of a newly-installed, depreciable, personal property in the first year of ownership rather than over the recovery period of *n* years. This means the basis of the asset is treated as a deductible business expense in the year of installation. The Section 179 limit changes frequently, depending upon the government's view of need for economic stimulus. Recently, the limit has varied widely from $24,000 in 2002 to a record high of $500,000 in 2010 and 2011 (based on special federal-level economic stimulus acts) to $139,000 in 2012. The amount of an asset's first cost above the Section 179 limit must be depreciated over the recovery period of the asset.

Tax law revisions occur often, and depreciation rules are changed from time to time, but the basic principles and relations are always applicable. The U.S. Department of the Treasury, Internal Revenue Service (IRS) website at www.irs.gov has all pertinent publications. Publication 946, *How to Depreciate Property,* is especially applicable.

12.2 STRAIGHT LINE (SL) DEPRECIATION

Straight line is considered the standard against which any depreciation method is compared. It derives its name from the fact that the book value decreases linearly with time. For *book depreciation* purposes, it offers an excellent representation of book value for any asset that is used regularly over a number of years.

Podcast available

The annual SL depreciation is determined by multiplying the first cost minus the salvage value by the depreciation rate.

$$D_t = (B - S)d$$
$$= \frac{B - S}{n} \qquad \textbf{[12.1]}$$

where D_t = depreciation charge for year t ($t = 1, 2, \ldots, n$)

$\quad\quad\ B$ = basis or first cost

$\quad\quad\ S$ = estimated salvage value

$\quad\quad\ n$ = recovery period

$\quad\quad\ d$ = depreciation rate = $1/n$

Since the asset is depreciated by the same amount each year, the book value after t years of service, denoted by BV_t, is the basis B minus the annual depreciation times t.

$$BV_t = B - tD_t \qquad [12.2]$$

The depreciation rate for a specific year is d_t. However, the SL model has the same rate for all years.

$$d = d_t = \frac{1}{n} \qquad [12.3]$$

The format for the spreadsheet function to display annual SL depreciation is

$$= SLN(B,S,n)$$

EXAMPLE 12.1 If an asset has a first cost of $50,000 with a $10,000 estimated salvage value after 5 years, calculate the annual SL depreciation and plot the yearly book value.

Solution

The depreciation each year for 5 years is

$$D_t = \frac{B - S}{n} = \frac{50{,}000 - 10{,}000}{5} = \$8000$$

The book values, computed using Equation [12.2], are plotted in Figure 12.2. For year 5, for example,

$$BV_5 = 50{,}000 - 5(8000) = \$10{,}000 = S$$

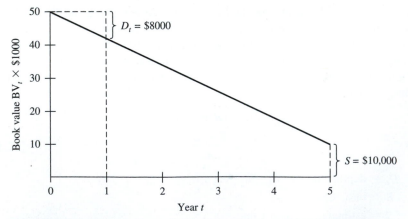

FIGURE 12.2   Book value of an asset depreciated using the straight line method, Example 12.1.

12.3 DECLINING BALANCE DEPRECIATION

Podcast available

Declining balance is also known as the fixed percentage or uniform percentage method. DB depreciation accelerates the write-off of asset value because the annual depreciation is determined by multiplying the *book value at the beginning of a year* by a fixed percentage d, expressed in decimal form. If $d = 0.2$, then 20% of the book value is removed each year. The amount of depreciation decreases each year.

The maximum annual depreciation rate for the DB method is twice the straight line rate.

$$d_{max} = 2/n \qquad\qquad [12.4]$$

This is called *double declining balance (DDB)*. If $n = 5$ years, the DDB rate is 0.4; so 40% of the book value is removed annually. Another commonly used percentage for the DB method is 150% of the SL rate, where $d = 1.5/n$.

The depreciation for year t is the fixed rate d times the book value at the end of the previous year.

$$D_t = (d)BV_{t-1} \qquad\qquad [12.5]$$

Book value in year t is determined by

$$BV_t = B(1 - d)^t \qquad\qquad [12.6]$$

The actual depreciation rate for each year t, relative to the first cost, is

$$d_t = d(1 - d)^{t-1} \qquad\qquad [12.7]$$

If BV_{t-1} is not known, the depreciation in year t can be calculated using B and d_t from Equation [12.7].

$$D_t = dB(1 - d)^{t-1} \qquad\qquad [12.8]$$

It is important to understand that the DB book value never goes to zero, because the book value is always decreased by a fixed percentage. The implied salvage value after n years is the BV_n amount.

$$\text{Implied } S = BV_n = B(1 - d)^n \qquad\qquad [12.9]$$

If a salvage value is initially estimated, this value is *not used* in the DB or DDB method. However, if the implied $S <$ estimated S, it is correct to stop charging further depreciation when the book value is at or below the estimated salvage value.

The spreadsheet functions DDB and DB display depreciation amounts for specific years. The formats are

$$= \mathbf{DDB}(B,S,n,t,d)$$

$$= \mathbf{DB}(B,S,n,t)$$

The d is a number between 1 and 2. If omitted, it is assumed to be 2 for DDB. The DDB function automatically checks to determine when the book value equals the estimated S value. No further depreciation is charged when this occurs. Consult Section 12.8 and Appendix A for further details on using the functions.

EXAMPLE 12.2 Albertus Natural Stone Quarry purchased a computer-controlled face-cutter saw for $80,000. The unit has an anticipated life of 5 years and a salvage value of $10,000. (*a*) Compare the schedules for annual depreciation and book value using two methods: DB at 150% of the straight line rate and at the DDB rate. (*b*) How is the estimated $10,000 salvage value used?

Solution

a. The DB depreciation rate is $d = 1.5/5 = 0.30$ while the DDB rate is $d_{max} = 2/5 = 0.40$. Table 12.1 and Figure 12.3 present the comparison of

TABLE 12.1 **Annual Depreciation and Book Value, Example 12.2**

Year, t	Declining Balance, $d = 0.30$		Double Declining Balance, $d = 0.40$	
	D_t	BV_t	D_t	BV_t
0		$80,000		$80,000
1	$24,000	56,000	$32,000	48,000
2	16,800	39,200	19,200	28,800
3	11,760	27,440	11,520	17,280
4	$8,232	19,208	6,912	10,368
5	5,762	13,446	368	10,000

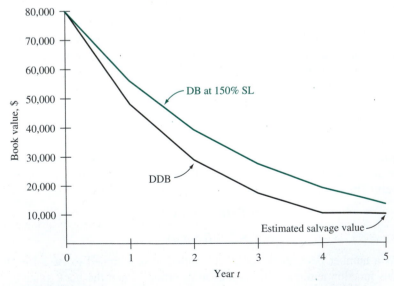

FIGURE 12.3 Plot of book values for two declining balance methods, Example 12.2.

depreciation and book value. Example calculations of depreciation and book value for each method follow.

150% DB for year 2 by Equation [12.5]
with $d = 0.30$
$$D_2 = 0.30(56,000) = \$16,800$$
by Equation [12.6] $BV_2 = 80,000(0.70)^2 = \$39,200$

DDB for year 3 by Equation [12.5]
with $d = 0.40$
$$D_3 = 0.40(28,800) = \$11,520$$
by Equation [12.6] $BV_3 = 80,000(0.60)^3 = \$17,280$

The DDB depreciation is considerably larger during the first years, causing the book values to decrease faster, as indicated in Figure 12.3.

b. The $10,000 salvage value is not utilized by the 150% DB method since the book value is not reduced this far. However, the DDB method reduces book value to $10,368 in year 4. Therefore, not all of the calculated depreciation for year 5, $D_5 = 0.40(10,368) = \$4147$, can be removed; only the $368 above S can be written off.

12.4 MODIFIED ACCELERATED COST RECOVERY SYSTEM (MACRS)

In the U.S., MACRS is the required *tax depreciation* method for all depreciable assets. MACRS rates take advantage of the accelerated DB and DDB methods. Corporations are still free to apply any of the classical methods for *book depreciation*.

MACRS determines annual depreciation amounts using the relation

$$D_t = d_t B \qquad [12.10]$$

where the depreciation rate is tabulated in Table 12.2. (Tab this page for future reference.) The book value in year t is determined by either subtracting the annual depreciation from the previous year's book value, or by subtracting the accumulated depreciation from the first cost.

$$BV_t = BV_{t-1} - D_t \qquad [12.11]$$

$$= B - \sum_{j=1}^{j=t} D_j \qquad [12.12]$$

The first cost is always completely depreciated, since MACRS assumes that $S = 0$, even though there may be an estimated positive salvage.

The MACRS recovery periods are standardized to the values of 3, 5, 7, 10, 15, and 20 years for personal property. Note that all MACRS depreciation rates (Table 12.2) are presented for 1 year longer than the recovery period, and that the extra-year rate is one-half of the previous year's rate. This is because the built-in *half-year convention* imposed by MACRS assumes that all property is

TABLE 12.2   **MACRS Depreciation Rates Applied to the Basis**

| Year | Depreciation Rate (%) | | | | | |
	$n = 3$	$n = 5$	$n = 7$	$n = 10$	$n = 15$	$n = 20$
1	33.33	20.00	14.29	10.00	5.00	3.75
2	44.45	32.00	24.49	18.00	9.50	7.22
3	14.81	19.20	17.49	14.40	8.55	6.68
4	7.41	11.52	12.49	11.52	7.70	6.18
5		11.52	8.93	9.22	6.93	5.71
6		5.76	8.92	7.37	6.23	5.29
7			8.93	6.55	5.90	4.89
8			4.46	6.55	5.90	4.52
9				6.56	5.91	4.46
10				6.55	5.90	4.46
11				3.28	5.91	4.46
12					5.90	4.46
13					5.91	4.46
14					5.90	4.46
15					5.91	4.46
16					2.95	4.46
17–20						4.46
21						2.23

placed in service at the midpoint of the tax year. Therefore, only 50% of the first-year DB depreciation applies for tax purposes. This removes some of the accelerated depreciation advantage and requires that leftover depreciation be taken in year $n + 1$.

MACRS depreciation rates incorporate the DDB method ($d = 2/n$) and switch to SL depreciation as an inherent component for *personal property* depreciation. The MACRS rates start at the DDB rate ($n = 3, 5, 7,$ and 10) or the 150% DB rate ($n = 15$ and 20) and switch when the SL method offers faster write-off.

For *real property (buildings)*, MACRS utilizes the SL method for $n = 39$ throughout the recovery period, and forces partial-year recovery in years 1 and 40. The MACRS real property rates in percentage amounts are

$$\begin{array}{ll} \text{Year 1} & d_1 = 1.391\% \\ \text{Years 2–39} & d_t = 2.564\% \\ \text{Year 40} & d_{40} = 1.177\% \end{array}$$

There is no specific spreadsheet function for the MACRS method; however, the VDB function can be altered slightly to work for MACRS. See Example 12.7 and Appendix A.

EXAMPLE 12.3

Baseline, a nationwide franchise for environmental engineering services, has acquired new workstations and 3-D modeling software for its 100 affiliate sites at a cost of $4000 per site. The estimated salvage for each system after 3 years is expected to be 5% of the first cost. The franchise manager in the home office in San Francisco wants to compare the depreciation for a 3-year MACRS model (tax depreciation) with that for a 3-year DDB model (book depreciation). To help,

a. determine which model offers the larger total depreciation after 2 years, and
b. determine the book value for each method at the end of 3 years.

Solution

The basis is $B = \$400,000$, and the estimated $S = 0.05(400,000) = \$20,000$. The MACRS rates for $n = 3$ are taken from Table 12.2, and the depreciation rate for DDB is $d_{max} = 2/3 = 0.6667$. Table 12.3 presents the depreciation and book values. Depreciation is calculated using Equations [12.10] (MACRS) and [12.5] (DDB). Year 3 depreciation for DDB would be $\$44,444(0.6667) = \$29,629$, except that this results in $BV_3 < \$20,000$. Only the remaining amount of $\$24,444$ can be removed.

a. The 2-year accumulated depreciation values are

MACRS:	$D_1 + D_2 = \$133,320 + 177,800 = \$311,120$
DDB:	$D_1 + D_2 = \$266,667 + 88,889 = \$355,556$

The DDB book depreciation is larger.

b. After 3 years, the MACRS book value is $29,640, while the DDB model indicates $BV_3 = \$20,000$. This occurs because MACRS removes the entire first cost, regardless of the estimated salvage value, but takes an extra year to do so. This is usually a tax advantage of the MACRS method.

TABLE 12.3 Comparing MACRS and DDB Depreciation, Example 12.3

		MACRS		DDB	
Year	Rate	Tax Depreciation	Book Value	Book Depreciation	Book Value
0			$400,000		$400,000
1	0.3333	$133,320	266,680	$266,667	133,333
2	0.4445	177,800	88,880	88,889	44,444
3	0.1481	59,240	29,640	24,444	20,000
4	0.0741	29,640	0		

TABLE 12.4   Example MACRS Recovery Periods

Asset Description	MACRS n Value, Years	
	GDS	ADS Range
Special manufacturing and handling devices, tractors, racehorses	3	3–5
Computers and peripherals, oil and gas drilling equipment, construction assets, autos, trucks, buses, cargo containers, some manufacturing equipment	5	6–9.5
Office furniture; some manufacturing equipment; railroad cars, engines, tracks; agricultural machinery; petroleum and natural gas equipment; *all property not in another class*	7	10–15
Equipment for water transportation, petroleum refining, agriculture product processing, durable-goods manufacturing, ship building	10	15–19
Land improvements, docks, roads, drainage, bridges, landscaping, pipelines, nuclear power production equipment, telephone distribution	15	20–24
Municipal sewers, farm buildings, telephone switching buildings, power production equipment (steam and hydraulic), water utilities	20	25–50
Residential rental property (house, mobile home)	27.5	40
Nonresidential real property attached to the land, but not the land itself	39	40

All depreciable property is classified into *property classes,* which identify their MACRS-allowed recovery periods. Table 12.4, a summary of material from IRS Publication 946, gives examples of assets and the MACRS n values. This table provides two MACRS n values for each property. The first is the *general depreciation system (GDS)* value, which we use in examples and problems. The depreciation rates in Table 12.2 correspond to the n values for the GDS column and provide the fastest write-off allowed. Note that any asset not in a stated class is automatically assigned a 7-year recovery period under GDS.

The far right column of Table 12.4 lists the *alternative depreciation system (ADS)* recovery period range. This alternative method, which uses *SL depreciation over a longer recovery period* than GDS, removes the early-life tax advantages of MACRS. Since it takes longer to depreciate the asset to zero, and since the SL model is required, ADS is usually not considered for an economic analysis.

It is worth noting that, in general, an economic comparison that includes depreciation may be performed more rapidly and usually without altering the final decision by applying the classical straight-line model in lieu of the MACRS method.

12.5 TAX DEPRECIATION SYSTEM IN CANADA

In most industrialized countries, depreciation is based on classical methods – SL, DDB, and others. Some countries require a standardized form of depreciation for tax purposes, much like the United States uses MACRS. Canada is a good example. It uses a system called *CCA (Capitalized Cost Allowance),* which has as a foundation the declining balance method. The CCA system prescribes only a fixed annual rate of write-off (the CCA rate), rather than the recovery period and a varying annual rate of depreciation like the MACRS rates. The annual CCA depreciation and book value, which is termed UCC (Undepreciated Capital Cost) are calculated using relations similar in several respects to the DB and MACRS methods. Additionally, only 50% of the first year's CCA can be claimed, thus preventing the purchase of depreciable assets at the end of the fiscal year and claiming an entire year's CCA charge. For years $t = 1, 2, \ldots$ and basis B, the CCA and UCC are calculated as follows.

$$\text{CCA}_t \begin{cases} = 0.5(\text{CCA rate})B & t = 1 \\ = (\text{CCA rate})\text{UCC}_{t-1} & t = 2, 3, \ldots \end{cases}$$

$$\text{UCC}_t = B - \sum_{j=1}^{j=t} \text{CCA}_j \qquad t = 1, 2, \ldots$$

It is easy to conclude that the relation for CCA is very similar to the DB method's annual depreciation, Equation [12.5]. The UCC relation is the same as the BV relation for any depreciation method, namely, the basis minus all accumulated depreciation to date.

In the Canadian system, assets are clustered into classes and the entire class is depreciated together. However, for the purposes of an economic analysis, the assets being evaluated are usually considered as a class of their own. A few sample classes, their description, and CCA rates are listed below.

Class Number	Description	CCA rate
8	Property not included in any other class	20%
10	Automobiles, vans, trucks, buses, tractors, etc.	30%
38	Most power-operated moveable equipment purchased after 1987 used for moving, excavating, and placing earth, concrete rock or asphalt	30%
52	Computer equipment and software purchased after Jan 27, 2009 and before February 2011	100%

There are recognizable similarities and differences between the depreciation systems of Canada, those of other countries, and MACRS of the U. S. In all cases, the national method of depreciation is a noncash book method by which the capital investment in depreciable assets can be removed from the corporate books, while providing a tax advantage to corporations.

12.6 SWITCHING BETWEEN CLASSICAL METHODS; RELATION TO MACRS RATES

The logic of switching between two methods over the depreciable life of an asset is of interest to understand how the MACRS rates are derived. MACRS includes switching from the DB to the SL model as an implicit property of the method. Additionally, applying the logic is necessary if tax law allows switching. Many countries other than the United States allow switching in order to accelerate the depreciation in the first years of life. The goal is to switch in order to *maximize* the present worth of total depreciation using the equation

$$PW_D = \sum_{t=1}^{t=n} D_t(P/F,i,t) \qquad [12.13]$$

By maximizing PW_D, the PW of taxes are also minimized, though the total amount of taxes paid through the life of the asset are not reduced.

Switching from a DB model to the SL method offers the best advantage, especially if the DB model is the DDB. General rules of switching are as follows:

1. Switch when the depreciation for year t by the current method is less than that for a new method. The selected depreciation is the larger amount.
2. Only one switch can take place during the recovery period.
3. Regardless of the method, $BV < S$ is not allowed.
4. If no switch is made in year t, the undepreciated amount, that is, BV_t, is used as the new basis to select the larger depreciation for the switching decision the next year, $t + 1$.

The specific procedure for step 1 (above) to switch from DDB to SL depreciation is as follows:

1. For each year t, compute the two depreciation charges.

   For DDB: $D_{DDB} = d(BV_{t-1})$ [12.14]

   For SL: $D_{SL} = \dfrac{BV_{t-1}}{n - t + 1}$ [12.15]

2. The depreciation for each year is

$$D_t = \max[D_{DDB}, D_{SL}] \qquad [12.16]$$

The VDB spreadsheet function determines annual depreciation for the DB-to-SL switch by applying this procedure. Refer to Section 12.8 for an example.

EXAMPLE 12.4 Hemisphere Bank purchased a $100,000 online document imaging system with a depreciation recovery period of 5 years. Determine the depreciation for (*a*) the SL method and (*b*) DDB-to-SL switching. (*c*) Use a rate of $i = 15\%$ per year to determine the PW_D values. (MACRS is not involved in this example.)

TABLE 12.5 Depreciation and Present Worth for DDB-to-SL Switching, Example 12.4

Year t	DDB Model D_{DDB}	DDB Model BV_t	SL Model D_{SL}	Larger D_t	$(P/F,15\%,t)$ Factor	Present Worth of D_t
0	—	$100,000				
1	$40,000	60,000	$20,000	$ 40,000	0.8696	$34,784
2	24,000	36,000	15,000	24,000	0.7561	18,146
3	14,400	21,600	12,000	14,400	0.6575	9,468
4*	8,640	12,960	10,800	10,800	0.5718	6,175
5	5,184	7,776		10,800	0.4972	5,370
Totals	$92,224			$100,000		$73,943

*Indicates year of switch from DDB to SL depreciation.

Solution

a. Equation [12.1] determines the annual SL depreciation, which is the same each year.

$$D_t = \frac{100,000 - 0}{5} = \$20,000$$

b. Use the DDB-to-SL switching procedure.

  1. The DDB depreciation rate is $d = 2/5 = 0.40$. D_t amounts for DDB (Table 12.5) are compared with the D_{SL} values from Equation [12.15]. The D_{SL} values change each year because BV_{t-1} is different. Only in year 1 is $D_{SL} = \$20,000$, the same as computed in part (a). For illustration, compute D_{SL} values for years 2 and 4. For $t = 2$, $BV_1 = \$60,000$ by the DDB method and

$$D_{SL} = \frac{60,000 - 0}{5 - 2 + 1} = \$15,000$$

  For $t = 4$, $BV_3 = \$21,600$ by the DDB method and

$$D_{SL} = \frac{21,600 - 0}{5 - 4 + 1} = \$10,800$$

  2. The column "Larger D_t" indicates a switch in year 4 with $D_4 = \$10,800$. Total depreciation with switching is $100,000 compared to the DDB amount of $92,224.

c. The present worth values for depreciation are calculated using Equation [12.13]. For SL, $D_t = \$20,000$ per year, and the P/A factor replaces P/F.

$$PW_D = 20,000(P/A,15\%,5) = 20,000(3.3522) = \$67,044$$

With switching, $PW_D = \$73,943$ as detailed in Table 12.5. This is an increase of $6899 over the SL PW_D.

12.7 DEPLETION METHODS

Podcast available

Up to this point, we have discussed depreciation for assets that can be replaced. Depletion is another method to write off investment that is applicable only to *natural resources*. When the resources are removed, they cannot be replaced or repurchased in the same manner as can a machine, computer, or building. Depletion is applicable to mines, wells, quarries, geothermal deposits, forests, and the like. There are two methods of depletion—*percentage* and *cost depletion.* (Details for U.S. taxes are provided in IRS Publication 535, *Business Expenses.*)

Percentage depletion is a special consideration given for natural resources. A constant, stated percentage of the resource's gross income may be depleted each year *provided it does not exceed 50% of the company's taxable income.* (The limit for oil and gas properties is 100% of taxable income.) The annual depletion amount is

Percentage depletion amount = percentage
$$\times \text{ gross income from property} \qquad [12.17]$$

Using percentage depletion, total depletion charges may exceed first cost with no limitation. The U.S. government does not generally allow percentage depletion to be applied to timber or oil and gas wells (except small independent producers). The annual percentage depletion for some common natural deposits is listed below.

Deposit	Percentage
Sulfur, uranium, lead, nickel, zinc, and some other ores and minerals	22%
Gold, silver, copper, iron ore, and some oil shale	15
Oil and natural gas wells (varies)	15–22
Coal, lignite, sodium chloride	10
Gravel, sand, some stones	5
Most other minerals, metallic ores	14

EXAMPLE 12.5   A gold mine was purchased for $10 million. It has an anticipated gross income of $8.0 million per year for years 1 to 5 and $5.0 million per year after year 5. Assume that depletion charges do not exceed 50% of taxable income. Compute the annual depletion amount and determine how long it will take to recover the initial investment.

Solution

A 15% depletion applies for gold.

Years 1 to 5:	0.15(8.0 million) = $1.2 million
Years thereafter:	0.15(5.0 million) = $750,000

A total of $6 million is written off in 5 years, and the remaining $4 million is written off at $750,000 per year. Total recovery is attained in

$$5 + \frac{\$4 \text{ million}}{\$750,000} = 5 + 5.3 = 10.3 \text{ years}$$

Cost depletion, also called factor depletion, is based on the level of activity or usage. It may be applied to most types of natural resources. The annual cost depletion factor p_t is the ratio of the first cost to the estimated number of units recoverable.

$$p_t = \frac{\textbf{first cost}}{\textbf{resource capacity}} \qquad \text{[12.18]}$$

The annual depletion charge is p_t times the year's usage. *The total cost depletion cannot exceed the first cost of the resource.* If the capacity of the property is reestimated as higher or lower in the future, a new p_t is determined based upon the undepleted amount.

Temple-Inland Corporation has negotiated the rights to cut timber on privately **EXAMPLE 12.6** held forest acreage for $700,000. An estimated 350 million board feet of lumber are harvestable. Determine the depletion amount for the first 2 years if 15 million and 22 million board feet are removed.

Solution

Use Equation [12.18] for p_t in dollars per million board feet.

$$p_t = \frac{\$700,000}{350} = \$2000 \text{ per million board feet}$$

Multiply p_t by the annual harvest to obtain depletion of $30,000 in year 1 and $44,000 in year 2. Continue using p_t until a total of $700,000 is written off or the remaining timber requires a reestimate of total board feet harvestable.

When allowed by law, the depletion each year may be determined using either the cost method or the percentage method. Usually, the percentage depletion amount is chosen because of the possibility of writing off more than the original investment. However, the law does require that the cost depletion amount be chosen if the percentage depletion is smaller in any year.

12.8 USING SPREADSHEETS FOR DEPRECIATION COMPUTATIONS

Spreadsheet functions calculate the annual depreciation for methods discussed and for switching between declining balance and straight line. All methods except MACRS have specific functions, but the VDB (variable declining balance) function may be adjusted to obtain the correct MACRS rates, including the initial year and extra year depreciation adjustments. The next example illustrates the functions in the same order in which the methods were discussed. The final part illustrates the use of an *x-y* scatter chart of book values to compare methods.

EXAMPLE 12.7   BA Aerospace purchased new aircraft engine diagnostics equipment for its maintenance support facility in France. Installed cost was $500,000 with a depreciable life of 5 years and an estimated 1% salvage. Use a spreadsheet to determine the following:

  a. Straight line (SL) depreciation and book value schedule.
  b. Double declining balance (DDB) depreciation and book value schedule.
  c. Declining balance (DB) at 150% of the SL rate depreciation and book value schedule.
  d. MACRS depreciation and book value schedule.
  e. Depreciation and book value schedule allowing a switch from DDB to SL depreciation.
  f. One graph that plots the book value curves for all five schedules.

Solution

Assume that any method can be used to depreciate this equipment. All five schedules and the graph may be developed using a single worksheet; however, several are developed here to show the progression through different methods. The functions for depreciation applied here—SLN, DDB, and VDB—are detailed in Appendix A and summarized inside the front cover.

  a. Figure 12.4: The SLN function, which determines SL depreciation in column B, has the same entry each year. The entire amount $B - S = \$495,000$ is depreciated. Yearly book value for this and all other methods is determined by subtraction of annual depreciation D_t from BV_{t-1}.
  b. Figure 12.5: Column B details double declining balance depreciation using the DDB function = DDB(500000,5000,5,t,2) in which the period value t changes each year. The last entry is optional; if omitted, 2 for DDB is assumed. $BV_5 = \$38,880$ does not reach the estimated salvage of $5000.

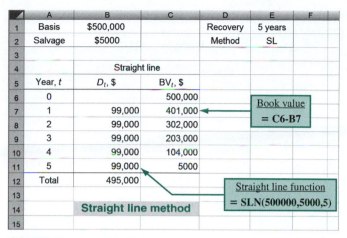

	A	B	C	D	E	F
1	Basis	$500,000		Recovery	5 years	
2	Salvage	$5000		Method	SL	
3						
4			Straight line			
5	Year, t	D_t, $	BV_t, $			
6	0		500,000			
7	1	99,000	401,000		Book value = C6-B7	
8	2	99,000	302,000			
9	3	99,000	203,000			
10	4	99,000	104,000			
11	5	99,000	5000			
12	Total	495,000				
13					Straight line function	
14		**Straight line method**			= SLN(500000,5000,5)	
15						

FIGURE 12.4 Straight line depreciation and book value schedule, Example 12.7*a*.

	A	B	C	D	E
1	First cost	$500,000	Recovery	5 years	
2	Salvage	$5000	Methods	DDB and DB at 150% SL	
3					
4		Double declining balance		Declining balance at 150% SL	
5	Year, t	D_t, $	BV_t, $	D_t, $	BV_t, $
6	0		500,000		500,000
7	1	200,000	300,000	150,000	350,000
8	2	120,000	180,000	105,000	245,000
9	3	72,000	108,000	73,500	171,500
10	4	43,200	64,800	51,450	120,050
11	5	25,920	38,880	36,015	84,035
12	Total	461,120		415,965	
13					
14		DDB function, year 5 = DDB(500000,5000,5,5,2)		DB function for 150% SL, year 5 = DDB(500000,5000,5,5,1.5)	
15					
16					
17					
18		**Declining balance methods**			
19					

FIGURE 12.5 DDB and DB at 150% SL rate depreciation and book value schedules, Example 12.7*b* and *c*.

c. Figure 12.5: Declining balance depreciation at 150% of the SL rate (column D) is best determined using the DDB function with the last field entered as 1.5. (Refer to the DB function description in Appendix A for more details.) The ending book value (column E) is now higher at $84,035 since the annual depreciation rate is 1.5/5 = 0.3, compared to the DDB rate of 0.4.

	A	B	C	D	E
1	First cost	$500,000	Recovery	5 years	
2	Salvage	$5000	Methods	MACRS and switching	
3					
4			MACRS	DDB-to-SL switch	
5	Year, t	D_t, $	BV_t, $	D_t, $	BV_t, $
6	0		500,000		500,000
7	1	100,000	400,000	200,000	300,000
8	2	160,000	240,000	120,000	180,000
9	3	96,000	144,000	72,000	108,000
10	4	57,600	86,400	51,500	56,500
11	5	57,600	28,800	51,500	5000
12	6	28,800	0		
13	Total	500,000		495,000	
14					
15					
16					

VDB function for switching,
years 3 to 4

$$= \text{VDB}(500000,5000,5,A9,A10)$$

MACRS using VDB function, year 6

$$= \text{VDB}(500000,0,5,\text{MAX}(0,A12\text{-}1.5),\text{MIN}(5,A12\text{-}0.5),2)$$

MACRS method and DDB-to-SL switching

FIGURE 12.6 MACRS and DDB-to-SL switch depreciation using the VDB function and book value schedules, Example 12.7*d* and *e*.

d. Figure 12.6: Column B displays the MACRS depreciation. The VDB function normally calculates depreciation when switching from the declining balance to the straight line method. However, to determine the correct MACRS rates it is necessary to embed MAX and MIN functions into the VDB function so that only one-half of the DDB depreciation is allowed in year 1 and an additional one-half of the final year depreciation is carried over to year $n + 1$. The VDB format for each year t is

$$= \text{VDB}(\textbf{cost, 0, life, MAX}(0, t - 1.5), \textbf{MIN}(\textbf{life}, t - 0.5), \textbf{factor})$$

The salvage entry is 0 since MACRS assumes $S = 0$. The factor entry is 2 for DDB (MACRS lives of 3, 5, 7, or 10) or 1.5 (lives of 15 or 20). Here, for example, when $t = 6$ (row 12), the function is $= \text{VDB}(500000,0,5, \text{MAX}(0,6 - 1.5), \text{MIN}(5,6 - 0.5),2)$ as shown in cell B12 and the associated cell tag.

e. Figure 12.6: Column D shows the larger of DDB or SL depreciation using the VDB function in its normal format (inside front cover and Appendix A). Now the estimated salvage of $5000 is entered. The factor entry (last field) of 2 is optional. When the SL amount of $51,500 is larger than the DDB

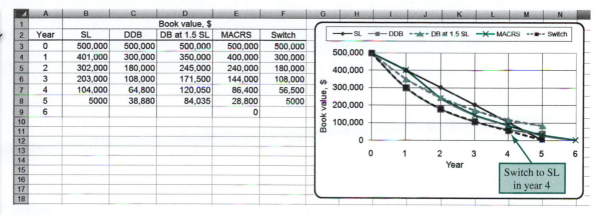

	A	B	C	D	E	F	G	H	I	J	K	L	M	N
1				Book value, $										
2	Year	SL	DDB	DB at 1.5 SL	MACRS	Switch								
3	0	500,000	500,000	500,000	500,000	500,000								
4	1	401,000	300,000	350,000	400,000	300,000								
5	2	302,000	180,000	245,000	240,000	180,000								
6	3	203,000	108,000	171,500	144,000	108,000								
7	4	104,000	64,800	120,050	86,400	56,500								
8	5	5000	38,880	84,035	28,800	5000								
9	6				0									
10														

FIGURE 12.7 Book value curves using an *x-y* scatter chart, Example 12.7*f*.

amount of $43,200 (see cell B10 in Figure 12.5) for the first time in year 4, the switch occurs.

f. Figure 12.7: Book values are copied from previous spreadsheets for construction of the *x-y* scatter chart. Some conclusions are: DDB and switching are the most accelerated (book value curves are coincident until year 4) compared to SL; the switch to SL from DDB in year 4 is indicated by the separation of the BV curves; DDB and switching do not come close to the estimated salvage; and, MACRS ignores salvage completely.

SUMMARY

This chapter discussed depreciation and depletion. Depreciation, which is not actual cash flow, writes off the investment in assets. For the purpose of income tax reduction, the MACRS method is applied (in the United States only). For book depreciation the classical methods of straight line and declining balance can always be utilized.

Information about each method follows, and formulas are summarized in Table 12.6.

Modified Accelerated Cost Recovery System (MACRS)

- It is the only approved tax depreciation system in the United States.
- The rates automatically switch from DDB or DB to SL.
- It always depreciates to zero; that is, it assumes $S = 0$.

- Recovery periods are specified by property classes.
- The actual recovery period is 1 year longer due to the imposed half-year convention.

Straight Line (SL)

- It writes off capital investment linearly over *n* years.
- The estimated salvage value is always considered.
- This is the classical, nonaccelerated depreciation model.

Declining Balance (DB)

- The model accelerates depreciation compared to straight line.
- The book value is reduced each year by a fixed percentage.

TABLE 12.6   **Summary of Depreciation Method Relations**

Model	MACRS	SL	DDB
Fixed depreciation rate d	Varies	$\dfrac{1}{n}$	$\dfrac{2}{n}$
Annual depreciation rate d_t	Table 12.2	$\dfrac{1}{n}$	$d(1-d)^{t-1}$
Annual depreciation D_t	$d_t B$	$\dfrac{B-S}{n}$	$d(\mathrm{BV}_{t-1})$
Book value BV_t	$\mathrm{BV}_{t-1} - D_t$	$B - tD_t$	$B(1-d)^t$

- The most used rate is twice the SL rate, which is called double declining balance (DDB).
- It has an implied salvage that may be different from the estimated salvage.
- Not approved for tax depreciation in the United States, but it is frequently used for book depreciation in many countries worldwide.

Depletion writes off investment in natural resources. Percentage depletion, which can recover more than the initial investment, reduces the value of the resource by a constant percentage of gross income each year. Alternatively, cost depletion is applied to the amount of resource removed. No more than the initial investment can be recovered with cost depletion.

PROBLEMS

Depreciation Terms and Computations

12.1 How does depreciation affect income taxes?

12.2 What are three depreciable costs that are included in an asset's basis?

12.3 What is the difference between book value and market value?

12.4 There are 3 different life values (recovery periods) associated with a depreciable asset. Identify each by name and explain how it is correctly used.

12.5 Cyber Manufacturing is purchasing a complete video borescope system for applications that require work in places that eyes cannot see. The purchase price is $8000, shipping and delivery is $300, installation cost is $1200, tax recovery period is 5 years, book depreciation period is 10 years, salvage value is estimated to be $500, operating cost (with technician) will be $45,000 per year. For MACRS depreciation of the system, what are the values of B, S, and n?

12.6 Jobe Concrete Products placed a new sand sifter into production 3 years ago. It had an installed cost of $100,000, a life of 5 years, and an anticipated salvage of $20,000. Book depreciation charges for the 3 years are $40,000, $24,000, and $14,000, respectively. (*a*) Determine the book value after 2 years. (*b*) If the sifter's market value today is $20,000, determine the difference between its current book value and its market value, and state which is lower. (*c*) Determine the total percentage of installed cost written off through year 3.

12.7 Quantum Electronic Services paid $P = \$40,000$ for its networked computer system. Both tax and book depreciation accounts are maintained. The annual tax depreciation rate is based on the previous year's book value (BV), while the book depreciation rate is based on the original first cost (P). Use the rates listed to plot annual depreciation and book values for each method. Develop the graphs using hand calculations or a spreadsheet, as directed by your instructor.

Year of Ownership	1	2	3	4
Tax depreciation rate, % of BV	40	40	40	40
Book depreciation rate, % of P	25	25	25	25

Straight Line Depreciation

12.8 A company that manufactures pulse Doppler insertion flow meters uses the straight line method for book depreciation purposes. Newly acquired equipment has a first cost of $170,000 with a 3-year life and $20,000 salvage value. Determine the depreciation charge and book value for year two.

12.9 Kobi Technologies book-depreciated an asset at $27,500 per year for four years using the straight line method. If the book value in year two is $65,000, what was (*a*) the asset's salvage value, and (*b*) its basis?

12.10 An asset owned by Photon Environmental was book depreciated by the straight line method over a 5-year period with book values of $296,000 and $224,000 in years two and three, respectively. Determine (*a*) the salvage value used in the calculation, and (*b*) the asset's basis.

12.11 Carl is curious about the original cost of the digital imaging equipment he uses at the First National Bank. Accounting cannot tell him the cost, but they know the annual depreciation over an 8-year period is $18,900 per year. If all items are straight-line depreciated and the salvage is always 25% of the first cost, estimate the original cost.

12.12 Butler Buildings purchased semi-automated assembly and riveting robotics equipment for constructing its modular warehouse buildings. The first cost was $475,000 and installation costs were $75,000; life is estimated at 10 years with a salvage of 15% of first cost. Use the SL method to determine (*a*) annual recovery rate, (*b*) annual depreciation, (*c*) book value after 5 years, and (*d*) book value after 10 years.

12.13 Columbia Construction purchased new equipment for its project to transform an existing, vacant facility into a milk and butter processing plant. For the equipment, $B = \$350,000$, and $S = \$50,000$. Book depreciation will use the SL method with $n = 5$ years. Use calculator or spreadsheet-based computations (or both, as directed) to plot annual depreciation, accumulated depreciation, and book value on one graph.

Declining Balance Depreciation

12.14 Halcrow Yolles purchased equipment for new highway construction in Manitoba, Canada, costing $500,000 Canadian. Estimated salvage at the end of the expected life of 5 years is $50,000. Various book depreciation methods are being studied currently. Determine the depreciation for year 2 using the DDB, 150% DB and SL methods.

12.15 Determine the original basis of a machine that is used for making spill containment pallets if its book value in year 3 is $25,000. The machine has a 5-year life and the double-declining balance method is applied.

12.16 If an asset is book-depreciated by the DDB method over a 5-year period, how long will it take to reach its salvage value, provided the estimated salvage is 25% of the first cost?

12.17 If the salvage value of an asset is nil and it is depreciated by the double-declining balance method, what percentage of the asset's first cost will remain after its 5-year life?

12.18 An engineer with Accenture Middle East BV in Dubai was asked by her client to help understand the difference between 150% DB and DDB depreciation. Answer the questions if $B = \$180,000$, $n = 12$ years, and $S = \$30,000$.

    a. What are the book values after 12 years for both methods?

    b. How do the estimated salvage and the two book values after 12 years compare in value?

    c. Which of the two methods, when calculated correctly considering $S = \$30,000$, writes off more of the first cost over the 12 years?

12.19 Exactly 10 years ago, Boyditch Professional Associates purchased $100,000 in depreciable assets with an estimated salvage of $10,000. For tax depreciation the SL method with $n = 10$ years was used, but for book depreciation, Boyditch applied the DDB method with $n = 7$ years and neglected the salvage estimate. The company sold the assets today for $12,500.

    a. Compare this sales amount with the book values using SL and DDB methods.

    b. If a salvage of $12,500 had been estimated exactly 10 years ago, determine the depreciation for each method in year 10.

12.20 Shirley is studying depreciation in her engineering management course. The instructor asked her to graphically compare the total percent of first cost depreciated for an asset costing B dollars over a life of $n = 5$ years for DDB and 125% DB depreciation. Help her by developing the plots of percent of B depreciated versus years. Use a spreadsheet unless instructed otherwise.

MACRS Depreciation

12.21 Del Norte Brick Co. is located near the intersection of Texas, New Mexico, and Mexico. Improved access to the company's property is via a small, privately owned bridge across the Rio Grande River.

The cost was $770,000 and it has a recovery period of 15 years. Determine the depreciation and book value for year three according to the MACRS method.

12.22 An automated assembly robot that cost $400,000 has a depreciable life of five years with a $100,000 salvage value. If the MACRS depreciation rates for years 1, 2, and 3 are 20.00, 32.00, and 19.20%, respectively, what is the book value of the robot at the end of year three?

12.23 The manager of a plant that manufactures stepper drives knows that MACRS and DDB are both accelerated depreciation methods, but out of curiosity, he wants to determine which one offers faster write-off in the first three years for equipment that has a first cost of $300,000, a 5-year life, and a $60,000 salvage value. Determine which method yields the lower book value and by how much.

12.24 Bison Gear and Engineering of St Charles, IL, makes sensorless and brushless dc gear motors suited for foodservice equipment, factory automation, alternative energy systems, and other specialty machinery applications. The company purchased an asset 2 years ago that has a 5-year recovery period. If the depreciation charge by the MACRS method for year 3 is $14,592, what was (a) the first cost of the asset, and (b) the depreciation charge in year 1?

12.25 A plant manager for a large cable company knows that the real value of certain types of cable-maintenance equipment is more closely approximated when the equipment is depreciated linearly by the SL method rather than with the more rapid write-off method MACRS. Therefore, he keeps two sets of books, one using MACRS for taxes and a second using SL for equipment-management purposes. For an asset that has a first cost of $80,000, a depreciable life of 5 years, and a salvage value equal to 25% of the first cost, determine the difference in the book values shown in the two sets of books at the end of year 4, and identify the method that has a lower BV after 4 years.

12.26 A 120-metric-ton telescoping crane that cost $320,000 is owned by Upper State Power. Salvage is estimated at $75,000. (a) Compare book values for MACRS and classical SL depreciation over a 7-year recovery period. (b) Explain how the estimated salvage is treated using MACRS.

12.27 Youngblood Shipbuilding Yard just purchased $800,000 in capital equipment for Ship-repairing functions on dry-docked ships. Estimated salvage is $150,000 for any year after 5 years of use. Compare the depreciation and book value for year 3 for each of the following depreciation methods:
a. GDS MACRS where the recovery period is 10 years
b. Double-declining balance with a recovery period of 15 years.
c. ADS SL, as an alternative to MACRS, with a recovery period of 15 years

12.28 Fairfield Properties owns real property that is MACRS depreciated with $n = 39$ years. They paid $3.4 million for the apartment complex and hope to sell it after 10 years of ownership for 50% more than the book value at that time. Compare the expected selling price with the amount that Fairfield paid for the property.

12.29 Blackwater Spring and Metal utilizes the same computerized spring forming machinery in its U.S. and Malaysian plants. Purchased in 2010, the first cost was $750,000 with $S = $150,000 after $n = 10$ years. MACRS depreciation with $n = 5$ years is applied in the United States and SL depreciation with $n = 10$ years is used by the Malaysian facility. (a) Develop and graph the book value curves for both plants. (b) If the equipment is sold after 6 years for $100,000, calculate the over or under depreciation amounts for each method.

12.30 Aaron Pipeline has the service contract for part of the Black Mesa slurry coal pipeline in Arizona. The company placed $500,000 worth of depreciable capital equipment into operation. Use the VDB spreadsheet function to calculate the MACRS depreciation and book value schedules for a 5-year recovery period.

Switching Between Methods

12.31 For a country that allows switching between declining balance (not DDB) and straight line, determine the difference in depreciation for year 2 between the two methods and determine if a switch is advisable. The asset has a first cost of $100,000, a 5-year recovery period, a $10,000 salvage value, and $d = 1/n$.

12.32 Henry has an assignment from his boss at Czech Glass and Wood Sculpting to evaluate depreciation methods for writing off the $200,000 first cost of a newly acquired Trotec

CO_2 laser system for engraving and cutting. Productive life is 8 years and salvage is estimated at $10,000. Henry wants to compare the PW of depreciation at $i = 10\%$ per year for DDB-to-SL switching with MACRS for $n = 7$ years to determine which is the preferred method. Perform the analysis using a table or a spreadsheet, as requested by the instructor.

12.33 ConocoPhillips alkylation processes are licensed to produce high-octane, low-sulphur blendstocks domestically and internationally. Halliburton Industries has newly licensed alkylation equipment costing $1 million per system at its Moscow, Houston, and Abu Dhabi refinery service operations. Russia requires a 10-year, straight line recovery with a 10% salvage value. The United States allows a 7-year MACRS recovery with no salvage considered. The United Arab Emirates allows a 7-year recovery with switching from DDB to SL method and no salvage considered. Which of the country's methods has the largest PW of depreciation at $i = 15\%$ per year?

12.34 To determine the MACRS rates in Table 12.2, the switching procedure in Section 12.5 must be altered slightly to accommodate the half-year convention imposed by MACRS. The first difference is in year 1, where the DDB rate is only one-half of the DDB rate. The second difference is in the denominator for the SL depreciation. The term is $(n - t + 1.5)$, instead of the $(n - t + 1)$ shown. The third and final change is that one-half of the SL depreciation allowed in year n is taken in the year $n + 1$. Use the switching procedure and three changes to verify the MACRS rates for $n = 5$.

Depletion Methods

12.35 What is the difference between depreciation and depletion?

12.36 A relatively small privately owned coal-mining company has the sales results summarized below. Determine the annual percentage depletion for the coal mine. Assume the company's taxable income is $140,000 each year.

Year	Sales, tons	Spot Sales Price, $/ton
1	34,300	9.68
2	50,100	10.50
3	71,900	11.23

12.37 Chaparral Sand and Gravel purchased a pit for $900,000 that is expected to yield 6000 tons of gravel and 7000 tons of sand per year. If the gravel will sell for $6 per ton and the sand for $9 per ton, determine the annual depletion charge according to the percentage depletion method.

12.38 Vesco Mineral Resources purchased mineral rights to land in the foothills of the Santa Cristo mountains. The cost of the purchase was $9 million. Vesco originally estimated that 200,000 tons of lignite coal was removable. However, further exploration during the second year of operation revealed that a total of 280,000 tons could be economically removed. If the company sold 20,000 tons in year 1 and 30,000 tons in year 2, what are the depletion charges each year according to the cost depletion method?

12.39 Carrolton Oil and Gas, an independent oil and gas producer, is approved to use a 20% of gross income depletion allowance. The write-off last year was $700,000 on its horizontal directional drill wells. Determine the estimated total reserves in barrels, if the volume pumped last year amounted to 1% of the total and the delivered-product price averaged $75 per barrel.

12.40 Ederly Quarry sells a wide variety of cut limestone for residential and commercial building construction. A recent quarry expansion cost $2.9 million and added an estimated 100,000 tons of reserves. (a) Estimate the cost depletion allowance for the next 5 years for these new reserves, using the projections made by John Ederly. (b) Will the depletion charge be limited in any year due to restrictions placed on the cost depletion method?

Year	1	2	3	4	5
Volume, 1000 tons	10	12	15	15	18
Price, $ per ton	85	90	90	95	95

12.41 For the last 10 years, Am-Mex Coal has used the cost depletion factor of $2500 per 100 tons to write off the investment of $35 million in its Pennsylvania anthracite coal mine. Depletion thus far totals $24.8 million. A new study to appraise mine reserves indicates that no more than 800,000 tons of salable coal remain. Determine next year's percentage and cost depletion amounts, if estimated gross income is expected to be between $6.125 and $8.50 million on a production level of 72,000 tons.

ADDITIONAL PROBLEMS AND FE EXAM REVIEW QUESTIONS

12.42 All of the following assets can be depreciated, except:
- **a.** a bulldozer.
- **b.** a copper mine.
- **c.** a surgical robot.
- **d.** a conveyor belt.

12.43 Classical straight line depreciation of a $100,000 asset takes place over a 5-year recovery period. If the salvage value is 20% of first cost, the depreciation charge for year 3 is closest to:
- **a.** $16,000
- **b.** $20,000
- **c.** $24,000
- **d.** $28,000

12.44 An asset with a first cost of $50,000 is to be depreciated by the straight line method over a 5-year period. The asset will have annual operating costs of $35,000 and a salvage value of $10,000. According to the straight line method, the book value at the end of year 3 will be closest to:
- **a.** $8000
- **b.** $20,000
- **c.** $24,000
- **d.** $26,000

12.45 A motorized cultivator with a first cost of $28,000 and salvage value of 25% of the first cost is depreciated by the DDB method over a 5-year period. If the operating cost is $43,000 per year, the depreciation charge for year 2 is closest to:
- **a.** $18,000
- **b.** $11,200
- **c.** $6700
- **d.** $4030

12.46 An assembly line conveyor system with a 5-year life is to be depreciated by the DDB method. The conveyor units had a first cost of $30,000 with a $9000 salvage value. The annual operating cost allocated to the conveyor is $7000 per year. The book value at the end of year 2 is closest to:
- **a.** $6,480
- **b.** $10,800
- **c.** $12,400
- **d.** $18,000

12.47 Gisele is performing a make/buy study involving the retention or disposal of a 4-year-old machine that was to be in production for 8 years. It cost $500,000 originally. Without a market value estimate, she decided to use the current book value plus 20%. If DDB depreciation is applied, the market value estimate is closest to:
- **a.** $158,200
- **b.** $253,125
- **c.** $217,900
- **d.** $189,800

12.48 A MACRS-depreciated asset has $B = \$100,000$, $S = \$40,000$, and a 10-year recovery period. (The d_t values for years $t = 1, 2, 3, 4$ and 5 are 10.00%, 18.00%, 14.40%, 11.52%, and 9.22%, respectively.) The depreciation charge for year 4 according to the MACRS method is closest to:
- **a.** $58,700
- **b.** $62,400
- **c.** $11,500
- **d.** $46,100

12.49 An industrial robot depreciated by the MACRS method has $B = \$60,000$ and a 5-year depreciable life. The MACRS d_t values for years 1, 2, 3, 4 and 5 are 10.00%, 18.00%, 14.40%, 11.52%, and 9.22%, respectively. If the depreciation charge in year 3 is $8640, the salvage value that was used in the depreciation calculation is closest to:
- **a.** $0
- **b.** $10,000
- **c.** $20,000
- **d.** $30,000

12.50 South African Gold Mines, Inc. is writing off its $210 million investment using the cost depletion method. An estimated 700,000 ounces of gold are available in its developed mines. This year 35,000 ounces were produced and sold at an average price of $1400 per ounce. Taxable income for the year is estimated at $4.8 million. The cost depletion for the year is closest to:
- **a.** $7.35 million, which is 15% of the gross income of $49.0 million.
- **b.** $2.4 million, which is 50% of estimated taxable income.
- **c.** $10.5 million.
- **d.** $42 million.

12.51 Rayonier Forrest Resources purchased a small tract of timber land for $70,000 that contained 25,000 trees. The value of the land was estimated to be $20,000. In the first year of operation, the lumber company cut down 7000 trees. According to the cost depletion method, the depletion deduction for year 1 is closest to: (Hint: Value of land cannot be depreciated or depleted.)

a. $2000
b. $7000
c. $10,000
d. $14,000

12.52 A stone and gravel quarry in Texas can use a percentage depletion rate of 5% of gross income, or

a cost depletion rate of $1.28 per ton. Quarry first cost = $3.2 million; estimated total tonnage = 2.5 million tons; tonnage this year = 65,000; gross income = $40 per ton. Of the two depletion charges, the method and larger amount are:

a. percentage at $83,200.
b. percentage at $130,000.
c. cost at $80,000.
d. cost at $130,000.

After-Tax Economic Analysis

Royalty-Free/CORBIS

This chapter provides an overview of tax terminology and equations pertinent to an after-tax economic analysis. Income tax computations for corporations and individuals are illustrated. The transfer from estimating cash flow before taxes (CFBT) to cash flow after taxes (CFAT) involves a consideration of significant tax effects that may alter the final decision, as well as estimate the magnitude of the tax effect on cash flow over the life of the alternative.

Mutually exclusive alternative comparisons using after-tax PW, AW, and ROR methods are explained with major tax implications considered. Replacement studies are discussed with tax effects that occur at the time that a defender is replaced. Additionally, the cost of debt and equity investment capital and its relation to the MARR is examined.

Additional information on U.S. federal taxes—tax law, and annually updated tax rates—is available through Internal Revenue Service publications and, more readily, on the IRS website www.irs.gov. Publications 542, *Corporations,* and 544, *Sales and Other Dispositions of Assets,* are especially applicable.

Purpose: Perform an economic evaluation of alternatives considering the effect of income taxes and pertinent tax regulations.

LEARNING OUTCOMES

Terminology and rates

1. Correctly use the basic terminology and income tax rates for corporations and individuals.

CFAT analysis

2. Calculate after-tax cash flow and evaluate alternatives.

Taxes and depreciation

3. Demonstrate the tax impact of depreciation recapture, accelerated depreciation, and a shortened recovery period.

After-tax replacement

4. Evaluate a defender and challenger in an after-tax replacement study.

Cost of capital

5. Determine the weighted average cost of capital (WACC) and its relation to MARR.

Spreadsheets

6. Use a spreadsheet to perform an after-tax PW, AW, or ROR analysis.

Value-added analysis

7. Evaluate alternatives using after-tax economic value-added analysis

13.1 INCOME TAX TERMINOLOGY AND RELATIONS

Podcast available

Some basic tax terms and relationships are explained here.

Gross income GI, also called *operating revenue R*, is the total income realized from all revenue-producing sources of the corporation, plus any income from other sources such as sale of assets, royalties, and license fees.

Income tax is the amount of taxes to be paid based on some form of income or profit levied by the federal (or lower-level) government. A large percentage of U.S. tax revenue is based upon taxation of corporate and personal income. Taxes are actual cash flows.

Net Operating Income NOI is the difference between gross income and operating expenses, that is NOI = GI − E. In financial terms, NOI represents the *earnings before interest and taxes* are removed; *EBIT* is another term for NOI.

Operating expenses E include all corporate costs incurred in the transaction of business. These expenses are tax deductible for corporations. Depreciation is not an operating expense. For engineering economy alternatives, these are the AOC (annual operating cost) and M&O (maintenance and operating) costs.

Taxable income TI is the amount upon which income taxes are based. For corporations, depreciation and operating expenses are tax-deductible.

$$\textbf{TI = gross income − expenses − depreciation}$$
$$\textbf{= GI − E − D} \qquad \text{[13.1]}$$

Tax rate T is a percentage, or decimal equivalent, of TI owed in taxes. The tax rate is graduated (or progressive); that is, higher rates apply as TI increases.

$$\textbf{Taxes = (taxable income) × (applicable tax rate)}$$
$$\textbf{= (TI)(T)} \qquad \text{[13.2]}$$

Effective tax rate T_e is a single-figure rate used in an economy study to estimate the effects of federal, state, and local taxes. The T_e rate, which usually ranges from 25% to 50% of TI, includes an allowance for state (and possibly local) taxes that are deductible when determining federal taxes.

$$\textbf{T_e = state and local rate + (1 − state and local rate)(federal rate)} \quad \text{[13.3]}$$
$$\textbf{Taxes = TI(T_e)} \qquad \text{[13.4]}$$

Different bases (and taxes) are used. The most common is a basis of income (and income taxes). Others are total sales (sales tax); appraised value of property (property tax); value-added tax (VAT); net capital investment (asset tax); winnings from gambling (part of income tax); and retail value of items imported (import tax).

TABLE 13.1 U.S. Corporate Federal Income Tax Rate Schedule (sample for 2012)

If Taxable Income ($) Is:			
Over	But Not over	Tax Is	Of the Amount over
0	50,000	15%	0
50,000	75,000	7,500 + 25%	50,000
75,000	100,000	13,750 + 34%	75,000
100,000	335,000	22,250 + 39%	100,000
335,000	10,000,000	113,900 + 34%	335,000
10,000,000	15,000,000	3,400,000 + 35%	10,000,000
15,000,000	18,333,333	5,150,000 + 38%	15,000,000
18,333,333	—	35%	0

(*Source:* Internal Revenue Service, Publication 542 – Corporations, March 2012, p. 17.)

The annual U.S. federal tax rate T for corporations and individuals is based upon the principle of *graduated tax rates,* which means that higher rates go with larger taxable incomes. Table 13.1 presents recent T values for corporations. The portion of each new dollar of TI is taxed at what is called the *marginal tax rate.* As an illustration, scan the tax rates in Table 13.1. A business with an annual TI of $50,000 has a marginal rate of 15%. However, a business with TI = $100,000 pays 15% for the first $50,000, 25% on the next $25,000, and 34% on the remainder.

$$\text{Taxes} = 0.15(50,000) + 0.25(75,000 - 50,000) + 0.34(100,000 - 75,000)$$
$$= \$22,250$$

To simplify tax computations, an average federal tax rate may be used in Equation [13.3] to determine the single-figure effective tax rate T_e.

If the video division of Marvel Comics has an annual gross income of $2,750,000 with expenses and depreciation totaling $1,950,000, (*a*) compute the company's exact federal income taxes. (*b*) Estimate total federal and state taxes if a state tax rate is 8% and a 34% federal average tax rate applies.

EXAMPLE 13.1

Solution

a. Compute the TI by Equation [13.1] and the income taxes using Table 13.1 rates.

$$\text{TI} = 2,750,000 - 1,950,000 = \$800,000$$
$$\text{Taxes} = 50,000(0.15) + 25,000(0.25) + 25,000(0.34)$$
$$+ 235,000(0.39) + (800,000 - 335,000)(0.34)$$
$$= 7500 + 6250 + 8500 + 91,650 + 158,100$$
$$= \$272,000$$

A faster approach uses the amount in the "Tax Is" column of Table 13.1 that is closest to the total TI and adds the tax for the next TI range.

Taxes = 113,900 + (800,000 − 335,000)(0.34) = $272,000

b. Equations [13.3] and [13.4] determine the effective tax rate and taxes.

$$T_e = 0.08 + (1 − 0.08)(0.34) = 0.3928$$

Taxes = (800,000)(0.3928) = $314,240

These two tax amounts are not comparable, because the tax in part (*a*) does not include state taxes.

How do corporate tax and individual tax computations compare? Gross income for an individual taxpayer is, for the most part, the total of salaries and wages. When determining an individual's taxable income, most of the expenses for living and working are not tax-deductible, whereas operating expenses are deductible for a corporation. For individual taxpayers,

GI = salaries + wages + interest and dividends + other income

TI = GI − personal exemption − deductions

Taxes = (taxable income)(applicable tax rate) = (TI)(*T*)

In the United States, the tax rates *T* for individuals, like those for corporations, are graduated by level of TI. The TI levels for each marginal rate level are adjusted annually to account for inflation and other factors. This process is called *indexing*. For the last several years, marginal rates have ranged from 10% to 35%; however, these marginal rates are the subject of significant debate depending upon the balance of power in federal congressional bodies and upon the economic conditions of the United States and other industrialized countries. As a result, marginal tax rates for individuals change more frequently than the rates for corporations. Table 13.2 is a sample schedule for the filing category "Married filing jointly." Other categories are: Single; Married filing separately; and Head of household. Current tax rates are available on the IRS website www.irs.gov in Publication 17—*Your Federal Income Tax (for individuals)*.

TABLE 13.2 Sample U.S. Federal Income Tax Rates for Individuals with a Filing Status of "Married Filing Jointly"

If your taxable income is: Over–	But not over–	The tax is:	of the amount over–
$0	$17,000	.... 10%	$0
17,000	69,000	$1,700.00 + 15%	17,000
69,000	139,350	9,500.00 + 25%	69,000
139,350	212,300	27,087.50 + 28%	139,350
212,300	379,150	47,513.50 + 33%	212,300
379,150	....	102,574.00 + 35%	379,150

(*Source:* Internal Revenue Service, Publication 17 — Your Federal Income Tax (for individuals) for use in preparing 2011 returns, 2011, p. 274.)

EXAMPLE 13.2

Josh and Allison, both consulting engineers, submit a married-filing-jointly return. During the year, their two jobs provided them with a combined income of $168,000. They had their second child during the year, and they plan to use the standard deduction of $9500 applicable for the year. Dividends, interest, and capital gains amounted to $8400. Personal exemptions are $3700 currently. Josh wants to estimate (*a*) the federal taxes, and (*b*) the percentages of gross income and taxable income paid in federal taxes for the year.

Solution

a. First, calculate GI and TI values using the relations for individuals with 4 exemptions.

$$GI = 168,000 + 8,400 = \$176,400$$
$$TI = 176,400 - 4(3700) - 9500 = \$152,100$$

Josh consulted the tax schedule (Table 13.2) for married filing jointly and a TI of $152,100. The tax amount is $27,087.50 + 28% of all over $139,350.

$$Taxes = 27,087.50 + 0.28(152,100 - 139,350)$$
$$= \$30,657.50$$

b. For the percentage of GI, taxes amount to 30,657.50/176,400 = 0.174 or 17.4%. Of the TI, taxes amount to 30,657.50/152,100 = 0.202 or 20.2%, which is considerably less than the marginal tax rate of 28% for the TI category.

13.2 BEFORE-TAX AND AFTER-TAX ALTERNATIVE EVALUATION

Early in the text, the term *net cash flow* (*NCF*) was identified as the best estimate of actual cash flow. The NCF, calculated as annual cash inflows minus cash outflows, is used to perform alternative evaluations via the PW, AW, ROR, and B/C methods. Now that the impact of taxes will be considered, it is time to expand our terminology. NCF is replaced by the terms *cash flow before taxes* (*CFBT*) and *cash flow after taxes* (*CFAT*). Both are actual cash flows and are related as

Podcast available

$$\textbf{CFAT = CFBT - taxes} \qquad \text{[13.5]}$$

The CFBT should include the initial investment *P* and the salvage value estimate *S* in the years they occur. *Depreciation D is included in TI, but not directly in the CFAT estimate since depreciation is not an actual cash flow.* This is very important since the engineering economy study must be based on actual cash flows. Using the effective tax rate, equations are:

CFBT = gross income − expenses − initial investment + salvage value

$$= GI − E − P + S \qquad [13.6]$$

CFAT = CFBT − TI(T_e)

$$= GI − E − P + S − (GI − E − D)(T_e) \qquad [13.7]$$

Suggested table column headings for CFBT and CFAT calculations are shown in Table 13.3. Numerical relations are shown in column numbers, with the effective tax rate T_e used for income taxes. Expenses E and initial investment P are negative. It is possible that in some years TI may be negative due to a depreciation amount that is larger than (GI − E). In this case, *the associated negative income tax is considered a tax savings for the year.* The assumption is that the negative tax is an advantage to the alternative and that it will offset taxes for the same year in other income-producing areas of the corporation.

Once the CFAT series is developed, apply an evaluation method—PW, AW, ROR, or B/C—and use exactly the same guidelines as in Chapters 4 through 7 to justify a single project or to select one mutually exclusive alternative. Example 13.3 illustrates CFAT computation and after-tax analysis.

If estimating the after-tax ROR is important, but after-tax detailed numbers are not of interest, or are too complicated to tackle, the before-tax ROR can be used to approximate the effects of taxation by using T_e and the relation

After-tax ROR = Before-tax ROR(1 − T_e) $\qquad$ [13.8]

Applying the same logic, the required after-tax MARR to use in a PW- or AW-based study is approximated as

After-tax MARR = Before-tax MARR(1 − T_e) $\qquad$ [13.9]

If an alternative's PW or AW value is close to zero, a more detailed analysis of the impact of taxes should be undertaken.

TABLE 13.3 Table Column Headings for Calculation of (*a*) CFBT and (*b*) CFAT

(*a*) CFBT table headings

Year	Gross Income GI	Operating Expenses E	Investment P and Salvage S	CFBT (4) = (1) + (2) + (3)
	(1)	(2)	(3)	

(*b*) CFAT table headings

Year	Gross Income GI	Operating Expenses E	Investment P and Salvage S	Depreciation D	Taxable Income TI (5) = (1) + (2) − (4)	Taxes (TI)(T_e) (6)	CFAT (7) = (1) + (2) + (3) − (6)
	(1)	(2)	(3)	(4)			

AMRO Engineering is evaluating a very large flood control program for several southern U.S. cities. One component is a 4-year project for a special-purpose transport ship-crane for use in building permanent storm surge protection against hurricanes on the New Orleans coastline. The estimates are $P = \$300,000$, $S = 0$, $n = 3$ years. MACRS depreciation with a 3-year recovery is indicated. Gross income and expenses are estimated at $200,000 and $80,000, respectively, for each of 4 years. (*a*) Perform before-tax and after-tax ROR analyses. AMRO uses $T_e = 35\%$ and a before-tax MARR of 15% per year. (*b*) Approximate the after-tax ROR with Equation [13.8] and comment on its accuracy.

EXAMPLE 13.3

Solution

a. Table 13.4 uses the format of Table 13.3 to determine the CFBT and CFAT series.

Before taxes: By Equation [13.6], CFBT = 200,000 − 80,000 = $120,000 for each year 1 to 4. The PW relation to estimate before-tax ROR per year is

PW = −300,000 + 120,000($P/A,i^*$,4)

This is easily solved using one of several methods (tabulated factors, a calculator's *i* function, or a spreadsheet's RATE or IRR function) to obtain

TABLE 13.4 CFBT, CFAT, and ROR Calculations Using MACRS and $T_e = 35\%$, Example 13.3

Year	GI	E	P and S	CFBT			
			Before-Tax ROR Analysis				
0			$−300,000	$−300,000			
1	$200,000	$−80,000		120,000			
2	200,000	−80,000		120,000			
3	200,000	−80,000	0	120,000			
4	200,000	−80,000		120,000			

Year	GI	E	P and S	D	TI	Taxes	CFAT
			After-Tax ROR Analysis				
0			$−300,000				$−300,000
1	$200,000	$−80,000		$ 99,990	$ 20,010	$ 7,003	112,997
2	200,000	−80,000		133,350	−13,350	−4,673	124,673
3	200,000	−80,000	0	44,430	75,570	26,450	93,551
4	200,000	−80,000		22,230	97,770	34,220	85,781

$i* = 21.86\%$. Comparing $i*$ with the before-tax MARR of 15%, the project is justified.

After taxes: Table 13.4 (bottom half) details MACRS depreciation (rates from Table 12.2), TI using Equation [13.1], and CFAT by Equation [13.7]. Because depreciation exceeds (GI − E) in year 2, the tax savings of $4673 will increase after-tax ROR. The PW relation and after-tax ROR are

$$PW = -300,000 + 112,997(P/F,i*,1) + \dots + 85,781(P/F,i*,4)$$

After-tax ROR $= i* = 15.54\%$

By Equation [13.9], the after-tax MARR is $15(1 − 0.35) = 9.75\%$. The project is economically justified since $i* > 9.75\%$.

b. Equation [13.8] yields the approximation

After-tax ROR $= 21.86(1 − 0.35) = 14.21\%$

This is an underestimate of the calculated amount 15.54%, in part due to the neglect of the tax savings in year 2, which is explicitly considered in the after-tax analysis.

After-tax evaluation of 2 or more alternatives follows the same guidelines detailed in Chapters 4 through 7. A couple of important points to remember follow.

PW: For different-life alternatives, the LCM of lives must be used to ensure an equal service comparison.

ROR: Incremental analysis of CFAT must be performed, since overall $i*$ values do not guarantee a correct selection (Section 6.5 provides details.)

Similar to the before-tax situation, the use of AW relations is preferable since neither of these complications is present. Additionally, spreadsheet usage is recommended for an after-tax analysis. Section 13.6 demonstrates several spreadsheet functions useful in after-tax PW, AW, and ROR analyses.

13.3 HOW DEPRECIATION CAN AFFECT AN AFTER-TAX STUDY

Podcast available

Several things may significantly alter the amount and timing of taxes and CFAT: depreciation method, length of recovery period, time at which the asset is disposed of, and relevant tax laws. A few of the tax implications should be considered in the economic evaluation (such as depreciation recovery). Others, such as capital gain or loss, can be neglected since they commonly occur toward the end of the asset's useful life and, consequently, are not reliably predictable at evaluation time. As discussed below, the key to estimating the potential tax effect is having a good sense of the selling price (salvage value) relative to the book value at disposal time.

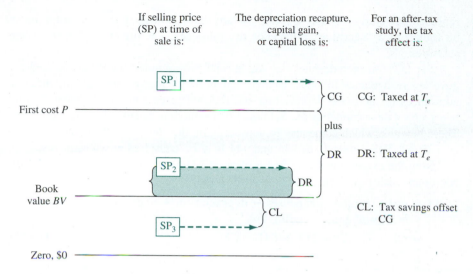

If selling price (SP) at time of sale is:	The depreciation recapture, capital gain, or capital loss is:	For an after-tax study, the tax effect is:

FIGURE 13.1
Summary of calculations and tax treatment for depreciation recapture and capital gains (losses).

Depreciation recapture DR, also called *ordinary gain*, occurs when a depreciable asset is sold for more than the current book value. As shown in Figure 13.1 (shaded area),

$$\text{Depreciation recapture} = \text{selling price} - \text{book value}$$

$$\mathbf{DR = SP - BV} \qquad \qquad \mathbf{[13.10]}$$

Depreciation recapture is often present in the after-tax study. In the United States, an amount equal to the estimated salvage value can always be anticipated as DR when the asset is disposed of at the end of or after the recovery period. This occurs because MACRS depreciates to zero in $n + 1$ years. The amount DR is treated as ordinary taxable income and taxed at the T_e rate.

When the selling price is expected to exceed first cost, a capital gain (discussed next) is also incurred and the TI due to the sale is the gain *plus* the depreciation recapture, as shown in Figure 13.1 for a selling price of SP_1. In this case, DR is the total depreciation taken thus far.

Capital gain CG is an amount incurred when the selling price exceeds its first cost. See Figure 13.1. At the time of asset disposal,

$$\text{Capital gain} = \text{selling price} - \text{first cost}$$

$$\mathbf{CG = SP - P} \qquad \qquad \mathbf{[13.11]}$$

Since future capital gains are difficult to predict, they are usually not detailed in an after-tax study. An exception is for assets that historically increase in value, such as buildings and land. *If included, the gain is taxed as ordinary TI.*

Capital loss CL occurs when a depreciable asset is disposed of for less than its current book value.

$$\text{Capital loss} = \text{book value} - \text{selling price}$$

$$\mathbf{CL = BV - SP} \qquad \qquad \mathbf{[13.12]}$$

An economic analysis does not commonly account for capital loss, simply because it is not estimable for a specific alternative. However, an after-tax replacement study should account for any capital loss if the defender must be traded at a "sacrifice" price. For tax purposes, CL offsets CG from other activities.

Tax law can include rules that are special, time-limited incentives offered by government agencies to boost capital, and possibly foreign investment, through allowances of increased depreciation and reduced taxes. One of the incentives to boost investment capital is the Section 179 Deduction discussed in Section 12.1. These benefits, which come and go depending on the "health of the economy," can alter CFAT. IRS publications 946 and 544 may be consulted at www.irs.gov.

Now the expression for TI in Equation [13.1] can be expanded to include the additional cash flow estimates for asset disposal.

$$\text{TI} = \text{gross income} - \text{expenses} - \text{depreciation} + \text{depreciation recapture} + \text{capital gain} - \text{capital loss}$$
$$= \text{GI} - E - D + \text{DR} + \text{CG} - \text{CL} \qquad \text{[13.13]}$$

EXAMPLE 13.4   Biotech, a medical imaging and modeling company, must purchase a bone cell analysis system for use by a team of bioengineers and mechanical engineers studying bone density in athletes. This particular part of a 3-year contract with the NBA will provide additional gross income of $100,000 per year. The effective tax rate is 35%.

	Analyzer 1	Analyzer 2
First cost, $	150,000	225,000
Operating expenses, $ per year	30,000	10,000
MACRS recovery, years	5	5

a. The Biotech president, who is very tax conscious, wishes to use a criterion of minimizing total taxes incurred over the 3 years of the contract. Which analyzer should be purchased?

b. Assume that after 3 years the company will sell the selected analyzer. Using the same total tax criterion, does either analyzer have an advantage? The selling prices are expected to be $130,000 for analyzer 1 and $225,000 for analyzer 2.

Solution

a. Table 13.5 details the tax computations. First, the yearly MACRS depreciation is determined. Equation [13.1], $\text{TI} = \text{GI} - E - D$, is used to calculate TI, after which the 35% tax rate is applied. Taxes for the 3-year period are summed.

Analyzer 1 tax total: $36,120   Analyzer 2 tax total: $38,430

The two analyzers are very close, but analyzer 1 wins with $2310 less in total taxes.

TABLE 13.5 **Comparison of Total Taxes for Two Alternatives, Example 13.4a**

Year	Gross Income GI	Operating Expenses E	First Cost P	MACRS Depreciation D	Book Value BV	Taxable Income TI	Taxes at 0.35TI
				Analyzer 1			
0			$150,000		$150,000		
1	$100,000	$30,000		$30,000	120,000	$40,000	$14,000
2	100,000	30,000		48,000	72,000	22,000	7,700
3	100,000	30,000		28,800	43,200	41,200	14,420
							$36,120
				Analyzer 2			
0			$225,000		$225,000		
1	$100,000	$10,000		$45,000	180,000	$45,000	$15,750
2	100,000	10,000		72,000	108,000	18,000	6,300
3	100,000	10,000		43,200	64,800	46,800	16,380
							$38,430

b. When the analyzer is sold after 3 years of service, there is a depreciation recapture (DR) that is taxed at the 35% rate. This tax is in addition to the third-year tax in Table 13.5. For each analyzer, account for the DR by Equation [13.10]; then determine the TI, using Equation [13.13], TI = GI − E − D + DR. Again, find the total taxes for 3 years, and select the analyzer with the smaller total.

Analyzer 1:   DR = SP − BV_3 = 130,000 − 43,200 = $86,800
   Year 3 TI = 100,000 − 30,000 − 28,800 + 86,800 = $128,000
   Year 3 taxes = 128,000(0.35) = $44,800
   Total taxes = 14,000 + 7700 + 44,800 = $66,500

Analyzer 2:   DR = 225,000 − 64,800 = $160,200
   Year 3 TI = 100,000 − 10,000 − 43,200 + 160,200 = $207,000
   Year 3 taxes = 207,000(0.35) = $72,450
   Total taxes = 15,750 + 6300 + 72,450 = $94,500

Now, analyzer 1 has a considerable advantage in total taxes.

It may be important to understand why accelerated depreciation rates and shortened recovery periods included in MACRS and DB methods give the corporation a tax advantage relative to that offered by the straight line method. Larger depreciation rates in earlier years require less in taxes due to the larger reductions in taxable income. The key is to choose the depreciation rates and n value that result in the *minimum present worth of taxes*.

$$\text{PW}_{\text{tax}} = \sum_{t=1}^{t=n} (\text{taxes in year } t)(P/F,i,t) \qquad [13.14]$$

To compare depreciation methods or recovery periods, assume the following: (1) There is a constant single-value tax rate, (2) CFBT exceeds the annual depreciation amount, and (3) the method reduces book value to the same salvage value. The following are then correct:

- Total taxes paid are *equal* for all depreciation methods or recovery periods.
- PW_{tax} is *less* for accelerated depreciation methods with the same n value.
- PW_{tax} is *less* for smaller n values with the same depreciation method.

MACRS is the prescribed tax depreciation model in the United States, and the only alternative is MACRS straight line depreciation with an extended recovery period. The accelerated write-off of MACRS always provides a smaller PW_{tax} compared to less accelerated models. If the DDB model were still allowed directly, it would not fare as well as MACRS, because DDB does not reduce the book value to zero.

EXAMPLE 13.5   An after-tax analysis for a new $50,000 machine proposed for a fiber optics manufacturing line is in process. The CFBT for the machine is estimated at $20,000. If a recovery period of 5 years applies, use the present-worth-of-taxes criterion, an effective tax rate of 35%, and a return of 8% per year to compare the following: straight line, DDB, and MACRS depreciation. Use a 6-year period to accommodate the MACRS half-year convention.

Solution

Table 13.6 presents a summary of depreciation, taxable income, and taxes for each model. For straight line depreciation with $n = 5$, $D = \$10,000$ for 5 years and $D_6 = 0$ (column 3). The CFBT of $20,000 is fully taxed at 35% in year 6.

The DDB percentage of $d = 2/n = 0.40$ is applied for 5 years. The implied salvage value is $50,000 - 46,112 = \$3888$, so not all $50,000 is tax deductible. The taxes using DDB are $3888 (0.35) = \$1361$ larger than for the SL model.

MACRS writes off the $50,000 in 6 years using the rates of Table 12.2. Total taxes are $24,500, the same as for SL depreciation.

The annual taxes (columns 5, 8, and 11) are accumulated year by year in Figure 13.2. Note the pattern of the curves, especially the lower total taxes relative to the SL model after year 1 for MACRS and in years 1 through 4 for DDB. These higher tax values for SL cause PW_{tax} for SL depreciation to be larger. The PW_{tax} values are at the bottom of Table 13.6. The MACRS PW_{tax} value is the smallest at $18,162.

TABLE 13.6 Comparison of Taxes and Present Worth of Taxes for Different Depreciation Methods

(1) Year	(2) CFBT	Straight Line			Double Declining Balance			MACRS		
		(3) D	(4) TI	(5) Taxes	(6) D	(7) TI	(8) Taxes	(9) D	(10) TI	(11) Taxes
1	+20,000	$10,000	$10,000	$ 3,500	$20,000	$ 0	$ 0	$10,000	$10,000	$ 3,500
2	+20,000	10,000	10,000	3,500	12,000	8,000	2,800	16,000	4,000	1,400
3	+20,000	10,000	10,000	3,500	7,200	12,800	4,480	9,600	10,400	3,640
4	+20,000	10,000	10,000	3,500	4,320	15,680	5,488	5,760	14,240	4,984
5	+20,000	10,000	10,000	3,500	2,592	17,408	6,093	5,760	14,240	4,984
6	+20,000	0	20,000	7,000	0	20,000	7,000	2,880	17,120	5,992
Totals		$50,000		$24,500	$46,112		$25,861*	$50,000		$24,500
PW$_{tax}$				$18,386			$18,549			$18,162

*Larger than other values since there is an implied salvage value of $3888 not recovered.

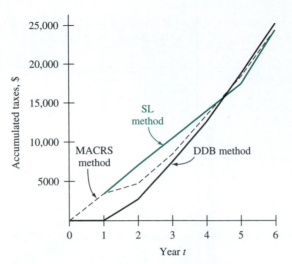

FIGURE 13.2   Taxes incurred by different depreciation rates for a 6-year comparison period, Example 13.5.

Comment: If a similar example is developed that applies only one depreciation method for different n values, the PW_{tax} value will be less for the smallest n.

13.4 AFTER-TAX REPLACEMENT STUDY

Podcast available

When a currently installed asset (the defender) is challenged with possible replacement, the effect of taxes can have an impact upon the decision of the replacement study. The final decision may not be reversed by taxes, but the difference between before-tax PW or AW values may be significantly different from the after-tax difference. There may be tax considerations in the year of the possible replacement due to *depreciation recapture* or *capital gain,* or there may be tax savings due to a sizable *capital loss,* if the trade of the defender is expected to occur at a sacrifice price. Additionally, the after-tax replacement study considers tax-deductible *depreciation* and *operating expenses* not accounted for in a before-tax analysis. The same procedure is applied for CFAT estimates as for the before-tax replacement study. A review of Sections 9.3 and 9.4 is recommended.

EXAMPLE 13.6   Savannah Power purchased railroad transport equipment for coal 3 years ago for $600,000. Management has discovered that it is technologically outdated now. New equipment has been identified. If the market value of $400,000 is offered as the trade-in for the current equipment, perform a replacement study using a 7% per year after-tax MARR. Assume an effective tax rate of 34% and straight line depreciation with $S = 0$ for both alternatives.

	Defender	Challenger
Market value, $	400,000	
First cost, $		−1,000,000
Annual cost, $/year	−100,000	−15,000
Recovery period, years	8 (originally)	5

Solution

For the defender *after-tax replacement study,* there are no tax effects other than income tax. The annual SL depreciation is 600,000/8 = $75,000, determined when the equipment was purchased 3 years ago. Now $P_D = \$-400,000$, which is the current market value.

Table 13.7 shows the TI and taxes at 34%, which are the same each year. The taxes are actually tax savings of $59,500 per year, as indicated by the minus sign. Since only costs are estimated, the annual CFAT is negative, but the $59,500 tax savings has reduced it to $−40,500. The CFAT and AW at 7% per year are

$$\text{CFAT} = \text{CFBT} - \text{taxes} = -100,000 - (-59,500) = \$-40,500$$
$$\text{AW}_D = -400,000(A/P,7\%,5) - 40,500 = \$-138,056$$

For the challenger, depreciation recapture on the defender occurs when it is replaced, because the trade-in amount of $400,000 is larger than the current

TABLE 13.7 After-Tax Replacement Analyses, Example 13.6

		Before Taxes			After Taxes			
Defender Age	Year	Expenses E	P	CFBT	Depreciation D	Taxable Income TI	Taxes* at 0.34TI	CFAT
			DEFENDER					
3	0		$−400,000	$−400,000				$−400,000
4–8	1–5	$−100,000		−100,000	$75,000	$−175,000	$−59,500	−40,500
AW at 7%								$−138,056
			CHALLENGER					
	0		$−1,000,000	$−1,000,000		$ +25,000[†]	$ 8,500	$−1,008,500
	1–5	$−15,000		−15,000	$200,000	−215,000	−73,100	+58,100
AW at 7%								$−187,863

*Minus sign indicates a tax savings for the year.
†DR₃, Depreciation recapture on defender trade-in.

book value. In year 0 for the challenger, Table 13.7 includes the following computations to arrive at a tax of $8500:

Defender book value, year 3: $BV_3 = 600,000 - 3(75,000) = \$375,000$

Depreciation recapture: $DR_3 = TI = 400,000 - 375,000 = \$25,000$

Taxes on the trade-in, year 0: $Taxes = 0.34(25,000) = \$8500$

The SL depreciation is $\$1,000,000/5 = \$200,000$ per year. This results in tax saving and CFAT as follows:

$$Taxes = (-15,000 - 200,000)(0.34) = \$-73,100$$
$$CFAT = CFBT - taxes = -15,000 - (-73,100) = \$+58,100$$

In year 5, it is assumed the challenger is sold for $0; there is no depreciation recapture. The AW for the challenger at the 7% after-tax MARR is

$$AW_C = -1,008,500(A/P,7\%,5) + 58,100 = \$-187,863$$

The defender is selected.

13.5 CAPITAL FUNDS AND THE COST OF CAPITAL

Podcast available

The funds that finance engineering projects are called *capital*, and the interest rate paid on these funds is called the *cost of capital*. The specific MARR used in an economic analysis is established such that the MARR exceeds the cost of capital, as discussed initially in Section 1.3 of this text. To understand how a MARR is set, it is necessary to understand the two types of capital.

Debt capital—These are funds borrowed from sources outside the corporation and its owners/stockholders. Debt financing includes loans, notes, mortgages, and bonds. The debt must be repaid at some stated interest rate using a specific time schedule, for example, over 15 years at 10% per year simple interest based on the declining loan balance. A corporation indicates outstanding debt financing in the liability section of its balance sheet. (See Appendix B, Table B.1.)

Equity capital—These are retained earnings previously kept within the corporation for future capital investment, and owners' funds obtained from stock sales (public corporations) or individual owner's funds (private corporations). Equity funds are indicated in the net worth section of the balance sheet. (Again, see Table B.1.)

Suppose a $55 million (U.S. dollars) project to expand the capacity for generated electricity by Mexico City's Luz y Fuerza del Centro is financed by $33 million from funds set aside over the last 5 years as retained earnings and

$22 million in municipal bond sales. The project's *capital pool* is developed as follows:

Equity capital: $33 million or 33/55 = 60% equity funds
Debt capital: $22 million or 22/55 = 40% debt funds

The *debt-to-equity (D-E) mix* is the ratio of debt to equity capital for a corporation or a project. In this example, the D-E mix is 40-60 with 40% debt from bonds and 60% equity from retained earnings.

Once the D-E mix is known, the *weighted average cost of capital (WACC)* can be calculated. The MARR is then established to exceed (or at least equal in the case of a not-for-profit project) this value. WACC is an estimate of the interest rate in percentage for all funds in the capital pool used to finance corporate projects. The relation is

$$\textbf{WACC} = \textbf{(equity fraction)(cost of equity capital)} \qquad \textbf{[13.15]}$$
$$\textbf{+ (debt fraction)(cost of debt capital)}$$

The cost terms, expressed as percentages, are the rates associated with each type of capital. For example, if invested retained earnings are making 12.5% per year, this is the cost of equity capital. If debt capital is acquired via loans that charge a 7% per year interest rate, the cost of debt capital in Equation [13.15] is 7%. The monetary amount of interest paid on debt capital is referred to as *debt service*. The D-E mix and WACC are illustrated in the next example.

Figure 13.3 indicates the usual shape of cost of capital curves. If 100% of the capital is derived from equity or 100% is from debt sources, the WACC equals the cost of capital for that source of funds. There is virtually always a mixture of capital

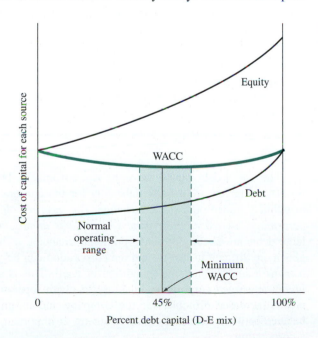

FIGURE 13.3
General shape of different cost of capital curves.

sources involved for any capitalization program. *As an illustration only,* Figure 13.3 indicates a minimum WACC at about 45% debt capital. Most firms operate over a range of D-E mixes for different projects.

EXAMPLE 13.7   A new program in genetics engineering at Gentex will require $10 million in capital. The chief financial officer (CFO) has estimated the following amounts of capital at the indicated interest rates per year.

Stock sales	$5 million at 13.7% per year
Use of retained earnings	$2 million at 8.9% per year
Debt financing through bonds	$3 million at 7.5% per year

Historically, Gentex has financed projects using a D-E mix of 40% from debt sources costing 7.5% per year and 60% from equity sources, which return 10% per year. (*a*) Compare the historical WACC value with that for this current genetics program. (*b*) Determine the MARR if a return of 5% per year is required by Gentex.

Solution

a. Equation [13.15] estimates the *historical* WACC.

$$\text{WACC} = 0.6(10) + 0.4(7.5) = 9.0\%$$

For the current program, the equity financing is comprised of two forms of equity capital—50% stock ($5 million out of $10 million) and 20% retained earnings. The remaining 30% is from debt sources. To calculate the *current* program's WACC, rewrite Equation [13.15] for equity funds.

$$\text{WACC} = \text{stock portion} + \text{retained earnings portion} + \text{debt portion}$$
$$= 0.5(13.7) + 0.2(8.9) + 0.3(7.5) = 10.88\%$$

This is higher than the 9% historical average.

b. The program should be evaluated using a MARR of $10.88 + 5.0 = 15.88\%$ per year.

The use of debt capital has some real advantages, but high D-E mixes, such as 50–50 and larger, are usually unhealthy for the corporation, be it privately owned or publicly traded on a stock exchange. *The leverage offered by larger debt capital percentages increases the riskiness of projects undertaken by the company.* When large debts are already present, additional financing using debt (or equity) sources gets more difficult to justify, and the corporation can be placed in a situation where it owns a smaller and smaller portion of itself. This is sometimes referred to as a *highly leveraged* corporation. Inability to obtain operating and investment capital means increased difficulty for the company and its projects. Thus, a reasonable balance between debt and equity financing is important for the financial health of a corporation.

Three companies that subcontract with automobile manufacturers have the following debt and equity capital amounts and D-E mixes. Assume all equity capital is in the form of stock.

EXAMPLE 13.8

	Amount of Capital		
Company	Debt ($ in millions)	Equity ($ in millions)	D-E Mix (% – %)
A	10	40	20 – 80
B	20	20	50 – 50
C	40	10	80 – 20

Assume the annual revenue is $15 million for each corporation and that, after debt service is considered, the net incomes are $14.4, $13.4, and $10.0 million, respectively. Compute the return on stock for each company, and comment on the return relative to the D-E mixes.

Solution

Divide the net income by the equity amount to compute the stock return. In million dollars,

$$\text{A:} \quad \text{Return} = \frac{14.4}{40} = 0.36 \quad (36\%)$$

$$\text{B:} \quad \text{Return} = \frac{13.4}{20} = 0.67 \quad (67\%)$$

$$\text{C:} \quad \text{Return} = \frac{10.0}{10} = 1.00 \quad (100\%)$$

As expected, the return is by far the largest for highly leveraged C, where only 20% of the company is in the hands of the ownership. The return on owner's equity is excellent, but the risk associated with this firm is high compared to A, where the D-E mix is only 20% debt.

The same principles discussed previously for corporations are applicable to governments and individuals. The individual who is highly leveraged has large debts in terms of credit card balances, personal loans, and house mortgages. As an example, assume two engineers each have take-home pay of $60,000 after all income tax, social security, and insurance premiums are deducted from their annual salaries. Further, assume that the cost (interest) on their debt (money borrowed via credit cards and loans) averages 15% per year and that the current debt principal is being repaid in equal amounts over 20 years. If Sherry has a total debt of $25,000 and Carlos owes $150,000, the remaining amounts of the annual take-home pay vary considerably, as shown on the next page. Sherry has $55,000, or 91.7% of her $60,000 take-home available, while Carlos has only 50% available.

Person	Total Debt, $	Annual Cost of Debt at 15%, $	Annual Repayment $	Amount Remaining from $60,000, $
Sherry	25,000	3,750	1,250	55,000
Carlos	150,000	22,500	7,500	30,000

The previous computations are correct for a before-tax analysis. If an after-tax analysis is performed, the tax advantage of debt capital should be considered. In the United States, and many other countries, interest paid on all forms of debt capital (loans, bonds, and mortgages) is considered a corporate expense, and is therefore, tax deductible. Equity capital does not have this advantage; dividends on stocks, for example, are not considered tax deductible. Though a detailed analysis can be performed, the easiest way to estimate an after-tax WACC via Equation [13.15] is to approximate the cost of debt capital using the effective tax rate T_e.

$$\textbf{After-tax cost of debt} = \textbf{(before-tax cost)}(1 - T_e) \qquad \textbf{[13.16]}$$

The cost of debt capital will be less after taxes are considered, but the cost of equity will remain the same in the after-tax WACC computation. Once determined, the after-tax MARR can be set. (Alternatively, the after-tax MARR can be approximated using the logic of Section 13.2.)

EXAMPLE 13.9   If Gentex's effective tax rate is 38%, determine the after-tax MARR to be applied when evaluating the program described in Example 13.7.

Solution

The after-tax WACC will *decrease* from the previous result of WACC = 10.88% based on the tax-advantaged cost of debt capital approximated by Equation [13.16].

After-tax cost of debt = $(7.5\%)(1 - 0.38) = 4.65\%$

After-tax WACC = $0.5(13.7) + 0.2(8.9) + 0.3(4.65) = 10.03\%$

Assuming that the expected 5% per year return is after taxes, the after-tax MARR is now 15.03% per year.

13.6 USING SPREADSHEETS FOR AFTER-TAX EVALUATION

The next example illustrates spreadsheet use for several after-tax evaluation techniques, including CFAT computation (using Table 13.3 format), PW and incremental ROR evaluations, and finally graphical breakeven ROR analysis. The solution stages build on each other in order to allow study of selected areas.

Two units offer similar features to perform bone density analysis for a skeletal diagnostic clinic for each of 10 NBA teams during a 3-year contract period. (The situation here is the same as that in Example 13.4.) Use MACRS 5-year recovery, $T_e = 35\%$, and an after-tax MARR of 10% per year to perform the following analyses for the 3-year contract:

EXAMPLE 13.10

a. Present worth and annual worth
b. Incremental ROR
c. Graphical determination of breakeven ROR to compare with MARR = 10%.

	Analyzer 1	Analyzer 2
First cost, $	−150,000	−225,000
Gross income, $/year	100,000	100,000
AOC, $/year	−30,000	−10,000
MACRS recovery, years	5	5
Estimated selling price after 3 years, $	130,000	225,000

Solution

Depreciation, income taxes, depreciation recapture, and salvage (selling price) are all included in this evaluation. The combination of Equations [13.7] and [13.13] define the CFAT relation for the spreadsheets. (No capital gains or losses are anticipated.)

$$CFAT = CFBT - taxes$$
$$= GI - E - P + S - (GI - E - D + DR)(T_e)$$

The spreadsheet and cell tags in Figure 13.4 detail CFAT calculation. At the end of the contract, year 3 computations of CFAT use the relation above. In $1000 units, they are

Analyzer 1: $CFAT_3 = 100 - 30 + 130 - [100 - 30 - 28.8 + (130 - 43.2)](0.35)$
$$= 200 - 128(0.35)$$
$$= \$155.20$$

Analyzer 2: $CFAT_3 = 100 - 10 + 225 - [100 - 10 - 43.2 + (225 - 64.8)](0.35)$
$$= 315 - 207(0.35)$$
$$= \$242.55$$

In practice, only one of the following analyses is necessary. For illustration purposes only, all four are presented here.

a. Figure 13.4: Since the 3-year contract period is the same for both alternatives, NPV functions at MARR = 10% are excellent for determining PW values for the CFAT series (column I). Likewise, the PMT function for 3 years is used to obtain AW from PW. (The minus sign on PMT ensures

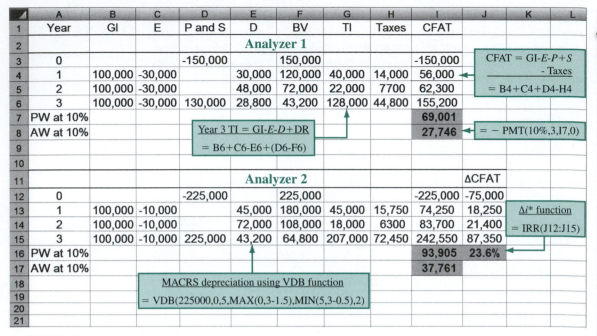

FIGURE 13.4   After-tax PW, AW, and incremental ROR analyses including depreciation recapture, Example 13.10*a* and *b*.

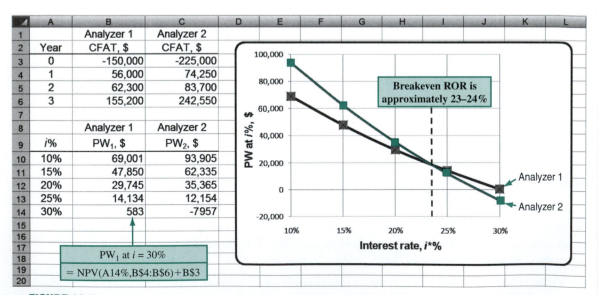

FIGURE 13.5   Graphical determination of breakeven ROR using PW at different interest rates, Example 13.10*c*.

that AW maintains the same sign as PW.) Analyzer 2 is selected with the larger PW and AW values.

Calculating AW is not necessary since the conclusion will always be the same as PW analysis for equal-life alternatives. Were lives unequal, the LCM years would be used in the NPV and PMT functions to ensure equal service comparison.

b. Figure 13.4: Column J presents the incremental CFAT of the larger investment in analyzer 2 over analyzer 1. The incremental $i^* = 23.6\%$ obtained using the IRR function (cell J16) shows that analyzer 2 significantly exceeds the 10% after-tax requirement.

c. Figure 13.5: The upper portion repeats the two CFAT series. In the lower portion, the NPV functions determine the PW at different i values. An x-y scatter chart shows that the PW curves cross in the range of 23 to 24% per year. (The exact value was determined previously to be 23.6%.) Since this breakeven value exceeds 10%, the larger investment in analyzer 2 is justified.

13.7 AFTER-TAX VALUE-ADDED ANALYSIS

Value added is a term that indicates a product or service has *added economic worth* from the perspective of the owner, an investor, or a consumer. It is possible to highly leverage the value-added activities of a process. For example, onions are sold at the farm level for cents per pound. They may be purchased in a store for $1 to $2 per pound. When onions are cut and coated with a special batter, they may be fried in hot oil and sold as onion rings for several dollars per pound. From the perspective of the consumer's willingness to pay, there has been great value added by the processing.

Value-added analysis of an alternative is performed in a slightly different way than CFAT analysis. But the decision about an alternative will be the same, because the AW of economic value added (EVA) is the same as the AW of CFAT.*

Value added analysis starts with the net operating profit after taxes (NOPAT). This is the amount remaining each year when income taxes are subtracted from taxable income.

$$\textbf{NOPAT = taxable income} - \textbf{taxes = TI} - (\textbf{TI})(T_e)$$
$$= (\textbf{TI})(1 - T_e) \qquad\qquad \textbf{[13.17]}$$

NOPAT implicitly includes depreciation accumulated thus far when TI is calculated. This is different from CFAT, where the depreciation is specifically removed so that only *actual* cash flows are used.

---

*W. G. Sullivan and K. L. Needy, "Determination of Economic Value Added for a Proposed Investment in New Manufacturing." *The Engineering Economist,* vol. 45, no. 2 (2000), pp. 166–181.

The annual EVA is the amount of NOPAT remaining after removing the *cost of invested capital* during the year. That is, EVA indicates the project's contribution to the *net worth of the corporation after taxes*. The cost of invested capital is the after-tax MARR multiplied by the book value of the asset during the year. This is the interest incurred by the current level of capital invested in the asset. Computationally, including Equation [13.17],

$$\textbf{EVA} = \textbf{NOPAT} - \textbf{cost of invested capital}$$
$$= \textbf{NOPAT} - \textbf{(after-tax interest rate)(book value in year } t - 1)$$
$$= \textbf{TI}(1 - T_e) - (i)(\textbf{BV}_{t-1}) \qquad\qquad [13.18]$$

Since both TI and BV consider depreciation, EVA is a measure of worth that mingles actual cash flows with noncash flows to determine financial worth. This financial worth is the amount used in public documents of the corporation (balance sheet, income statement, stock reports, etc.). Because corporations want to present the largest value possible to stockholders, the EVA method is often more appealing than the AW method.

The result of an EVA analysis is a series of annual EVA estimates. Calculate the AW of EVA estimates and select the alternative with the larger AW value. If only one project is evaluated, AW > 0 means the after-tax MARR is exceeded, thus making the project value adding. Since the final AW of EVA and the AW of CFAT values are equal, either method can be used. The annual EVA estimates indicate added worth to the corporation generated by the alternative, while the annual CFAT estimates describe how cash will flow. This comparison is made in the next (spreadsheet) example.

EXAMPLE 13.11   Electrical engineers and medical doctors at First Hope Health Center have developed a project for investing in new capital equipment with the expectation of increased revenue from its medical diagnostic services to cancer patients. The estimates are summarized below. (*a*) Use classical straight line depreciation, an after-tax MARR of 12%, and an effective tax rate of 40% to perform two annual worth after-tax analyses: EVA and CFAT. (*b*) Explain the fundamental difference between the results of the two analyses.

Initial investment	$-500,000
Gross income − expenses	$170,000 per year
Estimated life	4 years
Salvage value	None

Solution

a. *EVA evaluation:* All the necessary information for EVA estimation is in columns B through G of Figure 13.6. The NOPAT (column H) is calculated by Equation [13.17]. The book values (column E) resulting from straight line

	A	B	C	D	E	F	G	H	I	J	K	L
1									EVA analysis		CFAT analysis	
2									Cost of			
3	Year	GI - E	P	D	BV	TI	Taxes	NOPAT	invested capital	EVA	CFAT	
4	0		-500,000		500,000						-500,000	
5	1	170,000		125,000	375,000	45,000	18,000	27,000	-60,000	-33,000	152,000	
6	2	170,000		125,000	250,000	45,000	18,000	27,000	-45,000	-18,000	152,000	
7	3	170,000		125,000	125,000	45,000	18,000	27,000	-30,000	-3000	152,000	
8	4	170,000		125,000	0	45,000	18,000	27,000	-15,000	12,000	152,000	
9	AW value for EVA and CFAT analysis using PMT functions									-12,617	-12,617	

$$\text{NOPAT} = \text{TI - taxes}$$
$$= \text{F8 - G8}$$

$$\text{Cost of invested capital}$$
$$= \text{MARR} \times \text{BV}_{t-1}$$
$$= -0.12 * \text{E7}$$

$$\text{EVA in years 1 to 4 only}$$
$$= \text{column H + column I}$$

$$\text{CFAT} = \text{GI-}E\text{-}P\text{-taxes}$$
$$= \text{B8 + C8 - G8}$$

FIGURE 13.6 Project evaluation using EVA and CFAT approaches, Example 13.11.

depreciation are used to determine the cost of invested capital in column I by applying the second term in Equation [13.18], that is, $i(\text{BV}_{t-1})$, where i is 12%. This represents the amount of interest for the currently invested capital. The annual EVA is the sum of columns H and I. *Notice there is no EVA estimate for year 0,* since NOPAT and the cost of invested capital are estimated for years 1 through n. The AW of EVA (cell J9), determined using the PMT function, is negative, which indicates that plan A does not make the 12% return.

CFAT evaluation: The CFAT (column K) is calculated as $GI - E - P -$ taxes. The AW of CFAT again concludes that plan A does not return the 12%.

b. The series in columns J and K are clearly equivalent since the AW values are numerically the same. To explain the difference, consider, the constant CFAT of $152,000 per year. To obtain the AW of EVA estimate of $-12,617 for years 1 through 4, the initial investment of $500,000 is distributed over the 4-year life using the A/P factor at 12%. That is, an equivalent amount of $500,000($A/P$,12%,4) = $164,617 is "charged" against the cash inflows in each of years 1 through 4. In effect, the yearly CFAT is reduced by this charge.

$$\text{CFAT} - (\text{initial investment})(A/P,12\%,4) = \$152,000 - 500,000(A/P,12\%,4)$$
$$152,000 - 164,617 = \$-12,617$$

This is the AW value for both series, demonstrating that the two methods are economically equivalent. The EVA method indicates yearly contribution to the *value of the corporation,* which is negative for the first three years. The CFAT method estimates the *actual cash flows* to the corporation for all years 0 through 4.

SUMMARY

After-tax analysis does not usually change the decision to select one alternative over another; however, it does offer a much clearer estimate of the monetary impact of taxes. After-tax PW, AW, and ROR evaluations are performed on the CFAT series using exactly the same procedures as in previous chapters.

U.S. income tax rates are graduated—higher taxable incomes pay higher income taxes. A single-value, effective tax rate T_e is usually applied in an after-tax economic analysis. Taxes are reduced

because of tax-deductible items, such as depreciation and operating expenses.

If an alternative's estimated contribution to corporate financial worth is the economic measure, the economic value added (EVA) should be determined. Unlike CFAT, the EVA includes the effect of depreciation. The AW of CFAT and EVA are the same numerically; they interpret the annual cost of the capital investment in different, but equivalent manners.

PROBLEMS

Tax Terms and Computations

13.1 For a corporation that has a taxable income of $250,000 for federal tax computation, determine (*a*) the marginal tax rate, (*b*) the total taxes, and (*c*) the average tax rate.

13.2 In its first year of operation, Borsberry Construction had a total tax bill of $72,000. What was the company's taxable income?

13.3 Last year, an investor in rental property had gross income of $160,000 with the following expenses: maintenance $22,000, insurance $5000, management $10,000, utilities $16,000, and debt service (interest) $19,000. Income taxes totaled $8000. What was the net operating income for the year?

13.4 Helical Products makes machined springs with elastic redundant elements so that a broken spring will continue to function. The company has gross income of $450,000 with expenses of $230,000 and depreciation of $48,000. Approximate the company's total taxes for an effective tax rate of 38%.

13.5 In 2011, a married couple (both professionals) have a total annual income of $146,000. Their exemptions and deductions totaled $15,000. (*a*) Determine their federal tax liability. (*b*) Using the same TI for the couple, determine their tax liability using the most current tax rates found in Publication 17 (taken from www.irs.gov). Has the tax amount gone up or down for this couple?

13.6 Carl read the annual report of Harrison Engineering's 3-D Imaging Division. From it he deduced that sales revenue generated a GI of $4.9 million.

Other information was $E = \$2.1$ million, and $D = \$1.4$ million. If the average federal tax rate was 31% and state/local tax rates totaled 9.8%, estimate (*a*) federal income taxes, and (*b*) the percent of sales revenue that the federal government required in income taxes.

13.7 Last year, Marylynn opened Baron's Appliance Sales and Service. Her tax accountant provided the year's results.

Gross income = $320,000
Business expenses = $149,000
MACRS depreciation = $95,000
Average federal tax rate = 18.5%
Average state tax rate = 6%
City and county flat tax rates combined = 4.5%

Determine the following for Marylynn.
a. Taxable income.
b. Exact amount of federal income taxes.
c. Estimate of percent of the GI needed to pay all income taxes—federal, state, city and county.

Before-Tax and After-Tax Evaluation

13.8 Estimate the approximate after-tax rate of return (ROR) for a project that has a before-tax ROR of 28%. Assume the company has $T_e = 37\%$ and it uses 5-year MACRS depreciation for project assets.

13.9 Estimate the approximate before-tax rate of return (ROR) for a project that has a first cost of $750,000 and a salvage value of 25% of the first cost after three years. The project's NOI is $260,000. Assume an effective tax rate of 37%.

13.10 Approximate the after-tax ROR on a project that had a first cost of $500,000, a salvage value of 20% of the first cost after five years, and annual CFBT of $230,000. Assume the company had a 35% effective tax rate.

13.11 Estimate the CFAT for a company that has taxable income of $120,000, depreciation of $133,350, and an effective tax rate of 35%.

13.12 A dynamic measurement machine used by Renshaw Instruments had an annual cash flow before taxes of $550,000 over its life of 10 years. Its first cost was $1.7 million, it was straight-line depreciated with $n = 10$ years, and it was sold today for the estimated salvage of $200,000. The company's $T_e = 36\%$ and after-tax MARR = 15%. Was the purchase and use of the machine economically justified?

13.13 Fill in the missing values in the table below for the CFBT, D, TI, Taxes, and CFAT columns. Depreciation amounts are based on 3-year MACRS depreciation and the effective tax rate is 35%.

Year	GI, $	E, $	P and S, $	CFBT, $	D, $	TI, $	Taxes, $	CFAT, $
0			−1900	−1900				−1900
1	800	−100	0	700	633	67	23	677
2	950	−150	0			−45		816
3	600	−200	0	400	281		42	
4	300	−250	700	750		−91	−32	782

13.14 Estimate the gross income for Bling Enterprises, which reports a CFAT of $2.5 million, $900,000 in expenses, $900,000 in depreciation charges, and has an effective tax rate of 26.4%.

13.15 Your brother, who owns a chemical processing company, has complained about the company's high tax rate of 42% for federal and state taxes, comprised of a 34% federal rate and a state 8% flat rate. Determine his company's *average federal income tax rate* if the annual gross income is $4,000,000, cash flow after taxes is $2,250,000, expenses are $850,000, and depreciation is $650,000.

13.16 Advanced Anatomists, Inc., researchers in medical science, is contemplating a commercial venture concentrating on proteins based on the new X-ray technology of free-electron lasers. To recover the huge investment needed, an annual $2.5 million CFAT is needed. A favored average federal tax rate of 20% is expected; however, state taxing authorities will levy an 8% tax on TI. Over a 3-year period, the deductible expenses and depreciation are estimated to total $1.3 million the first year, increasing by $500,000 per year thereafter. Of this, 50% is expenses and 50% is depreciation. What is the required gross income each year?

13.17 The information shown below is for a marginally successful project. Use hand computations or a spreadsheet as requested by the instructor to do the following:

a. Determine the CFBT and CFAT series. The effective tax rate is 32%.

b. Obtain before-tax and after-tax i^* values to evaluate the effect of depreciation and taxes. (Assume that tax savings are used to offset taxes in other parts of the corporation.)

c. Determine the project's economic success if the after-tax MARR is 5%.

Year	Gross Income, $	Operating Expenses, $	Depreciation, $	First Cost and Salvage, $
0				−130,000
1	60,000	−55,000	35,000	
2	75,000	−50,000	35,000	
3	90,000	−45,000	35,000	
4	105,000	−40,000	35,000	10,000

13.18 Elias wants to perform an after-tax evaluation of equivalent methods to electrostatically remove airborne particulate matter from clean rooms used to package liquid pharmaceutical products. Using the information shown, MACRS depreciation with $n = 3$ years, a 5-year study period, after-tax MARR = 7% per year, a T_e of 34% and a spreadsheet, he obtained the results $AW_A = \$-2176$ and $AW_B = \$3545$. Any tax effects when the equipment is salvaged were neglected. Method B is the better method.

Use classical SL depreciation with $n = 5$ years to select the better method. Is the decision different from that reached using MACRS?

	Method A	Method B
First cost, $	−100,000	−150,000
Salvage value, $	10,000	20,000
Savings, $ per year	35,000	45,000
AOC, $ per year	−15,000	−6,000
Expected life, years	5	5

Depreciation Effects on Taxes

13.19 An asset that had a first cost of $80,000 was depreciated according to the MACRS method over a 5-year period. At the end of year 4, it was replaced with a more-advanced system and sold for $15,000. Determine if depreciation recapture or a capital loss was present and, if so, how much.

13.20 An automated assembly robot that cost $300,000 has a recovery period of five years with an expected $50,000 salvage value. If the MACRS depreciation rates for years 1, 2, and 3 are 20.0%, 32.0%, and 19.2%, respectively, what is the depreciation recapture, capital gain, or capital loss, provided the robot was sold after 3 years for $80,000?

13.21 A machine that had a first cost of $120,000 was depreciated by the MACRS method over a 3-year period. The machine was sold for $60,000 at the end of year 2, because the company decided to import the component made by the machine. If the company's gross income was $1.4 million with operating expenses of $500,000, what was the tax liability in year 2 for an effective tax rate of 35%?

13.22 A manufacturing company is considering the purchase of one of two material handling systems. The estimates are as follows:

System 1—$P = \$370,000$; $S = \$100,000$;
   $n = 3$ years
System 2—$P = \$490,000$; $S = \$150,000$;
   $n = 3$ years

The plant manager (who has a real desire to reduce corporate taxes) asked you to determine which system offers the bigger tax advantage over a 3-year recovery period. The company's annual gross income is projected to be $5.8 million with operating expenses of $1.6 million. Use a spreadsheet to determine the difference in taxes between the two systems if $T_e = 38\%$ and MACRS depreciation is applied.

13.23 Last month, a company specializing in wind power plant design and engineering made a large capital investment of $400,000 in physical simulation equipment that will be used for at least 5 years, then sold for approximately 25% of the first cost. By law, the assets are MACRS depreciated using a 3-year recovery period. (*a*) Explain why there is a predictable tax implication when the assets are sold. (*b*) By how much will the sale cause TI and taxes to change in year 5?

13.24 Cheryl, a EE student who is working on a business minor, is studying depreciation and finance in her engineering management course. The assignment is to demonstrate that shorter recovery periods require the same total taxes, but they offer a time value of taxes advantage for depreciable assets. Help her using asset estimates developed for a 6-year study period: $P = \$65,000$, $S = \$5000$ whenever it is sold, GI = $32,000 per year, AOC = $10,000 per year, SL depreciation, $i = 12\%$ per year, $T_e = 31\%$. The recovery period is either 3 or 6 years.

13.25 A bioengineer is evaluating methods used to apply an adhesive to microporous paper tape that is commonly used after surgery. The machinery costs $200,000, has no salvage value, and the CFBT estimate is $75,000 per year for up to 10 years. The $T_e = 38\%$ and $i = 8\%$ per year. The two depreciation methods to consider are: MACRS with $n = 5$ years and SL with $n = 8$ years (neglect the half-year convention effect). For a study period of 8 years, (*a*) determine which depreciation method and recovery period offers the better tax advantage, and (b) demonstrate that the same total taxes are paid for MACRS and SL depreciation.

13.26 Thomas completed a study of a $1 million 3-year old DNA analysis and modeling system that DynaScope Enterprises wants to keep for 1 more year or dispose of now. His table (in $1000 units) details the analysis, including an anticipated $100,000 selling price (SP) next year, SL depreciation, taxes at the all-inclusive rate of $T_e = 52\%$, and PW at the after-tax MARR of 5% per year. Thomas recommends retention since PW > 0. Critique the analysis to determine if he made the correct recommendation.

Year	CFBT	SP	Depreciation	TI	Taxes	CFAT
0	$-1000					$-1000
1	275		$250	$25	$13	262
2	275		250	25	13	262
3	275		250	25	13	262
4	275	$100	250	25	13	362
PW @ 5%						$11.3

After-Tax Replacement Analysis

13.27 The defender in a multiple-effect solar still manufacturing plant has a market value of $130,000 and expected annual operating costs of $70,000 with no salvage value after its remaining life of three years. The depreciation charges for the next three years will be $69,960, $49,960, and $35,720. Assume the company's effective tax rate is 35% and its after-tax MARR is 12%. A present worth equation for comparing the defender against a challenger that also has a 3-year life is being developed. Determine the after-tax cash flow for year 2 only used in the PW relation.

13.28 A 2-year old injection molding machine was expected to serve out its projected life of 5 years, but a challenger promises to be more efficient and have lower operating costs. You have been asked to perform an AW evaluation to determine if it is economically attractive to replace the defender now or keep it for 3 more years as originally planned. The defender had a first cost of $300,000, but its market value now is only $100,000. It has chargeable expenses of $120,000 per year and no expected salvage value. To simplify calculations, assume that SL depreciation was charged at $60,000 per year, and that it will continue at that rate for the next 3 years.

The challenger will cost $420,000; have no salvage value after its 3-year life; have chargeable expenses of $30,000 per year, and be SL-depreciated at $140,000 per year (again, for simplicity). Assume the company's effective tax rate is 35% and its after-tax MARR is 15% per year. Since GI is not estimated, all taxes are negative and considered "savings" to the alternative.

a. Determine the CFAT in year 0 for the challenger and defender. (Hint: Check for DR, CG, and CL.)

b. Determine the CFAT in years 1, 2, and 3 for the challenger and defender.

c. Conduct the replacement study to determine if the defender should be kept for 3 more years or replaced now.

13.29 Justyne needs assistance with the information shown below. The defender can be replaced now or kept for 4 more years. (Notes: All monetary values are in $1000 units. Assume that either asset is salvaged in the future at its original salvage estimate. Since no revenues are estimated, all taxes are negative and considered "savings" to the alternative. Neglect any capital gains or losses.)

	Defender	Challenger
First cost, $	−45	−24
Estimated S at purchase, $	5	0
Market value now, $	35	—
AOC, $ per year	−7	−8
Depreciation method	SL	MACRS
Recovery period, years	8	3
Useful life, years	8	5
Years owned	3	—

a. Perform a PW-based replacement study using an after-tax MARR = 12% per year and T_e = 35%. Do this using hand calculations.

b. Verify your results using a spreadsheet-based replacement study.

13.30 After 8 years of use, the heavy truck engine overhaul equipment at Pete's Truck Repair was evaluated for replacement. Pete's accountant used an after-tax MARR of 8% per year, T_e = 30%, and a current market value of $25,000 to determine AW_D = $2100. The new equipment costs $75,000, uses SL depreciation over a 10-year recovery period, and has a $15,000 salvage estimate. Estimated CFBT is $15,000 per year. Pete asked his engineer son Ramon to determine if the new equipment should replace what is owned currently. From the accountant, Ramon learned the current equipment cost was $20,000 when purchased and reached a zero book value several years ago. Help Ramon answer his father's question.

Capital and WACC

13.31 Nucor Corp manufactures generator coolers for nuclear and gas turbine power plants. The company completed a plant expansion through financing that had a debt/equity mix of 40–60. If $15 million came from mortgages and bond sales, what was the total amount of the financing?

13.32 Nano-Technologies bought out RT-Micro using financing as follows: $16 million from mortgages, $4 million from retained earnings, $12 million from cash on hand, and $30 million from bonds. Determine the debt-to-equity mix.

13.33 Master Bond Inc. makes a no-mix single component adhesive that cures at temperatures between 250 and 300°F and meets the low-outgassing requirements specified by NASA. Determine the weighted average cost of capital for financing and interest rates as follows: $4 million in stock sales at 12%, $6 million in bonds at 8%, and $5 million in retained earnings at 6% per year.

13.34 Alpha Engineering invested $30 million in a project that has a D-E mix of 65-35. Determine the return on the company's equity, if the net income is $4 million from revenue of $6 million.

13.35 Two public corporations, First Engineering and Midwest Development, each show capitalization of $175 million in their annual reports. The balance sheet for First indicates total debt of $87 million, and that of Midwest indicates net worth of $62 million. Determine the D-E mix for each company.

13.36 Oriental Motor reported a WACC of 10.4% in its recent report to stockholders. Their common stock has averaged a total return of 8% per year over the last three years. If projects within the corporation are 70% funded by its own capital, estimate the company's cost of debt capital.

13.37 Fruit Transgenics Engineering is contemplating the purchase of its rival. One of FTE's genetics engineers is interested in the financing strategy of the buyout. He learned of two plans. Plan A requires 50% equity funds from FTE retained earnings that currently earn 9% per year, with the balance borrowed externally at 6%, based on the company's excellent stock rating. Plan B requires only 20% equity funds with the balance borrowed at a higher rate of 8% per year.

a. Which plan has the lower average cost of capital?

b. If the current corporate WACC of 8.2% will not be exceeded, what is the maximum cost of debt capital allowed for each plan? Are these rates higher or lower than the current estimates?

13.38 Deavyanne Johnston, the engineering manager at TZO Chemicals, wants to complete an alternative evaluation study. She asked the finance manager for the corporate MARR. The finance manager gave her some data on the project and stated that all projects must clear their average (pooled) cost by at least 4%.

Funds Source	Amount, $	Cost, %
Retained earnings	4 million	7.4
Stock sales	6 million	4.8
Long-term loans	5 million	9.8
Budgeted funds for project	15 million	

a. Use the data to determine the minimum MARR.

b. The study is after-taxes and part (a) provided the before-tax MARR.

Determine the correct MARR to use if T_e was 32% last year and the finance manager meant that the 4% above the cost is for after-tax evaluations.

13.39 Dougherty Construction has worked on several international housing projects during the last year. The D-E mixes and rates are shown. Plot the WACC curve and identify the D-E mix that had the lowest WACC.

	Debt		Equity	
Project	Percent	Rate	Percent	Rate
203	100	10.9	0	0
206	50	7.0	50	8.5
306	65	11.6	35	7.5
367	25	8.2	75	6.0
456	0	0	100	8.9
913	10	5.5	90	7.2
914	75	11.4	25	8.4
987	80	10.5	20	8.1

13.40 Bow Chemical will invest $14 million this year to upgrade its ethylene glycol processes. This chemical is used to produce polyester resins to manufacture products varying from construction materials to aircraft, and from luggage to home appliances. Equity capital costs 14.5% per year and will supply 65% of the capital funds. Debt capital costs 10% per year before taxes. The effective tax rate is 36%.

a. Determine the amount of annual revenue after taxes that is consumed in covering the interest on the project's initial cost.

b. If the corporation does not want to use 65% of its own funds, the financing plan may include 75% debt capital. Determine the amount of annual revenue needed to cover the interest with this plan, and explain the effect it may have on the corporation's ability to borrow in the future.

Economic Value Analysis

13.41 While an engineering manager may prefer to use CFAT estimates to evaluate the AW of a project, a financial manager may select AW of EVA estimates. Why are these preferences predictable?

13.42 An asset with a first cost of $300,000 is depreciated by the MACRS method using a 5-year recovery period. Determine the monetary value added to the corporation by the asset in year two of its service, if the net operating profit after taxes is $70,000 and the company uses an after-tax interest rate of 15%. The MACRS depreciation rates for years 1 and 2 are 20% and 32%, respectively.

13.43 In conducting an EVA analysis for year two for a newly introduced product line, Bethune, Inc., which manufactures pre-assembled blower packages and other water treatment components, determined the EVA to be $28,000. Bethune's CEO knew that the gross income was $700,000, but he asked you to find out how much expense was associated with the new product line for year 2. The company uses an after-tax interest rate of 14% and a T_e of 35%. The initial investment capital required for the new product was $550,000 and all equipment is 3-year MACRS depreciated.

13.44 Triple Play Innovators Corporation (TPIC) plans to offer IPTV (Internet Protocol TV) service to North American customers starting soon. Perform an AW analysis of the EVA series for the two alternative suppliers that bid for hardware and software contracts. Let $T_e = 30\%$ and after-tax MARR = 8%; use SL depreciation (for simplicity) and a study period of 8 years. Use a spreadsheet, unless requested to perform hand calculations.

Bidder's country	United States	Malaysia
First cost, $	4.2 million	3.6 million
Recovery period, years	8	5
Salvage value, $	0	0
NOI, $ per year	1,500,000 in year 1; increasing by 300,000 per year up to 8 years	

ADDITIONAL PROBLEMS AND FE EXAM REVIEW QUESTIONS

13.45 If the after-tax rate of return for a cash flow series is 13.3% and the corporate effective tax rate is 39%, the approximated before-tax rate of return is closest to:
  a. 6.8%
  b. 15.4%
  c. 18.4%
  d. 21.8%

13.46 For a state tax rate of 8% and a federal tax rate of 34%, the effective tax rate is closest to:
  a. 41.7%
  b. 39.3%
  c. 36.4%
  d. 31.8%

13.47 If all values carry a + sign, cash flow before taxes is represented by the following equation:
  a. gross income − expenses − depreciation − initial investment + salvage value
  b. gross income − expenses − depreciation + salvage value
  c. gross income − expenses − initial investment + salvage value
  d. gross income − expenses + initial investment + salvage value

13.48 Depreciation recapture occurs when a depreciable asset is sold for:
  a. more that the current book value.
  b. more that the current market value.
  c. more that the estimated salvage value.
  d. more than the first cost.

13.49 A capital gain is calculated as:
  a. book value − selling price
  b. book value − first cost
  c. market value − selling price
  d. selling price − first cost

13.50 A small manufacturing company with a gross income of $360,000 has the following expenses: M&O $76,000, insurance $7000, labor $110,000, utilities $29,000. If debt service is $37,000 and taxes are $9000, the net operating income is closest to:
  a. $92,000
  b. $101,000
  c. $138,000
  d. $174,000

13.51 An after-market auto parts company just received $16,000 for a robot that has been depreciated to zero. If the company's effective tax rate is 36%, the sale will:
  a. increase the company's taxes by $16,000
  b. increase the company's taxes by $5760
  c. reduce the company's taxes by $16,000
  d. reduce the company's taxes by $5760

13.52 The after-tax analysis for a $60,000 investment with associated gross income minus expenses (GI − E) is shown below for only the first 2 years. If the effective tax rate is 40%, the values for depreciation (D), taxable income (TI), and taxes for year 1 are closest to:
  a. D = $5,000; TI = $25,000; Taxes = $10,000
  b. D = $30,000; TI = $30,000; Taxes = $4000
  c. D = $20,000; TI = $50,000; Taxes = $20,000
  d. D = $20,000; TI = $10,000; Taxes = $4000

Year	Investment	GI − E	D	TI	Taxes	CFAT
0	$−60,000					$−60,000
1		$30,000				26,000
2		35,000	$15,000	$6000		29,000

13.53 Huntsman Corp. completed a plant expansion using financing as follows: $9 million from mortgages, $3 million from retained earnings, and $4 million from cash on hand, The debt-to-equity mix was closest to:
  a. 56-44
  b. 75-25
  c. 44-56
  d. 25-75

13.54 A contractor with an effective tax rate of 35% has the following for a tax year: gross income of $155,000, other income of $4000, expenses of $45,000, and other deductions and exemptions of $12,000. The income tax due is closest to:
  a. $35,700
  b. $42,700
  c. $51,750
  d. $55,750

13.55 An asset purchased for $100,000 with S = $20,000 after 5 years was depreciated using the 5-year MACRS rates. Expenses average $18,000 per year and the effective tax rate is 30%. The asset is actually sold after 5 years of service for $22,000. MACRS rates in years 5 and 6 are 11.52% and

5.76%, respectively. The after-tax cash flow from the sale is closest to:

a. $27,760

b. $17,130

c. $26,870

d. $20,585

13.56 The Wilkins Company has maintained a 50-50 D-E mix for capital investments. Equity capital costs 11%; however, debt capital that historically costs 9% has now increased by 20%. If Wilkins does not want to exceed its past weighted average cost of capital (WACC) and it is forced to go to a D-E mix of 75-25, the maximum cost of equity capital that Wilkins can accept is closest to:

a. 9.8%

b. 10.9%

c. 7.6%

d. 9.2%

13.57 Net operating income (NOI) is defined as the difference between gross income and operating expenses. Of other ways to express the meaning of NOI, correct one(s) are:

1–corporate earnings before interest and taxes

2–same as net cash flow (NCF = cash inflow − cash outflow)

3–Net operating profit after taxes (NOPAT)

a. 1

b. 2

c. 3

d. 1 and 2

14 Chapter

Alternative Evaluation Considering Multiple Attributes and Risk

John Lund/Blend Images LLC

This chapter discusses techniques that expand the ability of a project's evaluation to include estimate variation, decision making under risk, and factors other than money. First, noneconomic attributes that may alter the straight economic decision are examined. Then the element of variation in parameter values is examined using simple probability and statistics. This allows the aspect of risk to be considered as the best alternative is selected. The use of spreadsheet-based simulation to account for variation in estimates and risk is introduced.

14.1 MULTIPLE ATTRIBUTE ANALYSIS

In all evaluations thus far, only one attribute—the economic one—has been relied upon in selecting the best alternative by maximizing the PW, AW, ROR, or B/C value. However, noneconomic factors are considered in most alternative evaluations. These factors are mostly intangible and usually difficult to quantify in economic terms.

Public sector projects are excellent examples of multiple-attribute selection. For example, a project to construct a dam forming a lake usually has several purposes, such as flood control, industrial use, commercial development, drinking water, recreation, and nature conservation. Noneconomic attributes, evaluated in different ways by different stakeholders, make selection of the best dam alternative very complex.

In evaluations, key noneconomic attributes can be considered directly using several techniques. A technique popular in most engineering disciplines, the *weighted attribute method,* is described here.

Once the decision is made to consider multiple attributes in an evaluation, the following must be accomplished:

1. Identify the key attributes.
2. Determine each attribute's importance and weight.
3. Rate (value) each alternative by attribute.
4. Calculate the evaluation measure and select the best alternative.

14.1.1 Key Attribute Identification

Attributes are identified by several methods, depending upon the situation. Seeking input from other people is important because it helps focus on key attributes as determined by those with experience and those who will use the selected system. Some identification approaches are:

- Comparison with similar studies that include multiple attributes.
- Input from experts with relevant experience.
- Survey of stakeholders (customers, employees, managers)
- Small group discussions (brainstorming, focus groups)
- Delphi method, a formal procedure that reaches consensus from people with different perspectives.

As an example of identifying key attributes, consider the purchase of a car for the Kerry family of four versus a car purchase for Clare, a university student. Economics (possibly with different slants) will be a key attribute for both, but other attributes will be considered. Examples follow.

Kerry Family

Economics—first cost and operating cost (mileage and maintenance)

Safety—airbags; rollover and crash factors; traction

Inside design—seating space, cargo space, etc.

Reliability—warranty coverage; breakdown record

Clare

Economics—first cost and mileage cost

Style—exterior design, color, sleek and modern look

Inside design—cargo room; seating

Dependability—pick up and speed factors; required maintenance

14.1.2 Importance and Weights of Attributes

An importance score is set by a person or group experienced with each attribute compared to alternative attributes. If a group is involved, consensus is required to arrive at one score for each attribute. The resulting score is used to determine a weight W_i for each attribute i.

$$W_i = \frac{\text{importance score}_i}{\text{sum of all scores}} = \frac{\text{importance score}_i}{S} \qquad [14.1]$$

This normalizes the weights, making their sum equal to 1.0 over all attributes $i = 1,$ $2, \ldots, m$. Table 14.1 is a tabular layout of attributes, weights, and alternatives used to implement the weighted attribute method. Attributes and weights from Equation [14.1] are entered on the left; value ratings for each alternative complete the table as discussed in the next step.

Three approaches to assigning weights are *equal, rank order,* and *weighted rank order. Equal weighting* means that all attributes are of the same importance, because there is no rationale or criterion to distinguish differences. This default approach sets all importance scores to 1, which makes each weight equal to $1/m$. To *rank order* attributes, place them in increasing importance, assigning a 1 to the least important and m to the most important attribute. This means that the difference between attribute importance is constant. Equation [14.1] results in weights of $1/S, 2/S, \ldots, m/S$.

A more practical and versatile approach is to assign importance using a *weighted rank order*. First, place the attributes in decreasing order of importance, then assign a score of 100 to the most important one(s), and score other attributes relative to this one using scores between 100 and 0. If s_i identifies the score for each

TABLE 14.1 Tabular Layout of Attributes and Alternatives Used for Multiple Attribute Evaluation

Attributes	Weights	Alternatives				
		1	2	3	...	n
1	W_1					
2	W_2					
3	W_3		Value ratings			
⋮	⋮					
m	W_m					

attribute, Equation [14.1] is rewritten to determine attribute weights. This auto-matically normalizes them to sum to 1.0.

$$W_i = \frac{s_i}{S} \qquad [14.2]$$

This approach is commonly applied because one or more attributes can be heav-ily weighted, while attributes of minor importance can be included in the analysis. As an example, suppose the 4 key attributes developed for the Kerry family car purchase are ordered as safety, economics, inside design, and reliability. If the eco-nomics attribute is half as important as safety, while the remaining 2 are half as important as economics, the attribute list, importance scores, and weights are as follows:

Attribute, i	Score, s_i	Weight, W_i
Safety	100	100/200 = 0.50
Economics	50	50/200 = 0.25
Inside design	25	25/200 = 0.125
Reliability	25	25/200 = 0.125
Sum	200	1.000

There are other weighting techniques, usually designed for group input where diverse opinions prevail. Some are utility functions, pairwise comparison, and the AHP (analytic hierarchy process). These techniques provide a significant advan-tage compared to those described here; they guarantee consistency between ranks, scores, and individual scorers.

14.1.3 Value Rating for Each Alternative

A decision maker evaluates each alternative (j) based on each attribute (i) to deter-mine a value rating V_{ij}. These are the right-side entries of Table 14.1. The value rating scale uses some numeric basis, such as, 0 to 100, 1 to 10, -1 to $+1$, or -3 to $+3$, with larger scores indicating a higher value rating. The last two scales allow an individual to give negative input on an alternative. A commonly used technique is the *Likert* scale, which defines several gradations with a range of numbers for each one. For example, a scale from 0 to 10 may be described as follows:

If You Value the Alternative as	Give it a Value Rating between the Numbers
Very poor	0–2
Poor	3–5
Good	6–8
Very good	9–10

TABLE 14.2   Value Ratings of Four Attributes and Three
Alternatives for Multiple-Attribute Evaluation

		Alternative		
Attribute	Weight	1	2	3
Safety	0.50	6	4	8
Economics	0.25	9	3	1
Inside Design	0.125	5	6	6
Reliability	0.125	5	9	7

Each evaluator uses this scale to rate each alternative on each attribute, thus generating the V_{ij} values. Likert scales with an even number of choices, say, 4, are
preferred so that the central tendency of "fair" is not overrated.

   Continuing the car purchase example, assume that the father has value rated
3 car alternatives using a 0 to 10 scale against the 4 key attributes. The result may
look like Table 14.2 with all V_{ij} values entered. Mrs. Kelly and each child will have
a table with (probably) different V_{ij} entries.

14.1.4 Evaluation Measure for Each Alternative

Alternative evaluation by the weighted attribute method results in a measure R_j for
each alternative, which is the sum of each attribute's weight multiplied by a corresponding alternative's value rating.

$$R_j = \text{sum of (weights} \times \text{value rating)}$$
$$= \sum_{i=1}^{m} W_i \times V_{ij} \qquad\qquad [14.3]$$

The selection guideline is:

   Select the alternative with the largest R_j value.

When several decision makers are involved, a different R_j can be calculated for
each person and some resolution made if different alternatives are indicated to be
the best. Alternatively, resolution to agree on one set of V_{ij} values can be reached
before a single R_j is determined. As always, sensitivity analysis of scores, weights,
and value ratings offers insight into the sensitivity of the final selection for different people and groups.

   A final note on the evaluation measure R_j is in order. It is a single-dimension
number that quite effectively combines the different dimensions addressed by
the attributes, people's importance scores, and evaluators' value ratings. This
type of aggregate measure, often called a *rank-and-rate method*, removes the
complexity of balancing different attributes, but it also eliminates much of the
robust information captured in ranking attributes and rating alternatives against
attributes.

EXAMPLE 14.1 Two vendors offer chlorine gas distribution systems in 1-ton cylinders for use in industrial water cooling systems. Hartmix, Inc. engineers have completed a present worth analysis of both alternatives using a 5-year study period. The values are $PW_A = \$-432{,}500$ and $PW_B = \$-378{,}750$, leading to a recommendation of vendor B. This result and some noneconomic factors were considered by the regional manager, Herb, and district superintendent, Charlotte. They independently defined attributes and assigned importance scores (0 to 100) and value ratings (1 to 10) for the two vendor proposals. Higher scores and values are considered better. Use their results (Table 14.3) to determine if vendor B is the better choice when these key attributes are considered.

Solution

Of the four steps in the weighted attribute method, the first (attribute definition) and third (value rating) are completed. Finish the other two steps using each evaluator's scores and ratings and compare the selections to determine if Herb, Charlotte, and the engineers all selected vendor B.

TABLE 14.3 Importance Scores and Value Ratings Determined by Herb and Charlotte, Example 14.1

Attribute	Importance Score		Herb's Values		Charlotte's Values	
	Herb	Charlotte	Vendor A	Vendor B	Vendor A	Vendor B
Safety	100	80	10	9	7	9
Economics	35	100	3	10	5	5
Flexibility	20	10	10	9	5	8
Maintainability	20	50	2	10	8	4
Total	175	240				

TABLE 14.4 Evaluation Measure for Multiple Attribute Comparison, Example 14.1

Attribute	Herb's Evaluation			Charlotte's Evaluation		
	Weights	A	B	Weights	A	B
Safety	0.5714	10	9	0.3333	7	9
Economics	0.2000	3	10	0.4167	5	5
Flexibility	0.1143	10	9	0.0417	5	8
Maintainability	0.1143	2	10	0.2083	8	4
Totals and R values	1.0000	7.69	9.31	1.0000	6.29	6.25

Equation [14.2] calculates the weights for each attribute for both evaluators. The totals are $S = 175$ for Herb and $S = 240$ for Charlotte.

Herb   Safety: $100/175 = 0.5714$ Economics: $35/175 = 0.20$
         Flexibility: $20/175 = 0.1143$ Maintainability: $20/175 = 0.1143$

Charlotte   Safety: $80/240 = 0.3333$ Economics: $100/240 = 0.4167$
         Flexibility: $10/240 = 0.0417$ Maintainability: $50/240 = 0.2083$

Table 14.4 details the weights and value ratings as well as R_A and R_B values (last row). As an illustration, the measure for vendor B for Charlotte is calculated by Equation [14.3].

Charlotte: $R_B = 0.3333(9) + 0.4167(5) + 0.0417(8) + 0.2083(4) = 6.25$

Herb selects vendor B conclusively and Charlotte selects A by a very small margin. Since the engineers selected B on purely economic terms, some negotiated agreement is necessary; however, B is likely the better choice.

14.2 ECONOMIC EVALUATION WITH RISK CONSIDERED

All things in the world vary from one situation or environment to another and over time. We are guaranteed that variation will occur in engineering economy due to its emphasis on future estimates. All analyses thus far have used estimates (e.g., AOC = $-45,000 per year) and computations (PW, ROR, and other measures) that are *certain*, that is, no variation. We can observe and estimate outcomes with a high degree of certainty, but even this depends upon the precision and accuracy of our skills and tools.

When variation enters into estimation and evaluation, it is called *risk*. When there may be two or more observable values for a parameter *and* it is possible to estimate the chance that each value may occur, risk is considered. Decision making under risk is present, for example, if a cash flow estimate has a 50-50 chance of being $10,000 or $5,000. In reality, virtually all decisions are made under risk, but the risk may not be explicitly taken into account. This section introduces *decision making under risk*.

> **Decision making under certainty**—Deterministic estimates are made and entered into engineering economy computations for PW, AW, FW, ROR, or B/C equivalency values. Estimates are the *most likely value*, also called *single-value estimates*. All previous examples and problems in this book are of this type. In fact, sensitivity analysis is simply another form of analysis with certainty, except that the computations are repeated with different values, *each estimated with certainty*.

> **Decision making under risk**—The element of chance is formally taken into account; however, a clear-cut decision is harder to make because variation in estimates is allowed. One or more parameters (P, A, S, AOC, i, n, etc.) in an

alternative can vary. Two fundamental ways to consider risk in the economic evaluation are expected value analysis and simulation analysis.

- *Expected value analysis* uses the possible values of an estimate and the chance of occurrence associated with each value to calculate simple statistical measures such as the expected value and standard deviation of a single estimate or an entire alternative's PW (or other measure). The alternative with the most favorable "statistics" is indicated as best.
- *Simulation analysis* uses the chance and parameter estimates to generate repeated computations of the PW (or other measure) by random sampling. Large enough samples are generated to conclude which alternative is best with a reasonable degree of confidence. Spreadsheets and software are generally used to conduct the sampling and plotting needed to reach a statistical conclusion.

The remainder of this section concentrates on decision making under risk using the expected value approach applied to the engineering economy methods we have learned in previous chapters. First, however, it is important to understand how to use several fundamentals of probability and statistics.

Random Variable (or Variable)   A random variable can take on any one of several values. It is classified as either *discrete* or *continuous* and is identified by a letter. Discrete variables have several isolated values, while continuous variables can assume any value between two stated limits. A project's life n that is estimated to have a value of, say, 4 or 6 or 10 years is a discrete variable, and is written $n = 4, 6, 10$. The rate of return i is a continuous variable that may range from -100% to ∞, that is, $-100\% \leq i < \infty$.

Probability   A probability is a number between 0 and 1.0 that expresses the chance in decimal form that a variable may take on any value identified as possible for it. Simply stated, it is the *chance that something specific will occur,* divided by 100. Probability for a variable X, for example, is identified by $P(X = 6)$ and is interpreted as the probability that X will equal the specific value 6. If the chance is 25%, the probability that $X = 6$ is written $P(X = 6) = 0.25$. In all probability statements, the $P(X)$ values over all possible X values must total to 1.0.

Probability Distribution   This is a graphical representation describing how probability is distributed over all possible values of a variable. Figure 14.1a is a plot of the *discrete variable* estimated asset life n, which has possible values of 4, 6, and 10 years, all with equal probability of 1/3. The distribution for a *continuous variable* is described by a continuous curve over the range of the variable. Figure 14.1b indicates an equal probability of 0.01 for all values of a uniform gross income GI between \$400 and \$500 per year. These are called *uniform distributions*.

Random Sample (or Sample)   A sample is a collection of N values drawn in a random fashion from all possible values of the variable using the probability distribution of the variable. Suppose a random sample of size

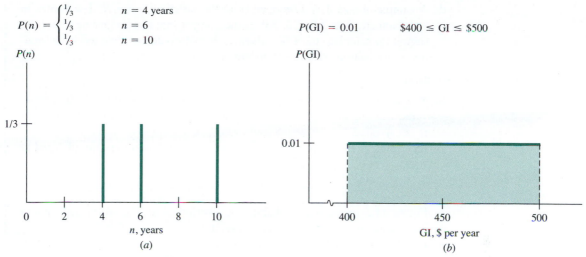

FIGURE 14.1 Probability distribution of (a) discrete variable and (b) continuous variable.

$N = 10$ is taken from the salvage value variable S, where $S = \$500$ or $\$800$, each with an equal probability of 0.5. If a coin is tossed with heads representing $\$500$ and tails $\$800$, the sample outcome may be HHTHTTTHTT, which translates to four $\$500$s and six $\$800$s. As larger samples are taken, the equal probability for each value of S will be clearly represented in the sample. Samples are used to "estimate" properties for the population of the variable. As discussed below, these estimates are used to perform engineering economy calculations necessary to select the best alternative. If the data is assumed to be uniformly distributed between any two numbers, the spreadsheet functions RAND or RANDBETWEEN can be used to generate random numbers (RN) for a random sample. These functions are described in Appendix A and applied in Example 14.6.

EXAMPLE 14.2

Carlos, a cost estimator for Deblack Chemicals, LLC, has estimated the annual net revenue R for a newly developed antifungal spray applied to post-harvest citrus fruit. He predicts the expected revenue to be $\$3.1$ million per year for the next 5 years. This expected amount is based on his estimate that R could be $\$2.6$, $\$2.8$, $\$3.0$, $\$3.2$, $\$3.4$, or $\$3.6$ million per year, all with the same probability. Do the following for Carlos:

a. Identify the variable R as discrete or continuous; describe it in engineering economy terms.
b. Write the probability statements for the estimates identified.
c. Plot the probability distribution of R.

d. A sample of size 4 is developed from the distribution for R. The values in $ million are: 2.6; 3.0; 3.2; 3.0. If the interest rate is 12% per year, use the sample to calculate the PW values of R that could be included in an economic evaluation with risk considered.

Solution

a. R is discrete since, as estimated by Carlos, it can take on only 6 specific values. In engineering economy terms, R is a uniform series amount, that is, an A value.

b. Probability statements for 6 estimated values (in $ million) are all equal at 1/6.

$$P(R = 2.6) = P(R = 2.8) = \ldots = P(R = 3.6) = 1/6 \text{ or } 0.16667$$

c. Figure 14.2 shows that probability is distributed as a uniform distribution with equal probability for each value of R.

d. Let the symbol $PW(R_i)$ indicate the present worth of the ith sample value. Since R (in $ million) is a uniform series, the P/A factor is used.

$$PW(R_1) = R_1(P/A,12\%,5) = 2.6(3.6048) = \$9,372,480$$
$$PW(R_2) = 3.0(P/A,12\%,5) = \$10,814,400$$
$$PW(R_3) = 3.2(P/A,12\%,5) = \$11,535,360$$
$$PW(R_4) = 3.0(P/A,12\%,5) = \$10,814,400$$

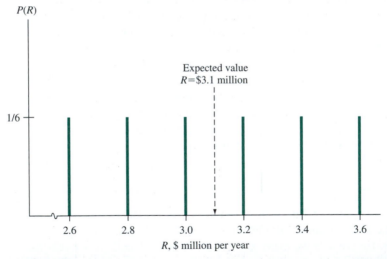

FIGURE 14.2   Probability distribution of estimated annual revenue, Example 14.2.

Discrete variables use summation operations and continuous variables use integration operations to calculate the properties discussed next. The development here is for discrete variables only.

The expected value of a variable X is the long-term average if a variable is sampled many times. For a complete project, the expected value of a measure, say,

PW, is the long-run average of PW, were the project repeated many times. The expected value of the variable X is identified by $E(X)$. The *sample average* $\overline{X}$ estimates the expected value by summing all values in the sample and dividing by the sample size N.

E(X) estimate: $$\overline{X} = \sum_{i=1}^{N} X_i/N \qquad [14.4]$$

In Example 14.2, the variable is PW and $E(\text{PW})$ is estimated by $\overline{\text{PW}}$ from Equation [14.4].

$$\overline{\text{PW}} = (9{,}372{,}480 + \ldots + 10{,}814{,}400)/4 = \$10{,}634{,}160$$

The spreadsheet function = AVERAGE() calculates the sample mean, which is the same as Equation [14.4], from a list of entries in spreadsheet cells or up to 30 individually entered values.

If the probability distribution is known or estimated, the expected value is calculated as

$$E(X) = \sum_{\text{all } i} X_i P(X_i) \qquad [14.5]$$

An expected value can be utilized in one of two ways in engineering economy. The first calculates $E(\text{parameter estimate})$, for example $E(P)$ or $E(\text{AOC})$, over all possible values of the variable. If an entire alternative is to be evaluated, once $E()$ is calculated, the evaluation is performed in the same way discussed in previous chapters. The second way determines $E(\text{measure})$, for example $E(\text{PW})$ or $E(\text{ROR})$ of the entire alternative. Now the best alternative is selected from the best $E(\text{measure})$ value. The next two examples illustrate these two approaches in turn.

EXAMPLE 14.3

Plumb Electric Cooperative is experiencing a difficult time obtaining natural gas for electricity generation. Fuels other than natural gas are purchased at an extra cost, which is transferred to the customer. Total monthly fuel expenses are now averaging $7,750,000. An engineer with this city-owned utility has calculated the average revenue for the past 24 months using three fuel-mix situations—gas plentiful, less than 30% other fuels purchased, and 30% or more other fuels. Table 14.5 indicates the number of months that each fuel-mix situation occurred. Can the utility expect to meet future monthly expenses based on the 24 months of data, if a similar fuel-mix pattern continues?

TABLE 14.5 Revenue and Fuel-Mix Data, Example 14.3

Fuel-Mix Situation	Months in Past 24	Average Revenue, $ per Month
Gas plentiful	12	5,270,000
<30% other	6	7,850,000
≥30% other	6	12,130,000

Solution

Using the 24 months of data, estimate a probability for each fuel mix.

Fuel-Mix Situation	Probability of Occurrence
Gas plentiful	12/24 = 0.50
<30% other	6/24 = 0.25
≥30% other	6/24 = 0.25

Let R represent average monthly revenue. Use Equation [14.5] to determine expected revenue per month.

$$E(R) = 5{,}270{,}000(0.50) + 7{,}850{,}000(0.25) + 12{,}130{,}000(0.25)$$
$$= \$7{,}630{,}000$$

With expenses averaging $7,750,000, the average monthly revenue shortfall is $120,000. To break even, other sources of revenue must be generated, or the additional costs may be transferred to the customer.

EXAMPLE 14.4 Lite-Weight Wheelchair Company has a substantial investment in tubular steel bending equipment. A new piece of equipment costs $5000 and has a life of 3 years. Estimated cash flows (Table 14.6) depend on economic conditions classified as receding, stable, or expanding. A probability is estimated that each of the economic conditions will prevail during the 3-year period. Apply expected value and PW analysis to determine if the equipment should be purchased. Use a MARR of 15% per year.

TABLE 14.6 Equipment Cash Flow and Probabilities, Example 14.4

Year	Receding (Prob. = 0.2)	Stable (Prob. = 0.6)	Expanding (Prob. = 0.2)
	Annual Cash Flow Estimates, $ per year		
0	$-5000	$-5000	$-5000
1	+2500	+2000	+2000
2	+2000	+2000	+3000
3	+1000	+2000	+3500

Solution

First determine the PW of the cash flows in Table 14.6 for each economic condition, and then calculate $E(PW)$, using Equation [14.5]. Define subscripts R for receding economy, S for stable, and E for expanding. The PW values for the three scenarios are

$$PW_R = -5000 + 2500(P/F,15\%,1) + 2000(P/F,15\%,2) + 1000(P/F,15\%,3)$$
$$= -5000 + 4344 = \$-656$$
$$PW_S = -5000 + 4566 = \$-434$$
$$PW_E = -5000 + 6309 = \$+1309$$

Only in an expanding economy will the cash flows return the 15% and justify the investment. Using the probabilities estimated for each economic condition, the expected present worth is

$$E(PW) = \sum_{j=R,S,E} PW_j[P(j)]$$
$$= -656(0.2) - 434(0.6) + 1309(0.2)$$
$$= \$-130$$

At 15%, $E(PW) < 0$; the equipment is not justified using an expected value analysis.

The *average* is a *measure of central tendency* of data. Another measure, the *standard deviation* identified by the small letter s, is a *measure of the spread* of the data points around the average. By definition, s is the dispersion or spread about the expected value $E(X)$ or sample average $\overline{X}$. The standard deviation for a sample is calculated by taking each X minus the sample average $\overline{X}$, squaring it, adding all terms, dividing by $N-1$, then extracting the square root of the result.

$$s = \left[\frac{\sum_{i=1}^{N} (X_i - \overline{X})^2}{N - 1} \right]^{1/2} \qquad [14.6]$$

An equivalent, easier way to compute s eliminates the subtractions.

$$s = \left[\frac{\sum_{i=1}^{N} X_i^2}{N - 1} - \frac{N}{N - 1}\overline{X}^2 \right]^{1/2} \qquad [14.7]$$

From the random sample, the central tendency, estimated by $\overline{X}$, and dispersion, estimated by s, measures are calculated and combined to determine the fraction or percentage of the values that are expected to be within ± 1, ± 2, and ± 3 standard deviations of the average.

$$\overline{X} \pm ts \quad \text{for } t = 1, 2, 3 \qquad [14.8]$$

In probability terms, this can be stated as

$$P(\bar{X} - ts \leq X \leq \bar{X} + ts) \qquad \text{[14.9]}$$

This is a very good measure of the clustering of the data. The tighter the clustering about the expected value, the more confidence that the decision maker can place on the selection of one alternative. This logic is illustrated in Example 14.5, which considers risk.

The function = STDEV() calculates the sample standard deviation using Equation [14.6] from a series of spreadsheet entries or up to 30 individually entered values.

EXAMPLE 14.5   Jerry, an electrical engineer with TGS Nationwide Utility Services, is analyzing electric bills for 1-bedroom apartments in Atlanta and Chicago. In both cities, TGS charges apartment complex owners a flat rate of $125 per month, with this cost passed on to the resident through the rent. This eliminates the expensive alternative of metering and billing each resident monthly. Initially, small random samples of monthly bills from both cities will assist in decision making under risk about the future of "blanket" versus "individual" billing. Help Jerry by answering several questions using the sample data.

a. Estimate the average monthly bill for each city. Do they appear to be approximately equal or significantly different?
b. Estimate the standard deviations. How does the clustering about the average compare for the two cities?
c. Determine the number of bills within the limits $\bar{X} \pm 1s$ for each sample. Based on the sample results, what decision seems reasonable concerning the flat rate of $125 per month versus individual billing in the future?

Sample point	1	2	3	4	5	6	7
Atlanta sample, A, $	65	66	73	92	117	159	225
Chicago sample, C, $	84	90	104	140	157		

Solution

The following solution uses manual calculations. It is followed by a spreadsheet-based solution.

a. Equation [14.4] estimates the expected values, which are quite close in amount.

Atlanta: $N = 7$ $\bar{X}_A = 797/7 = \$113.86$ per month
Chicago: $N = 5$ $\bar{X}_C = 575/5 = \$115.00$ per month

b. Only for illustration purposes, calculate the standard deviation using both formats; Equation [14.6] for Atlanta and Equation [14.7] for Chicago. Tables 14.7

TABLE 14.7 **Calculation of Standard Deviation using Equation [14.6] for Atlanta Sample, Example 14.5**

X	$(X - \bar{X})$	$(X - \bar{X})^2$
65	−48.86	2,387.30
66	−47.86	2,290.58
73	−40.86	1,669.54
92	−21.86	477.86
117	3.14	9.86
159	45.14	2,037.62
225	111.14	12,352.10
Total		21,224.86

$$\bar{X}_A = 113.86$$
$$s_A = [(21,224.86)/(7 - 1)]^{1/2} = 59.48$$

TABLE 14.8 **Calculation of Standard Deviation using Equation [14.7] for Chicago Sample, Example 14.5**

X	X^2
84	7,056
90	8,100
104	10,816
140	19,600
157	24,649
Total	70,221

$$\bar{X}_C = 115$$
$$s_c = [70,221/4 - 5/4(115)^2]^{1/2} = 32$$

and 14.8 summarize the calculations, resulting in $s_A = \$59.48$ and $s_C = \$32.00$. In terms of the sample averages, these vary considerably for the two cities.

Atlanta: s_A is 52% of the average $\bar{X}_A = \$113.86$

Chicago: s_C is 28% of the average $\bar{X}_C = \$115.00$

The clustering about the average for the Chicago sample is tighter than for Atlanta.

	A	B	C	D	E	F	G	H	I	J	K	L
1	Sample point	1	2	3	4	5	6	7	Average	Standard deviation	Avg - 1s	Avg + 1s
2	Atlanta	65	66	73	92	117	159	225	$ 113.86	$ 59.48	$ 54.38	$ 173.33
3												
4	Chicago	84	90	104	140	157			$ 115.00	$ 32.00	$ 83.00	$ 147.00
5												
6												
7					These sample values			= AVERAGE(B4:F4)				
8					are outside ± 1s limits							
9									= STDEV(B4:F4)			
10												

FIGURE 14.3 Spreadsheet functions display sample average, standard deviation, and ± 1s limits, Example 14.5.

c. One standard deviation from the mean is calculated using $t = 1$ in Equation [14.8]. The number of sample points within the $\pm 1s$ is then determined.

> Atlanta: 113.86 ± 59.48 results in limits of $54.38 and $173.34. One data point is outside these limits.
>
> Chicago: 115.00 ± 32.00 results in limits of $83.00 and $147.00. Again, one data point is outside these limits.

Based on these quite small samples, the clustering is tighter for Chicago bills, but the same pattern of bills is experienced within the $\pm 1s$ range. Therefore, the average charge of $125 per month adequately covers the expected cost. However, larger samples may prove this conclusion incorrect.

Solution using a spreadsheet is presented in Figure 14.3. The AVERAGE and STDEV functions are used to determine $\overline{X}$ and s. The results about clustering are the same as discussed above.

14.3 ALTERNATIVE EVALUATION USING SAMPLING AND SIMULATION

Variation and error are always present to some degree in estimation and economic evaluation; these cause risk in making the initial alternative selection and throughout the life of the alternative. Besides the expected value approach discussed in the previous section, other approaches consider risk. However, they all require more time and resources to apply correctly. Given this, only parameters considered of prime importance to the economic decision should be analyzed using techniques that evaluate risk beyond the expected value approach.

The development of a random sample using Monte Carlo sampling is a very common approach to account for variation. With Monte Carlo, it is first necessary to formulate the probability distribution of the parameter(s) chosen for risk analysis. Once random sampling is complete, a simulation performs the economic evaluation using a measure of worth in order to develop a conclusion about economic viability. One important assumption is that *all parameters are considered independent of each other*. In other words, each parameter is a random variable and it is not affected by any other variable in the project.

The sampling and simulation procedure follows these basic steps:

1. **Formulate alternative(s).** Formulate the alternative(s) and associated parameters. Select the measure of worth (PW, AW, ROR, B/C, etc.) and develop the relations for economic evaluation.
2. **Select parameters.** Select the parameters that will vary and are important to the economic decision.
3. **Formulate probability distributions.** Determine the distribution for each parameter. Where possible, use standard, simple discrete or continuous distributions to make the sampling and simulation results faster to obtain and easier to interpret.
4. **Take random sample.** Use a hand or (preferably) spreadsheet-based technique to select a sample of size N for each varying parameter. (Commonly, $N = 50$, 100, 500, 1000, or more trials using Monte Carlo sampling, as illustrated in Example 14.6.)
5. **Calculate measure of worth.** Use parameters estimated with certainty and with variation to calculate N values of the selected measure's relation; one value of the measure for each random sample trial.
6. **Describe measure of worth.** In order to better understand the simulated results, determine the average, standard deviation, and probability statements for the measure of worth.
7. **Conclusions.** Make decisions on the economic viability of the project or select the best mutually exclusive alternative. If the conclusion is not clear, a larger sample size may be helpful or, more likely, additional information about troublesome parameters may be necessary.

A relatively easy spreadsheet-based sampling and simulation is illustrated next. In addition to spreadsheet functions used previously, others are introduced here, including RAND, RANDBETWEEN, VLOOKUP, and COUNTIF. These greatly simplify the building of probability distributions and sampling from them. RAND and RANDBETWEEN are described in Appendix A. If needed, details about VLOOKUP and COUNTIF beyond that explained here are available on the help system of your spreadsheet.

EXAMPLE 14.6

In the past, JAGBA Glass of Memphis has purchased several versions of automated glass scrolling and cutting machines. They have had productive lives varying from 5 to 9 years. Depending on the quality of the machine itself and the economy in Memphis, machine life and annual gross income can vary significantly. Expenses are quite well known based on years of experience. Estimates for the current-technology machine are:

First cost, P = $-50,000$

Expenses, E = $-10,000$ per year

Salvage value, S = $10,000$ whenever it is sold or traded

Life, n = 5 to 9 years, with equal chance for each integer value

GI = $10,000$ to $40,000$ per year, with equal chance for all numbers in between

MARR = 10% per year

The owner, who took engineering economy and kept his textbook after graduating, performed an evaluation at the extremes of n and GI. Because of the results, he determined that they are not sufficient for a reliable economic decision. His general AW relation and computations follow.

$$AW = P(A/P,MARR,n) + GI - E + S(A/F,MARR,n)$$

Low: $n = 5$, GI = $10,000$

$$AW = -50,000(A/P,10\%,5) + 10,000 - 10,000 + 10,000(A/F,10\%,5)$$
$$= \$-11,552$$

High: $n = 9$, GI = $40,000$

$$AW = -50,000(A/P,10\%,9) + 40,000 - 10,000 + 10,000(A/F,10\%,9)$$
$$= \$22,054$$

Using the AW measure of worth, uniform discrete and continuous probability distributions for n and GI, respectively, select a random sample of size 50 to estimate the following and make a recommendation to JAGBA's owner about the purchase of the new machine.

Average AW value, $\overline{AW}$

Standard deviation of AW, s_{AW}

$P(AW \geq 0)$

Solution

The life estimate has a uniform probability distribution with integer-only values from 5 to 9, each with a probability of 0.20. The gross income estimate also varies according to a uniform distribution; however, it is continuous from $10,000$ to $40,000$. The probability distributions for these two variables are similar to those in Figure 14.1a and b, with the axis scales changed to match the estimates here.

Figure 14.4 details the sampling for 10 of 50 n and GI values (columns C and I). The resulting AW for each trial is shown in column J, with the average,

	A	B	C	D	E	F	G	H	I	J	K	L	M
1	SAMPLING FOR LIFE VALUES							GROSS INCOME		EVALUATION			
2	Sample	Random	Sample					Sample	Sample	AW			
3	Trial	Number	n Value		Distribution for life			Trial	GI Value	Value, $		AW statistics	
4	1	0.3525	6		Lower cutoff	n, years		1	20,298	114	Average	$	5211
5	2	0.2037	6		0.0	5		2	19,841	-343	Std dev	$	9161
6	3	0.6085	8		0.2	6		3	23,327	4829	AW > 0		31
7	4	0.0376	5		0.4	7		4	32,785	11,233	AW < 0		19
8	5	0.0248	5		0.6	8		5	30,662	9110			
9	6	0.8518	9		0.8	9		6	10,395	-7551			
10	7	0.3586	6					7	13,648	-6536			
11	8	0.8864	9					8	39,356	21,410			
12	9	0.6651	8					9	33,524	15,026			
13	10	0.9197	9					10	33,981	16,035			

FIGURE 14.4 Sampling and simulation results for AW analysis when life and gross income vary, Example 14.6.

standard deviation, and number of positive and negative AW values shown in column M for all 50 trials. Sample functions for the random sample and AW results are shown below.

Result	Column	Sample Function
RN between 0 and 1 for n	B	= RAND()
Random value of n (trial 1)	C	= VLOOKUP(B4,E$5:F$9,2)
Random value of GI	I	= RANDBETWEEN(10000,40000)
AW computation (trial 1)	J	= −PMT(10%,C4,−50000,10000) + I4 − 10000
Average of 50 AW values	M	= AVERAGE(J4:J53)
Std. dev. of 50 AW values	M	= STDEV(J4:J53)
Number of AW values ≥ 0	M	= COUNTIF(J4:J53,">=0")
Number of AW values < 0	M	= COUNTIF(J4:J53,"<0")

For the life values, RAND() generates a random number (RN) between 0 and 1 in column B, then the VLOOKUP function (column C) uses this RN in the lookup table in columns E and F to obtain a corresponding n value between 5 and 9 years. For GI, the RANDBETWEEN function generates a GI estimate directly in whole dollars. Once these two values are available, the PMT function calculates the AW value (column J). Finally, the requested statistics for the sample of 50 trials are developed *for this simulation* (column M).

$$\overline{AW} = \$5211$$
$$s_{AW} = \$9161$$
$$P(AW \geq 0) = 31/50 = 0.62$$

The good news is that the average AW is positive over 60% of the time, which indicates that the machine should be purchased. However, the standard deviation is very large compared to the average — $9161 compared to $5211 *for this simulation*. If this comparison is similar for other and larger samples, it means that, though AW is generally expected to be positive, a negative variation of only one standard deviation below the average puts AW in negative territory, that is, $5211 - 9161 = \$-3950$.

Comment: It is important to remember that this is only a small sample of 50; larger samples may alter the statistics considerably. Also, realize that for RAND() and other RN generators, the resulting RN values change every time any function on the worksheet is activated. Therefore, results will vary slightly from one simulation to another. The help system on Excel has an elementary, but excellent introduction to spreadsheet-based simulation using Monte Carlo sampling. It is a very powerful tool for engineering economy and other applications.

SUMMARY

When multiple attributes are involved in decision making, noneconomic and economic factors are included. The weighted attribute method is a straightforward approach to take several factors into consideration. It is useful for committee-based decision making, especially when diverse opinions are present and significantly different value systems are represented by committee members.

To peform decision making under risk means that some parameters of a project or alternative are considered random variables, for which probability distributions must be established. Measures such as expected value, standard deviation and probability statements are helpful in understanding how the varying parameters behave and the risk involved.

Monte Carlo sampling is combined with time-value-of-money measures such as PW, AW and ROR to implement a simulation approach to risk analysis. The results of the sampling and simulation can be compared with an analysis made under certainty, thereby improving the economic decision.

PROBLEMS

Multiple Attributes

14.1 Three alternatives are being evaluated based on six different attributes, all of which are considered of equal importance. Determine the weight to assign to each attribute.

14.2 A consultant asked a corporate president to assign importance values from 0 to 100 to five attributes that will be included in an alternative evaluation. Determine the weight of each attribute using the president's importance scores.

Attribute	Importance Score
1. Safety	40
2. Cost	60
3. Impact	70
4. Environmental	30
5. Acceptability	50

14.3 Ten attributes were rank-ordered in terms of increasing importance and were identified as A, B, C,..., and J. Determine the weight of (*a*) attribute D, and (*b*) attribute J.

14.4 Different types and capacities of crawler hoes are being considered for a pipe-laying project across the country. Several supervisors who served on similar projects in the past have identified attributes and their views of relative importance. Determine (a) the importance scores using a 10 (high) to 0 scale, and (b) the normalized weights using the weighted rank order approach.

Attribute	Comment
1 Truck vs. hoe height	90% as important as trenching speed
2 Type of topsoil	Only 10% of most important attribute
3 Type of subsoil	30% as important as trenching speed
4 Hoe cycle time	Twice as important as type of subsoil
5 Hoe trenching speed	Most important attribute
6 Pipe-laying speed	80% as important as hoe cycle time

14.5 The Athlete's Shop owner evaluated two proposals for exercise equipment. A present worth analysis at $i = 15\%$ resulted in $PW_A = \$340,000$ and $PW_B = \$290,000$. Independently, the lead trainer assigned a relative importance score from 0 to 100 to three noneconomic attributes, plus the economic attribute.

Attribute	Trainer's Score
Economics	80
Durability	35
Flexibility	30
Maintainability	50

Separately, you have used the four attributes to value rate the proposals on a scale of 0 to 1.0 as shown in the following table.

Attribute	Value Ratings Proposal A	Proposal B
Economics	1.00	0.90
Durability	0.35	1.00
Flexibility	1.00	0.90
Maintainability	0.25	1.00

Select the better proposal using (a) present worth, and (b) the weighted attribute method.

14.6 A public project proposal evaluated at MARR = 6% per year resulted in AW = $5600 per year and a B/C ratio of 1.7, indicating it is clearly economically justified. The Public Utility Commission (PUC), which must approve the proposal, uses the weighted attribute method with four attributes – economic, safety, environmental, and legal – equally weighted. It is well known that the PUC has never approved a proposal with a weighted attribute score of less than 80, based on averaged value ratings from each of the three PUC members. They value rate a proposal using a scale of 0 to 100. The PUC chair displayed a PowerPoint slide showing the result and rejected the proposal.

$$
\begin{aligned}
R_{prop} &= (\text{economic weight}) \times (\text{PUC rating}_{eco}) \\
&+ (\text{safety weight}) \times (\text{PUC rating}_{saf}) \\
&+ (\text{environmental weight}) \times (\text{PUC rating}_{env}) \\
&+ (\text{legal weight}) \times (\text{PUC rating}_{leg}) \\
&= (0.25) \times (20) + (0.25) \times (40) \\
&+ (0.25) \times (90) + (0.25) \times (80) \\
&= 57.5
\end{aligned}
$$

Since both AW and B/C indicate that approval is warranted, is it possible for the proposal to be approved by increasing the economic weight determined by the PUC members?

Probability and Samples

14.7 Identify a fundamental reason why all decisions in engineering economics are performed under risk.

14.8 For each situation below, determine (a) if the variable is discrete or continuous and (b) if the information involves certainty or risk.
(1) The first cost of a new front-end loader is $34,000 or $38,000 depending on the size purchased.
(2) The raises for engineers and technical staff employees will be 3%, or 5%, with half getting 3% and half getting 5%.
(3) Revenue from a new product line is expected to be between $350,000 and $475,000 per year with a larger chance that it is low in the range.
(4) The salvage value for an old machine will be $500 (the asking price) or $0 (it will be scrapped), depending upon who is selected by the manager to take it.
(5) Profits are expected to be up by anywhere between 25% and 60% this year, with equal probability for all estimates.

14.9 Royalty income received by an investor in an oil well will vary according to the price of oil. Data collected from stripper wells in an established oilfield were used to develop a probability-royalty relationship. Calculate (*a*) the expected value of royalties per year, and (*b*) the probability that the royalties will be at least $12,600 per year.

Royalties R, $/year	6200	8500	9600	10,300	12,600	15,500
Probability, P(R)	0.10	0.21	0.32	0.24	0.09	0.04

14.10 The Car Buyer's Guild surveyed 1000 households to determine the number of operating vehicles that are owned by residents at the address surveyed. Use the results below to determine the percentage of households that own (*a*) 1 or less vehicles; (*b*) 1 or 2 vehicles, and (*c*) more than 3 vehicles.

Number of cars, N	Number of Households
0	120
1	560
2	260
3	32
4	22
5 or more	6

14.11 An engineer was asked to determine whether the average air quality in a volatile chemical mixing room was within OSHA guidelines. The following air quality readings were collected: 81, 86, 80, 91, 83, 83, 96, 85, 89.
  a. Determine the arithmetic mean.
  b. Calculate the standard deviation.
  c. Determine the percent of readings that fall within ± 1 standard deviation from the mean.
  d. Verify your answers using spreadsheet functions.

14.12 Karen has collected monthly operating expenses for the past 3 years for a micro-finishing department. (*a*) She wants to know the probability that a month's expense may be above $50,000. (*b*) Provide Karen with the expected value calculated by hand in two forms—using the number of months and using the probability for each range. Use the midpoints of each range in preparing the answers.

Expense range, $1000	Midpoint, $1000	Number of months
1–10	5	2
10–20	15	5
20–30	25	8
30–40	35	7
40–50	45	6
50–60	55	5
60–70	65	3

14.13 A sample of 100 monthly maintenance costs for automated soldering machines was collected. The costs are clustered into $200 cells with midpoints ranging from $600 to $2000. The number of times (frequency) each cell value was observed and its probability are shown below. (*a*) Find the expected value of the expenses for use in a PW analysis. (*b*) Develop the discrete probability distribution of the monthly maintenance costs similar to Figure 14.1*a* and indicate the expected value on it.

Cell Midpoint	Frequency	Probability
600	6	0.06
800	10	0.10
1000	9	0.09
1200	15	0.15
1400	28	0.28
1600	15	0.15
1800	7	0.07
2000	10	0.10

14.14 A newsstand manager is tracking *Y*, the number of weekly magazines left on the shelf when the new edition is delivered. Data collected over a 30-week period are summarized by a discrete probability distribution. A friend determined that the expected value and standard deviation of these results are $E(Y) = 7.08$ and $s_Y = 3.23$, respectively. Plot the distribution and indicate the expected value and one standard deviation on either side of it.

Copies, Y	3	7	10	12
P(Y)	1/3	1/4	1/3	1/12

Sampling and Simulation

14.15 When using Monte Carlo sampling to obtain random numbers from the probability distributions of varying parameters, what is a fundamental assumption that must be made?

14.16 Use the RAND()*100 spreadsheet function to generate 100 values from a uniform distribution with the range of 0 to 100. Then use other spreadsheet functions to calculate (*a*) the average and compare the sample value to 50, and (*b*) the standard deviation and compare the sample value to 28.87. These two values are correct for a continuous uniform distribution with limits of 0 and 100.

14.17 (*a*) Explain the meaning of a sample standard deviation. (*b*) Use the Excel© help utility to determine the formula used by the STDEV function.

14.18 Use the Excel© help utility to determine what the following functions are designed to do: (*a*) VLOOKUP, and (*b*) RANDBETWEEN.

14.19 Carl, a colleague in Europe working for the same company as you, estimated annual cash flow after taxes (CFAT) for a project he is working on.

Year	CFAT, $
0	−35,000
1–10	5,500

The PW value at the current MARR of 8% per year is:

PW = −35,000 + 5500(*P/A*,8%,10) = $1905

Because of the changing economic scene, Carl believes the MARR will vary over a relatively narrow range, as will the CFAT. He is willing to accept the other estimates as certain. Use the following probability distribution assumptions for MARR and CFAT to perform a simulation. Take a sample of 30 or more simulation trials. Is the project economically justified using decision making under certainty? Under risk?

MARR	Discrete uniform distribution over the range 7% to 10%
CFAT	Continuous uniform distribution over the range $4000 to $7000 for each year

14.20 Janice is a process engineer for Upland Chemicals. Yesterday, she was handed the following information about a piece of air quality sampling equipment that she earlier requested be purchased for her department. The request was denied by Jim, the Corporate Finance Manager based on the large negative PW value that he calculated.

P = $−150,000
GI = $50,000 per year
E = 10% of P or $−15,000 per year
n = 10 years
S = 20% of P or $30,000
PW = $−21,811

Janice was quite aggravated when she determined the interest rate used in the analysis was 25%, well above the published corporate MARR of 10% to justify required product or environmental quality equipment. In a meeting with you, Janice shared some of her own estimates about first cost and expected life. Using the following estimates and distribution assumptions, perform a simulation for Janice that may help her in an effort to obtain approval of the equipment's purchase.

Accepted as certain: GI = $50,000 per year
E = 10% of P
S = 20% of P

Varying: P	Continuous uniform distribution over the range $−100,000 to $−150,000.
n	Discrete uniform distribution over the range 8 to 15 years.
MARR	Discrete uniform distribution; values from 10% through 15% have a 15% chance each; values from 16% to 25% have a 1% chance each.

ADDITIONAL PROBLEMS

14.21 All of the following are excellent attribute identification approaches except:
a. Employing small group discussions
b. Using the same attributes that competing entities use
c. Getting input from experts with relevant experience
d. Surveying the stakeholders

14.22 If attributes of first cost, safety, and environmental concerns had scores of 100, 75, and 50, respectively, the weight for environmental concerns will be closest to:
- **a.** 0.44
- **b.** 0.33
- **c.** 0.22
- **d.** 0.11

14.23 Alternative locations for a new drinking water treatment plant are being evaluated using four attributes identified as A, B, C, and D with weights of 0.4, 0.3, 0.2, and 0.1, respectively. If the value rating scale is from 1 to 10, the evaluation measure of the weighted attribute method for ratings of 3, 7, 2 and 10 for attributes A, B, C, and D, respectively, is closest to:
- **a.** 3.3
- **b.** 3.9
- **c.** 4.1
- **d.** 4.7

14.24 The president of Bullnose Shoes estimated the chances of different levels of earnings for this calendar year. They are:

Earnings	Chance
$260,000	1 in 10
$400,000	6 in 10
$800,000	3 in 10

The expected value of the earnings is closest to:
- **a.** $506,000
- **b.** $493,000
- **c.** $461,000
- **d.** $402,000

14.25 A prime difference between a discrete and continuous random variable is:
- **a.** a discrete variable can take on only integer values.
- **b.** a continuous variable cannot have negative values.
- **c.** a discrete variable can assume only isolated values.
- **d.** in the highest and lowest values that they can assume.

14.26 A fundamental assumption made when obtaining a random sample from the probability distribution of a variable is that:
- **a.** the distribution must be discrete or continuous uniform.
- **b.** once set, the sample size cannot be increased.
- **c.** the sample size must be large enough to estimate the actual value within $\pm 5\%$.
- **d.** no other variable affects the one being sampled.

14.27 You were informed that the highest salary of engineers in your department was $78,000 per year and that this was two standard deviations above the average. If you discover that the standard deviation is $3000, the average salary is closest to:
- **a.** $66,000
- **b.** $72,000
- **c.** $75,000
- **d.** $81,000

14.28 The historic average and standard deviation of the percentage moisture content MC of a product is $E(MC) = 78.5\%$ and $s_{MC} = 10.5\%$, respectively. The results of a sample of size 5 taken over the last 5 hours included the following 4 data points and the sample average: 81, 74, 83, 66, $\overline{X} = 79$. If the guideline is to maintain the product's moisture within 1 standard deviation of the historic average, the number of sample values outside these limits is:
- **a.** 3
- **b.** 2
- **c.** 1
- **d.** none

Using Spreadsheets and Microsoft Excel©

(UPDATED FOR EXCEL 2010)

This appendix explains the layout of a spreadsheet and the use of Microsoft Excel (hereafter called Excel) functions in engineering economy. Since new versions provide improved capabilities, refer to the Excel help system for your particular computer and version of Excel for current detail.

A.1 INTRODUCTION TO USING EXCEL

Enter a Formula or Use a Function

The = sign is necessary to perform any formula or function computation in a cell. The formulas and functions on the worksheet can be displayed by pressing Ctrl and `. The symbol ` is usually in the upper left of the keyboard with the ~ (tilde) symbol. Pressing Ctrl+` a second time hides the formulas and functions. Some examples of entries follow.

1. Run Excel.
2. Move to cell C3.
3. Type = PV(5%,12,10) and <Enter>. This function will calculate the present value of 12 payments of $10 at a 5% per year interest rate.

Another example: To calculate the future value of 12 payments of $10 at 6% per year interest, do the following:

1. Move to cell B3, and type INTEREST.
2. Move to cell C3, and type 6% or = 6/100.
3. Move to cell B4, and type PAYMENT.
4. Move to cell C4, and type 10 (to represent the size of each payment).
5. Move to cell B5, and type NUMBER OF PAYMENTS.
6. Move to cell C5, and type 12 (to represent the number of payments).
7. Move to cell B7, and type FUTURE VALUE.

8. Move to cell C7, and type = FV(C3,C5,C4) and hit <Enter>. The answer will appear in cell C7.

To edit the values in cells,

1. Move to cell C3 and type 5% (the previous value will be replaced).
2. The value in cell C7 will update.

Cell References in Formulas and Functions

If a cell reference is used in lieu of a specific number, it is possible to change the number once and perform sensitivity analysis on any variable (entry) that is referenced by the cell number, such as C5. This approach defines the referenced cell as a *global variable* for the worksheet. There are two types of cell references—relative and absolute.

Relative References

If a cell reference is entered, for example, A1, into a formula or function that is copied or dragged into another cell, the reference is changed relative to the movement of the original cell. If the formula in C5 is = A1, and it is copied into cell C6, the formula is changed to = A2. This feature is used when dragging a function through several cells, and the source entries must change with the column or row.

Absolute References

If adjusting cell references is not desired, place a $ sign in front of the part of the cell reference that is not to be adjusted—the column, row, or both. For example, = A1 will retain the formula when it is moved anywhere on the worksheet. Similarly, = $A1 will retain the column A, but the relative reference on 1 will adjust the row number upon movement around the worksheet.

Absolute references are used in engineering economy for sensitivity analysis of parameters such as MARR, first cost, and annual cash flows. In these cases, a change in the absolute-reference cell entry can help determine the sensitivity of a measure such as PW or AW.

Print the Spreadsheet

First define the portion (or all) of the spreadsheet to be printed.

1. Move the mouse pointer to the top left corner of your spreadsheet.
2. Hold down the left click button. (Do not release the left click button.)
3. Drag the pointer to the lower right corner of your spreadsheet or to wherever you want to stop printing.
4. Release the left click button. (It is ready to print.)
5. Left-click the File tab in the upper left of the screen.
6. Move the pointer down to select Print. The worksheet image will appear with choices of what to print.
7. In the Settings area, left-click on the option Print Selection, then left-click the option Print.
8. Left-click the Print button to start printing.

Create an (*x-y*) Scatter Chart

There is a large variety of chart types available in Excel – scatter, column, pie, bar, area, radar, and more. The (*x-y*) scatter chart is one of the most commonly used in scientific analysis, including engineering economy. It plots pairs of data and can place multiple series of entries on the *y* axis. The *x-y* scatter chart is especially useful for results such as breakeven analysis and the PW vs. *i* graph that displays the results of the NPV function for different interest rates.

1. Run Excel.
2. Enter the following in columns A, B, and C cells, respectively.
   Column A, A1 through A7: Year, 0, 1, 2, 3, 4, 5
   Column B, B1 through B7: Project 1, −50, 40, 30, 20, 10, 5
   Column C, C1 through C7: Project 2, −50, 5, 10, 20, 30, 40
3. Move the pointer to A1, left click, and hold while dragging to cell C7. All cells will be highlighted, including the title cell for each column. (Note that for the chart legend to identify the data series correctly, only the bottom row of a title should be highlighted. If no title is highlighted, the data series are identified as Series 1, Series 2, etc.)
4. If all the columns for the chart are not adjacent to one another, press and hold the Control key on the keyboard *after* dragging over the first (leftmost) column of data. Momentarily release the left click, then move to the top of the next (nonadjacent) column for the chart. Do not release the Control key until all columns to be plotted have been highlighted.
5. Left-click on the Insert button on the upper toolbar.
6. Select the Scatter option and choose a subtype of (*x-y*) scatter chart.

Once selected, the initial chart will appear to the right of the spreadsheet data in a fashion similar to Figure A.1. Most elements of the chart can be altered; these include chart area, plot area, data series, gridlines, titles, legend, axis, and others. There are two ways to begin to enhance the chart's appearance. First, by selecting an item from the Chart Tools-Layout options shown at the top of the screen;

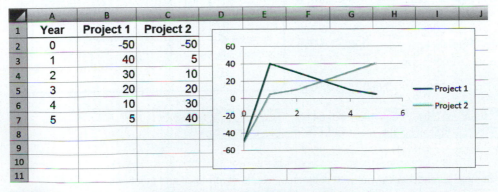

FIGURE A.1   Initial view of (*x-y*) scatter chart for cash flows of two projects.

second, by *right* clicking on the area to be changed. View the menu for each element by *left* clicking on the bottom entry entitled Format.

A.2 ORGANIZATION (LAYOUT) OF THE SPREADSHEET

A spreadsheet can be used in several ways to obtain answers to numerical questions. The first is as a rapid solution tool, often with the entry of only a few numbers or one predefined function. For example, to find the future worth in a single cell operation, move the pointer to cell B4 and type = FV(8%,5,−2500). The number $14,666.50 is displayed as the 8% per year future worth at the end of the fifth year of five payments of $2500 each. A second use is more formal; it may present data, answers, graphs and tables that identify what problem(s) the spreadsheet solves. Some fundamental guidelines useful in setting up the spreadsheet follow. A very simple layout is presented in Figure A.2. As the solutions become more complex, an orderly arrangement of information makes the spreadsheet easier to read and use by you and others.

Cluster the data and the answers. It is advisable to organize the given or estimated data in the top left of the spreadsheet. A very brief label should be used to identify the data, for example, MARR = in cell A1 and the value, 12%, in cell B1. Then B1 can be the referenced cell for all entries requiring the MARR. Additionally, it may be worthwhile to cluster the answers into one area and frame it using the Outside Border button on the toolbar. Often, the answers are best placed at the bottom or top of the column of entries used in the formula or predefined function.

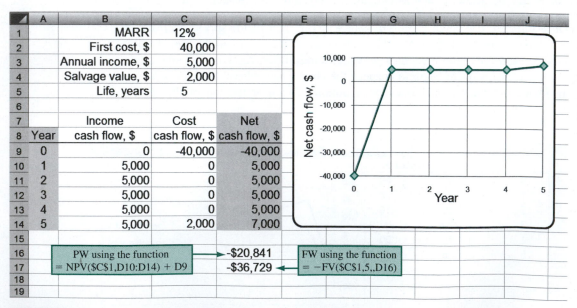

FIGURE A.2   Sample spreadsheet layout with estimates, results of formulas and functions, and an *xy* scatter chart.

Enter titles for columns and rows. Each column or row should be labeled so its entries are clear to the reader. It is very easy to select from the wrong column or row when no brief title is present.

Enter income and cost cash flows separately. When there are both income and cost cash flows involved, it is strongly recommended that the cash flow estimates for revenue (usually positive) and first cost, salvage value, and annual costs (usually negative, with salvage a positive number) be entered into two adjacent columns. Then a formula combining them in a third column displays the net cash flow. There are two immediate advantages to this practice: this reduces errors made when performing the summation and subtraction mentally, and changes for sensitivity analysis are more easily made.

Use cell references. The use of absolute and relative cell references is a must when any changes in entries are expected. For example, suppose the MARR is entered in cell C1, and three separate references are made to the MARR in functions on the spreadsheet. The absolute cell reference entry C1 in the three functions allows the MARR to be changed one time, not three.

Obtain a final answer through summing and embedding. When the formulas and functions are kept relatively simple, the final answer can be obtained using the SUM function. For example, if the present worth values (PW) of two columns of cash flows are determined separately, then the total PW is the SUM of the subtotals. This practice is especially useful when the cash flow series are complex.

Prepare for a chart. If a chart (graph) will be developed, plan ahead by leaving sufficient room on the right of the data and answers. Using the Move Chart-Location button on the upper right, the graph can be located on the same worksheet or a new one. Leaving the graph on the same worksheet is recommended, especially when the results of sensitivity analysis are plotted.

A.3 EXCEL FUNCTIONS USEFUL IN ENGINEERING ECONOMY (alphabetical order)

DB (Declining Balance)

Calculates the depreciation amount for an asset for a specified period n using the declining balance method. The depreciation rate d used in the computation is determined from asset values S (salvage value) and B (basis or first cost) as $d = 1 - (S/B)^{1/n}$. Three-decimal-place accuracy is used for d.

= **DB(cost, salvage, life, period, month)**

cost   First cost or basis of the asset.
salvage   Salvage value.
life   Recovery period.
period   The period, year, for which the depreciation is to be calculated.
month   (optional entry) If this entry is omitted, a full year is assumed for the first year.

Example   A new machine costs $100,000 and is expected to last 10 years. At the end of 10 years, the salvage value of the machine is $50,000. What is the depreciation of the machine in the first year and the fifth year?

$$\text{Depreciation for the first year:} = \text{DB}(100000,50000,10,1)$$
$$\text{Depreciation for the fifth year:} = \text{DB}(100000,50000,10,5)$$

DDB (Double Declining Balance)

Calculates the depreciation of an asset for a specified period n using the double declining balance method. A factor can also be entered for some other declining balance depreciation method by specifying a factor value in the function.

= DDB(cost, salvage, life, period, factor)

cost	First cost or basis of the asset.
salvage	Salvage value of the asset.
life	Recovery period.
period	The period, year, for which the depreciation is to be calculated.
factor	(optional entry) If this entry is omitted, the function will use a double declining method with 2 times the straight line rate. If, for example, the entry is 1.5, the 150% declining balance method will be used.

Example   A new machine costs $200,000 and is expected to last 10 years. The salvage value is $10,000. Calculate the depreciation of the machine for the first and the eighth years. Finally, calculate the depreciation for the fifth year using the 175% declining balance method.

Depreciation for the first year: = DDB(200000,10000,10,1)

Depreciation for the eighth year: = DDB(200000,10000,10,8)

Depreciation for the fifth year using 175% DB: = DDB(200000,10000,10,5,1.75)

Because of the manner in which the DB function determines the fixed percentage d and the accuracy of the computations, it is recommended that the DDB function be used for all declining balance depreciation rates. Simply use the optional factor entry for rates other than $d = 2/n$.

EFFECT (Effective Interest Rate)

Calculates the effective annual interest rate for a stated nominal annual rate and a given number of compounding periods per year. Excel uses Equation [3.2] to calculate the effective rate.

= EFFECT(nominal, npery)

nominal	Nominal interest rate for the year.
npery	Number of times that interest is compounded per year.

Example   Claude has applied for a $10,000 loan. The bank office told him the interest rate is 8% per year and that interest is compounded monthly to conveniently match his monthly payments. What effective annual rate will Claude pay?

Effective annual rate: = EFFECT(8%,12)

EFFECT can also be used to find effective rates other than annually. Enter the nominal rate for the time period of the required effective rate; npery is the number of times compounding occurs during the time period of the effective rate.

Example   Interest is stated as 3.5% per quarter with quarterly compounding. Find the effective semiannual rate.

The 6-month nominal rate is 7% and compounding is 2 times per 6 months.

Effective semiannual rate: = EFFECT(7%,2)

FV (Future Value)

Calculates the future value (worth) based on periodic payments at a specific interest rate.

= FV(rate, nper, pmt, pv, type)

rate	Interest rate per compounding period.
nper	Number of compounding periods.
pmt	Constant payment amount.
pv	The present value amount. If pv is not specified, the function will assume it to be 0.
type	(optional entry) Either 0 or 1. A 0 represents payments made at the end of the period, and 1 represents payments at the beginning of the period. If omitted, 0 is assumed.

Example   Jack wants to start a savings account that can be increased as desired. He will deposit $12,000 to start the account and plans to add $500 to the account at the beginning of each month for the next 24 months. The bank pays 0.25% per month. How much will be in Jack's account at the end of 24 months?

Future value in 24 months: = FV(0.25%,24,500,12000,1)

IF (IF Logical Function)

Determines which of two entries are entered into a cell based on the outcome of a logical check on the outcome of another cell. The logical test can be a function or a simple value check, but it must use an equality or inequality sense. If the response is a text string, place it between quote marks (" "). The responses can themselves be IF functions. Up to seven IF functions can be nested for very complex logical tests.

= IF(logical_test,value_if_true,value_if_false)

logical_test	any worksheet function can be used here, including a mathematical operation.

value_if_true   result if the logical_test argument is true.

value_if_false   result if the logical_test argument is false.

Example   The entry in cell B4 should be "selected" if the PW value in cell B3 is greater than or equal to zero, and "rejected" if PW < 0.

Entry in cell B4: = IF(B3 >=0,"selected", "rejected")

Example   The entry in cell C5 should be "selected" if the PW value in cell C4 is greater than or equal to zero, "rejected" if PW < 0, and "fantastic" if PW ≥ 200.

Entry in cell C5: = IF(C4<0, "rejected",IF(C4>=200,"fantastic", "selected"))

IRR (Internal Rate of Return)

Calculates the internal rate of return between −100% and infinity for a series of cash flows at regular periods.

= IRR(values, guess)

values   A set of numbers in a spreadsheet column (or row) for which the rate of return will be calculated. The set of numbers must consist of at least *one* positive and *one* negative number.

guess   (optional entry) To reduce the number of iterations, a *guessed rate of return* can be entered. In most cases, a guess is not required, and a 10% rate of return is initially assumed. If the #NUM! error appears, try using different values for guess. Inputting different guess values makes it possible to determine the multiple roots for the rate of return equation of a nonconventional cash flow series.

Example   John wants to start a printing business. He will need $25,000 in capital and anticipates that the business will generate the following incomes during the first 5 years. Calculate his rate of return.

Year 1	$5,000
Year 2	$7,500
Year 3	$8,000
Year 4	$10,000
Year 5	$15,000

Set up an array in the spreadsheet.

In cell A1, type −25000 (negative for payment).

In cell A2, type 5000 (positive for income).

In cell A3, type 7500.

In cell A4, type 8000.

In cell A5, type 10000.

In cell A6, type 15000.

Note that any years with a zero cash flow must have a zero entered to ensure that the year value is correctly maintained for computation purposes.

To calculate the internal rate of return after 5 years and specify a guess value of 5%, type = IRR(A1:A6,5%).

MIRR (Modified Internal Rate of Return)

Calculates the modified internal rate of return for a series of cash flows with a given reinvestment rate on incomes and stated interest (finance) rate on borrowed funds.

= MIRR(values, finance_rate, reinvest_rate)

values	Refers to an array in the spreadsheet. The series must occur at regular periods and must contain at least *one* positive number and *one* negative number.
finance_rate	Interest rate on borrowed money for years with negative cash flows.
reinvest_rate	Rate of return for reinvestment on positive cash flows.

Example Jane opened a hobby store 4 years ago. When she started the business, Jane borrowed $50,000 from a bank at 6% per year interest. If additional loans are necessary, the same 6% rate will apply. Since then, the business has yielded $10,000 the first year, $15,000 the second year, $18,000 the third year, and $21,000 the fourth year. Jane reinvests her profits, earning 8% per year. What is the modified rate of return after 4 years?

In cells A1 through A5, type −50000, 10000, 15000, 18000, and 21000.

Modified rate of return after 4 years: = MIRR(A1:A5,6%,8%).

NOMINAL (Nominal Interest Rate)

Calculates the nominal *annual* interest rate for a stated effective annual rate and a given number of compounding periods per year.

= NOMINAL(effective, npery)

effective	Effective interest rate for the year.
npery	Number of times interest is compounded per year.

Example Last year, a corporate stock earned an effective return of 12.55% per year. Calculate the nominal annual rate, if interest is compounded quarterly and if interest is compounded continuously.

Nominal annual rate, quarterly compounding: = NOMINAL(12.55%,4)

Nominal annual rate, continuous compounding: = NOMINAL(12.55%,100000)

NPER (Number of Periods)

Calculates the number of periods for the present worth of an investment to equal the future value specified, based on uniform regular payments and a stated interest rate.

= NPER(rate, pmt, pv, fv, type)

rate — Interest rate per compounding period.

pmt — Amount paid during each compounding period.

pv — Present value (lump-sum amount).

fv — (optional entry) Future value or cash balance after the last payment. If fv is omitted, the function will assume a value of 0.

type — (optional entry) Enter 0 if payments are due at the end of the compounding period, and 1 if payments are due at the beginning of the period. If omitted, 0 is assumed.

Example Sally plans to open a savings account which pays 0.25% per month. Her initial deposit is $3000, and she plans to deposit $250 at the beginning of every month. How many payments does she have to make to accumulate $15,000 to buy a used car?

Number of payments: $= \text{NPER}(0.25\%, -250, -3000, 15000, 1)$

NPV (Net Present Value)

Calculates the net present value, that is, the PW, of a series of future cash flows at a stated interest rate.

= NPV(rate, series)

rate — Interest rate per compounding period.

series — Series of costs and incomes set up in a range of cells in the spreadsheet. Any cash flow in year 0 (now) is not included in the series entry, since it is already a present value.

Example Mark is considering buying a sports franchise for $100,000 and expects to receive the following income during the next 6 years of business: $25,000, $40,000, $42,000, $44,000, $48,000, $50,000. The interest rate is 8% per year.

In cells A1 through A7, enter −100000, followed by the six estimated annual incomes.

Present worth: $= \text{NPV}(8\%, \text{A2:A7}) + \text{A1}$

PMT (Payments)

Calculates equivalent periodic amounts based on present value and/or future value at a stated interest rate.

= PMT(rate, nper, pv, fv, type)

rate — Interest rate per compounding period.

nper — Total number of periods.

pv — Present value.

fv   Future value.

type   (optional entry) Enter 0 for payments due at the end of the compounding period, and 1 if payment is due at the start of the compounding period. If omitted, 0 is assumed.

Example   Jim plans to take a $15,000 loan to help him buy a new car. The interest rate is 7%. He wants to pay the loan off in 5 years (60 months). What are his monthly payments?

$$\text{Monthly payments:} = PMT(7\%/12,60,15000)$$

PV (Present Value)

Calculates the present value, that is, the PW value of a future series of equal cash flows and a single lump sum in the last period at a constant interest rate.

    = PV(rate, nper, pmt, fv, type)

rate   Interest rate per compounding period.

nper   Total number of periods.

pmt   Cash flow at regular intervals. Negative numbers represent payments (cash outflows), and positive numbers represent income.

fv   Future value or cash balance at the end of the last period.

type   (optional entry) Enter 0 if payments are due at the end of the compounding period, and 1 if payments are due at the start of each compounding period. If omitted, 0 is assumed.

There are two primary differences between the PV function and the NPV function: PV allows for end or beginning of period cash flows, and PV requires that all amounts have the same value, whereas they may vary for the NPV function.

Example   Jose is considering leasing a car for $300 a month for 3 years (36 months). After the 36-month lease, he can purchase the car for $12,000. Using an interest rate of 8% per year, find the present worth of this option.

$$\text{Present worth:} = PV(8\%/12,36,-300,-12000)$$

Note the minus signs on the pmt and fv amounts.

RAND and RANDBETWEEN (Random Numbers)

Returns evenly distributed numbers that are (1) ≥ 0 and < 1; (2) ≥ 0 and < 100; or (3) between two specified numbers.

= RAND()	**for range 0 to 1**
= RAND()*100	**for range 0 to 100**
= RAND()*(b-a) + a	**for range a to b (decimals)**
= RANDBETWEEN(a,b)	**for range a to b (integers only)**

a = minimum integer to be generated
b = maximum integer to be generated

Example   Abby needs random numbers between 5 and 15. What spreadsheet functions are available to her? Here a = 5 and b = 15. There are two functions available.

Random number: = RAND()*10 + 5   (includes decimal numbers)
                  = RANDBETWEN(5,15)   (integer values only)

RATE (Interest Rate)

Calculates the interest rate per compounding period for a series of equal cash flows.

= RATE(nper, pmt, pv, fv, type, guess)

nper   Total number of periods.

pmt   Payment amount made each compounding period.

pv   Present value.

fv   Future value (not including the pmt amount).

type   (optional entry) Enter 0 for payments due at the end of the compounding period, and 1 if payments are due at the start of each compounding period. If omitted, 0 is assumed.

guess   (optional entry) To minimize computing time, include a guessed interest rate. If a value of guess is not specified, the function will assume a rate of 10%. This function usually converges to a solution, if the rate is between 0% and 100%.

Example   Mary wants to start a savings account at a bank. She will make an initial deposit of $1000 to open the account and plans to deposit $100 at the beginning of each month. She plans to do this for the next 3 years (36 months). At the end of 3 years, she wants to have at least $5000. What is the minimum interest required to achieve this result?

Interest rate: = RATE(36,−100,−1000,5000,1)

SLN (Straight Line Depreciation)

Calculates the straight line depreciation of an asset for a given year.

= SLN(cost, salvage, life)

cost   First cost or basis of the asset.

salvage   Salvage value.

life   Recovery period.

Example   Marisco, Inc., purchased a printing machine for $100,000. The machine has a life of 8 years and an estimated salvage value of $15,000. What is the depreciation each year?

Depreciation: = SLN(100000,15000,8)

VDB (Variable Declining Balance)

Calculates the depreciation using the declining balance method with a switch to straight line depreciation in the year in which straight line has a larger depreciation amount. This function automatically implements the switch from DB to SL depreciation, unless specifically instructed to not switch.

   $= \text{VDB (cost, salvage, life, start\_period, end\_period, factor, no\_switch)}$

cost	First cost of the asset.
salvage	Salvage value.
life	Recovery period.
start_period	First period for depreciation to be calculated.
end_period	Last period for depreciation to be calculated.
factor	(optional entry) If omitted, the function will use the double declining rate of $2/n$.
no_switch	(optional entry) If omitted or entered as FALSE, the function will switch from declining balance to straight line depreciation when the latter is greater than DB depreciation. If entered as TRUE, the function will not switch to SL depreciation at any time.

Example   Newly purchased equipment with a first cost of $300,000 has a depreciable life of 10 years with no salvage value. Calculate the 175% declining balance depreciation for the first year and the ninth year if switching to SL depreciation is acceptable, and if switching is not permitted.

Depreciation for first year, with switching: $= \text{VDB}(300000,0,10,0,1,1.75)$
Depreciation for ninth year, with switching: $= \text{VDB}(300000,0,10,8,9,1.75)$
Depreciation for first year, no switching: $= \text{VDB}(300000,0,10,0,1,1.75,\text{TRUE})$
Depreciation for ninth year, no switching: $= \text{VDB}(300000,0,10,8,9,1.75,\text{TRUE})$

VDB (for MACRS Depreciation)

The VDB function can be adapted to generate annual MACRS depreciation amounts by replacing the start_period and end_period with the MAX and MIN functions, respectively. As above, the factor option should be entered if other than DDB rates start the MACRS depreciation. The VDB format is

   $= \text{VDB(cost, 0, life, MAX(0,t}-1.5), \text{MIN(life,t}-0.5), \text{factor)}$

Example   Determine the MACRS depreciation for year 4 for a $350,000 asset that has a 20% salvage and a MACRS recovery period of 3 years. $D_4 = \$25,926$ is the display.

   For year 4: $= \text{VDB}(350000,0,3,\text{MAX}(0,4-1.5),\text{MIN}(3,4-.5),2)$

As a second example, if the MACRS recovery period is $n = 15$ years, $D_{16} = \$10,334$.

For year 16: $= \text{VDB}(350000,0,15,\text{MAX}(0,16-1.5),\text{MIN}(15,16-0.5),1.5)$

The optional factor 1.5 is required here, since MACRS starts with 150% DB for $n = 15$ and 20 year recoveries.

A.4 GOAL SEEK—A SPREADSHEET TOOL FOR BREAKEVEN AND SENSITIVITY ANALYSES

GOAL SEEK is found on the Excel toolbar labeled Data, followed by What-if Analysis. It changes the value in a specific cell (called the set or target cell) based on a numerical value for another (changing) cell as input by the user. It is a good tool for sensitivity analysis, 'what-if' analysis, as well as breakeven. The initial GOAL SEEK template is pictured in Figure A.3. One of the cells (set or changing cell) must contain an equation or function that uses the other cell to determine a numeric value. Only a single cell can be identified as the changing cell; however, this limitation can be avoided by using equations rather than specific numerical inputs in all additional cells to be changed.

Example   A proposed asset will cost $25,000, generate an annual cash flow of $6000 over its 5-year life, and then have an estimated $500 salvage value. The rate of return using the IRR function is 6.94%. Determine the annual cash flow necessary to raise the return to 12% per year.

Figure A.4 (left side) shows the cash flows and return displayed using the function $= \text{IRR(B4:B9)}$ prior to the use of GOAL SEEK. Note that the initial $6000 is input in cell B5, but other years' cash flows are input as equations that equate them to B5. The $500 salvage is added for the last year. This format allows GOAL SEEK to change only cell B5 while forcing the other cash flows to the same value. The tool finds the required cash flow of $6857 to obtain the 12% per year return. The GOAL SEEK Status inset indicates that a solution is found. Clicking OK saves all changed cells; clicking Cancel returns to the original values.

More complicated analysis can be performed using the SOLVER tool. For example, SOLVER can handle multiple changing cells. Additionally, equality and inequality constraints can be developed. See the Excel Help system to utilize this

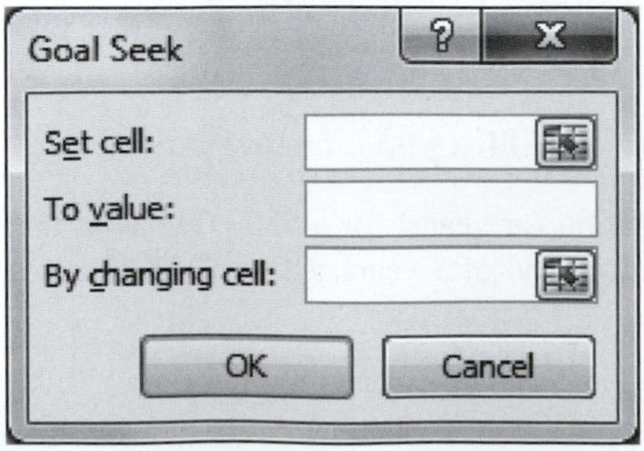

FIGURE A.3
GOAL SEEK template used to specify a 'set' cell and value and the 'changing' cell.

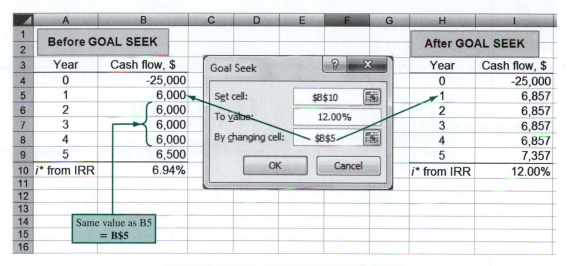

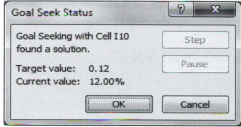

FIGURE A.4 Use of GOAL SEEK tool to determine an annual cash flow to increase the rate of return from 6.94% to 12%.

more powerful tool. If SOLVER is not shown on the Tools toolbar, click Add-ins and install the SOLVER Add-In.

A.5 ERROR MESSAGES

If Excel is unable to complete a formula or function computation, an error message is displayed.

#DIV/0!	Requires division by zero.
#N/A	Refers to a value that is not available.
#NAME?	Uses a name that Excel doesn't recognize.
#NULL!	Specifies an invalid intersection of two areas.
#NUM!	Uses a number incorrectly.
#REF!	Refers to a cell that is not valid.
#VALUE!	Uses an invalid argument or operand.
#####	Produces a result, or includes a constant numeric value, that is too long to fit in the cell. (Widen the column.)

Accounting Reports and Business Ratios

This appendix provides a fundamental description of financial statements. The documents discussed here will assist in reviewing or understanding basic financial statements and in gathering information useful in an engineering economy study.

B.1 THE BALANCE SHEET

The fiscal year and the tax year are defined identically for a corporation or an individual—12 months in length. The corporate fiscal year (FY) is commonly not the calendar year (CY). The U.S. government uses October through September as its FY. For example, October 2014 through September 2015 is FY2015. The fiscal or tax year is always the calendar year for an individual.

At the end of each fiscal year, a company publishes a *balance sheet*. A sample balance sheet for Delta Corporation is presented in Table B.1. This is a yearly presentation of the state of the firm at a particular time, for example, December 31, 2015; however, a balance sheet is also usually prepared quarterly and monthly. There are three main categories.

Assets. This section is a summary of all resources owned by or owed to the company. *Current assets* represent shorter-lived working capital (cash, accounts receivable, etc.), which is more easily converted to cash, usually within 1 year. Longer-lived assets are referred to as *fixed assets* (land, equipment, etc.). Conversion of these holdings to cash in a short period of time would require a major corporate reorientation.

Liabilities. This section is a summary of all financial obligations (debts, bonds, mortgages, loans, etc.) of a corporation.

Net worth. Also called *owner's equity,* this section provides a summary of the financial value of ownership, including stocks issued and earnings retained by the corporation.

The balance sheet is constructed using the relation

$$\text{Assets} = \text{liabilities} + \text{net worth}$$

TABLE B.1 **Sample Balance Sheet**

DELTA CORPORATION Balance Sheet December 31, 2015			
Assets		**Liabilities**	
Current			
Cash	$10,500	Accounts payable	$19,700
Accounts receivable	18,700	Dividends payable	7,000
Interest accrued receivable	500	Long-term notes payable	16,000
Inventories	52,000	Bonds payable	20,000
Total current assets	$81,700	Total liabilities	$62,700
Fixed		**Net Worth**	
Land	$25,000	Common stock	$275,000
Building and equipment	438,000	Preferred stock	100,000
Less: Depreciation		Retained earnings	25,000
allowance $82,000	356,000		
Total fixed assets	381,000	Total net worth	400,000
Total assets	$462,700	Total liabilities and net worth	$462,700

In Table B.1 each major category is further divided into standard subcategories. For example, current assets is comprised of cash, accounts receivable, etc. Each subdivision has a specific interpretation, say, accounts receivable, which represents all money owed to the company by its customers.

B.2 INCOME STATEMENT AND COST OF GOODS SOLD STATEMENT

A second important financial statement is the *income statement* (Table B.2), which summarizes the profits or losses for a stated period of time. Income statements always accompany balance sheets. The major categories of an income statement are

Revenues. This includes all sales and revenues that the company has received in the past accounting period.

Expenses. This is a summary of all expenses for the period. Some expense amounts are detailed in other statements, for example, cost of goods sold.

The income statement, published at the same time as the balance sheet, uses the basic equation

$$\text{Revenues} - \text{expenses} = \text{profit (or loss)}$$

The *cost of goods sold* is an important accounting term for a manufacturing company. It represents the net cost of producing the products marketed by the firm.

TABLE B.2 **Sample Income Statement**

DELTA CORPORATION Income Statement Year Ended December 31, 2015		
Revenues		
Sales	$505,000	
Interest revenue	3,500	
Total revenues		$508,500
Expenses		
Cost of goods sold (from Table B.3)	$290,000	
Selling	28,000	
Administrative	35,000	
Other	12,000	
Total expenses		365,000
Income before taxes		143,500
Taxes for year		64,575
Net profit for year		**$ 78,925**

TABLE B.3 **Sample Cost of Goods Sold Statement**

DELTA CORPORATION Statement of Cost of Goods Sold Year Ended December 31, 2015		
Materials		
Inventory, January 1, 2015	$ 54,000	
Purchases during year	174,500	
Total	$228,500	
Less: Inventory December 31, 2015	50,000	
Cost of materials		$178,500
Direct labor		110,000
Prime cost		288,500
Indirect costs		7,000
Factory cost		295,500
Less: Increase in finished goods inventory during year		5,500
Cost of goods sold (into Table B.2)		$290,000

Cost of goods sold may also be called *factory cost*. Note that the total of the cost of goods sold statement in Table B.3 is entered as an expense item on the income statement. This total is determined using the relations

$$\text{Cost of goods sold} = \text{prime cost} + \text{indirect cost}$$
$$\text{Prime cost} = \text{direct materials} + \text{direct labor}$$

[B.1]

Indirect costs include all indirect and overhead charges made to a product, process, service, or cost center. Indirect cost allocation is discussed in Chapter 11.

B.3 BUSINESS RATIOS

Accountants, financial analysts, and engineering economists frequently utilize business ratio analysis to evaluate the financial health of a company over time and in relation to industry norms. Because the engineering economist must continually communicate with others, it is helpful to have a basic understanding of several ratios. For comparison purposes, it is necessary to compute the ratios for several companies in the same industry. Industrywide median ratio values are published annually by firms such as Dun and Bradstreet in *Industry Norms and Key Business Ratios*. The ratios are classified according to their role in measuring different dimensions of the corporation.

Solvency ratios. Assess ability to meet short-term and long-term financial obligations.

Efficiency ratios. Measure management's ability to use and control assets.

Profitability ratios. Evaluate the ability to earn a return for the owners of the corporation.

Numerical data for several important ratios are discussed here and are extracted from the Delta balance sheet and income statement, Tables B.1 and B.2.

Current Ratio

This ratio is utilized to analyze the company's working capital condition.

$$\text{Current ratio} = \frac{\text{current assets}}{\text{current liabilities}}$$

Current liabilities include all short-term debts, such as accounts and dividends payable. Note that only balance sheet data are utilized in the current ratio; that is, no association with revenues or expenses is made. For the balance sheet of Table B.1, current liabilities amount to $19,700 + $7000 = $26,700 and

$$\text{Current ratio} = \frac{81,700}{26,700} = 3.06$$

Since current liabilities are those debts payable in the next year, the current ratio value of 3.06 means that the current assets would cover short-term debts approximately 3 times. Current ratio values of 2 to 3 usually indicate that corporate solvency is acceptable.

The current ratio assumes that the working capital invested in inventory can be converted to cash quite rapidly. Often, however, a better idea of a company's *immediate* financial position can be obtained by using the acid test ratio.

Acid Test Ratio (Quick Ratio)

$$\text{Acid test ratio} = \frac{\text{quick assets}}{\text{current liabilities}} = \frac{\text{current assets} - \text{inventories}}{\text{current liabilities}}$$

It is meaningful for the emergency situation when the firm must cover short-term debts using its readily convertible assets. For Delta Corporation,

$$\text{Acid test ratio} = \frac{81,700 - 52,000}{26,700} = 1.11$$

Comparison of this and the current ratio shows that approximately 2 times the current debts of the company are invested in inventories. However, an acid test ratio of approximately 1.0 is generally regarded as a strong current position, regardless of the amount of assets in inventories.

Debt Ratio

This ratio is a measure of financial strength.

$$\text{Debt ratio} = \frac{\text{total liabilities}}{\text{total assets}}$$

For Delta Corporation,

$$\text{Debt ratio} = \frac{62,700}{462,700} = 0.135$$

Delta is 13.5% creditor-owned and 86.5% stockholder-owned. A debt ratio in the range of 20% or less usually indicates a sound financial condition, with little fear of forced reorganization because of unpaid liabilities. However, a company with virtually no debts is inexperienced in dealing with short-term and long-term debt financing. The debt-equity (D-E) mix, discussed in Chapter 13, is another measure of financial strength.

Return on Sales Ratio

This often quoted ratio indicates the profit margin for the company.

$$\text{Return on sales} = \frac{\text{net profit}}{\text{net sales}} (100\%)$$

Net profit is the after-tax value from the income statement. This ratio measures profit earned per sales dollar and indicates how well the corporation can sustain adverse conditions over time, such as falling prices, rising costs, and declining sales. For Delta Corporation,

$$\text{Return on sales} = \frac{78,925}{505,000} (100\%) = 15.6\%$$

Corporations may point to small return on sales ratios, say, 2.5% to 4.0%, as indications of sagging economic conditions. In truth, for a relatively large-volume, high-turnover business, an income ratio of 3% is quite healthy.

Return on Assets Ratio

This is the key indicator of profitability since it evaluates the ability of the corporation to transfer assets into operating profit. The definition and value for Delta are

$$\text{Return on assets} = \frac{\text{net profit}}{\text{total assets}}(100\%)$$

$$= \frac{78,925}{462,700}(100\%) = 17.1\%$$

Efficient use of assets indicates that the company should earn a high return, while low returns usually accompany lower values of this ratio compared to the industry group norms.

Inventory Turnover Ratio

This ratio indicates the number of times the average inventory value passes through the operations of the company.

$$\text{Net sales to inventory} = \frac{\text{net sales}}{\text{average inventory}}$$

where average inventory is the figure recorded in the balance sheet. For Delta Corporation this ratio is

$$\text{Net sales to inventory} = \frac{505,000}{52,000} = 9.71$$

This means that the average value of the inventory has been sold 9.71 times during the year. Values of this ratio vary greatly from one industry to another.

EXAMPLE B.1

Sample values for financial ratios or percentages for different industry sectors are presented below. Compare the corresponding Delta Corporation values with these norms.

Ratio or Percentage	Motor Vehicles and Auto Parts 336105*	Air Transportation 481000	Industrial Machinery Manufacturing 333200	Home Furnishings 442000
Current ratio	1.3	1.7	2.1	2.0
Quick ratio	0.4	0.7	1.2	0.7
Inventory turnover ratio	5.4	240.9	5.4	3.0
Debt ratio	90.3%	87.7%	65.8%	74.3%

*North American Industry Classification System (NAICS) code for this industry sector.
Source: L. Troy, *Almanac of Business and Industrial Financial Ratios,* 43d annual edition, CCH, Riverwoods, IL, 2012.

Solution

It is not correct to compare business ratios for a company in one industry with norms from other categories, that is, different industry sectors. Let's assume that Delta is a manufacturer of industrial machinery (NAICS code 333200). The values for Delta are:

$$\text{Current ratio} = 3.06$$
$$\text{Quick ratio} = 1.11$$
$$\text{Inventory turnover ratio} = 9.71$$
$$\text{Debt ratio} = 13.5\%$$

Comparing these results with the norms in the table for sector 333200, Delta has a significantly larger current ratio than the norm; 3.06 versus 2.1. Delta can cover current liabilities 3 times compared to the industry average of slightly over 2 times. Delta turns its average inventory value through the company's operations about 2 times as often as the industry norm; 9.71 versus 5.4. Finally, we can determine that Delta is a financially strong corporation as its debt ratio is only 13.5% compared with an industry standard of 65.8%. This means that Delta is largely stockholder-owned (86.5%), while its industry sector, in general, is only about 1/3 stockholder-owned. Based on these few measures, Delta appears to be in excellent financial health.

Final Answers to Selected Problems

This appendix provides the final answer to every third end-of-chapter problem. The complete solution for each problem included here is available on the website of the book at www.mhhe.com/blank.

Chapter 1

1.1 It is not possible to select the best alternative if it is not recognized as an alternative.

1.4 See Sec 1.2.2

1.7 See Sec 1.2.4

1.10 Money's worth changes as a function of time.

1.13 See Sec 1.3

1.16 See Sec 1.3

1.19 (*a*) Equivalent cost = $41,800; (*b*) Later

1.22 ROR = 4.7%/year

1.25 ROR = 38.3%/year

1.28 Earnings = $100,000,000

1.31 Wrong, unless MARR = cost of capital

1.34 $n = 990$ years

1.37 $P = \$13,888.89$

1.40 $n = 10$ years

1.43 i and n

1.46 $P = \$50,000$; $F = ?$; $i = 15\%$; $n = 3$

1.49 $F = \$400,000$; $n = 2$; $i = 20\%$; $P = ?$

1.52 $P = \$16,000,000$; $A = \$3,800,000$; $i = 18\%$; $n = ?$

1.55 Net cash flow or NCF

1.58 Down arrow of $40,000, year 5; up arrow, year 0 has $P = ?$; $i = 15\%$ per year

1.61 (*a*) F; (*b*) A; (*c*) n; (*d*) i; (*e*) P

1.64 See Sec. 1.8

1.67 After discussion, consult website at www.mhhe.com/blank

1.70 (a)
1.73 (b)
1.76 (d)

Chapter 2

2.1   (a) 6.7275; (b) 0.10853; (c) 9.8181; (d) 0.20122; (e) 167.2863
2.4   (a) $F = \$1,018,000$; (b) $= -FV(10\%,3,,100000) + 885000$
2.7   (a) $F = \$20,133.20$, (b, c) $F = \$20,133.23$
2.10 $A = \$3,309,747$
2.13 $A = \$4,592,200$
2.16   (a) $A = \$4,470,900$; (b) PMT(8,10,$-30000000$,0);
        (c) $= -$PMT(8%,10,30000000)
2.19   (a) $A = \$78,811$; (b) Recalls $= \$788,110$/year
2.22 $P = \$720,252$
2.25 $P = \$1,735,440$
2.28 $P = \$67,621,000$
2.31   (a) $P = \$21,245.30$;
        (b) $-$PV(10,2,0,7000) $-$ PV(10,4,0,9000) $-$ PV(10,5,0,15000)
2.34 $A = \$1,627,500$
2.37 $A = \$858,780$
2.40 $A = \$5696.25$
2.43   (a) $CF_4 = \$65,497.71$; (b) $\$65,495.05$
2.46 $n = 12$ years
2.49   (a) $P = \$112,284$; (b) Spreadsheet: enter costs, use NPV; calculator: use
        PV on each cash flow and add
2.52 $A = \$7986$
2.55 $F = \$721,018$
2.58 $G = \$37,805.65$
2.61 $A = \$6,094,950$
2.64 $P = \$90,405$
2.67 $A = \$197.986$ billion (factor); $\$197.984$ billion (spreadsheet)
2.70 $A = \$60,164$
2.73   Cost, year $3 = \$106,146$
2.76 $A_1 = \$1457.94$
2.79 $F = \$1,543,120$
2.82 $F_8 = \$1709.52$
2.85   Due, year $4 = \$6,593,776$
2.88 $F = \$2349.20$ million
2.91   (a) $P = \$143,278$; (b) $A = \$24,932$ (factors)
2.94 $F = \$20.6084$ billion
2.97 $A = \$16,819$
2.100 $P_0 = \$205,099$
2.103 $F = \$98,777$
2.106   (a) $x = \$70,730$; (b) $x = \$70,726$ (GOAL SEEK)

2.109 (*a*) $F = \$68,436,684$; (*b*) $F = \$68,436,701$
2.112 (*a*)
2.115 (*a*)
2.118 (*b*)
2.121 (*a*)
2.124 (*d*)

Chapter 3

3.1   (*a*) 4%/6-months; (*b*) 8%/year; (*c*) 16%/2-years
3.4   (*a*) year; (*b*) quarter; (*c*) day; (*d*) continuous; (*e*) hour
3.7   (*a*) 2%/quarter; (*b*) 1.33%/2-months; (*c*) 16%/2-years; (*d*) 8%/year;
          (*e*) 4%/6-months
3.10 (*a*) 0.262%/week; (*b*) effective
3.13 PP = day; CP = quarter
3.16 (*a*) 3%/quarter; (*b*) 6.09%/6-months; (*c*) 12.55%/year
3.19 $F = \$93,256,800$
3.22 $F = \$494,527$
3.25 $P = \$249,573$
3.28 $F = \$123,907$
3.31 $P = \$65,950,140$
3.34 $P = \$203,340$; $A = \$56,409$
3.37 $P = \$113.68$
3.40 $P = \$1023.98$
3.43 Cost of treatment = $\$126,825$; $A = \$2821$/month
3.46 $P = \$1,669,282$
3.49 $F = \$242,762$
3.52 $A = \$44,698$
3.55 $G = \$60.36$
3.58 $i = 1.26\%$/month; $F = \$2,782,130$
3.61 $A = \$4500$/3-months; $F = \$104,057$
3.64 $A = \$361.60$/month
3.67 Fuel savings = $\$4800$/year; $P = \$11,529$
3.70 (*a*)
3.73 (*b*)
3.76 (*a*)

Chapter 4

4.1   See Sec 4.1
4.4   (*a*) *A, B* and *C* are mutually exclusive; *D* and *E* are independent;
          (*b*) X, XD, XE, and XDE
4.7 $\text{PW}_{\text{solar}} = \$-16,364$; $\text{PW}_{\text{line}} = \$-13,902$; select line
4.10 $\text{PW}_{\text{manual}} = \$-781,502$; $\text{PW}_{\text{robotic}} = \$-693,119$; select robotic
4.13 $\text{PW} = \$1157.53$; justified

4.16 (a) PW$_A$ = \$−107,359; PW$_B$ = \$−130,742, select A;
(b) A: −PV(12,3,−20000,15000) − 70000;
   B: −PV(12,3,−8000,40000) − 140000

4.19 (a) PW$_{pull}$ = \$−5,187,780; PW$_{push}$ = \$−5,803,265; select pull;
(b) PW$_{pull}$: = −PV(10%,8,−700000,100000) −1500000;
   PW$_{push}$: = −PV(10%,8,−600000,50000) −2250000 −
   PV(10%,3,,−500000)

4.22 PW$_A$ = \$−259,271; PW$_J$ = \$−150,592; select Joshua

4.25 (a) PW = \$−59.74, no; (b) PW = \$−59.72 (spreadsheet)

4.28 PW = \$−155; no

4.31 LCM = 48 months; PW$_P$ = \$−31,996; PW$_F$ = \$−23,558; select fiber optic

4.34 (a) LCM = 4 years; PW$_{CFRP}$ = \$−463,320; PW$_{FRC}$ = \$−306,927; select FRC

4.37 (a) PW$_C$ = \$−375,922; PW$_A$ = \$−386,958; select concrete;
(b) PW$_C$ = \$−375,000; PW$_A$ = \$−257,667; select asphalt

4.40 FW$_C$ = \$−907,336; FW$_M$ = \$−1,075,244; select concrete

4.43 LCC$_A$ = \$−2,121,421; LCC$_B$ = \$−1,658,323; select B

4.46 CC = \$−88,875,000

4.49 CC = \$−3,532,308

4.52 CC = \$1,350,000

4.55 CC = \$−561,088

4.58 CC$_E$ = \$4,775,350; CC$_F$ = \$−1,527,217; CC$_G$ = \$3,333,333; select E

4.61

Projects	PW at 10%
X	\$−60,770
Y	−21,539
Z	−6,215
XY	−82,309
XZ	−66,985
DN	0

All PW < 0; select DN

4.64 (d)
4.67 (b)
4.70 (d)

Chapter 5

5.1 Revenue = \$−3,925,860/year
5.4 Revenue = \$5,522,900 per year
5.7 AW = \$−64,428,476
5.10 (a) CR = \$−603,112; (b) AW = \$−1,057,855;
(c) Example, CR: = −PMT(12%,12,−3800000,250000);
AW: gradient in B3:B14; = −PMT(12%,12,−3800000,250000)
−PMT(12%,12,NPV(12%,B3:B14))
5.13 (a) AW$_{plastic}$ = \$−439,222; AW$_{rubber}$ = \$−936,403; select plastic

5.16   Factor: (*a*) AW = $45.9 million, justified; (*b*) AW goes − between 30% and 40%
Spreadsheet:

	A	B	C	D
1	*i*%	CR	A of AOC	AW
2	5%	-13,641,240	65,875,000	52,233,760
3	10%	-19,968,136	65,875,000	45,906,864
4	20%	-34,910,610	65,875,000	30,964,390
5	30%	-51,269,770	65,875,000	14,605,230
6	35%	-59,647,543	65,875,000	6,227,457
7	40%	-68,081,371	65,875,000	-2,206,371

5.19   (*a*) $AW_{D103} = \$-152,760$; $AW_{490G} = \$-135,143$; select 490G;
(*b*) $P = \$-300,000$ changes decision to D103

5.22   AW = $615,650

5.25   (*a*) A = $6556/year; (*b*) A = $5904/year; (*c*) $652/year less

5.28   AW = $−144,198 (factor); AW = $−144,194 (spreadsheet)

5.31   Effective *i* = 10.381%/year; $AW_A = \$-4.35$ million; $AW_B = \$-4.23$ million; select *B*

5.34   (*b*)

5.37   (*d*)

5.40   (*d*)

Chapter 6

6.1   (*a*) Infinity; (*b*) −100%

6.4   Interest, year 2 = $562,350

6.7   (*a*) $i^* = 15.32\%$; (*b*) Function = IRR(B1:B4); $i^* = 15.32\%$/year

6.10   (*a*) $i^* = 5.32\%$ (interpolation); (*b*) FV function and GOAL SEEK, $i^* = 5.29\%$/year

6.13 $i^* = 23.38\%$/year

6.16 $i^* = 6.86\%$/year

6.19 $i^* = 10\%$/year

6.22 $i^* = 19.17\%$/year

6.25   (*a*) $\Delta i^* > 18\%$; (*b*) $\Delta i^* < 10\%$

6.28   (*a*) $\Delta CF_0 = \$-10,000$; (*b*) $\Delta CF_3 = \$+13,200$; (*c*) $\Delta CF_6 = \$+4200$

6.31 $\Delta i^* = 13.14\%$; select P

6.34 $\Delta i^* = 16.83\%$; select Y

6.37 $i_1^* = 18.52\%$, accept; $i_2^* = 8.23\%$, reject; $i_3^* = 15.49\%$, accept; $i_4^* = 10.35\%$, reject

6.40   Order is DN, D, A, B, E, C; (*a*) Select C; (*b*) Select C

6.43   (*a*) Accept A and C; (*b*) Order is DN, A, B, C, D; select D; (*c*) Select A

6.46   (*a*) Maximum i^* are two; (*b*) One positive i^* indicated

6.49   (*a*) Maximum i^* are two; (*b*) Norstrom's not satisfied; maximum of two i^*

6.52   (*a*) Maximum i^* are three; (*b*) Per Norstrom's, one positive i^*; $i^* = 17.39\%$

6.55   (*a*) Cash flow sign test: max of 3 i^*; cumulative cash flow test: one positive i^*;

(*b*) $i^* = 6.29\%$/year;

(*c*) Hand: $i' = 7.91\%$; spreadsheet: = MIRR(B1:B11,8%,20%), $i' = 7.90\%$;

(*d*) Hand: same as IRR, $i'' = i^* = 6.29\%$; spreadsheet follows:

	A	B	C
1	Year, t	NCF$_t$, $	FW$_t$, $
2	0	0	0
3	1	-5,000	-5,000
4	2	-5,000	-10,315
5	3	-5,000	-15,963
6	4	-5,000	-21,968
7	5	-5,000	-28,350
8	6	-5,000	-35,133
9	7	9,000	-28,344
10	8	-5,000	-35,127
11	9	-5,000	-42,337
12	10	45,000	0
13			
14	Reinvestment rate, i_r		20.0%
15	Resulting ROIC rate, i''		**6.29%**

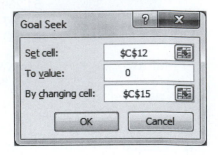

6.58   (*a*) Hand: $F_4 = -3901.48(1 + i'') + 4000 = 0$; $i'' = 2.53\%$;
spreadsheet: $i'' = 2.53\%$;

(*b*) Hand and spreadsheet: $i' = 8.76\%$

6.61   (*b*)

6.64   (*a*)

6.67   (*c*)

6.70   (*c*)

Chapter 7

7.1   See Sec 7.1

7.4   See Sec 7.1

7.7   B/C = 1.18

7.10   (*a*) B/C = 1.91; (*b*) Minimum P = $27,941

7.13   Modified B/C = 1.71

7.16 P = $2,547,825

7.19   (*a*) B/C = 1.98, justified; (*b*) Modified B/C = 4.59, justified

7.22   B/C = 22.86

7.25   Expand to Pond: ΔB/C = 0.90, select pond

7.28   S to N: ΔB/C = 0.98, select N

7.31   NS to EC: ΔB/C = 0.88, select EC

7.34 Order: DN, G, J, H, I, L, K; L-to-I ΔB/C is needed, select survivor
7.37 (c)
7.40 (a)
7.43 (a)
7.46 (b)
7.49 (d)

Chapter 8

8.1 (a) Q_{BE} =185,185 units; (b) Profit = $-190,000 (loss);
(c) Profit = $1,592,000
8.4 x = 1246 hours/month
8.7 (a) Graph: profit is about $20,000 at 1350 units; (b) Q_p = 1355 units;
max profit = $20,175/month
8.10 (a) i = 8%: n = 8.2; i =15%: n = 19.8 years; (b) NPER results in range
from 8.3 to 19.8 years
8.13 P_{asp} = $1,569,912/mile
8.16 x = 8070 square yards/year
8.19 x = 3701 ads per year
8.22 x = 790 units
8.25 (a) T = 951.2 tons/year, select 1; (b) Equate AW relations, can use
SOLVER

% change	Cost, $/hour	Breakeven T value, tons
−15%	20.40	1119
−5%	22.80	1001
0%	24.00	951
+5%	25.20	906
+15%	27.60	827

8.28 (a) Q = 2416 filters/year, make inhouse; (b) Q = 7248 filters/year,
outsource; (c) Inhouse: Profit = $77,520; outsource: Profit = $100,000
8.31 AW_{cont} = $-62,849; AW_{batch} = $-64,565 at i = 5% and n = 10;
AW_{batch} always > AW_{cont}
8.34 $AW_{10,000}$ = $-63,151; $AW_{18,000}$ = $-60,847; decision is sensitive
8.37 Decision is sensitive

MARR	AW$_1$	AW$_2$	Select
4%	$-2318	$-2234	2
6	−2444	−2448	1
8	−2573	−2673	1

8.40 Selection changes at +40% of best savings estimate

	A	B	C	D	E	F
1	**Percent**	**Company A**		**Company B**		
2	variation	Savings, $	AW, $	Savings, $	AW, $	Selection
3	-40%	9,000	-10,871	7,800	-9,486	B
4	-20%	12,000	-7,871	10,400	-6,886	B
5	0%	15,000	-4,871	13,000	-4,286	B
6	20%	18,000	-1,871	15,600	-1,686	B
7	40%	21,000	1,129	18,200	914	A

8.43 (a) Select D103 only for optimistic scenario

	A	B	C	D	E	F
1	Strategy	P	S	AOC	n	AW$_{D103}$
2	Pess	-500,000	50,000	-4,000	1	-$504,000
3	ML	-400,000	40,000	-4,000	3	-$152,761
4	Opt	-300,000	30,000	-4,000	5	-$78,225
5			= -PMT(10%,E4,B4,C4)+D4			
6						

8.46 (a) Machines 1 and 2: n_p = 4 and 6 years; machine 1 only; (b) Machines 1 and 2: n_p > 5 and > 9 years; neither
8.49 n_p = 51.1 months
8.52 n_p = 7.7 years; lease
8.55 (a) i_1^* = 38.6%; i_2^* = 26.55%; select 1; (b) ΔROR analysis over LCM required; analysis assumes return of 26.55% for years 6–10, which may not be correct
8.58 (a)
8.61 (a)
8.64 (c)
8.67 (b)
8.70 (d)

Chapter 9

9.1 See Sec 9.1
9.4 P = $-39,000; n = 4 years; S = $25,000; AOC = $-17,000/year
9.7 Defender ESL is 3 years
9.10 ESL is 4 years; AW = $-83,857
9.13 (a) ESL is n = 2 years; AW = $-77,929
9.16 AW$_C$ = $-49,345; $10,000 trade: AW$_D$ = $-47,496, select D; $20,000 trade: AW$_D$ = $-51,660, select C
9.19 AW$_D$ = $-201,900; AW$_C$ = $-158,177; replace now
9.22 RV = $12,734
9.25 RV = $21,954

9.28 $AW_D = \$-124{,}538$; $AW_C = \$-146{,}957$; keep defender
9.31 (a)
9.34 (b)
9.37 (a)

Chapter 10

10.1 See Sec 10.1
10.4 (a) Cost = \$20,986; (b) Cost = \$25,886
10.7 $f = 3.99\%$/year
10.10 (a) 2017 salary = \$69,678; (b) 2017 salary = \$92,853
10.13 Transportation cost = 12.1¢; labor cost = 46.8¢; 2012: profit = 4.50%; 2017: profit = 3.99%
10.16 $i = 1.49\%$/month
10.19 PW = \$−325,630
10.22 $PW_A = \$1{,}547{,}806$; $PW_B = \$1{,}700{,}000$; select A
10.25 Purchasing power = \$171,511
10.28 (a) $F = \$16{,}105$; (b) Purchasing power = \$12,619; (c) $i = 4.76\%$
10.31 Buying power = \$853,740
10.34 $F = \$46{,}304$
10.37 $P_{6000} = \$5547$; $P_{9000} = \$8001$; $P_{5000} = \$3952$
10.40 $AW_X = \$-62{,}094$; $AW_Y = \$-63{,}268$; select X
10.43 AW = \$−39,607/year
10.46 (a)
10.49 (c)
10.52 (c)
10.55 (c)

Chapter 11

11.1 See Sec 11.1
11.4 Cost = \$4,911,500
11.7 Cost = \$2,467,742
11.10 Cost = \$219,904
11.13 $C_t = \$41{,}628{,}000$
11.16 2002 cost = \$243
11.19 1913 rate = \$0.19/hour
11.22 $C_2 = \$11{,}466$
11.25 $x = -0.0566$
11.28 Cost = \$9570
11.31 $h = 5.87$
11.34 $T_{12} = 31.3$ months
11.37 A: \$6000; B: \$18,000; C: \$10,000
11.40 $AW_{make} = \$-3{,}793{,}075$; $AW_{buy} = \$-4{,}934{,}400$; select make
11.43 Percent changes: DFW is +4.1%; YYZ is +16.5%; MEX is −29.7%

11.46 (*c*)
11.49 (*b*)
11.52 (*a*)
11.55 (*d*)
11.58 (*c*)

Chapter 12

12.1 Depreciation is tax deductible
12.4 See Sec 12.1
12.7 Taxes: $D_t = \text{Rate} \times BV_{t-1}$; book: $D_t = \text{Rate} \times 40{,}000$

Year, *t*	Tax Depreciation		Book Depreciation	
	D_t	BV_t	D_t	BV_t
0		40,000		40,000
1	16,000	24,000	10,000	30,000
2	9,600	14,400	10,000	20,000
3	5,760	8,640	10,000	10,000
4	3,456	5,184	10,000	0

12.10 (*a*) $S = \$80{,}000$; (*b*) $B = \$440{,}000$
12.13 $D_t = \$60{,}000$ per year

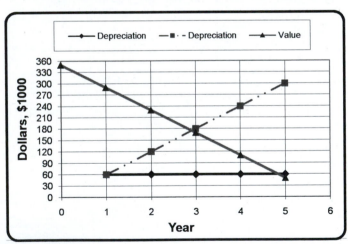

12.16 $t = 2.71$; *S* reached by end of 3rd year
12.19 (*a*) SL: \$10,000; DDB: \$9486; both are less; (*b*) SL: \$8750; DDB: 0
12.22 $BV_3 = \$115{,}200$
12.25 SL: $BV_4 = \$32{,}000$; MACRS: $BV_4 = \$13{,}824$; MACRS BV is lower
12.28 Selling price = \$3,852,183; 13.3% more than paid
12.31 SL larger by \$1500; switch
12.34 Switch to SL in year 4; for year 2: DDB, $d_2 = 0.32$; SL, $d_2 = 0.178$; for year 4: SL and DDB, $d_4 = 0.1152$

12.37 Percentage depletion = $4950
12.40 (*a*) In $1000 units: $290; 348; 435; 435; 522; (*b*) No
12.43 (*a*)
12.46 (*b*)
12.49 (*a*)
12.52 (*b*)

Chapter 13

13.1 (*a*) Rate = 39%; (*b*) Taxes = $80,750; (*c*) Average rate = 32.3%
13.4 Approximate taxes = $65,360
13.7 (*a*) TI = $76,000; (*b*) Taxes = $14,090; (*c*) 6.43% of GI
13.10 Approximated after-tax ROR = 25.01%
13.13 $CFBT_2$ = $800; D_2 = $845; D_4 = $141; TI_3 = $119; $Taxes_2$ = -16;
      $CFAT_3$ = $358
13.16 GI_1 = $3,813,587; GI_2 = $3,973,913; GI_3 = $4,134,239
13.19 Yes; DR = $1176
13.22 System 2 taxes are lower by $26,600
13.25 (*a*) MACRS: PW_{tax} = $102,119; SL: PW_{tax} = $109,185; MACRS is lower;
      (*b*) Taxes = $152,000 for both
13.28 (*a*) $CFAT_C$ = $-392,000$; $CFAT_D$ = $-100,000$;
      (*b*) $CFAT_D$ = $-57,000$; $CFAT_C$ = $29,500$;
      (*c*) AW_D = $-100,798$; AW_C = $-142,188$; retain defender
13.31 Total = $37,500,000
13.34 Return = 38.1%
13.37 (*a*) $WACC_A$ = 7.5%, $WACC_B$ = 8.2%; A is lower; (*b*) A: 7.4%, goes up;
      B: 8.0%, same
13.40 (*a*) After-tax WACC = 11.665%; interest is $1,633,100;
      (*b*) After-tax WACC = 8.425%; interest is $1,179,500
13.43 TI = $122,055; E = $333,470
13.46 (*b*)
13.49 (*d*)
13.52 (*d*)
13.55 (*b*)

Chapter 14

14.1 W_i = 0.1667
14.4 (*a, b*) Attributes (score; weight) are: #1 (9; 0.27); #2 (1; 0.03); #3 (3; 0.09);
      #4 (6; 0.18); #5 (10; 0.30); #6 (4.8; 0.14)
14.7 See Sec 14.2
14.10 (*a*) $P(N = 0 \text{ or } 1) \times 100$ = 68%;
      (*b*) $P(N = 1 \text{ or } 2) \times 100$ = 82%;
      (*c*) $P(N > 3) \times 100$ = 2.8%
14.13 (*a*) $E(X)$ = $1344
14.16 A result: (*a*) AVERAGE(A1:A100) of 49.2532; (*b*) STDEV(A1:A100)
      of 28.08

14.19 Sample of first 8 of 30 trials shown; project viable under certainty and risk

	A	B	C	D	E	F	G	H	I	J	K	L	M
1		MARR SAMPLE							CFAT SAMPLE	EVALUATION			
2	Sample	Random	Sample					Sample	Sample	PW			
3	Trial	Number	MARR Value		Distribution for MARR			Trial	CFAT Value	Value, $		PW statistics	
4	1	0.2324	7		Lower cutoff	MARR, %		1	6,968	13,940	Average	$	3,839
5	2	0.1339	7		0.00	7		2	5,241	1,811	Std dev	$	5,809
6	3	0.2193	7		0.25	8		3	6,599	11,349	PW > 0		23
7	4	0.3654	8		0.50	9		4	6,113	6,019	PW < 0		7
8	5	0.0398	7		0.75	10		5	6,076	7,675			
9	6	0.0145	7					6	5,658	4,739			
10	7	0.1787	7					7	5,433	3,159			
11	8	0.7683	10					8	6,931	7,588			

14.22 (c)

14.25 (c)

14.28 (b)

Reference Materials

TEXTBOOKS

Blank, L. T., A. Tarquin: *Engineering Economy,* 7th ed., McGraw-Hill, New York, 2012.

Blank, L. T., A. Tarquin, and S. Iverson: *Engineering Economy,* Canadian 2d ed., McGraw-Hill Ryerson, Whitby, ON, 2012.

Bowman, M. S.: *Applied Economic Analysis for Technologists, Engineers, and Managers,* 2d ed., Pearson Prentice-Hall, Upper Saddle River, NJ, 2003.

Canada, J. R., W. G. Sullivan, D. Kulonda, and J. A. White: *Capital Investment Analysis for Engineering and Management*, 3d ed., Pearson Prentice-Hall, Upper Saddle River, NJ, 2005.

Eschenbach, T. G.: *Engineering Economy: Applying Theory to Practice*, 3d ed., Oxford University Press, New York, 2011.

Fraser, N. M., E. M. Lewkes, I. Bernhardt, and M. Tajima: *Engineering Economics in Canada,* 3d ed., Pearson Prentice-Hall, Upper Saddle River, NJ, 2006.

Hartman, J. C.: *Engineering Economy and the Decision-Making Process,* Pearson Prentice-Hall, Upper Saddle River, NJ, 2007.

Newnan, D. G., J. P. Lavelle, and T. G. Eschenbach: *Essentials of Engineering Economic Analysis,* 2d ed., Oxford University Press, New York, 2002.

Newnan, D. G., T. G. Eschenbach, and J. P. Lavelle: *Engineering Economic Analysis,* 11th ed., Oxford University Press, New York, 2012.

Ostwald, P. F.: *Construction Cost Analysis and Estimating,* Pearson Prentice-Hall, Upper Saddle River, NJ, 2001.

Ostwald, P. F., and T. S. McLaren: *Cost Analysis and Estimating for Engineering and Management*, Pearson Prentice-Hall, Upper Saddle River, NJ, 2004.

Park, C. S.: *Contemporary Engineering Economics,* 5th ed., Pearson Prentice-Hall, Upper Saddle River, NJ, 2011.

Park, C. S.: *Fundamentals of Engineering Economics,* 3rd ed., Pearson Prentice-Hall, Upper Saddle River, NJ, 2013.

Peters, M. S., K. D. Timmerhaus, and R. E. West: *Plant Design and Economics for Chemical Engineers,* 5th ed., McGraw-Hill, New York, 2003.

Peurifoy, R. L., and G. D. Oberlender: *Estimating Construction Costs,* 5th ed., McGraw-Hill, New York, 2002.

Sullivan, W. G., E. Wicks, and C. P. Koelling: *Engineering Economy,* 15th ed., Pearson Prentice-Hall, Upper Saddle River, NJ, 2012.

Thuesen, G. J., and W. J. Fabrycky: *Engineering Economy,* 9th ed., Pearson Prentice-Hall, Upper Saddle River, NJ, 2001.

White, J. A., K. E. Case, and D. B. Pratt: *Principles of Engineering Economic Analysis,* 6th ed., John Wiley & Sons, Hoboken, New Jersey, 2012.

PUBLICATIONS (WITH WEBSITE ADDRESS)

Chemical Engineering, Access Intelligence, New York, monthly, www.che.com.

Corporations, Publication 542, Department of the Treasury, Internal Revenue Service, Government Printing Office, Washington, DC, annually, www.irs.gov.

Engineering News-Record, McGraw-Hill, New York, monthly, www.enr.construction.com.

How to Depreciate Property, Publication 946, U.S. Department of the Treasury, Internal Revenue Service, Government Printing Office, Washington, DC, annually, www.irs.gov.

Introduction to Monte Carlo Simulation, Microsoft Corporation, 2012, www.office.microsoft.com/en-us/excel-help/introduction-to-monte-carlo-simulation.

Sales and Other Dispositions of Assets, Publication 542, Department of the Treasury, Internal Revenue Service, Government Printing Office, Washington, DC, annually, www.irs.gov.

The Engineering Economist, joint publication of the Engineering Economy Divisions of ASEE and IIE, published by Taylor and Francis, Philadelphia, PA, quarterly, www.tandfonline.com.

Your Federal Income Tax, Publication 17, Department of the Treasury, Internal Revenue Service, Government Printing Office, Washington, DC, annually, www.irs.gov.

0.25%			TABLE 1	Discrete Cash Flow: Compound Interest Factors			0.25%	
	Single Payments		Uniform Series Payments				Arithmetic Gradients	
	F/P Compound Amount	P/F Present Worth	A/F Sinking Fund	F/A Compound Amount	A/P Capital Recovery	P/A Present Worth	P/G Gradient Present Worth	A/G Gradient Uniform Series
n								
1	1.0025	0.9975	1.00000	1.0000	1.00250	0.9975		
2	1.0050	0.9950	0.49938	2.0025	0.50188	1.9925	0.9950	0.4994
3	1.0075	0.9925	0.33250	3.0075	0.33500	2.9851	2.9801	0.9983
4	1.0100	0.9901	0.24906	4.0150	0.25156	3.9751	5.9503	1.4969
5	1.0126	0.9876	0.19900	5.0251	0.20150	4.9627	9.9007	1.9950
6	1.0151	0.9851	0.16563	6.0376	0.16813	5.9478	14.8263	2.4927
7	1.0176	0.9827	0.14179	7.0527	0.14429	6.9305	20.7223	2.9900
8	1.0202	0.9802	0.12391	8.0704	0.12641	7.9107	27.5839	3.4869
9	1.0227	0.9778	0.11000	9.0905	0.11250	8.8885	35.4061	3.9834
10	1.0253	0.9753	0.09888	10.1133	0.10138	9.8639	44.1842	4.4794
11	1.0278	0.9729	0.08978	11.1385	0.09228	10.8368	53.9133	4.9750
12	1.0304	0.9705	0.08219	12.1664	0.08469	11.8073	64.5886	5.4702
13	1.0330	0.9681	0.07578	13.1968	0.07828	12.7753	76.2053	5.9650
14	1.0356	0.9656	0.07028	14.2298	0.07278	13.7410	88.7587	6.4594
15	1.0382	0.9632	0.06551	15.2654	0.06801	14.7042	102.2441	6.9534
16	1.0408	0.9608	0.06134	16.3035	0.06384	15.6650	116.6567	7.4469
17	1.0434	0.9584	0.05766	17.3443	0.06016	16.6235	131.9917	7.9401
18	1.0460	0.9561	0.05438	18.3876	0.05688	17.5795	148.2446	8.4328
19	1.0486	0.9537	0.05146	19.4336	0.05396	18.5332	165.4106	8.9251
20	1.0512	0.9513	0.04882	20.4822	0.05132	19.4845	183.4851	9.4170
21	1.0538	0.9489	0.04644	21.5334	0.04894	20.4334	202.4634	9.9085
22	1.0565	0.9466	0.04427	22.5872	0.04677	21.3800	222.3410	10.3995
23	1.0591	0.9442	0.04229	23.6437	0.04479	22.3241	243.1131	10.8901
24	1.0618	0.9418	0.04048	24.7028	0.04298	23.2660	264.7753	11.3804
25	1.0644	0.9395	0.03881	25.7646	0.04131	24.2055	287.3230	11.8702
26	1.0671	0.9371	0.03727	26.8290	0.03977	25.1426	310.7516	12.3596
27	1.0697	0.9348	0.03585	27.8961	0.03835	26.0774	335.0566	12.8485
28	1.0724	0.9325	0.03452	28.9658	0.03702	27.0099	360.2334	13.3371
29	1.0751	0.9301	0.03329	30.0382	0.03579	27.9400	386.2776	13.8252
30	1.0778	0.9278	0.03214	31.1133	0.03464	28.8679	413.1847	14.3130
36	1.0941	0.9140	0.02658	37.6206	0.02908	34.3865	592.4988	17.2306
40	1.1050	0.9050	0.02380	42.0132	0.02630	38.0199	728.7399	19.1673
48	1.1273	0.8871	0.01963	50.9312	0.02213	45.1787	1040.06	23.0209
50	1.1330	0.8826	0.01880	53.1887	0.02130	46.9462	1125.78	23.9802
52	1.1386	0.8782	0.01803	55.4575	0.02053	48.7048	1214.59	24.9377
55	1.1472	0.8717	0.01698	58.8819	0.01948	51.3264	1353.53	26.3710
60	1.1616	0.8609	0.01547	64.6467	0.01797	55.6524	1600.08	28.7514
72	1.1969	0.8355	0.01269	78.7794	0.01519	65.8169	2265.56	34.4221
75	1.2059	0.8292	0.01214	82.3792	0.01464	68.3108	2447.61	35.8305
84	1.2334	0.8108	0.01071	93.3419	0.01321	75.6813	3029.76	40.0331
90	1.2520	0.7987	0.00992	100.7885	0.01242	80.5038	3446.87	42.8162
96	1.2709	0.7869	0.00923	108.3474	0.01173	85.2546	3886.28	45.5844
100	1.2836	0.7790	0.00881	113.4500	0.01131	88.3825	4191.24	47.4216
108	1.3095	0.7636	0.00808	123.8093	0.01058	94.5453	4829.01	51.0762
120	1.3494	0.7411	0.00716	139.7414	0.00966	103.5618	5852.11	56.5084
132	1.3904	0.7192	0.00640	156.1582	0.00890	112.3121	6950.01	61.8813
144	1.4327	0.6980	0.00578	173.0743	0.00828	120.8041	8117.41	67.1949
240	1.8208	0.5492	0.00305	328.3020	0.00555	180.3109	19399	107.5863
360	2.4568	0.4070	0.00172	582.7369	0.00422	237.1894	36264	152.8902
480	3.3151	0.3016	0.00108	926.0595	0.00358	279.3418	53821	192.6699

0.5%				TABLE 2	Discrete Cash Flow: Compound Interest Factors			0.5%
	Single Payments		Uniform Series Payments				Arithmetic Gradients	
	F/P	*P/F*	*A/F*	*F/A*	*A/P*	*P/A*	*P/G*	*A/G*
	Compound Amount	Present Worth	Sinking Fund	Compound Amount	Capital Recovery	Present Worth	Gradient Present Worth	Gradient Uniform Series
n								
1	1.0050	0.9950	1.00000	1.0000	1.00500	0.9950		
2	1.0100	0.9901	0.49875	2.0050	0.50375	1.9851	0.9901	0.4988
3	1.0151	0.9851	0.33167	3.0150	0.33667	2.9702	2.9604	0.9967
4	1.0202	0.9802	0.24813	4.0301	0.25313	3.9505	5.9011	1.4938
5	1.0253	0.9754	0.19801	5.0503	0.20301	4.9259	9.8026	1.9900
6	1.0304	0.9705	0.16460	6.0755	0.16960	5.8964	14.6552	2.4855
7	1.0355	0.9657	0.14073	7.1059	0.14573	6.8621	20.4493	2.9801
8	1.0407	0.9609	0.12283	8.1414	0.12783	7.8230	27.1755	3.4738
9	1.0459	0.9561	0.10891	9.1821	0.11391	8.7791	34.8244	3.9668
10	1.0511	0.9513	0.09777	10.2280	0.10277	9.7304	43.3865	4.4589
11	1.0564	0.9466	0.08866	11.2792	0.09366	10.6770	52.8526	4.9501
12	1.0617	0.9419	0.08107	12.3356	0.08607	11.6189	63.2136	5.4406
13	1.0670	0.9372	0.07464	13.3972	0.07964	12.5562	74.4602	5.9302
14	1.0723	0.9326	0.06914	14.4642	0.07414	13.4887	86.5835	6.4190
15	1.0777	0.9279	0.06436	15.5365	0.06936	14.4166	99.5743	6.9069
16	1.0831	0.9233	0.06019	16.6142	0.06519	15.3399	113.4238	7.3940
17	1.0885	0.9187	0.05651	17.6973	0.06151	16.2586	128.1231	7.8803
18	1.0939	0.9141	0.05323	18.7858	0.05823	17.1728	143.6634	8.3658
19	1.0994	0.9096	0.05030	19.8797	0.05530	18.0824	160.0360	8.8504
20	1.1049	0.9051	0.04767	20.9791	0.05267	18.9874	177.2322	9.3342
21	1.1104	0.9006	0.04528	22.0840	0.05028	19.8880	195.2434	9.8172
22	1.1160	0.8961	0.04311	23.1944	0.04811	20.7841	214.0611	10.2993
23	1.1216	0.8916	0.04113	24.3104	0.04613	21.6757	233.6768	10.7806
24	1.1272	0.8872	0.03932	25.4320	0.04432	22.5629	254.0820	11.2611
25	1.1328	0.8828	0.03765	26.5591	0.04265	23.4456	275.2686	11.7407
26	1.1385	0.8784	0.03611	27.6919	0.04111	24.3240	297.2281	12.2195
27	1.1442	0.8740	0.03469	28.8304	0.03969	25.1980	319.9523	12.6975
28	1.1499	0.8697	0.03336	29.9745	0.03836	26.0677	343.4332	13.1747
29	1.1556	0.8653	0.03213	31.1244	0.03713	26.9330	367.6625	13.6510
30	1.1614	0.8610	0.03098	32.2800	0.03598	27.7941	392.6324	14.1265
36	1.1967	0.8356	0.02542	39.3361	0.03042	32.8710	557.5598	16.9621
40	1.2208	0.8191	0.02265	44.1588	0.02765	36.1722	681.3347	18.8359
48	1.2705	0.7871	0.01849	54.0978	0.02349	42.5803	959.9188	22.5437
50	1.2832	0.7793	0.01765	56.6452	0.02265	44.1428	1035.70	23.4624
52	1.2961	0.7716	0.01689	59.2180	0.02189	45.6897	1113.82	24.3778
55	1.3156	0.7601	0.01584	63.1258	0.02084	47.9814	1235.27	25.7447
60	1.3489	0.7414	0.01433	69.7700	0.01933	51.7256	1448.65	28.0064
72	1.4320	0.6983	0.01157	86.4089	0.01657	60.3395	2012.35	33.3504
75	1.4536	0.6879	0.01102	90.7265	0.01602	62.4136	2163.75	34.6679
84	1.5204	0.6577	0.00961	104.0739	0.01461	68.4530	2640.66	38.5763
90	1.5666	0.6383	0.00883	113.3109	0.01383	72.3313	2976.08	41.1451
96	1.6141	0.6195	0.00814	122.8285	0.01314	76.0952	3324.18	43.6845
100	1.6467	0.6073	0.00773	129.3337	0.01273	78.5426	3562.79	45.3613
108	1.7137	0.5835	0.00701	142.7399	0.01201	83.2934	4054.37	48.6758
120	1.8194	0.5496	0.00610	163.8793	0.01110	90.0735	4823.51	53.5508
132	1.9316	0.5177	0.00537	186.3226	0.01037	96.4596	5624.59	58.3103
144	2.0508	0.4876	0.00476	210.1502	0.00976	102.4747	6451.31	62.9551
240	3.3102	0.3021	0.00216	462.0409	0.00716	139.5808	13416	96.1131
360	6.0226	0.1660	0.00100	1004.52	0.00600	166.7916	21403	128.3236
480	10.9575	0.0913	0.00050	1991.49	0.00550	181.7476	27588	151.7949

0.75%			TABLE 3	Discrete Cash Flow: Compound Interest Factors				0.75%
	Single Payments		Uniform Series Payments				Arithmetic Gradients	
	F/P Compound Amount	**P/F** Present Worth	**A/F** Sinking Fund	**F/A** Compound Amount	**A/P** Capital Recovery	**P/A** Present Worth	**P/G** Gradient Present Worth	**A/G** Gradient Uniform Series
n								
1	1.0075	0.9926	1.00000	1.0000	1.00750	0.9926		
2	1.0151	0.9852	0.49813	2.0075	0.50563	1.9777	0.9852	0.4981
3	1.0227	0.9778	0.33085	3.0226	0.33835	2.9556	2.9408	0.9950
4	1.0303	0.9706	0.24721	4.0452	0.25471	3.9261	5.8525	1.4907
5	1.0381	0.9633	0.19702	5.0756	0.20452	4.8894	9.7058	1.9851
6	1.0459	0.9562	0.16357	6.1136	0.17107	5.8456	14.4866	2.4782
7	1.0537	0.9490	0.13967	7.1595	0.14717	6.7946	20.1808	2.9701
8	1.0616	0.9420	0.12176	8.2132	0.12926	7.7366	26.7747	3.4608
9	1.0696	0.9350	0.10782	9.2748	0.11532	8.6716	34.2544	3.9502
10	1.0776	0.9280	0.09667	10.3443	0.10417	9.5996	42.6064	4.4384
11	1.0857	0.9211	0.08755	11.4219	0.09505	10.5207	51.8174	4.9253
12	1.0938	0.9142	0.07995	12.5076	0.08745	11.4349	61.8740	5.4110
13	1.1020	0.9074	0.07352	13.6014	0.08102	12.3423	72.7632	5.8954
14	1.1103	0.9007	0.06801	14.7034	0.07551	13.2430	84.4720	6.3786
15	1.1186	0.8940	0.06324	15.8137	0.07074	14.1370	96.9876	6.8606
16	1.1270	0.8873	0.05906	16.9323	0.06656	15.0243	110.2973	7.3413
17	1.1354	0.8807	0.05537	18.0593	0.06287	15.9050	124.3887	7.8207
18	1.1440	0.8742	0.05210	19.1947	0.05960	16.7792	139.2494	8.2989
19	1.1525	0.8676	0.04917	20.3387	0.05667	17.6468	154.8671	8.7759
20	1.1612	0.8612	0.04653	21.4912	0.05403	18.5080	171.2297	9.2516
21	1.1699	0.8548	0.04415	22.6524	0.05165	19.3628	188.3253	9.7261
22	1.1787	0.8484	0.04198	23.8223	0.04948	20.2112	206.1420	10.1994
23	1.1875	0.8421	0.04000	25.0010	0.04750	21.0533	224.6682	10.6714
24	1.1964	0.8358	0.03818	26.1885	0.04568	21.8891	243.8923	11.1422
25	1.2054	0.8296	0.03652	27.3849	0.04402	22.7188	263.8029	11.6117
26	1.2144	0.8234	0.03498	28.5903	0.04248	23.5422	284.3888	12.0800
27	1.2235	0.8173	0.03355	29.8047	0.04105	24.3595	305.6387	12.5470
28	1.2327	0.8112	0.03223	31.0282	0.03973	25.1707	327.5416	13.0128
29	1.2420	0.8052	0.03100	32.2609	0.03850	25.9759	350.0867	13.4774
30	1.2513	0.7992	0.02985	33.5029	0.03735	26.7751	373.2631	13.9407
36	1.3086	0.7641	0.02430	41.1527	0.03180	31.4468	524.9924	16.6946
40	1.3483	0.7416	0.02153	46.4465	0.02903	34.4469	637.4693	18.5058
48	1.4314	0.6986	0.01739	57.5207	0.02489	40.1848	886.8404	22.0691
50	1.4530	0.6883	0.01656	60.3943	0.02406	41.5664	953.8486	22.9476
52	1.4748	0.6780	0.01580	63.3111	0.02330	42.9276	1022.59	23.8211
55	1.5083	0.6630	0.01476	67.7688	0.02226	44.9316	1128.79	25.1223
60	1.5657	0.6387	0.01326	75.4241	0.02076	48.1734	1313.52	27.2665
72	1.7126	0.5839	0.01053	95.0070	0.01803	55.4768	1791.25	32.2882
75	1.7514	0.5710	0.00998	100.1833	0.01748	57.2027	1917.22	33.5163
84	1.8732	0.5338	0.00859	116.4269	0.01609	62.1540	2308.13	37.1357
90	1.9591	0.5104	0.00782	127.8790	0.01532	65.2746	2578.00	39.4946
96	2.0489	0.4881	0.00715	139.8562	0.01465	68.2584	2853.94	41.8107
100	2.1111	0.4737	0.00675	148.1445	0.01425	70.1746	3040.75	43.3311
108	2.2411	0.4462	0.00604	165.4832	0.01354	73.8394	3419.90	46.3154
120	2.4514	0.4079	0.00517	193.5143	0.01267	78.9417	3998.56	50.6521
132	2.6813	0.3730	0.00446	224.1748	0.01196	83.6064	4583.57	54.8232
144	2.9328	0.3410	0.00388	257.7116	0.01138	87.8711	5169.58	58.8314
240	6.0092	0.1664	0.00150	667.8869	0.00900	111.1450	9494.12	85.4210
360	14.7306	0.0679	0.00055	1830.74	0.00805	124.2819	13312	107.1145
480	36.1099	0.0277	0.00021	4681.32	0.00771	129.6409	15513	119.6620

1%			TABLE 4	Discrete Cash Flow: Compound Interest Factors				1%
	Single Payments		Uniform Series Payments				Arithmetic Gradients	
	F/P	P/F	A/F	F/A	A/P	P/A	P/G	A/G
n	Compound Amount	Present Worth	Sinking Fund	Compound Amount	Capital Recovery	Present Worth	Gradient Present Worth	Gradient Uniform Series
1	1.0100	0.9901	1.00000	1.0000	1.01000	0.9901		
2	1.0201	0.9803	0.49751	2.0100	0.50751	1.9704	0.9803	0.4975
3	1.0303	0.9706	0.33002	3.0301	0.34002	2.9410	2.9215	0.9934
4	1.0406	0.9610	0.24628	4.0604	0.25628	3.9020	5.8044	1.4876
5	1.0510	0.9515	0.19604	5.1010	0.20604	4.8534	9.6103	1.9801
6	1.0615	0.9420	0.16255	6.1520	0.17255	5.7955	14.3205	2.4710
7	1.0721	0.9327	0.13863	7.2135	0.14863	6.7282	19.9168	2.9602
8	1.0829	0.9235	0.12069	8.2857	0.13069	7.6517	26.3812	3.4478
9	1.0937	0.9143	0.10674	9.3685	0.11674	8.5660	33.6959	3.9337
10	1.1046	0.9053	0.09558	10.4622	0.10558	9.4713	41.8435	4.4179
11	1.1157	0.8963	0.08645	11.5668	0.09645	10.3676	50.8067	4.9005
12	1.1268	0.8874	0.07885	12.6825	0.08885	11.2551	60.5687	5.3815
13	1.1381	0.8787	0.07241	13.8093	0.08241	12.1337	71.1126	5.8607
14	1.1495	0.8700	0.06690	14.9474	0.07690	13.0037	82.4221	6.3384
15	1.1610	0.8613	0.06212	16.0969	0.07212	13.8651	94.4810	6.8143
16	1.1726	0.8528	0.05794	17.2579	0.06794	14.7179	107.2734	7.2886
17	1.1843	0.8444	0.05426	18.4304	0.06426	15.5623	120.7834	7.7613
18	1.1961	0.8360	0.05098	19.6147	0.06098	16.3983	134.9957	8.2323
19	1.2081	0.8277	0.04805	20.8109	0.05805	17.2260	149.8950	8.7017
20	1.2202	0.8195	0.04542	22.0190	0.05542	18.0456	165.4664	9.1694
21	1.2324	0.8114	0.04303	23.2392	0.05303	18.8570	181.6950	9.6354
22	1.2447	0.8034	0.04086	24.4716	0.05086	19.6604	198.5663	10.0998
23	1.2572	0.7954	0.03889	25.7163	0.04889	20.4558	216.0660	10.5626
24	1.2697	0.7876	0.03707	26.9735	0.04707	21.2434	234.1800	11.0237
25	1.2824	0.7798	0.03541	28.2432	0.04541	22.0232	252.8945	11.4831
26	1.2953	0.7720	0.03387	29.5256	0.04387	22.7952	272.1957	11.9409
27	1.3082	0.7644	0.03245	30.8209	0.04245	23.5596	292.0702	12.3971
28	1.3213	0.7568	0.03112	32.1291	0.04112	24.3164	312.5047	12.8516
29	1.3345	0.7493	0.02990	33.4504	0.03990	25.0658	333.4863	13.3044
30	1.3478	0.7419	0.02875	34.7849	0.03875	25.8077	355.0021	13.7557
36	1.4308	0.6989	0.02321	43.0769	0.03321	30.1075	494.6207	16.4285
40	1.4889	0.6717	0.02046	48.8864	0.03046	32.8347	596.8561	18.1776
48	1.6122	0.6203	0.01633	61.2226	0.02633	37.9740	820.1460	21.5976
50	1.6446	0.6080	0.01551	64.4632	0.02551	39.1961	879.4176	22.4363
52	1.6777	0.5961	0.01476	67.7689	0.02476	40.3942	939.9175	23.2686
55	1.7285	0.5785	0.01373	72.8525	0.02373	42.1472	1032.81	24.5049
60	1.8167	0.5504	0.01224	81.6697	0.02224	44.9550	1192.81	26.5333
72	2.0471	0.4885	0.00955	104.7099	0.01955	51.1504	1597.87	31.2386
75	2.1091	0.4741	0.00902	110.9128	0.01902	52.5871	1702.73	32.3793
84	2.3067	0.4335	0.00765	130.6723	0.01765	56.6485	2023.32	35.7170
90	2.4486	0.4084	0.00690	144.8633	0.01690	59.1609	2240.57	37.8724
96	2.5993	0.3847	0.00625	159.9273	0.01625	61.5277	2459.43	39.9727
100	2.7048	0.3697	0.00587	170.4814	0.01587	63.0289	2605.78	41.3426
108	2.9289	0.3414	0.00518	192.8926	0.01518	65.8578	2898.42	44.0103
120	3.3004	0.3030	0.00435	230.0387	0.01435	69.7005	3334.11	47.8349
132	3.7190	0.2689	0.00368	271.8959	0.01368	73.1108	3761.69	51.4520
144	4.1906	0.2386	0.00313	319.0616	0.01313	76.1372	4177.47	54.8676
240	10.8926	0.0918	0.00101	989.2554	0.01101	90.8194	6878.60	75.7393
360	35.9496	0.0278	0.00029	3494.96	0.01029	97.2183	8720.43	89.6995
480	118.6477	0.0084	0.00008	11765	0.01008	99.1572	9511.16	95.9200

1.25% **TABLE 5** Discrete Cash Flow: Compound Interest Factors **1.25%**

	Single Payments		Uniform Series Payments				Arithmetic Gradients	
n	F/P Compound Amount	P/F Present Worth	A/F Sinking Fund	F/A Compound Amount	A/P Capital Recovery	P/A Present Worth	P/G Gradient Present Worth	A/G Gradient Uniform Series
1	1.0125	0.9877	1.00000	1.0000	1.01250	0.9877		
2	1.0252	0.9755	0.49680	2.0125	0.50939	1.9631	0.9755	0.4969
3	1.0380	0.9634	0.32920	3.0377	0.34170	2.9265	2.9023	0.9917
4	1.0509	0.9515	0.24536	4.0756	0.25786	3.8781	5.7569	1.4845
5	1.0641	0.9398	0.19506	5.1266	0.20756	4.8178	9.5160	1.9752
6	1.0774	0.9282	0.16153	6.1907	0.17403	5.7460	14.1569	2.4638
7	1.0909	0.9167	0.13759	7.2680	0.15009	6.6627	19.6571	2.9503
8	1.1045	0.9054	0.11963	8.3589	0.13213	7.5681	25.9949	3.4348
9	1.1183	0.8942	0.10567	9.4634	0.11817	8.4623	33.1487	3.9172
10	1.1323	0.8832	0.09450	10.5817	0.10700	9.3455	41.0973	4.3975
11	1.1464	0.8723	0.08537	11.7139	0.09787	10.2178	49.8201	4.8758
12	1.1608	0.8615	0.07776	12.8604	0.09026	11.0793	59.2967	5.3520
13	1.1753	0.8509	0.07132	14.0211	0.08382	11.9302	69.5072	5.8262
14	1.1900	0.8404	0.06581	15.1964	0.07831	12.7706	80.4320	6.2982
15	1.2048	0.8300	0.06103	16.3863	0.07353	13.6005	92.0519	6.7682
16	1.2199	0.8197	0.05685	17.5912	0.06935	14.4203	104.3481	7.2362
17	1.2351	0.8096	0.05316	18.8111	0.06566	15.2299	117.3021	7.7021
18	1.2506	0.7996	0.04988	20.0462	0.06238	16.0295	130.8958	8.1659
19	1.2662	0.7898	0.04696	21.2968	0.05946	16.8193	145.1115	8.6277
20	1.2820	0.7800	0.04432	22.5630	0.05682	17.5993	159.9316	9.0874
21	1.2981	0.7704	0.04194	23.8450	0.05444	18.3697	175.3392	9.5450
22	1.3143	0.7609	0.03977	25.1431	0.05227	19.1306	191.3174	10.0006
23	1.3307	0.7515	0.03780	26.4574	0.05030	19.8820	207.8499	10.4542
24	1.3474	0.7422	0.03599	27.7881	0.04849	20.6242	224.9204	10.9056
25	1.3642	0.7330	0.03432	29.1354	0.04682	21.3573	242.5132	11.3551
26	1.3812	0.7240	0.03279	30.4996	0.04529	22.0813	260.6128	11.8024
27	1.3985	0.7150	0.03137	31.8809	0.04387	22.7963	279.2040	12.2478
28	1.4160	0.7062	0.03005	33.2794	0.04255	23.5025	298.2719	12.6911
29	1.4337	0.6975	0.02882	34.6954	0.04132	24.2000	317.8019	13.1323
30	1.4516	0.6889	0.02768	36.1291	0.04018	24.8889	337.7797	13.5715
36	1.5639	0.6394	0.02217	45.1155	0.03467	28.8473	466.2830	16.1639
40	1.6436	0.6084	0.01942	51.4896	0.03192	31.3269	559.2320	17.8515
48	1.8154	0.5509	0.01533	65.2284	0.02783	35.9315	759.2296	21.1299
50	1.8610	0.5373	0.01452	68.8818	0.02702	37.0129	811.6738	21.9295
52	1.9078	0.5242	0.01377	72.6271	0.02627	38.0677	864.9409	22.7211
55	1.9803	0.5050	0.01275	78.4225	0.02525	39.6017	946.2277	23.8936
60	2.1072	0.4746	0.01129	88.5745	0.02379	42.0346	1084.84	25.8083
72	2.4459	0.4088	0.00865	115.6736	0.02115	47.2925	1428.46	30.2047
75	2.5388	0.3939	0.00812	123.1035	0.02062	48.4890	1515.79	31.2605
84	2.8391	0.3522	0.00680	147.1290	0.01930	51.8222	1778.84	34.3258
90	3.0588	0.3269	0.00607	164.7050	0.01857	53.8461	1953.83	36.2855
96	3.2955	0.3034	0.00545	183.6411	0.01795	55.7246	2127.52	38.1793
100	3.4634	0.2887	0.00507	197.0723	0.01757	56.9013	2242.24	39.4058
108	3.8253	0.2614	0.00442	226.0226	0.01692	59.0865	2468.26	41.7737
120	4.4402	0.2252	0.00363	275.2171	0.01613	61.9828	2796.57	45.1184
132	5.1540	0.1940	0.00301	332.3198	0.01551	64.4781	3109.35	48.2234
144	5.9825	0.1672	0.00251	398.6021	0.01501	66.6277	3404.61	51.0990
240	19.7155	0.0507	0.00067	1497.24	0.01317	75.9423	5101.53	67.1764
360	87.5410	0.0114	0.00014	6923.28	0.01264	79.0861	5997.90	75.8401
480	388.7007	0.0026	0.00003	31016	0.01253	79.7942	6284.74	78.7619

1.5%				TABLE 6	Discrete Cash Flow: Compound Interest Factors			1.5%
	Single Payments		Uniform Series Payments				Arithmetic Gradients	
	F/P	P/F	A/F	F/A	A/P	P/A	P/G	A/G
n	Compound Amount	Present Worth	Sinking Fund	Compound Amount	Capital Recovery	Present Worth	Gradient Present Worth	Gradient Uniform Series
1	1.0150	0.9852	1.00000	1.0000	1.01500	0.9852		
2	1.0302	0.9707	0.49628	2.0150	0.51128	1.9559	0.9707	0.4963
3	1.0457	0.9563	0.32838	3.0452	0.34338	2.9122	2.8833	0.9901
4	1.0614	0.9422	0.24444	4.0909	0.25944	3.8544	5.7098	1.4814
5	1.0773	0.9283	0.19409	5.1523	0.20909	4.7826	9.4229	1.9702
6	1.0934	0.9145	0.16053	6.2296	0.17553	5.6972	13.9956	2.4566
7	1.1098	0.9010	0.13656	7.3230	0.15156	6.5982	19.4018	2.9405
8	1.1265	0.8877	0.11858	8.4328	0.13358	7.4859	25.6157	3.4219
9	1.1434	0.8746	0.10461	9.5593	0.11961	8.3605	32.6125	3.9008
10	1.1605	0.8617	0.09343	10.7027	0.10843	9.2222	40.3675	4.3772
11	1.1779	0.8489	0.08429	11.8633	0.09929	10.0711	48.8568	4.8512
12	1.1956	0.8364	0.07668	13.0412	0.09168	10.9075	58.0571	5.3227
13	1.2136	0.8240	0.07024	14.2368	0.08524	11.7315	67.9454	5.7917
14	1.2318	0.8118	0.06472	15.4504	0.07972	12.5434	78.4994	6.2582
15	1.2502	0.7999	0.05994	16.6821	0.07494	13.3432	89.6974	6.7223
16	1.2690	0.7880	0.05577	17.9324	0.07077	14.1313	101.5178	7.1839
17	1.2880	0.7764	0.05208	19.2014	0.06708	14.9076	113.9400	7.6431
18	1.3073	0.7649	0.04881	20.4894	0.06381	15.6726	126.9435	8.0997
19	1.3270	0.7536	0.04588	21.7967	0.06088	16.4262	140.5084	8.5539
20	1.3469	0.7425	0.04325	23.1237	0.05825	17.1686	154.6154	9.0057
21	1.3671	0.7315	0.04087	24.4705	0.05587	17.9001	169.2453	9.4550
22	1.3876	0.7207	0.03870	25.8376	0.05370	18.6208	184.3798	9.9018
23	1.4084	0.7100	0.03673	27.2251	0.05173	19.3309	200.0006	10.3462
24	1.4295	0.6995	0.03492	28.6335	0.04992	20.0304	216.0901	10.7881
25	1.4509	0.6892	0.03326	30.0630	0.04826	20.7196	232.6310	11.2276
26	1.4727	0.6790	0.03173	31.5140	0.04673	21.3986	249.6065	11.6646
27	1.4948	0.6690	0.03032	32.9867	0.04532	22.0676	267.0002	12.0992
28	1.5172	0.6591	0.02900	34.4815	0.04400	22.7267	284.7958	12.5313
29	1.5400	0.6494	0.02778	35.9987	0.04278	23.3761	302.9779	12.9610
30	1.5631	0.6398	0.02664	37.5387	0.04164	24.0158	321.5310	13.3883
36	1.7091	0.5851	0.02115	47.2760	0.03615	27.6607	439.8303	15.9009
40	1.8140	0.5513	0.01843	54.2679	0.03343	29.9158	524.3568	17.5277
48	2.0435	0.4894	0.01437	69.5652	0.02937	34.0426	703.5462	20.6667
50	2.1052	0.4750	0.01357	73.6828	0.02857	34.9997	749.9636	21.4277
52	2.1689	0.4611	0.01283	77.9249	0.02783	35.9287	796.8774	22.1794
55	2.2679	0.4409	0.01183	84.5296	0.02683	37.2715	868.0285	23.2894
60	2.4432	0.4093	0.01039	96.2147	0.02539	39.3803	988.1674	25.0930
72	2.9212	0.3423	0.00781	128.0772	0.02281	43.8447	1279.79	29.1893
75	3.0546	0.3274	0.00730	136.9728	0.02230	44.8416	1352.56	30.1631
84	3.4926	0.2863	0.00602	166.1726	0.02102	47.5786	1568.51	32.9668
90	3.8189	0.2619	0.00532	187.9299	0.02032	49.2099	1709.54	34.7399
96	4.1758	0.2395	0.00472	211.7202	0.01972	50.7017	1847.47	36.4381
100	4.4320	0.2256	0.00437	228.8030	0.01937	51.6247	1937.45	37.5295
108	4.9927	0.2003	0.00376	266.1778	0.01876	53.3137	2112.13	39.6171
120	5.9693	0.1675	0.00302	331.2882	0.01802	55.4985	2359.71	42.5185
132	7.1370	0.1401	0.00244	409.1354	0.01744	57.3257	2588.71	45.1579
144	8.5332	0.1172	0.00199	502.2109	0.01699	58.8540	2798.58	47.5512
240	35.6328	0.0281	0.00043	2308.85	0.01543	64.7957	3870.69	59.7368
360	212.7038	0.0047	0.00007	14114	0.01507	66.3532	4310.72	64.9662
480	1269.70	0.0008	0.00001	84580	0.01501	66.6142	4415.74	66.2883

2%				TABLE 7	Discrete Cash Flow: Compound Interest Factors				2%
	Single Payments		**Uniform Series Payments**				**Arithmetic Gradients**		
n	*F/P* Compound Amount	*P/F* Present Worth	*A/F* Sinking Fund	*F/A* Compound Amount	*A/P* Capital Recovery	*P/A* Present Worth	*P/G* Gradient Present Worth	*A/G* Gradient Uniform Series	
1	1.0200	0.9804	1.00000	1.0000	1.02000	0.9804			
2	1.0404	0.9612	0.49505	2.0200	0.51505	1.9416	0.9612	0.4950	
3	1.0612	0.9423	0.32675	3.0604	0.34675	2.8839	2.8458	0.9868	
4	1.0824	0.9238	0.24262	4.1216	0.26262	3.8077	5.6173	1.4752	
5	1.1041	0.9057	0.19216	5.2040	0.21216	4.7135	9.2403	1.9604	
6	1.1262	0.8880	0.15853	6.3081	0.17853	5.6014	13.6801	2.4423	
7	1.1487	0.8706	0.13451	7.4343	0.15451	6.4720	18.9035	2.9208	
8	1.1717	0.8535	0.11651	8.5830	0.13651	7.3255	24.8779	3.3961	
9	1.1951	0.8368	0.10252	9.7546	0.12252	8.1622	31.5720	3.8681	
10	1.2190	0.8203	0.09133	10.9497	0.11133	8.9826	38.9551	4.3367	
11	1.2434	0.8043	0.08218	12.1687	0.10218	9.7868	46.9977	4.8021	
12	1.2682	0.7885	0.07456	13.4121	0.09456	10.5753	55.6712	5.2642	
13	1.2936	0.7730	0.06812	14.6803	0.08812	11.3484	64.9475	5.7231	
14	1.3195	0.7579	0.06260	15.9739	0.08260	12.1062	74.7999	6.1786	
15	1.3459	0.7430	0.05783	17.2934	0.07783	12.8493	85.2021	6.6309	
16	1.3728	0.7284	0.05365	18.6393	0.07365	13.5777	96.1288	7.0799	
17	1.4002	0.7142	0.04997	20.0121	0.06997	14.2919	107.5554	7.5256	
18	1.4282	0.7002	0.04670	21.4123	0.06670	14.9920	119.4581	7.9681	
19	1.4568	0.6864	0.04378	22.8406	0.06378	15.6785	131.8139	8.4073	
20	1.4859	0.6730	0.04116	24.2974	0.06116	16.3514	144.6003	8.8433	
21	1.5157	0.6598	0.03878	25.7833	0.05878	17.0112	157.7959	9.2760	
22	1.5460	0.6468	0.03663	27.2990	0.05663	17.6580	171.3795	9.7055	
23	1.5769	0.6342	0.03467	28.8450	0.05467	18.2922	185.3309	10.1317	
24	1.6084	0.6217	0.03287	30.4219	0.05287	18.9139	199.6305	10.5547	
25	1.6406	0.6095	0.03122	32.0303	0.05122	19.5235	214.2592	10.9745	
26	1.6734	0.5976	0.02970	33.6709	0.04970	20.1210	229.1987	11.3910	
27	1.7069	0.5859	0.02829	35.3443	0.04829	20.7069	244.4311	11.8043	
28	1.7410	0.5744	0.02699	37.0512	0.04699	21.2813	259.9392	12.2145	
29	1.7758	0.5631	0.02578	38.7922	0.04578	21.8444	275.7064	12.6214	
30	1.8114	0.5521	0.02465	40.5681	0.04465	22.3965	291.7164	13.0251	
36	2.0399	0.4902	0.01923	51.9944	0.03923	25.4888	392.0405	15.3809	
40	2.2080	0.4529	0.01656	60.4020	0.03656	27.3555	461.9931	16.8885	
48	2.5871	0.3865	0.01260	79.3535	0.03260	30.6731	605.9657	19.7556	
50	2.6916	0.3715	0.01182	84.5794	0.03182	31.4236	642.3606	20.4420	
52	2.8003	0.3571	0.01111	90.0164	0.03111	32.1449	678.7849	21.1164	
55	2.9717	0.3365	0.01014	98.5865	0.03014	33.1748	733.3527	22.1057	
60	3.2810	0.3048	0.00877	114.0515	0.02877	34.7609	823.6975	23.6961	
72	4.1611	0.2403	0.00633	158.0570	0.02633	37.9841	1034.06	27.2234	
75	4.4158	0.2265	0.00586	170.7918	0.02586	38.6771	1084.64	28.0434	
84	5.2773	0.1895	0.00468	213.8666	0.02468	40.5255	1230.42	30.3616	
90	5.9431	0.1683	0.00405	247.1567	0.02405	41.5869	1322.17	31.7929	
96	6.6929	0.1494	0.00351	284.6467	0.02351	42.5294	1409.30	33.1370	
100	7.2446	0.1380	0.00320	312.2323	0.02320	43.0984	1464.75	33.9863	
108	8.4883	0.1178	0.00267	374.4129	0.02267	44.1095	1569.30	35.5774	
120	10.7652	0.0929	0.00205	488.2582	0.02205	45.3554	1710.42	37.7114	
132	13.6528	0.0732	0.00158	632.6415	0.02158	46.3378	1833.47	39.5676	
144	17.3151	0.0578	0.00123	815.7545	0.02123	47.1123	1939.79	41.1738	
240	115.8887	0.0086	0.00017	5744.44	0.02017	49.5686	2374.88	47.9110	
360	1247.56	0.0008	0.00002	62328	0.02002	49.9599	2482.57	49.7112	
480	13430	0.0001			0.02000	49.9963	2498.03	49.9643	

			TABLE 8	Discrete Cash Flow: Compound Interest Factors				3%

	Single Payments		Uniform Series Payments				Arithmetic Gradients	
n	F/P Compound Amount	P/F Present Worth	A/F Sinking Fund	F/A Compound Amount	A/P Capital Recovery	P/A Present Worth	P/G Gradient Present Worth	A/G Gradient Uniform Series
1	1.0300	0.9709	1.00000	1.0000	1.03000	0.9709		
2	1.0609	0.9426	0.49261	2.0300	0.52261	1.9135	0.9426	0.4926
3	1.0927	0.9151	0.32353	3.0909	0.35353	2.8286	2.7729	0.9803
4	1.1255	0.8885	0.23903	4.1836	0.26903	3.7171	5.4383	1.4631
5	1.1593	0.8626	0.18835	5.3091	0.21835	4.5797	8.8888	1.9409
6	1.1941	0.8375	0.15460	6.4684	0.18460	5.4172	13.0762	2.4138
7	1.2299	0.8131	0.13051	7.6625	0.16051	6.2303	17.9547	2.8819
8	1.2668	0.7894	0.11246	8.8923	0.14246	7.0197	23.4806	3.3450
9	1.3048	0.7664	0.09843	10.1591	0.12843	7.7861	29.6119	3.8032
10	1.3439	0.7441	0.08723	11.4639	0.11723	8.5302	36.3088	4.2565
11	1.3842	0.7224	0.07808	12.8078	0.10808	9.2526	43.5330	4.7049
12	1.4258	0.7014	0.07046	14.1920	0.10046	9.9540	51.2482	5.1485
13	1.4685	0.6810	0.06403	15.6178	0.09403	10.6350	59.4196	5.5872
14	1.5126	0.6611	0.05853	17.0863	0.08853	11.2961	68.0141	6.0210
15	1.5580	0.6419	0.05377	18.5989	0.08377	11.9379	77.0002	6.4500
16	1.6047	0.6232	0.04961	20.1569	0.07961	12.5611	86.3477	6.8742
17	1.6528	0.6050	0.04595	21.7616	0.07595	13.1661	96.0280	7.2936
18	1.7024	0.5874	0.04271	23.4144	0.07271	13.7535	106.0137	7.7081
19	1.7535	0.5703	0.03981	25.1169	0.06981	14.3238	116.2788	8.1179
20	1.8061	0.5537	0.03722	26.8704	0.06722	14.8775	126.7987	8.5229
21	1.8603	0.5375	0.03487	28.6765	0.06487	15.4150	137.5496	8.9231
22	1.9161	0.5219	0.03275	30.5368	0.06275	15.9369	148.5094	9.3186
23	1.9736	0.5067	0.03081	32.4529	0.06081	16.4436	159.6566	9.7093
24	2.0328	0.4919	0.02905	34.4265	0.05905	16.9355	170.9711	10.0954
25	2.0938	0.4776	0.02743	36.4593	0.05743	17.4131	182.4336	10.4768
26	2.1566	0.4637	0.02594	38.5530	0.05594	17.8768	194.0260	10.8535
27	2.2213	0.4502	0.02456	40.7096	0.05456	18.3270	205.7309	11.2255
28	2.2879	0.4371	0.02329	42.9309	0.05329	18.7641	217.5320	11.5930
29	2.3566	0.4243	0.02211	45.2189	0.05211	19.1885	229.4137	11.9558
30	2.4273	0.4120	0.02102	47.5754	0.05102	19.6004	241.3613	12.3141
31	2.5001	0.4000	0.02000	50.0027	0.05000	20.0004	253.3609	12.6678
32	2.5751	0.3883	0.01905	52.5028	0.04905	20.3888	265.3993	13.0169
33	2.6523	0.3770	0.01816	55.0778	0.04816	20.7658	277.4642	13.3616
34	2.7319	0.3660	0.01732	57.7302	0.04732	21.1318	289.5437	13.7018
35	2.8139	0.3554	0.01654	60.4621	0.04654	21.4872	301.6267	14.0375
40	3.2620	0.3066	0.01326	75.4013	0.04326	23.1148	361.7499	15.6502
45	3.7816	0.2644	0.01079	92.7199	0.04079	24.5187	420.6325	17.1556
50	4.3839	0.2281	0.00887	112.7969	0.03887	25.7298	477.4803	18.5575
55	5.0821	0.1968	0.00735	136.0716	0.03735	26.7744	531.7411	19.8600
60	5.8916	0.1697	0.00613	163.0534	0.03613	27.6756	583.0526	21.0674
65	6.8300	0.1464	0.00515	194.3328	0.03515	28.4529	631.2010	22.1841
70	7.9178	0.1263	0.00434	230.5941	0.03434	29.1234	676.0869	23.2145
75	9.1789	0.1089	0.00367	272.6309	0.03367	29.7018	717.6978	24.1634
80	10.6409	0.0940	0.00311	321.3630	0.03311	30.2008	756.0865	25.0353
84	11.9764	0.0835	0.00273	365.8805	0.03273	30.5501	784.5434	25.6806
85	12.3357	0.0811	0.00265	377.8570	0.03265	30.6312	791.3529	25.8349
90	14.3005	0.0699	0.00226	443.3489	0.03226	31.0024	823.6302	26.5667
96	17.0755	0.0586	0.00187	535.8502	0.03187	31.3812	858.6377	27.3615
108	24.3456	0.0411	0.00129	778.1863	0.03129	31.9642	917.6013	28.7072
120	34.7110	0.0288	0.00089	1123.70	0.03089	32.3730	963.8635	29.7737

4%			TABLE 9	Discrete Cash Flow: Compound Interest Factors				4%
	Single Payments		**Uniform Series Payments**				**Arithmetic Gradients**	
	F/P Compound Amount	*P/F* Present Worth	*A/F* Sinking Fund	*F/A* Compound Amount	*A/P* Capital Recovery	*P/A* Present Worth	*P/G* Gradient Present Worth	*A/G* Gradient Uniform Series
n								
1	1.0400	0.9615	1.00000	1.0000	1.04000	0.9615		
2	1.0816	0.9246	0.49020	2.0400	0.53020	1.8861	0.9246	0.4902
3	1.1249	0.8890	0.32035	3.1216	0.36035	2.7751	2.7025	0.9739
4	1.1699	0.8548	0.23549	4.2465	0.27549	3.6299	5.2670	1.4510
5	1.2167	0.8219	0.18463	5.4163	0.22463	4.4518	8.5547	1.9216
6	1.2653	0.7903	0.15076	6.6330	0.19076	5.2421	12.5062	2.3857
7	1.3159	0.7599	0.12661	7.8983	0.16661	6.0021	17.0657	2.8433
8	1.3686	0.7307	0.10853	9.2142	0.14853	6.7327	22.1806	3.2944
9	1.4233	0.7026	0.09449	10.5828	0.13449	7.4353	27.8013	3.7391
10	1.4802	0.6756	0.08329	12.0061	0.12329	8.1109	33.8814	4.1773
11	1.5395	0.6496	0.07415	13.4864	0.11415	8.7605	40.3772	4.6090
12	1.6010	0.6246	0.06655	15.0258	0.10655	9.3851	47.2477	5.0343
13	1.6651	0.6006	0.06014	16.6268	0.10014	9.9856	54.4546	5.4533
14	1.7317	0.5775	0.05467	18.2919	0.09467	10.5631	61.9618	5.8659
15	1.8009	0.5553	0.04994	20.0236	0.08994	11.1184	69.7355	6.2721
16	1.8730	0.5339	0.04582	21.8245	0.08582	11.6523	77.7441	6.6720
17	1.9479	0.5134	0.04220	23.6975	0.08220	12.1657	85.9581	7.0656
18	2.0258	0.4936	0.03899	25.6454	0.07899	12.6593	94.3498	7.4530
19	2.1068	0.4746	0.03614	27.6712	0.07614	13.1339	102.8933	7.8342
20	2.1911	0.4564	0.03358	29.7781	0.07358	13.5903	111.5647	8.2091
21	2.2788	0.4388	0.03128	31.9692	0.07128	14.0292	120.3414	8.5779
22	2.3699	0.4220	0.02920	34.2480	0.06920	14.4511	129.2024	8.9407
23	2.4647	0.4057	0.02731	36.6179	0.06731	14.8568	138.1284	9.2973
24	2.5633	0.3901	0.02559	39.0826	0.06559	15.2470	147.1012	9.6479
25	2.6658	0.3751	0.02401	41.6459	0.06401	15.6221	156.1040	9.9925
26	2.7725	0.3607	0.02257	44.3117	0.06257	15.9828	165.1212	10.3312
27	2.8834	0.3468	0.02124	47.0842	0.06124	16.3296	174.1385	10.6640
28	2.9987	0.3335	0.02001	49.9676	0.06001	16.6631	183.1424	10.9909
29	3.1187	0.3207	0.01888	52.9663	0.05888	16.9837	192.1206	11.3120
30	3.2434	0.3083	0.01783	56.0849	0.05783	17.2920	201.0618	11.6274
31	3.3731	0.2965	0.01686	59.3283	0.05686	17.5885	209.9556	11.9371
32	3.5081	0.2851	0.01595	62.7015	0.05595	17.8736	218.7924	12.2411
33	3.6484	0.2741	0.01510	66.2095	0.05510	18.1476	227.5634	12.5396
34	3.7943	0.2636	0.01431	69.8579	0.05431	18.4112	236.2607	12.8324
35	3.9461	0.2534	0.01358	73.6522	0.05358	18.6646	244.8768	13.1198
40	4.8010	0.2083	0.01052	95.0255	0.05052	19.7928	286.5303	14.4765
45	5.8412	0.1712	0.00826	121.0294	0.04826	20.7200	325.4028	15.7047
50	7.1067	0.1407	0.00655	152.6671	0.04655	21.4822	361.1638	16.8122
55	8.6464	0.1157	0.00523	191.1592	0.04523	22.1086	393.6890	17.8070
60	10.5196	0.0951	0.00420	237.9907	0.04420	22.6235	422.9966	18.6972
65	12.7987	0.0781	0.00339	294.9684	0.04339	23.0467	449.2014	19.4909
70	15.5716	0.0642	0.00275	364.2905	0.04275	23.3945	472.4789	20.1961
75	18.9453	0.0528	0.00223	448.6314	0.04223	23.6804	493.0408	20.8206
80	23.0498	0.0434	0.00181	551.2450	0.04181	23.9154	511.1161	21.3718
85	28.0436	0.0357	0.00148	676.0901	0.04148	24.1085	526.9384	21.8569
90	34.1193	0.0293	0.00121	827.9833	0.04121	24.2673	540.7369	22.2826
96	43.1718	0.0232	0.00095	1054.30	0.04095	24.4209	554.9312	22.7236
108	69.1195	0.0145	0.00059	1702.99	0.04059	24.6383	576.8949	23.4146
120	110.6626	0.0090	0.00036	2741.56	0.04036	24.7741	592.2428	23.9057
144	283.6618	0.0035	0.00014	7066.55	0.04014	24.9119	610.1055	24.4906

5%			TABLE 10	Discrete Cash Flow: Compound Interest Factors				5%
	Single Payments		**Uniform Series Payments**				**Arithmetic Gradients**	
	F/P Compound Amount	*P/F* Present Worth	*A/F* Sinking Fund	*F/A* Compound Amount	*A/P* Capital Recovery	*P/A* Present Worth	*P/G* Gradient Present Worth	*A/G* Gradient Uniform Series
n								
1	1.0500	0.9524	1.00000	1.0000	1.05000	0.9524		
2	1.1025	0.9070	0.48780	2.0500	0.53780	1.8594	0.9070	0.4878
3	1.1576	0.8638	0.31721	3.1525	0.36721	2.7232	2.6347	0.9675
4	1.2155	0.8227	0.23201	4.3101	0.28201	3.5460	5.1028	1.4391
5	1.2763	0.7835	0.18097	5.5256	0.23097	4.3295	8.2369	1.9025
6	1.3401	0.7462	0.14702	6.8019	0.19702	5.0757	11.9680	2.3579
7	1.4071	0.7107	0.12282	8.1420	0.17282	5.7864	16.2321	2.8052
8	1.4775	0.6768	0.10472	9.5491	0.15472	6.4632	20.9700	3.2445
9	1.5513	0.6446	0.09069	11.0266	0.14069	7.1078	26.1268	3.6758
10	1.6289	0.6139	0.07950	12.5779	0.12950	7.7217	31.6520	4.0991
11	1.7103	0.5847	0.07039	14.2068	0.12039	8.3064	37.4988	4.5144
12	1.7959	0.5568	0.06283	15.9171	0.11283	8.8633	43.6241	4.9219
13	1.8856	0.5303	0.05646	17.7130	0.10646	9.3936	49.9879	5.3215
14	1.9799	0.5051	0.05102	19.5986	0.10102	9.8986	56.5538	5.7133
15	2.0789	0.4810	0.04634	21.5786	0.09634	10.3797	63.2880	6.0973
16	2.1829	0.4581	0.04227	23.6575	0.09227	10.8378	70.1597	6.4736
17	2.2920	0.4363	0.03870	25.8404	0.08870	11.2741	77.1405	6.8423
18	2.4066	0.4155	0.03555	28.1324	0.08555	11.6896	84.2043	7.2034
19	2.5270	0.3957	0.03275	30.5390	0.08275	12.0853	91.3275	7.5569
20	2.6533	0.3769	0.03024	33.0660	0.08024	12.4622	98.4884	7.9030
21	2.7860	0.3589	0.02800	35.7193	0.07800	12.8212	105.6673	8.2416
22	2.9253	0.3418	0.02597	38.5052	0.07597	13.1630	112.8461	8.5730
23	3.0715	0.3256	0.02414	41.4305	0.07414	13.4886	120.0087	8.8971
24	3.2251	0.3101	0.02247	44.5020	0.07247	13.7986	127.1402	9.2140
25	3.3864	0.2953	0.02095	47.7271	0.07095	14.0939	134.2275	9.5238
26	3.5557	0.2812	0.01956	51.1135	0.06956	14.3752	141.2585	9.8266
27	3.7335	0.2678	0.01829	54.6691	0.06829	14.6430	148.2226	10.1224
28	3.9201	0.2551	0.01712	58.4026	0.06712	14.8981	155.1101	10.4114
29	4.1161	0.2429	0.01605	62.3227	0.06605	15.1411	161.9126	10.6936
30	4.3219	0.2314	0.01505	66.4388	0.06505	15.3725	168.6226	10.9691
31	4.5380	0.2204	0.01413	70.7608	0.06413	15.5928	175.2333	11.2381
32	4.7649	0.2099	0.01328	75.2988	0.06328	15.8027	181.7392	11.5005
33	5.0032	0.1999	0.01249	80.0638	0.06249	16.0025	188.1351	11.7566
34	5.2533	0.1904	0.01176	85.0670	0.06176	16.1929	194.4168	12.0063
35	5.5160	0.1813	0.01107	90.3203	0.06107	16.3742	200.5807	12.2498
40	7.0400	0.1420	0.00828	120.7998	0.05828	17.1591	229.5452	13.3775
45	8.9850	0.1113	0.00626	159.7002	0.05626	17.7741	255.3145	14.3644
50	11.4674	0.0872	0.00478	209.3480	0.05478	18.2559	277.9148	15.2233
55	14.6356	0.0683	0.00367	272.7126	0.05367	18.6335	297.5104	15.9664
60	18.6792	0.0535	0.00283	353.5837	0.05283	18.9293	314.3432	16.6062
65	23.8399	0.0419	0.00219	456.7980	0.05219	19.1611	328.6910	17.1541
70	30.4264	0.0329	0.00170	588.5285	0.05170	19.3427	340.8409	17.6212
75	38.8327	0.0258	0.00132	756.6537	0.05132	19.4850	351.0721	18.0176
80	49.5614	0.0202	0.00103	971.2288	0.05103	19.5965	359.6460	18.3526
85	63.2544	0.0158	0.00080	1245.09	0.05080	19.6838	366.8007	18.6346
90	80.7304	0.0124	0.00063	1594.61	0.05063	19.7523	372.7488	18.8712
95	103.0347	0.0097	0.00049	2040.69	0.05049	19.8059	377.6774	19.0689
96	108.1864	0.0092	0.00047	2143.73	0.05047	19.8151	378.5555	19.1044
98	119.2755	0.0084	0.00042	2365.51	0.05042	19.8323	380.2139	19.1714
100	131.5013	0.0076	0.00038	2610.03	0.05038	19.8479	381.7492	19.2337

6%			TABLE 11	Discrete Cash Flow: Compound Interest Factors				6%
	Single Payments		**Uniform Series Payments**				**Arithmetic Gradients**	
	F/P	*P/F*	*A/F*	*F/A*	*A/P*	*P/A*	*P/G*	*A/G*
n	Compound Amount	Present Worth	Sinking Fund	Compound Amount	Capital Recovery	Present Worth	Gradient Present Worth	Gradient Uniform Series
1	1.0600	0.9434	1.00000	1.0000	1.06000	0.9434		
2	1.1236	0.8900	0.48544	2.0600	0.54544	1.8334	0.8900	0.4854
3	1.1910	0.8396	0.31411	3.1836	0.37411	2.6730	2.5692	0.9612
4	1.2625	0.7921	0.22859	4.3746	0.28859	3.4651	4.9455	1.4272
5	1.3382	0.7473	0.17740	5.6371	0.23740	4.2124	7.9345	1.8836
6	1.4185	0.7050	0.14336	6.9753	0.20336	4.9173	11.4594	2.3304
7	1.5036	0.6651	0.11914	8.3938	0.17914	5.5824	15.4497	2.7676
8	1.5938	0.6274	0.10104	9.8975	0.16104	6.2098	19.8416	3.1952
9	1.6895	0.5919	0.08702	11.4913	0.14702	6.8017	24.5768	3.6133
10	1.7908	0.5584	0.07587	13.1808	0.13587	7.3601	29.6023	4.0220
11	1.8983	0.5268	0.06679	14.9716	0.12679	7.8869	34.8702	4.4213
12	2.0122	0.4970	0.05928	16.8699	0.11928	8.3838	40.3369	4.8113
13	2.1329	0.4688	0.05296	18.8821	0.11296	8.8527	45.9629	5.1920
14	2.2609	0.4423	0.04758	21.0151	0.10758	9.2950	51.7128	5.5635
15	2.3966	0.4173	0.04296	23.2760	0.10296	9.7122	57.5546	5.9260
16	2.5404	0.3936	0.03895	25.6725	0.09895	10.1059	63.4592	6.2794
17	2.6928	0.3714	0.03544	28.2129	0.09544	10.4773	69.4011	6.6240
18	2.8543	0.3503	0.03236	30.9057	0.09236	10.8276	75.3569	6.9597
19	3.0256	0.3305	0.02962	33.7600	0.08962	11.1581	81.3062	7.2867
20	3.2071	0.3118	0.02718	36.7856	0.08718	11.4699	87.2304	7.6051
21	3.3996	0.2942	0.02500	39.9927	0.08500	11.7641	93.1136	7.9151
22	3.6035	0.2775	0.02305	43.3923	0.08305	12.0416	98.9412	8.2166
23	3.8197	0.2618	0.02128	46.9958	0.08128	12.3034	104.7007	8.5099
24	4.0489	0.2470	0.01968	50.8156	0.07968	12.5504	110.3812	8.7951
25	4.2919	0.2330	0.01823	54.8645	0.07823	12.7834	115.9732	9.0722
26	4.5494	0.2198	0.01690	59.1564	0.07690	13.0032	121.4684	9.3414
27	4.8223	0.2074	0.01570	63.7058	0.07570	13.2105	126.8600	9.6029
28	5.1117	0.1956	0.01459	68.5281	0.07459	13.4062	132.1420	9.8568
29	5.4184	0.1846	0.01358	73.6398	0.07358	13.5907	137.3096	10.1032
30	5.7435	0.1741	0.01265	79.0582	0.07265	13.7648	142.3588	10.3422
31	6.0881	0.1643	0.01179	84.8017	0.07179	13.9291	147.2864	10.5740
32	6.4534	0.1550	0.01100	90.8898	0.07100	14.0840	152.0901	10.7988
33	6.8406	0.1462	0.01027	97.3432	0.07027	14.2302	156.7681	11.0166
34	7.2510	0.1379	0.00960	104.1838	0.06960	14.3681	161.3192	11.2276
35	7.6861	0.1301	0.00897	111.4348	0.06897	14.4982	165.7427	11.4319
40	10.2857	0.0972	0.00646	154.7620	0.06646	15.0463	185.9568	12.3590
45	13.7646	0.0727	0.00470	212.7435	0.06470	15.4558	203.1096	13.1413
50	18.4202	0.0543	0.00344	290.3359	0.06344	15.7619	217.4574	13.7964
55	24.6503	0.0406	0.00254	394.1720	0.06254	15.9905	229.3222	14.3411
60	32.9877	0.0303	0.00188	533.1282	0.06188	16.1614	239.0428	14.7909
65	44.1450	0.0227	0.00139	719.0829	0.06139	16.2891	246.9450	15.1601
70	59.0759	0.0169	0.00103	967.9322	0.06103	16.3845	253.3271	15.4613
75	79.0569	0.0126	0.00077	1300.95	0.06077	16.4558	258.4527	15.7058
80	105.7960	0.0095	0.00057	1746.60	0.06057	16.5091	262.5493	15.9033
85	141.5789	0.0071	0.00043	2342.98	0.06043	16.5489	265.8096	16.0620
90	189.4645	0.0053	0.00032	3141.08	0.06032	16.5787	268.3946	16.1891
95	253.5463	0.0039	0.00024	4209.10	0.06024	16.6009	270.4375	16.2905
96	268.7590	0.0037	0.00022	4462.65	0.06022	16.6047	270.7909	16.3081
98	301.9776	0.0033	0.00020	5016.29	0.06020	16.6115	271.4491	16.3411
100	339.3021	0.0029	0.00018	5638.37	0.06018	16.6175	272.0471	16.3711

7%	TABLE 12	Discrete Cash Flow: Compound Interest Factors					7%
	Single Payments		**Uniform Series Payments**				**Arithmetic Gradients**	

n	F/P Compound Amount	P/F Present Worth	A/F Sinking Fund	F/A Compound Amount	A/P Capital Recovery	P/A Present Worth	P/G Gradient Present Worth	A/G Gradient Uniform Series
1	1.0700	0.9346	1.00000	1.0000	1.07000	0.9346		
2	1.1449	0.8734	0.48309	2.0700	0.55309	1.8080	0.8734	0.4831
3	1.2250	0.8163	0.31105	3.2149	0.38105	2.6243	2.5060	0.9549
4	1.3108	0.7629	0.22523	4.4399	0.29523	3.3872	4.7947	1.4155
5	1.4026	0.7130	0.17389	5.7507	0.24389	4.1002	7.6467	1.8650
6	1.5007	0.6663	0.13980	7.1533	0.20980	4.7665	10.9784	2.3032
7	1.6058	0.6227	0.11555	8.6540	0.18555	5.3893	14.7149	2.7304
8	1.7182	0.5820	0.09747	10.2598	0.16747	5.9713	18.7889	3.1465
9	1.8385	0.5439	0.08349	11.9780	0.15349	6.5152	23.1404	3.5517
10	1.9672	0.5083	0.07238	13.8164	0.14238	7.0236	27.7156	3.9461
11	2.1049	0.4751	0.06336	15.7836	0.13336	7.4987	32.4665	4.3296
12	2.2522	0.4440	0.05590	17.8885	0.12590	7.9427	37.3506	4.7025
13	2.4098	0.4150	0.04965	20.1406	0.11965	8.3577	42.3302	5.0648
14	2.5785	0.3878	0.04434	22.5505	0.11434	8.7455	47.3718	5.4167
15	2.7590	0.3624	0.03979	25.1290	0.10979	9.1079	52.4461	5.7583
16	2.9522	0.3387	0.03586	27.8881	0.10586	9.4466	57.5271	6.0897
17	3.1588	0.3166	0.03243	30.8402	0.10243	9.7632	62.5923	6.4110
18	3.3799	0.2959	0.02941	33.9990	0.09941	10.0591	67.6219	6.7225
19	3.6165	0.2765	0.02675	37.3790	0.09675	10.3356	72.5991	7.0242
20	3.8697	0.2584	0.02439	40.9955	0.09439	10.5940	77.5091	7.3163
21	4.1406	0.2415	0.02229	44.8652	0.09229	10.8355	82.3393	7.5990
22	4.4304	0.2257	0.02041	49.0057	0.09041	11.0612	87.0793	7.8725
23	4.7405	0.2109	0.01871	53.4361	0.08871	11.2722	91.7201	8.1369
24	5.0724	0.1971	0.01719	58.1767	0.08719	11.4693	96.2545	8.3923
25	5.4274	0.1842	0.01581	63.2490	0.08581	11.6536	100.6765	8.6391
26	5.8074	0.1722	0.01456	68.6765	0.08456	11.8258	104.9814	8.8773
27	6.2139	0.1609	0.01343	74.4838	0.08343	11.9867	109.1656	9.1072
28	6.6488	0.1504	0.01239	80.6977	0.08239	12.1371	113.2264	9.3289
29	7.1143	0.1406	0.01145	87.3465	0.08145	12.2777	117.1622	9.5427
30	7.6123	0.1314	0.01059	94.4608	0.08059	12.4090	120.9718	9.7487
31	8.1451	0.1228	0.00980	102.0730	0.07980	12.5318	124.6550	9.9471
32	8.7153	0.1147	0.00907	110.2182	0.07907	12.6466	128.2120	10.1381
33	9.3253	0.1072	0.00841	118.9334	0.07841	12.7538	131.6435	10.3219
34	9.9781	0.1002	0.00780	128.2588	0.07780	12.8540	134.9507	10.4987
35	10.6766	0.0937	0.00723	138.2369	0.07723	12.9477	138.1353	10.6687
40	14.9745	0.0668	0.00501	199.6351	0.07501	13.3317	152.2928	11.4233
45	21.0025	0.0476	0.00350	285.7493	0.07350	13.6055	163.7559	12.0360
50	29.4570	0.0339	0.00246	406.5289	0.07246	13.8007	172.9051	12.5287
55	41.3150	0.0242	0.00174	575.9286	0.07174	13.9399	180.1243	12.9215
60	57.9464	0.0173	0.00123	813.5204	0.07123	14.0392	185.7677	13.2321
65	81.2729	0.0123	0.00087	1146.76	0.07087	14.1099	190.1452	13.4760
70	113.9894	0.0088	0.00062	1614.13	0.07062	14.1604	193.5185	13.6662
75	159.8760	0.0063	0.00044	2269.66	0.07044	14.1964	196.1035	13.8136
80	224.2344	0.0045	0.00031	3189.06	0.07031	14.2220	198.0748	13.9273
85	314.5003	0.0032	0.00022	4478.58	0.07022	14.2403	199.5717	14.0146
90	441.1030	0.0023	0.00016	6287.19	0.07016	14.2533	200.7042	14.0812
95	618.6697	0.0016	0.00011	8823.85	0.07011	14.2626	201.5581	14.1319
96	661.9766	0.0015	0.00011	9442.52	0.07011	14.2641	201.7016	14.1405
98	757.8970	0.0013	0.00009	10813	0.07009	14.2669	201.9651	14.1562
100	867.7163	0.0012	0.00008	12382	0.07008	14.2693	202.2001	14.1703

8%		TABLE **13**	Discrete Cash Flow: Compound Interest Factors					8%
	Single Payments		**Uniform Series Payments**				**Arithmetic Gradients**	
	F/P	*P/F*	*A/F*	*F/A*	*A/P*	*P/A*	*P/G*	*A/G*
n	Compound Amount	Present Worth	Sinking Fund	Compound Amount	Capital Recovery	Present Worth	Gradient Present Worth	Gradient Uniform Series
1	1.0800	0.9259	1.00000	1.0000	1.08000	0.9259		
2	1.1664	0.8573	0.48077	2.0800	0.56077	1.7833	0.8573	0.4808
3	1.2597	0.7938	0.30803	3.2464	0.38803	2.5771	2.4450	0.9487
4	1.3605	0.7350	0.22192	4.5061	0.30192	3.3121	4.6501	1.4040
5	1.4693	0.6806	0.17046	5.8666	0.25046	3.9927	7.3724	1.8465
6	1.5869	0.6302	0.13632	7.3359	0.21632	4.6229	10.5233	2.2763
7	1.7138	0.5835	0.11207	8.9228	0.19207	5.2064	14.0242	2.6937
8	1.8509	0.5403	0.09401	10.6366	0.17401	5.7466	17.8061	3.0985
9	1.9990	0.5002	0.08008	12.4876	0.16008	6.2469	21.8081	3.4910
10	2.1589	0.4632	0.06903	14.4866	0.14903	6.7101	25.9768	3.8713
11	2.3316	0.4289	0.06008	16.6455	0.14008	7.1390	30.2657	4.2395
12	2.5182	0.3971	0.05270	18.9771	0.13270	7.5361	34.6339	4.5957
13	2.7196	0.3677	0.04652	21.4953	0.12652	7.9038	39.0463	4.9402
14	2.9372	0.3405	0.04130	24.2149	0.12130	8.2442	43.4723	5.2731
15	3.1722	0.3152	0.03683	27.1521	0.11683	8.5595	47.8857	5.5945
16	3.4259	0.2919	0.03298	30.3243	0.11298	8.8514	52.2640	5.9046
17	3.7000	0.2703	0.02963	33.7502	0.10963	9.1216	56.5883	6.2037
18	3.9960	0.2502	0.02670	37.4502	0.10670	9.3719	60.8426	6.4920
19	4.3157	0.2317	0.02413	41.4463	0.10413	9.6036	65.0134	6.7697
20	4.6610	0.2145	0.02185	45.7620	0.10185	9.8181	69.0898	7.0369
21	5.0338	0.1987	0.01983	50.4229	0.09983	10.0168	73.0629	7.2940
22	5.4365	0.1839	0.01803	55.4568	0.09803	10.2007	76.9257	7.5412
23	5.8715	0.1703	0.01642	60.8933	0.09642	10.3711	80.6726	7.7786
24	6.3412	0.1577	0.01498	66.7648	0.09498	10.5288	84.2997	8.0066
25	6.8485	0.1460	0.01368	73.1059	0.09368	10.6748	87.8041	8.2254
26	7.3964	0.1352	0.01251	79.9544	0.09251	10.8100	91.1842	8.4352
27	7.9881	0.1252	0.01145	87.3508	0.09145	10.9352	94.4390	8.6363
28	8.6271	0.1159	0.01049	95.3388	0.09049	11.0511	97.5687	8.8289
29	9.3173	0.1073	0.00962	103.9659	0.08962	11.1584	100.5738	9.0133
30	10.0627	0.0994	0.00883	113.2832	0.08883	11.2578	103.4558	9.1897
31	10.8677	0.0920	0.00811	123.3459	0.08811	11.3498	106.2163	9.3584
32	11.7371	0.0852	0.00745	134.2135	0.08745	11.4350	108.8575	9.5197
33	12.6760	0.0789	0.00685	145.9506	0.08685	11.5139	111.3819	9.6737
34	13.6901	0.0730	0.00630	158.6267	0.08630	11.5869	113.7924	9.8208
35	14.7853	0.0676	0.00580	172.3168	0.08580	11.6546	116.0920	9.9611
40	21.7245	0.0460	0.00386	259.0565	0.08386	11.9246	126.0422	10.5699
45	31.9204	0.0313	0.00259	386.5056	0.08259	12.1084	133.7331	11.0447
50	46.9016	0.0213	0.00174	573.7702	0.08174	12.2335	139.5928	11.4107
55	68.9139	0.0145	0.00118	848.9232	0.08118	12.3186	144.0065	11.6902
60	101.2571	0.0099	0.00080	1253.21	0.08080	12.3766	147.3000	11.9015
65	148.7798	0.0067	0.00054	1847.25	0.08054	12.4160	149.7387	12.0602
70	218.6064	0.0046	0.00037	2720.08	0.08037	12.4428	151.5326	12.1783
75	321.2045	0.0031	0.00025	4002.56	0.08025	12.4611	152.8448	12.2658
80	471.9548	0.0021	0.00017	5886.94	0.08017	12.4735	153.8001	12.3301
85	693.4565	0.0014	0.00012	8655.71	0.08012	12.4820	154.4925	12.3772
90	1018.92	0.0010	0.00008	12724	0.08008	12.4877	154.9925	12.4116
95	1497.12	0.0007	0.00005	18702	0.08005	12.4917	155.3524	12.4365
96	1616.89	0.0006	0.00005	20199	0.08005	12.4923	155.4112	12.4406
98	1885.94	0.0005	0.00004	23562	0.08004	12.4934	155.5176	12.4480
100	2199.76	0.0005	0.00004	27485	0.08004	12.4943	155.6107	12.4545

9%			TABLE 14	Discrete Cash Flow: Compound Interest Factors				9%
	Single Payments		Uniform Series Payments				Arithmetic Gradients	
n	*F/P* Compound Amount	*P/F* Present Worth	*A/F* Sinking Fund	*F/A* Compound Amount	*A/P* Capital Recovery	*P/A* Present Worth	*P/G* Gradient Present Worth	*A/G* Gradient Uniform Series
1	1.0900	0.9174	1.00000	1.0000	1.09000	0.9174		
2	1.1881	0.8417	0.47847	2.0900	0.56847	1.7591	0.8417	0.4785
3	1.2950	0.7722	0.30505	3.2781	0.39505	2.5313	2.3860	0.9426
4	1.4116	0.7084	0.21867	4.5731	0.30867	3.2397	4.5113	1.3925
5	1.5386	0.6499	0.16709	5.9847	0.25709	3.8897	7.1110	1.8282
6	1.6771	0.5963	0.13292	7.5233	0.22292	4.4859	10.0924	2.2498
7	1.8280	0.5470	0.10869	9.2004	0.19869	5.0330	13.3746	2.6574
8	1.9926	0.5019	0.09067	11.0285	0.18067	5.5348	16.8877	3.0512
9	2.1719	0.4604	0.07680	13.0210	0.16680	5.9952	20.5711	3.4312
10	2.3674	0.4224	0.06582	15.1929	0.15582	6.4177	24.3728	3.7978
11	2.5804	0.3875	0.05695	17.5603	0.14695	6.8052	28.2481	4.1510
12	2.8127	0.3555	0.04965	20.1407	0.13965	7.1607	32.1590	4.4910
13	3.0658	0.3262	0.04357	22.9534	0.13357	7.4869	36.0731	4.8182
14	3.3417	0.2992	0.03843	26.0192	0.12843	7.7862	39.9633	5.1326
15	3.6425	0.2745	0.03406	29.3609	0.12406	8.0607	43.8069	5.4346
16	3.9703	0.2519	0.03030	33.0034	0.12030	8.3126	47.5849	5.7245
17	4.3276	0.2311	0.02705	36.9737	0.11705	8.5436	51.2821	6.0024
18	4.7171	0.2120	0.02421	41.3013	0.11421	8.7556	54.8860	6.2687
19	5.1417	0.1945	0.02173	46.0185	0.11173	8.9501	58.3868	6.5236
20	5.6044	0.1784	0.01955	51.1601	0.10955	9.1285	61.7770	6.7674
21	6.1088	0.1637	0.01762	56.7645	0.10762	9.2922	65.0509	7.0006
22	6.6586	0.1502	0.01590	62.8733	0.10590	9.4424	68.2048	7.2232
23	7.2579	0.1378	0.01438	69.5319	0.10438	9.5802	71.2359	7.4357
24	7.9111	0.1264	0.01302	76.7898	0.10302	9.7066	74.1433	7.6384
25	8.6231	0.1160	0.01181	84.7009	0.10181	9.8226	76.9265	7.8316
26	9.3992	0.1064	0.01072	93.3240	0.10072	9.9290	79.5863	8.0156
27	10.2451	0.0976	0.00973	102.7231	0.09973	10.0266	82.1241	8.1906
28	11.1671	0.0895	0.00885	112.9682	0.09885	10.1161	84.5419	8.3571
29	12.1722	0.0822	0.00806	124.1354	0.09806	10.1983	86.8422	8.5154
30	13.2677	0.0754	0.00734	136.3075	0.09734	10.2737	89.0280	8.6657
31	14.4618	0.0691	0.00669	149.5752	0.09669	10.3428	91.1024	8.8083
32	15.7633	0.0634	0.00610	164.0370	0.09610	10.4062	93.0690	8.9436
33	17.1820	0.0582	0.00556	179.8003	0.09556	10.4644	94.9314	9.0718
34	18.7284	0.0534	0.00508	196.9823	0.09508	10.5178	96.6935	9.1933
35	20.4140	0.0490	0.00464	215.7108	0.09464	10.5668	98.3590	9.3083
40	31.4094	0.0318	0.00296	337.8824	0.09296	10.7574	105.3762	9.7957
45	48.3273	0.0207	0.00190	525.8587	0.09190	10.8812	110.5561	10.1603
50	74.3575	0.0134	0.00123	815.0836	0.09123	10.9617	114.3251	10.4295
55	114.4083	0.0087	0.00079	1260.09	0.09079	11.0140	117.0362	10.6261
60	176.0313	0.0057	0.00051	1944.79	0.09051	11.0480	118.9683	10.7683
65	270.8460	0.0037	0.00033	2998.29	0.09033	11.0701	120.3344	10.8702
70	416.7301	0.0024	0.00022	4619.22	0.09022	11.0844	121.2942	10.9427
75	641.1909	0.0016	0.00014	7113.23	0.09014	11.0938	121.9646	10.9940
80	986.5517	0.0010	0.00009	10951	0.09009	11.0998	122.4306	11.0299
85	1517.93	0.0007	0.00006	16855	0.09006	11.1038	122.7533	11.0551
90	2335.53	0.0004	0.00004	25939	0.09004	11.1064	122.9758	11.0726
95	3593.50	0.0003	0.00003	39917	0.09003	11.1080	123.1287	11.0847
96	3916.91	0.0003	0.00002	43510	0.09002	11.1083	123.1529	11.0866
98	4653.68	0.0002	0.00002	51696	0.09002	11.1087	123.1963	11.0900
100	5529.04	0.0002	0.00002	61423	0.09002	11.1091	123.2335	11.0930

10%			TABLE 15	Discrete Cash Flow: Compound Interest Factors				10%
	Single Payments		**Uniform Series Payments**				**Arithmetic Gradients**	
	F/P Compound Amount	*P/F* Present Worth	*A/F* Sinking Fund	*F/A* Compound Amount	*A/P* Capital Recovery	*P/A* Present Worth	*P/G* Gradient Present Worth	*A/G* Gradient Uniform Series
n								
1	1.1000	0.9091	1.00000	1.0000	1.10000	0.9091		
2	1.2100	0.8264	0.47619	2.1000	0.57619	1.7355	0.8264	0.4762
3	1.3310	0.7513	0.30211	3.3100	0.40211	2.4869	2.3291	0.9366
4	1.4641	0.6830	0.21547	4.6410	0.31547	3.1699	4.3781	1.3812
5	1.6105	0.6209	0.16380	6.1051	0.26380	3.7908	6.8618	1.8101
6	1.7716	0.5645	0.12961	7.7156	0.22961	4.3553	9.6842	2.2236
7	1.9487	0.5132	0.10541	9.4872	0.20541	4.8684	12.7631	2.6216
8	2.1436	0.4665	0.08744	11.4359	0.18744	5.3349	16.0287	3.0045
9	2.3579	0.4241	0.07364	13.5795	0.17364	5.7590	19.4215	3.3724
10	2.5937	0.3855	0.06275	15.9374	0.16275	6.1446	22.8913	3.7255
11	2.8531	0.3505	0.05396	18.5312	0.15396	6.4951	26.3963	4.0641
12	3.1384	0.3186	0.04676	21.3843	0.14676	6.8137	29.9012	4.3884
13	3.4523	0.2897	0.04078	24.5227	0.14078	7.1034	33.3772	4.6988
14	3.7975	0.2633	0.03575	27.9750	0.13575	7.3667	36.8005	4.9955
15	4.1772	0.2394	0.03147	31.7725	0.13147	7.6061	40.1520	5.2789
16	4.5950	0.2176	0.02782	35.9497	0.12782	7.8237	43.4164	5.5493
17	5.0545	0.1978	0.02466	40.5447	0.12466	8.0216	46.5819	5.8071
18	5.5599	0.1799	0.02193	45.5992	0.12193	8.2014	49.6395	6.0526
19	6.1159	0.1635	0.01955	51.1591	0.11955	8.3649	52.5827	6.2861
20	6.7275	0.1486	0.01746	57.2750	0.11746	8.5136	55.4069	6.5081
21	7.4002	0.1351	0.01562	64.0025	0.11562	8.6487	58.1095	6.7189
22	8.1403	0.1228	0.01401	71.4027	0.11401	8.7715	60.6893	6.9189
23	8.9543	0.1117	0.01257	79.5430	0.11257	8.8832	63.1462	7.1085
24	9.8497	0.1015	0.01130	88.4973	0.11130	8.9847	65.4813	7.2881
25	10.8347	0.0923	0.01017	98.3471	0.11017	9.0770	67.6964	7.4580
26	11.9182	0.0839	0.00916	109.1818	0.10916	9.1609	69.7940	7.6186
27	13.1100	0.0763	0.00826	121.0999	0.10826	9.2372	71.7773	7.7704
28	14.4210	0.0693	0.00745	134.2099	0.10745	9.3066	73.6495	7.9137
29	15.8631	0.0630	0.00673	148.6309	0.10673	9.3696	75.4146	8.0489
30	17.4494	0.0573	0.00608	164.4940	0.10608	9.4269	77.0766	8.1762
31	19.1943	0.0521	0.00550	181.9434	0.10550	9.4790	78.6395	8.2962
32	21.1138	0.0474	0.00497	201.1378	0.10497	9.5264	80.1078	8.4091
33	23.2252	0.0431	0.00450	222.2515	0.10450	9.5694	81.4856	8.5152
34	25.5477	0.0391	0.00407	245.4767	0.10407	9.6086	82.7773	8.6149
35	28.1024	0.0356	0.00369	271.0244	0.10369	9.6442	83.9872	8.7086
40	45.2593	0.0221	0.00226	442.5926	0.10226	9.7791	88.9525	9.0962
45	72.8905	0.0137	0.00139	718.9048	0.10139	9.8628	92.4544	9.3740
50	117.3909	0.0085	0.00086	1163.91	0.10086	9.9148	94.8889	9.5704
55	189.0591	0.0053	0.00053	1880.59	0.10053	9.9471	96.5619	9.7075
60	304.4816	0.0033	0.00033	3034.82	0.10033	9.9672	97.7010	9.8023
65	490.3707	0.0020	0.00020	4893.71	0.10020	9.9796	98.4705	9.8672
70	789.7470	0.0013	0.00013	7887.47	0.10013	9.9873	98.9870	9.9113
75	1271.90	0.0008	0.00008	12709	0.10008	9.9921	99.3317	9.9410
80	2048.40	0.0005	0.00005	20474	0.10005	9.9951	99.5606	9.9609
85	3298.97	0.0003	0.00003	32980	0.10003	9.9970	99.7120	9.9742
90	5313.02	0.0002	0.00002	53120	0.10002	9.9981	99.8118	9.9831
95	8556.68	0.0001	0.00001	85557	0.10001	9.9988	99.8773	9.9889
96	9412.34	0.0001	0.00001	94113	0.10001	9.9989	99.8874	9.9898
98	11389	0.0001	0.00001		0.10001	9.9991	99.9052	9.9914
100	13781	0.0001	0.00001		0.10001	9.9993	99.9202	9.9927

11%				TABLE 16 Discrete Cash Flow: Compound Interest Factors				11%

	Single Payments		Uniform Series Payments				Arithmetic Gradients	
	F/P Compound Amount	*P/F* Present Worth	*A/F* Sinking Fund	*F/A* Compound Amount	*A/P* Capital Recovery	*P/A* Present Worth	*P/G* Gradient Present Worth	*A/G* Gradient Uniform Series
n								
1	1.1100	0.9009	1.00000	1.0000	1.11000	0.9009		
2	1.2321	0.8116	0.47393	2.1100	0.58393	1.7125	0.8116	0.4739
3	1.3676	0.7312	0.29921	3.3421	0.40921	2.4437	2.2740	0.9306
4	1.5181	0.6587	0.21233	4.7097	0.32233	3.1024	4.2502	1.3700
5	1.6851	0.5935	0.16057	6.2278	0.27057	3.6959	6.6240	1.7923
6	1.8704	0.5346	0.12638	7.9129	0.23638	4.2305	9.2972	2.1976
7	2.0762	0.4817	0.10222	9.7833	0.21222	4.7122	12.1872	2.5863
8	2.3045	0.4339	0.08432	11.8594	0.19432	5.1461	15.2246	2.9585
9	2.5580	0.3909	0.07060	14.1640	0.18060	5.5370	18.3520	3.3144
10	2.8394	0.3522	0.05980	16.7220	0.16980	5.8892	21.5217	3.6544
11	3.1518	0.3173	0.05112	19.5614	0.16112	6.2065	24.6945	3.9788
12	3.4985	0.2858	0.04403	22.7132	0.15403	6.4924	27.8388	4.2879
13	3.8833	0.2575	0.03815	26.2116	0.14815	6.7499	30.9290	4.5822
14	4.3104	0.2320	0.03323	30.0949	0.14323	6.9819	33.9449	4.8619
15	4.7846	0.2090	0.02907	34.4054	0.13907	7.1909	36.8709	5.1275
16	5.3109	0.1883	0.02552	39.1899	0.13552	7.3792	39.6953	5.3794
17	5.8951	0.1696	0.02247	44.5008	0.13247	7.5488	42.4095	5.6180
18	6.5436	0.1528	0.01984	50.3959	0.12984	7.7016	45.0074	5.8439
19	7.2633	0.1377	0.01756	56.9395	0.12756	7.8393	47.4856	6.0574
20	8.0623	0.1240	0.01558	64.2028	0.12558	7.9633	49.8423	6.2590
21	8.9492	0.1117	0.01384	72.2651	0.12384	8.0751	52.0771	6.4491
22	9.9336	0.1007	0.01231	81.2143	0.12231	8.1757	54.1912	6.6283
23	11.0263	0.0907	0.01097	91.1479	0.12097	8.2664	56.1864	6.7969
24	12.2392	0.0817	0.00979	102.1742	0.11979	8.3481	58.0656	6.9555
25	13.5855	0.0736	0.00874	114.4133	0.11874	8.4217	59.8322	7.1045
26	15.0799	0.0663	0.00781	127.9988	0.11781	8.4881	61.4900	7.2443
27	16.7386	0.0597	0.00699	143.0786	0.11699	8.5478	63.0433	7.3754
28	18.5799	0.0538	0.00626	159.8173	0.11626	8.6016	64.4965	7.4982
29	20.6237	0.0485	0.00561	178.3972	0.11561	8.6501	65.8542	7.6131
30	22.8923	0.0437	0.00502	199.0209	0.11502	8.6938	67.1210	7.7206
31	25.4104	0.0394	0.00451	221.9132	0.11451	8.7331	68.3016	7.8210
32	28.2056	0.0355	0.00404	247.3236	0.11404	8.7686	69.4007	7.9147
33	31.3082	0.0319	0.00363	275.5292	0.11363	8.8005	70.4228	8.0021
34	34.7521	0.0288	0.00326	306.8374	0.11326	8.8293	71.3724	8.0836
35	38.5749	0.0259	0.00293	341.5896	0.11293	8.8552	72.2538	8.1594
40	65.0009	0.0154	0.00172	581.8261	0.11172	8.9511	75.7789	8.4659
45	109.5302	0.0091	0.00101	986.6386	0.11101	9.0079	78.1551	8.6763
50	184.5648	0.0054	0.00060	1668.77	0.11060	9.0417	79.7341	8.8185
55	311.0025	0.0032	0.00035	2818.20	0.11035	9.0617	80.7712	8.9135
60	524.0572	0.0019	0.00021	4755.07	0.11021	9.0736	81.4461	8.9762
65	883.0669	0.0011	0.00012	8018.79	0.11012	9.0806	81.8819	9.0172
70	1488.02	0.0007	0.00007	13518	0.11007	9.0848	82.1614	9.0438
75	2507.40	0.0004	0.00004	22785	0.11004	9.0873	82.3397	9.0610
80	4225.11	0.0002	0.00003	38401	0.11003	9.0888	82.4529	9.0720
85	7119.56	0.0001	0.00002	64714	0.11002	9.0896	82.5245	9.0790

12%			TABLE **17**	Discrete Cash Flow: Compound Interest Factors				12%
	Single Payments		**Uniform Series Payments**				**Arithmetic Gradients**	
	F/P Compound Amount	*P/F* Present Worth	*A/F* Sinking Fund	*F/A* Compound Amount	*A/P* Capital Recovery	*P/A* Present Worth	*P/G* Gradient Present Worth	*A/G* Gradient Uniform Series
n								
1	1.1200	0.8929	1.00000	1.0000	1.12000	0.8929		
2	1.2544	0.7972	0.47170	2.1200	0.59170	1.6901	0.7972	0.4717
3	1.4049	0.7118	0.29635	3.3744	0.41635	2.4018	2.2208	0.9246
4	1.5735	0.6355	0.20923	4.7793	0.32923	3.0373	4.1273	1.3589
5	1.7623	0.5674	0.15741	6.3528	0.27741	3.6048	6.3970	1.7746
6	1.9738	0.5066	0.12323	8.1152	0.24323	4.1114	8.9302	2.1720
7	2.2107	0.4523	0.09912	10.0890	0.21912	4.5638	11.6443	2.5512
8	2.4760	0.4039	0.08130	12.2997	0.20130	4.9676	14.4714	2.9131
9	2.7731	0.3606	0.06768	14.7757	0.18768	5.3282	17.3563	3.2574
10	3.1058	0.3220	0.05698	17.5487	0.17698	5.6502	20.2541	3.5847
11	3.4785	0.2875	0.04842	20.6546	0.16842	5.9377	23.1288	3.8953
12	3.8960	0.2567	0.04144	24.1331	0.16144	6.1944	25.9523	4.1897
13	4.3635	0.2292	0.03568	28.0291	0.15568	6.4235	28.7024	4.4683
14	4.8871	0.2046	0.03087	32.3926	0.15087	6.6282	31.3624	4.7317
15	5.4736	0.1827	0.02682	37.2797	0.14682	6.8109	33.9202	4.9803
16	6.1304	0.1631	0.02339	42.7533	0.14339	6.9740	36.3670	5.2147
17	6.8660	0.1456	0.02046	48.8837	0.14046	7.1196	38.6973	5.4353
18	7.6900	0.1300	0.01794	55.7497	0.13794	7.2497	40.9080	5.6427
19	8.6128	0.1161	0.01576	63.4397	0.13576	7.3658	42.9979	5.8375
20	9.6463	0.1037	0.01388	72.0524	0.13388	7.4694	44.9676	6.0202
21	10.8038	0.0926	0.01224	81.6987	0.13224	7.5620	46.8188	6.1913
22	12.1003	0.0826	0.01081	92.5026	0.13081	7.6446	48.5543	6.3514
23	13.5523	0.0738	0.00956	104.6029	0.12956	7.7184	50.1776	6.5010
24	15.1786	0.0659	0.00846	118.1552	0.12846	7.7843	51.6929	6.6406
25	17.0001	0.0588	0.00750	133.3339	0.12750	7.8431	53.1046	6.7708
26	19.0401	0.0525	0.00665	150.3339	0.12665	7.8957	54.4177	6.8921
27	21.3249	0.0469	0.00590	169.3740	0.12590	7.9426	55.6369	7.0049
28	23.8839	0.0419	0.00524	190.6989	0.12524	7.9844	56.7674	7.1098
29	26.7499	0.0374	0.00466	214.5828	0.12466	8.0218	57.8141	7.2071
30	29.9599	0.0334	0.00414	241.3327	0.12414	8.0552	58.7821	7.2974
31	33.5551	0.0298	0.00369	271.2926	0.12369	8.0850	59.6761	7.3811
32	37.5817	0.0266	0.00328	304.8477	0.12328	8.1116	60.5010	7.4586
33	42.0915	0.0238	0.00292	342.4294	0.12292	8.1354	61.2612	7.5302
34	47.1425	0.0212	0.00260	384.5210	0.12260	8.1566	61.9612	7.5965
35	52.7996	0.0189	0.00232	431.6635	0.12232	8.1755	62.6052	7.6577
40	93.0510	0.0107	0.00130	767.0914	0.12130	8.2438	65.1159	7.8988
45	163.9876	0.0061	0.0074	1358.23	0.12074	8.2825	66.7342	8.0572
50	289.0022	0.0035	0.00042	2400.02	0.12042	8.3045	67.7624	8.1597
55	509.3206	0.0020	0.00024	4236.01	0.12024	8.3170	68.4082	8.2251
60	897.5969	0.0011	0.00013	7471.64	0.12013	8.3240	68.8100	8.2664
65	1581.87	0.0006	0.00008	13174	0.12008	8.3281	69.0581	8.2922
70	2787.80	0.0004	0.00004	23223	0.12004	8.3303	69.2103	8.3082
75	4913.06	0.0002	0.00002	40934	0.12002	8.3316	69.3031	8.3181
80	8658.48	0.0001	0.00001	72146	0.12001	8.3324	69.3594	8.3241
85	15259	0.0001	0.00001		0.12001	8.3328	69.3935	8.3278

14%			TABLE 18	Discrete Cash Flow: Compound Interest Factors			14%	
	Single Payments		**Uniform Series Payments**				**Arithmetic Gradients**	
	F/P Compound Amount	*P/F* Present Worth	*A/F* Sinking Fund	*F/A* Compound Amount	*A/P* Capital Recovery	*P/A* Present Worth	*P/G* Gradient Present Worth	*A/G* Gradient Uniform Series
n								
1	1.1400	0.8772	1.00000	1.0000	1.14000	0.8772		
2	1.2996	0.7695	0.46729	2.1400	0.60729	1.6467	0.7695	0.4673
3	1.4815	0.6750	0.29073	3.4396	0.43073	2.3216	2.1194	0.9129
4	1.6890	0.5921	0.20320	4.9211	0.34320	2.9137	3.8957	1.3370
5	1.9254	0.5194	0.15128	6.6101	0.29128	3.4331	5.9731	1.7399
6	2.1950	0.4556	0.11716	8.5355	0.25716	3.8887	8.2511	2.1218
7	2.5023	0.3996	0.09319	10.7305	0.23319	4.2883	10.6489	2.4832
8	2.8526	0.3506	0.07557	13.2328	0.21557	4.6389	13.1028	2.8246
9	3.2519	0.3075	0.06217	16.0853	0.20217	4.9464	15.5629	3.1463
10	3.7072	0.2697	0.05171	19.3373	0.19171	5.2161	17.9906	3.4490
11	4.2262	0.2366	0.04339	23.0445	0.18339	5.4527	20.3567	3.7333
12	4.8179	0.2076	0.03667	27.2707	0.17667	5.6603	22.6399	3.9998
13	5.4924	0.1821	0.03116	32.0887	0.17116	5.8424	24.8247	4.2491
14	6.2613	0.1597	0.02661	37.5811	0.16661	6.0021	26.9009	4.4819
15	7.1379	0.1401	0.02281	43.8424	0.16281	6.1422	28.8623	4.6990
16	8.1372	0.1229	0.01962	50.9804	0.15962	6.2651	30.7057	4.9011
17	9.2765	0.1078	0.01692	59.1176	0.15692	6.3729	32.4305	5.0888
18	10.5752	0.0946	0.01462	68.3941	0.15462	6.4674	34.0380	5.2630
19	12.0557	0.0829	0.01266	78.9692	0.15266	6.5504	35.5311	5.4243
20	13.7435	0.0728	0.01099	91.0249	0.15099	6.6231	36.9135	5.5734
21	15.6676	0.0638	0.00954	104.7684	0.14954	6.6870	38.1901	5.7111
22	17.8610	0.0560	0.00830	120.4360	0.14830	6.7429	39.3658	5.8381
23	20.3616	0.0491	0.00723	138.2970	0.14723	6.7921	40.4463	5.9549
24	23.2122	0.0431	0.00630	158.6586	0.14630	6.8351	41.4371	6.0624
25	26.4619	0.0378	0.00550	181.8708	0.14550	6.8729	42.3441	6.1610
26	30.1666	0.0331	0.00480	208.3327	0.14480	6.9061	43.1728	6.2514
27	34.3899	0.0291	0.00419	238.4993	0.14419	6.9352	43.9289	6.3342
28	39.2045	0.0255	0.00366	272.8892	0.14366	6.9607	44.6176	6.4100
29	44.6931	0.0224	0.00320	312.0937	0.14320	6.9830	45.2441	6.4791
30	50.9502	0.0196	0.00280	356.7868	0.14280	7.0027	45.8132	6.5423
31	58.0832	0.0172	0.00245	407.7370	0.14245	7.0199	46.3297	6.5998
32	66.2148	0.0151	0.00215	465.8202	0.14215	7.0350	46.7979	6.6522
33	75.4849	0.0132	0.00188	532.0350	0.14188	7.0482	47.2218	6.6998
34	86.0528	0.0116	0.00165	607.5199	0.14165	7.0599	47.6053	6.7431
35	98.1002	0.0102	0.00144	693.5727	0.14144	7.0700	47.9519	6.7824
40	188.8835	0.0053	0.00075	1342.03	0.14075	7.1050	49.2376	6.9300
45	363.6791	0.0027	0.00039	2590.56	0.14039	7.1232	49.9963	7.0188
50	700.2330	0.0014	0.00020	4994.52	0.14020	7.1327	50.4375	7.0714
55	1348.24	0.0007	0.00010	9623.13	0.14010	7.1376	50.6912	7.1020
60	2595.92	0.0004	0.00005	18535	0.14005	7.1401	50.8357	7.1197
65	4998.22	0.0002	0.00003	35694	0.14003	7.1414	50.9173	7.1298
70	9623.64	0.0001	0.00001	68733	0.14001	7.1421	50.9632	7.1356
75	18530	0.0001	0.00001		0.14001	7.1425	50.9887	7.1388
80	35677				0.14000	7.1427	51.0030	7.1406
85	68693				0.14000	7.1428	51.0108	7.1416

15%			TABLE 19	Discrete Cash Flow: Compound Interest Factors				15%

	Single Payments		Uniform Series Payments				Arithmetic Gradients	
	F/P Compound Amount	**P/F** Present Worth	**A/F** Sinking Fund	**F/A** Compound Amount	**A/P** Capital Recovery	**P/A** Present Worth	**P/G** Gradient Present Worth	**A/G** Gradient Uniform Series
n								
1	1.1500	0.8696	1.00000	1.0000	1.15000	0.8696		
2	1.3225	0.7561	0.46512	2.1500	0.61512	1.6257	0.7561	0.4651
3	1.5209	0.6575	0.28798	3.4725	0.43798	2.2832	2.0712	0.9071
4	1.7490	0.5718	0.20027	4.9934	0.35027	2.8550	3.7864	1.3263
5	2.0114	0.4972	0.14832	6.7424	0.29832	3.3522	5.7751	1.7228
6	2.3131	0.4323	0.11424	8.7537	0.26424	3.7845	7.9368	2.0972
7	2.6600	0.3759	0.09036	11.0668	0.24036	4.1604	10.1924	2.4498
8	3.0590	0.3269	0.07285	13.7268	0.22285	4.4873	12.4807	2.7813
9	3.5179	0.2843	0.05957	16.7858	0.20957	4.7716	14.7548	3.0922
10	4.0456	0.2472	0.04925	20.3037	0.19925	5.0188	16.9795	3.3832
11	4.6524	0.2149	0.04107	24.3493	0.19107	5.2337	19.1289	3.6549
12	5.3503	0.1869	0.03448	29.0017	0.18448	5.4206	21.1849	3.9082
13	6.1528	0.1625	0.02911	34.3519	0.17911	5.5831	23.1352	4.1438
14	7.0757	0.1413	0.02469	40.5047	0.17469	5.7245	24.9725	4.3624
15	8.1371	0.1229	0.02102	47.5804	0.17102	5.8474	26.6930	4.5650
16	9.3576	0.1069	0.01795	55.7175	0.16795	5.9542	28.2960	4.7522
17	10.7613	0.0929	0.01537	65.0751	0.16537	6.0472	29.7828	4.9251
18	12.3755	0.0808	0.01319	75.8364	0.16319	6.1280	31.1565	5.0843
19	14.2318	0.0703	0.01134	88.2118	0.16134	6.1982	32.4213	5.2307
20	16.3665	0.0611	0.00976	102.4436	0.15976	6.2593	33.5822	5.3651
21	18.8215	0.0531	0.00842	118.8101	0.15842	6.3125	34.6448	5.4883
22	21.6447	0.0462	0.00727	137.6316	0.15727	6.3587	35.6150	5.6010
23	24.8915	0.0402	0.00628	159.2764	0.15628	6.3988	36.4988	5.7040
24	28.6252	0.0349	0.00543	184.1678	0.15543	6.4338	37.3023	5.7979
25	32.9190	0.0304	0.00470	212.7930	0.15470	6.4641	38.0314	5.8834
26	37.8568	0.0264	0.00407	245.7120	0.15407	6.4906	38.6918	5.9612
27	43.5353	0.0230	0.00353	283.5688	0.15353	6.5135	39.2890	6.0319
28	50.0656	0.0200	0.00306	327.1041	0.15306	6.5335	39.8283	6.0960
29	57.5755	0.0174	0.00265	377.1697	0.15265	6.5509	40.3146	6.1541
30	66.2118	0.0151	0.00230	434.7451	0.15230	6.5660	40.7526	6.2066
31	76.1435	0.0131	0.00200	500.9569	0.15200	6.5791	41.1466	6.2541
32	87.5651	0.0114	0.00173	577.1005	0.15173	6.5905	41.5006	6.2970
33	100.6998	0.0099	0.00150	664.6655	0.15150	6.6005	41.8184	6.3357
34	115.8048	0.0086	0.00131	765.3654	0.15131	6.6091	42.1033	6.3705
35	133.1755	0.0075	0.00113	881.1702	0.15113	6.6166	42.3586	6.4019
40	267.8635	0.0037	0.00056	1779.09	0.15056	6.6418	43.2830	6.5168
45	538.7693	0.0019	0.00028	3585.13	0.15028	6.6543	43.8051	6.5830
50	1083.66	0.0009	0.00014	7217.72	0.15014	6.6605	44.0958	6.6205
55	2179.62	0.0005	0.00007	14524	0.15007	6.6636	44.2558	6.6414
60	4384.00	0.0002	0.00003	29220	0.15003	6.6651	44.3431	6.6530
65	8817.79	0.0001	0.00002	58779	0.15002	6.6659	44.3903	6.6593
70	17736	0.0001	0.00001		0.15001	6.6663	44.4156	6.6627
75	35673				0.15000	6.6665	44.4292	6.6646
80	71751				0.15000	6.6666	44.4364	6.6656
85					0.15000	6.6666	44.4402	6.6661

16%		TABLE 20	Discrete Cash Flow: Compound Interest Factors				16%
	Single Payments			Uniform Series Payments			Arithmetic Gradients	

	F/P	P/F	A/F	F/A	A/P	P/A	P/G	A/G
n	Compound Amount	Present Worth	Sinking Fund	Compound Amount	Capital Recovery	Present Worth	Gradient Present Worth	Gradient Uniform Series
1	1.1600	0.8621	1.00000	1.0000	1.16000	0.8621		
2	1.3456	0.7432	0.46296	2.1600	0.62296	1.6052	0.7432	0.4630
3	1.5609	0.6407	0.28526	3.5056	0.44526	2.2459	2.0245	0.9014
4	1.8106	0.5523	0.19738	5.0665	0.35738	2.7982	3.6814	1.3156
5	2.1003	0.4761	0.14541	6.8771	0.30541	3.2743	5.5858	1.7060
6	2.4364	0.4104	0.11139	8.9775	0.27139	3.6847	7.6380	2.0729
7	2.8262	0.3538	0.08761	11.4139	0.24761	4.0386	9.7610	2.4169
8	3.2784	0.3050	0.07022	14.2401	0.23022	4.3436	11.8962	2.7388
9	3.8030	0.2630	0.05708	17.5185	0.21708	4.6065	13.9998	3.0391
10	4.4114	0.2267	0.04690	21.3215	0.20690	4.8332	16.0399	3.3187
11	5.1173	0.1954	0.03886	25.7329	0.19886	5.0286	17.9941	3.5783
12	5.9360	0.1685	0.03241	30.8502	0.19241	5.1971	19.8472	3.8189
13	6.8858	0.1452	0.02718	36.7862	0.18718	5.3423	21.5899	4.0413
14	7.9875	0.1252	0.02290	43.6720	0.18290	5.4675	23.2175	4.2464
15	9.2655	0.1079	0.01936	51.6595	0.17936	5.5755	24.7284	4.4352
16	10.7480	0.0930	0.01641	60.9250	0.17641	5.6685	26.1241	4.6086
17	12.4677	0.0802	0.01395	71.6730	0.17395	5.7487	27.4074	4.7676
18	14.4625	0.0691	0.01188	84.1407	0.17188	5.8178	28.5828	4.9130
19	16.7765	0.0596	0.01014	98.6032	0.17014	5.8775	29.6557	5.0457
20	19.4608	0.0514	0.00867	115.3797	0.16867	5.9288	30.6321	5.1666
22	26.1864	0.0382	0.00635	157.4150	0.16635	6.0113	32.3200	5.3765
24	35.2364	0.0284	0.00467	213.9776	0.16467	6.0726	33.6970	5.5490
26	47.4141	0.0211	0.00345	290.0883	0.16345	6.1182	34.8114	5.6898
28	63.8004	0.0157	0.00255	392.5028	0.16255	6.1520	35.7073	5.8041
30	85.8499	0.0116	0.00189	530.3117	0.16189	6.1772	36.4234	5.8964
32	115.5196	0.0087	0.00140	715.7475	0.16140	6.1959	36.9930	5.9706
34	155.4432	0.0064	0.00104	965.2698	0.16104	6.2098	37.4441	6.0299
35	180.3141	0.0055	0.00089	1120.71	0.16089	6.2153	37.6327	6.0548
36	209.1643	0.0048	0.00077	1301.03	0.16077	6.2201	37.8000	6.0771
38	281.4515	0.0036	0.00057	1752.82	0.16057	6.2278	38.0799	6.1145
40	378.7212	0.0026	0.00042	2360.76	0.16042	6.2335	38.2992	6.1441
45	795.4438	0.0013	0.00020	4965.27	0.16020	6.2421	38.6598	6.1934
50	1670.70	0.0006	0.00010	10436	0.16010	6.2463	38.8521	6.2201
55	3509.05	0.0003	0.00005	21925	0.16005	6.2482	38.9534	6.2343
60	7370.20	0.0001	0.00002	46058	0.16002	6.2492	39.0063	6.2419

18%			TABLE 21	Discrete Cash Flow: Compound Interest Factors				18%
	Single Payments		**Uniform Series Payments**				**Arithmetic Gradients**	
	F/P Compound Amount	*P/F* Present Worth	*A/F* Sinking Fund	*F/A* Compound Amount	*A/P* Capital Recovery	*P/A* Present Worth	*P/G* Gradient Present Worth	*A/G* Gradient Uniform Series
n								
1	1.1800	0.8475	1.00000	1.0000	1.18000	0.8475		
2	1.3924	0.7182	0.45872	2.1800	0.63872	1.5656	0.7182	0.4587
3	1.6430	0.6086	0.27992	3.5724	0.45992	2.1743	1.9354	0.8902
4	1.9388	0.5158	0.19174	5.2154	0.37174	2.6901	3.4828	1.2947
5	2.2878	0.4371	0.13978	7.1542	0.31978	3.1272	5.2312	1.6728
6	2.6996	0.3704	0.10591	9.4420	0.28591	3.4976	7.0834	2.0252
7	3.1855	0.3139	0.08236	12.1415	0.26236	3.8115	8.9670	2.3526
8	3.7589	0.2660	0.06524	15.3270	0.24524	4.0776	10.8292	2.6558
9	4.4355	0.2255	0.05239	19.0859	0.23239	4.3030	12.6329	2.9358
10	5.2338	0.1911	0.04251	23.5213	0.22251	4.4941	14.3525	3.1936
11	6.1759	0.1619	0.03478	28.7551	0.21478	4.6560	15.9716	3.4303
12	7.2876	0.1372	0.02863	34.9311	0.20863	4.7932	17.4811	3.6470
13	8.5994	0.1163	0.02369	42.2187	0.20369	4.9095	18.8765	3.8449
14	10.1472	0.0985	0.01968	50.8180	0.19968	5.0081	20.1576	4.0250
15	11.9737	0.0835	0.01640	60.9653	0.19640	5.0916	21.3269	4.1887
16	14.1290	0.0708	0.01371	72.9390	0.19371	5.1624	22.3885	4.3369
17	16.6722	0.0600	0.01149	87.0680	0.19149	5.2223	23.3482	4.4708
18	19.6733	0.0508	0.00964	103.7403	0.18964	5.2732	24.2123	4.5916
19	23.2144	0.0431	0.00810	123.4135	0.18810	5.3162	24.9877	4.7003
20	27.3930	0.0365	0.00682	146.6280	0.18682	5.3527	25.6813	4.7978
22	38.1421	0.0262	0.00485	206.3448	0.18485	5.4099	26.8506	4.9632
24	53.1090	0.0188	0.00345	289.4945	0.18345	5.4509	27.7725	5.0950
26	73.9490	0.0135	0.00247	405.2721	0.18247	5.4804	28.4935	5.1991
28	102.9666	0.0097	0.00177	566.4809	0.18177	5.5016	29.0537	5.2810
30	143.3706	0.0070	0.00126	790.9480	0.18126	5.5168	29.4864	5.3448
32	199.6293	0.0050	0.00091	1103.50	0.18091	5.5277	29.8191	5.3945
34	277.9638	0.0036	0.00065	1538.69	0.18065	5.5356	30.0736	5.4328
35	327.9973	0.0030	0.00055	1816.65	0.18055	5.5386	30.1773	5.4485
36	387.0368	0.0026	0.00047	2144.65	0.18047	5.5412	30.2677	5.4623
38	538.9100	0.0019	0.00033	2988.39	0.18033	5.5452	30.4152	5.4849
40	750.3783	0.0013	0.00024	4163.21	0.18024	5.5482	30.5269	5.5022
45	1716.68	0.0006	0.00010	9531.58	0.18010	5.5523	30.7006	5.5293
50	3927.36	0.0003	0.00005	21813	0.18005	5.5541	30.7856	5.5428
55	8984.84	0.0001	0.00002	49910	0.18002	5.5549	30.8268	5.5494
60	20555			114190	0.18001	5.5553	30.8465	5.5526

	Single Payments		Uniform Series Payments				Arithmetic Gradients	
	F/P	P/F	A/F	F/A	A/P	P/A	P/G	A/G
	Compound Amount	Present Worth	Sinking Fund	Compound Amount	Capital Recovery	Present Worth	Gradient Present Worth	Gradient Uniform Series
n								
1	1.2000	0.8333	1.00000	1.0000	1.20000	0.8333		
2	1.4400	0.6944	0.45455	2.2000	0.65455	1.5278	0.6944	0.4545
3	1.7280	0.5787	0.27473	3.6400	0.47473	2.1065	1.8519	0.8791
4	2.0736	0.4823	0.18629	5.3680	0.38629	2.5887	3.2986	1.2742
5	2.4883	0.4019	0.13438	7.4416	0.33438	2.9906	4.9061	1.6405
6	2.9860	0.3349	0.10071	9.9299	0.30071	3.3255	6.5806	1.9788
7	3.5832	0.2791	0.07742	12.9159	0.27742	3.6046	8.2551	2.2902
8	4.2998	0.2326	0.06061	16.4991	0.26061	3.8372	9.8831	2.5756
9	5.1598	0.1938	0.04808	20.7989	0.24808	4.0310	11.4335	2.8364
10	6.1917	0.1615	0.03852	25.9587	0.23852	4.1925	12.8871	3.0739
11	7.4301	0.1346	0.03110	32.1504	0.23110	4.3271	14.2330	3.2893
12	8.9161	0.1122	0.02526	39.5805	0.22526	4.4392	15.4667	3.4841
13	10.6993	0.0935	0.02062	48.4966	0.22062	4.5327	16.5883	3.6597
14	12.8392	0.0779	0.01689	59.1959	0.21689	4.6106	17.6008	3.8175
15	15.4070	0.0649	0.01388	72.0351	0.21388	4.6755	18.5095	3.9588
16	18.4884	0.0541	0.01144	87.4421	0.21144	4.7296	19.3208	4.0851
17	22.1861	0.0451	0.00944	105.9306	0.20944	4.7746	20.0419	4.1976
18	26.6233	0.0376	0.00781	128.1167	0.20781	4.8122	20.6805	4.2975
19	31.9480	0.0313	0.00646	154.7400	0.20646	4.8435	21.2439	4.3861
20	38.3376	0.0261	0.00536	186.6880	0.20536	4.8696	21.7395	4.4643
22	55.2061	0.0181	0.00369	271.0307	0.20369	4.9094	22.5546	4.5941
24	79.4968	0.0126	0.00255	392.4842	0.20255	4.9371	23.1760	4.6943
26	114.4755	0.0087	0.00176	567.3773	0.20176	4.9563	23.6460	4.7709
28	164.8447	0.0061	0.00122	819.2233	0.20122	4.9697	23.9991	4.8291
30	237.3763	0.0042	0.00085	1181.88	0.20085	4.9789	24.2628	4.8731
32	341.8219	0.0029	0.00059	1704.11	0.20059	4.9854	24.4588	4.9061
34	492.2235	0.0020	0.00041	2456.12	0.20041	4.9898	24.6038	4.9308
35	590.6682	0.0017	0.00034	2948.34	0.20034	4.9915	24.6614	4.9406
36	708.8019	0.0014	0.00028	3539.01	0.20028	4.9929	24.7108	4.9491
38	1020.67	0.0010	0.00020	5098.37	0.20020	4.9951	24.7894	4.9627
40	1469.77	0.0007	0.00014	7343.86	0.20014	4.9966	24.8469	4.9728
45	3657.26	0.0003	0.00005	18281	0.20005	4.9986	24.9316	4.9877
50	9100.44	0.0001	0.00002	45497	0.20002	4.9995	24.9698	4.9945
55	22645		0.00001		0.20001	4.9998	24.9868	4.9976

20%   TABLE 22   Discrete Cash Flow: Compound Interest Factors   **20%**

22%		TABLE 23	Discrete Cash Flow: Compound Interest Factors					22%
	Single Payments		**Uniform Series Payments**				**Arithmetic Gradients**	
n	*F/P* Compound Amount	*P/F* Present Worth	*A/F* Sinking Fund	*F/A* Compound Amount	*A/P* Capital Recovery	*P/A* Present Worth	*P/G* Gradient Present Worth	*A/G* Gradient Uniform Series
1	1.2200	0.8197	1.00000	1.0000	1.22000	0.8197		
2	1.4884	0.6719	0.45045	2.2200	0.67045	1.4915	0.6719	0.4505
3	1.8158	0.5507	0.26966	3.7084	0.48966	2.0422	1.7733	0.8683
4	2.2153	0.4514	0.18102	5.5242	0.40102	2.4936	3.1275	1.2542
5	2.7027	0.3700	0.12921	7.7396	0.34921	2.8636	4.6075	1.6090
6	3.2973	0.3033	0.09576	10.4423	0.31576	3.1669	6.1239	1.9337
7	4.0227	0.2486	0.07278	13.7396	0.29278	3.4155	7.6154	2.2297
8	4.9077	0.2038	0.05630	17.7623	0.27630	3.6193	9.0417	2.4982
9	5.9874	0.1670	0.04411	22.6700	0.26411	3.7863	10.3779	2.7409
10	7.3046	0.1369	0.03489	28.6574	0.25489	3.9232	11.6100	2.9593
11	8.9117	0.1122	0.02781	35.9620	0.24781	4.0354	12.7321	3.1551
12	10.8722	0.0920	0.02228	44.8737	0.24228	4.1274	13.7438	3.3299
13	13.2641	0.0754	0.01794	55.7459	0.23794	4.2028	14.6485	3.4855
14	16.1822	0.0618	0.01449	69.0100	0.23449	4.2646	15.4519	3.6233
15	19.7423	0.0507	0.01174	85.1922	0.23174	4.3152	16.1610	3.7451
16	24.0856	0.0415	0.00953	104.9345	0.22953	4.3567	16.7838	3.8524
17	29.3844	0.0340	0.00775	129.0201	0.22775	4.3908	17.3283	3.9465
18	35.8490	0.0279	0.00631	158.4045	0.22631	4.4187	17.8025	4.0289
19	43.7358	0.0229	0.00515	194.2535	0.22515	4.4415	18.2141	4.1009
20	53.3576	0.0187	0.00420	237.9893	0.22420	4.4603	18.5702	4.1635
22	79.4175	0.0126	0.00281	356.4432	0.22281	4.4882	19.1418	4.2649
24	118.2050	0.0085	0.00188	532.7501	0.22188	4.5070	19.5635	4.3407
26	175.9364	0.0057	0.00126	795.1653	0.22126	4.5196	19.8720	4.3968
28	261.8637	0.0038	0.00084	1185.74	0.22084	4.5281	20.0962	4.4381
30	389.7579	0.0026	0.00057	1767.08	0.22057	4.5338	20.2583	4.4683
32	580.1156	0.0017	0.00038	2632.34	0.22038	4.5376	20.3748	4.4902
34	863.4441	0.0012	0.00026	3920.20	0.22026	4.5402	20.4582	4.5060
35	1053.40	0.0009	0.00021	4783.64	0.22021	4.5411	20.4905	4.5122
36	1285.15	0.0008	0.00017	5837.05	0.22017	4.5419	20.5178	4.5174
38	1912.82	0.0005	0.00012	8690.08	0.22012	4.5431	20.5601	4.5256
40	2847.04	0.0004	0.00008	12937	0.22008	4.5439	20.5900	4.5314
45	7694.71	0.0001	0.00003	34971	0.22003	4.5449	20.6319	4.5396
50	20797		0.00001	94525	0.22001	4.5452	20.6492	4.5431
55	56207				0.22000	4.5454	20.6563	4.5445

	Single Payments		Uniform Series Payments				Arithmetic Gradients	
	F/P	*P/F*	*A/F*	*F/A*	*A/P*	*P/A*	*P/G*	*A/G*
	Compound Amount	Present Worth	Sinking Fund	Compound Amount	Capital Recovery	Present Worth	Gradient Present Worth	Gradient Uniform Series
n								
1	1.2400	0.8065	1.00000	1.0000	1.24000	0.8065		
2	1.5376	0.6504	0.44643	2.2400	0.68643	1.4568	0.6504	0.4464
3	1.9066	0.5245	0.26472	3.7776	0.50472	1.9813	1.6993	0.8577
4	2.3642	0.4230	0.17593	5.6842	0.41593	2.4043	2.9683	1.2346
5	2.9316	0.3411	0.12425	8.0484	0.36425	2.7454	4.3327	1.5782
6	3.6352	0.2751	0.09107	10.9801	0.33107	3.0205	5.7081	1.8898
7	4.5077	0.2218	0.06842	14.6153	0.30842	3.2423	7.0392	2.1710
8	5.5895	0.1789	0.05229	19.1229	0.29229	3.4212	8.2915	2.4236
9	6.9310	0.1443	0.04047	24.7125	0.28047	3.5655	9.4458	2.6492
10	8.5944	0.1164	0.03160	31.6434	0.27160	3.6819	10.4930	2.8499
11	10.6571	0.0938	0.02485	40.2379	0.26485	3.7757	11.4313	3.0276
12	13.2148	0.0757	0.01965	50.8950	0.25965	3.8514	12.2637	3.1843
13	16.3863	0.0610	0.01560	64.1097	0.25560	3.9124	12.9960	3.3218
14	20.3191	0.0492	0.01242	80.4961	0.25242	3.9616	13.6358	3.4420
15	25.1956	0.0397	0.00992	100.8151	0.24992	4.0013	14.1915	3.5467
16	31.2426	0.0320	0.00794	126.0108	0.24794	4.0333	14.6716	3.6376
17	38.7408	0.0258	0.00636	157.2534	0.24636	4.0591	15.0846	3.7162
18	48.0386	0.0208	0.00510	195.9942	0.24510	4.0799	15.4385	3.7840
19	59.5679	0.0168	0.00410	244.0328	0.24410	4.0967	15.7406	3.8423
20	73.8641	0.0135	0.00329	303.6006	0.24329	4.1103	15.9979	3.8922
22	113.5735	0.0088	0.00213	469.0563	0.24213	4.1300	16.4011	3.9712
24	174.6306	0.0057	0.00138	723.4610	0.24138	4.1428	16.6891	4.0284
26	268.5121	0.0037	0.00090	1114.63	0.24090	4.1511	16.8930	4.0695
28	412.8642	0.0024	0.00058	1716.10	0.24058	4.1566	17.0365	4.0987
30	634.8199	0.0016	0.00038	2640.92	0.24038	4.1601	17.1369	4.1193
32	976.0991	0.0010	0.00025	4062.91	0.24025	4.1624	17.2067	4.1338
34	1500.85	0.0007	0.00016	6249.38	0.24016	4.1639	17.2552	4.1440
35	1861.05	0.0005	0.00013	7750.23	0.24013	4.1664	17.2734	4.1479
36	2307.71	0.0004	0.00010	9611.28	0.24010	4.1649	17.2886	4.1511
38	3548.33	0.0003	0.00007	14781	0.24007	4.1655	17.3116	4.1560
40	5455.91	0.0002	0.00004	22729	0.24004	4.1659	17.3274	4.1593
45	15995	0.0001	0.00002	66640	0.24002	4.1664	17.3483	4.1639
50	46890		0.00001		0.24001	4.1666	17.3563	4.1653
55					0.24000	4.1666	17.3593	4.1663

25%			TABLE 25	Discrete Cash Flow: Compound Interest Factors				25%
	Single Payments		Uniform Series Payments				Arithmetic Gradients	
	F/P Compound Amount	P/F Present Worth	A/F Sinking Fund	F/A Compound Amount	A/P Capital Recovery	P/A Present Worth	P/G Gradient Present Worth	A/G Gradient Uniform Series
n								
1	1.2500	0.8000	1.00000	1.0000	1.25000	0.8000		
2	1.5625	0.6400	0.44444	2.2500	0.69444	1.4400	0.6400	0.4444
3	1.9531	0.5120	0.26230	3.8125	0.51230	1.9520	1.6640	0.8525
4	2.4414	0.4096	0.17344	5.7656	0.42344	2.3616	2.8928	1.2249
5	3.0518	0.3277	0.12185	8.2070	0.37185	2.6893	4.2035	1.5631
6	3.8147	0.2621	0.08882	11.2588	0.33882	2.9514	5.5142	1.8683
7	4.7684	0.2097	0.06634	15.0735	0.31634	3.1611	6.7725	2.1424
8	5.9605	0.1678	0.05040	19.8419	0.30040	3.3289	7.9469	2.3872
9	7.4506	0.1342	0.03876	25.8023	0.28876	3.4631	9.0207	2.6048
10	9.3132	0.1074	0.03007	33.2529	0.28007	3.5705	9.9870	2.7971
11	11.6415	0.0859	0.02349	42.5661	0.27349	3.6564	10.8460	2.9663
12	14.5519	0.0687	0.01845	54.2077	0.26845	3.7251	11.6020	3.1145
13	18.1899	0.0550	0.01454	68.7596	0.26454	3.7801	12.2617	3.2437
14	22.7374	0.0440	0.01150	86.9495	0.26150	3.8241	12.8334	3.3559
15	28.4217	0.0352	0.00912	109.6868	0.25912	3.8593	13.3260	3.4530
16	35.5271	0.0281	0.00724	138.1085	0.25724	3.8874	13.7482	3.5366
17	44.4089	0.0225	0.00576	173.6357	0.25576	3.9099	14.1085	3.6084
18	55.5112	0.0180	0.00459	218.0446	0.25459	3.9279	14.4147	3.6698
19	69.3889	0.0144	0.00366	273.5558	0.25366	3.9424	14.6741	3.7222
20	86.7362	0.0115	0.00292	342.9447	0.25292	3.9539	14.8932	3.7667
22	135.5253	0.0074	0.00186	538.1011	0.25186	3.9705	15.2326	3.8365
24	211.7582	0.0047	0.00119	843.0329	0.25119	3.9811	15.4711	3.8861
26	330.8722	0.0030	0.00076	1319.49	0.25076	3.9879	15.6373	3.9212
28	516.9879	0.0019	0.00048	2063.95	0.25048	3.9923	15.7524	3.9457
30	807.7936	0.0012	0.00031	3227.17	0.25031	3.9950	15.8316	3.9628
32	1262.18	0.0008	0.00020	5044.71	0.25020	3.9968	15.8859	3.9746
34	1972.15	0.0005	0.00013	7884.61	0.25013	3.9980	15.9229	3.9828
35	2465.19	0.0004	0.00010	9856.76	.025010	3.9984	15.9367	3.9858
36	3081.49	0.0003	0.00008	12322	0.25008	3.9987	15.9481	3.9883
38	4814.82	0.0002	0.00005	19255	0.25005	3.9992	15.9651	3.9921
40	7523.16	0.0001	0.00003	30089	0.25003	3.9995	15.9766	3.9947
45	22959		0.00001	91831	0.25001	3.9998	15.9915	3.9980
50	70065				0.25000	3.9999	15.9969	3.9993
55					0.25000	4.0000	15.9989	3.9997

30%   TABLE 26 Discrete Cash Flow: Compound Interest Factors   **30%**

	Single Payments		Uniform Series Payments				Arithmetic Gradients	
	F/P Compound Amount	*P/F* Present Worth	*A/F* Sinking Fund	*F/A* Compound Amount	*A/P* Capital Recovery	*P/A* Present Worth	*P/G* Gradient Present Worth	*A/G* Gradient Uniform Series
n								
1	1.3000	0.7692	1.00000	1.0000	1.30000	0.7692		
2	1.6900	0.5917	0.43478	2.3000	0.73478	1.3609	0.5917	0.4348
3	2.1970	0.4552	0.25063	3.9900	0.55063	1.8161	1.5020	0.8271
4	2.8561	0.3501	0.16163	6.1870	0.46163	2.1662	2.5524	1.1783
5	3.7129	0.2693	0.11058	9.0431	0.41058	2.4356	3.6297	1.4903
6	4.8268	0.2072	0.07839	12.7560	0.37839	2.6427	4.6656	1.7654
7	6.2749	0.1594	0.05687	17.5828	0.35687	2.8021	5.6218	2.0063
8	8.1573	0.1226	0.04192	23.8577	0.34192	2.9247	6.4800	2.2156
9	10.6045	0.0943	0.03124	32.0150	0.33124	3.0190	7.2343	2.3963
10	13.7858	0.0725	0.02346	42.6195	0.32346	3.0915	7.8872	2.5512
11	17.9216	0.0558	0.01773	56.4053	0.31773	3.1473	8.4452	2.6833
12	23.2981	0.0429	0.01345	74.3270	0.31345	3.1903	8.9173	2.7952
13	30.2875	0.0330	0.01024	97.6250	0.31024	3.2233	9.3135	2.8895
14	39.3738	0.0254	0.00782	127.9125	0.30782	3.2487	9.6437	2.9685
15	51.1859	0.0195	0.00598	167.2863	0.30598	3.2682	9.9172	3.0344
16	66.5417	0.0150	0.00458	218.4722	0.30458	3.2832	10.1426	3.0892
17	86.5042	0.0116	0.00351	285.0139	0.30351	3.2948	10.3276	3.1345
18	112.4554	0.0089	0.00269	371.5180	0.30269	3.3037	10.4788	3.1718
19	146.1920	0.0068	0.00207	483.9734	0.30207	3.3105	10.6019	3.2025
20	190.0496	0.0053	0.00159	630.1655	0.30159	3.3158	10.7019	3.2275
22	321.1839	0.0031	0.00094	1067.28	0.30094	3.3230	10.8482	3.2646
24	542.8008	0.0018	0.00055	1806.00	0.30055	3.3272	10.9433	3.2890
25	705.6410	0.0014	0.00043	2348.80	0.30043	3.3286	10.9773	3.2979
26	917.3333	0.0011	0.00033	3054.44	0.30033	3.3297	11.0045	3.3050
28	1550.29	0.0006	0.00019	5164.31	0.30019	3.3312	11.0437	3.3153
30	2620.00	0.0004	0.00011	8729.99	0.30011	3.3321	11.0687	3.3219
32	4427.79	0.0002	0.00007	14756	0.30007	3.3326	11.0845	3.3261
34	7482.97	0.0001	0.00004	24940	0.30004	3.3329	11.0945	3.3288
35	9727.86	0.0001	0.00003	32423	0.30003	3.3330	11.0980	3.3297

Index

A

Absolute cell referencing, 404
Accelerated write-off, 320, 323,
    325–28, 355–56
Accounting
    ratios, 421–24
    statements, 418–20
Accuracy of estimates, 293–94
Acid-test ratio, 422
Activity-based costing (ABC),
    309–11
A/F factor, 42–43
After-tax
    cash flow, 349–52
    debt *versus* equity financing,
        360–63
    and depreciation, 352–56
    evaluation, 349–52
    MARR, 6, 350
    rate or return, 350
    WACC, 364
A/G factor, 45. *See also* Gradients,
    arithmetic
Alternative depreciation system
    (ADS), 328
Alternatives
    and breakeven analysis, 213–16
    cost, 97
    defined, 4, 96–98
    incremental rate-of-return, 149–56
    independent, 96 (*See also*
        Independent projects)
    infinite life, 104–08, 130–32
    mutually exclusive, 96, 100–10,
        127–30, 152–56, 191–96
    revenue, 96

and simulation, 394–98
    types, 96–97
Amortization. *See* Depreciation
Annual interest rate
    effective, 73–77
    nominal, 73–77
Annual operating costs (AOC), 96,
    126, 346
Annual percentage rate (APR), 73
Annual percentage yield (APY), 73
Annual worth
    advantages, 125
    after-tax analysis, 352
    of annual operating costs, 126
    and B/C analysis, 189, 193–95
    and breakeven analysis, 213–16
    equivalent uniform, 125–30
    and EVA, 367–69
    and future worth, 126
    and incremental rate of
        return, 156
    of infinite-life projects, 130–32
    and inflation, 125, 278–79
    and present worth, 126
    and replacement analysis, 244–50
    and sensitivity analysis, 216,
        221–23
A/P factor, 40–41
Arithmetic gradient. *See* Gradient,
    arithmetic
Attributes
    evaluating multiple, 380–85
    weighting, 381–82
Average, 389–91
Average cost per unit, 212
AVERAGE function, Excel, 389, 394

B

Balance sheet, 418–19
Base amount, and gradients, 44, 46, 50–52
Basis, 319
B/C. *See* Benefit/cost ratio
Before-tax rate of return, 145–47
  and after-taxes, 350
Benefit and cost difference, 189–90
Benefit/cost ratio
  conventional, 189
  incremental analysis, 191–95
  independent project evaluation, 195
  modified, 189
Benefits
  direct *versus* usage costs, 191
  in public projects, 184, 189
Bonds
  and debt financing, 360, 362
  description, 99
  present worth, 99–100
  rate of return, 147
Book depreciation, 319, 325
Book value
  by declining balance method, 323
  defined, 319
  by double declining balance method, 323
  and EVA, 368–69
  by MACRS, 325
  versus market value, 319, 358–60
  by straight line method, 322
Borrowing rate, 161, 165
Bottom-up approach, 292–93
Breakeven analysis. *See also* Sensitivity analysis
  average cost per unit, 212
  description, 208–13
  fixed costs, 209
  and GOAL SEEK, 229–30
  and make-buy decisions, 207, 215–16
  PW *versus* i graph, 168
  and replacement value, 250
  versus sensitivity analysis, 216
  single project, 208–13
  and time value of money, 213
  total cost, 208–13
  two alternatives, 213–16
  variable costs, 209
Breakeven value, 208–09
Bundles. *See* Independent projects

C

Calculator. *See* Financial calculator
Canadian depreciation system, 329
Capital
  cost of (*See* Cost of capital)
  debt *versus* equity, 360
  recovery and AW, 126
Capital budgeting. *See* Independent projects
Capitalized cost, 14, 104–08, 130–32
  and annual worth, 130–32
  computation, 104
  and present worth, 104–08
Capital financing
  debt, 360
  debt *versus* equity, 360–62
  equity, 360
Capital gain, 353, 358
Capital loss, 353, 354
Capital recovery
  in annual worth of costs, 126–27
  defined, 126
  and inflation, 278–79
  and replacement analysis, 244–45
Capital recovery factor. *See* A/P factor
Capital recovery for assets. *See* Depreciation
Cash flow
  after-tax, 349–51
  and EVA, 367–69
  before-tax, 349
  continuous, 76
  conventional series, 156–57
  defined, 4
  diagramming, 18–20

discounted, 94
estimating, 16–17, 291–305
incremental, 149
net, 17, 349
nonconventional, 156–57
and payback analysis, 223–24
positive net, and ROR, 160, 162
and public sector projects, 184
recurring and nonrecurring, 104
and replacement analysis, 250
revenue *versus* cost, 96–97
sign test, 156
zero, 53–54, 411
Cash flow after taxes (CFAT),
      349–52
Cash flow before taxes (CFBT), 349
Cash flow diagram, 18–20
Cell references, in Excel, 404
Challenger, in replacement analysis,
      243, 246–50, 358
Class life, 328
Compound amount factors
   single payment (F/P), 35–37
   uniform series (F/A), 42–43
Compound interest, 10–14
Compounding
   continuous, 76–77
   frequency, 75
   interperiod, 82–84
   versus simple interest, 9–12
Compounding period
   continuous, 76
   defined, 73
   and payment period, 77–78
Constant value dollars, 266,
      269–70
Consultant's perspective, 243
Consumer Price Index, 267, 296
Continuous compounding, 76–77
Contracts, types, 187
Conventional benefit/cost ratio, 189
Conventional cash flow series,
      156–57
Conventional gradient, 45

Cost, life-cycle, 103
Cost alternative, 97, 150, 152
Cost-capacity equations, 299–301
Cost centers, 309
Cost depletion. *See* Depletion
Cost driver, 309–11
Cost-estimating relationships,
      299–305
Cost estimation
   approaches, 292–93
   cost-capacity method, 299–301
   cost indexes, 296–99
   factor method, 301–03
   and inflation, 101, 266
   learning curve, 303–05
   unit method, 294–95
Cost indexes, 296–99
Cost of capital
   and debt-equity mix, 361–63
   for debt financing, 362, 364
   defined, 6–7, 361
   for equity financing, 360
   and MARR, 6–7, 360
   weighted average, 361–62
Cost-of-goods-sold, 307, 419–20
Cost of invested capital, 368
Cost(s)
   of asset ownership, 126
   direct, estimating, 291–305
   fixed, 209
   indirect, estimating, 305–11
   of invested capital, 368
   life-cycle, 103
   linear, 208–10
   nonlinear, 211
   in public projects, 184
   sign convention, 18, 189
   sunk, 243
   variable, 209
Cumulative cash flow sign
      test, 157
Current assets, 418
Current liabilities, 418
Current ratio, 421

D

DB function, Excel, 323, 407–08
DDB function, Excel, 323, 408
Debt capital, 360–64
Debt-equity mix, 361
Debt financing
  on balance sheet, 418
  cost of, 362–64
  leveraging, 362–63
Debt ratio, 422
Debt service, 361
Decision making
  attributes, 380
  under certainty, 385
  engineering economy role, 3
  under risk, 4, 385–96
Declining balance depreciation,
    323–25, 407
Decreasing gradients, 45, 46
Defender, in replacement analysis,
    243, 246–50
Deflation, 268–69
Depletion, 332–33
Depreciation
  accelerated, 320, 323, 325–28,
    355–56
  alternative system (ADS), 328
  book, 319, 325
  class life, 328
  Canadian system, 329
  declining balance, 323–25, 407
  defined, 319
  double declining balance, 323–25,
    327, 408
  and EVA, 368
  half-year convention, 320,
    325, 336
  MACRS, 325–28
  and Excel functions, 407–08,
    414–16
  recovery period for, 328
  present worth of, 330
  recapture, 353–54
  straight line, 321–22
  straight line alternative, 328
  switching methods, 330–31, 415
Descartes' rule, 156
Design stages, 103, 293
Design-to-cost approach, 292–93
Direct benefits, 191
Direct costs, 291–305
Disbenefits, 184, 189
Discount rate, 94, 185
Discounted cash flow, 14, 94
Discounted payback, 223
Dollars, today *versus* future, 266
Do-nothing alternative
  and B/C analysis, 191–93
  defined, 4, 96–97
  and independent projects, 108–10
  and rate of return, 152, 154–56
Double declining balance, 223–25
  in Excel, 323, 408
  in switching, 330–31, 415
  and taxes, 356–58
Dumping, 268

E

Economic service life (ESL), 244–46,
    247–49
Economic value added (EVA),
    367–69
Effective interest rate
  and APY, 73
  and compounding frequency, 75
  for continuous compounding, 76–77
  defined, 75
  and nominal rate, 73–74
EFFECT function, Excel, 84–86,
    408–09
Effective tax rate, 346, 350
Efficiency ratios, 421
End-of-period convention, 18
Engineering economy
  defined, 3
  role in decision-making, 3

Equal-service alternatives, 98, 101–03, 125, 152

Equity financing, 360–63
  cost of, 361, 363

Equivalence, 8–9
  compounding period *versus* payment period, 77–84

Equivalent uniform annual cost, 123. *See also* Annual worth

Equivalent uniform annual worth. *See* Annual worth

Estimation
  accuracy, 293–94
  of cash flow, 4, 291–94
  of interest rates, 267–68
  and sensitivity analysis, 216–23
  before and after tax MARR, 350
  before and after tax ROR, 350
  using cost-estimating relations, 299–305

Ethics
  code, for engineers, 25–26
  and morals, 23
  and public sector projects, 187–88

Evaluation criteria, 4–5

Excel. *See also* Spreadsheets; *specific functions*
  after-tax analysis, 364–67
  and B/C analysis, 197–98
  and breakeven analysis, 225–26
  charts, 405–06
  and depreciation, 334–37
  embedded function, 134, 265
  error messages, 417
  functions listing, 21, 407–16
  GOAL SEEK, Excel, 165, 228–30, 256–57, 416–17
  and inflation, 279–82
  introduction, 21–23
  and Monte Carlo, 394–98
  and rate of return, 166–69
  and sensitivity analysis, 227–28
  spreadsheet layout, 406–07

Expected value, 389

Expenses, 346, 419–20. *See also* Cost estimation; Cost(s)

Experience curve, 304

External rate of return. *See also* Rate of return
  description, 160
  by MIRR method, 161, 163–65
  by ROIC method, 162, 163–65

F

F/A factor, 42–43

Face value, of bonds, 99

Factor method estimation, 301–03

Factors
  formulas and use, 35–52
  and non-tabulated *i* or *n*, 57–58
  and tabulated values, 439–64

Factory cost, 307, 420

Financial calculator
  and ESL analysis, 254
  functions, 22
  versus factors, 21–22, 53–56
  formula used, 37
  and PW analysis, 111
  and ROR analysis, 146, 166–67

Financial worth of corporations, 368

First cost, 17, 96, 319

Fiscal year, 418

Fixed assets, 418

Fixed cost, 209

Fixed percentage method. *See* Declining balance depreciation

F/P factor, 35

Future dollars, 266

Future worth
  calculation, 103, 126
  evaluation by, 103, 189
  and inflation, 274–78

FV function, Excel, 21, 53–54, 409

G

Gains and losses. *See* Capital gain;
    Capital loss
General depreciation system (GDS),
    328
GOAL SEEK, Excel, 416–17
    for breakeven, 225–26
    for external ROR, 162, 165
    for replacement value, 256–57
Gradient, arithmetic, 44–45, 50–51
Gradient, geometric, 43–47, 50–52
Graduated tax rates, 347–49
Gross income, 332, 346

H

Half-year convention, 320, 335, 336
Highly leveraged corporations,
    362–63
Hurdle rate. *See* Minimum attractive
    rate of return
Hyperinflation, 267, 278

I

IF function, Excel, 162, 165
Income
    gross, 346
    net, 346, 368
    as revenue, 346, 419
    taxable, 346–49
Income statement, 419–20
Income tax, 346–60
    average tax rate, 347
    and cash flow, 349–52
    corporate, 347
    defined, 346
    and depreciation, 352–58
    effective rates, 346
    individual taxpayers, 348–49
    present worth of, 356–58
    and rate of return, 350
    rates, 347–48
    and replacement studies, 358–60
    tax savings, 350

Incremental benefit/cost analysis,
    191–97
Incremental cash flow, 149–51
Incremental costs
    and benefit/cost analysis, 191–92
    definition, 149–51
    and rate of return, 152–56
Incremental rate of return analysis,
    152–56
Independent projects
    B/C evaluation of, 195–96
    bundles of, 109
    formation, 96–97
    PW evaluation of, 108–10
    ROR evaluation of, 152–53
Index creep, 298
Indexing, income taxes, 348
Indirect costs
    activity-based costing, 309–11
    definition, 291, 305
    factor method, 301–03
    rates, 306–08
Infinite life, 104–08, 130–32, 183
Inflation, 266–78
    assumption in PW and AW,
        102, 125
    and capital recovery, 278–79
    definition, 266
    and future worth, 274–78
    hyper, 267, 278
    and interest rates, 267–68
    and MARR, 277
    and present worth, 269–73
Inhouse-outsource decisions. *See*
    Make-or-buy decisions
Initial investment. *See* First cost
Installation costs, in depreciation, 319
Intangible factors, 4. *See also* Multiple
    attribute evaluation;
    Noneconomic factors
Interest
    compound, 10–14
    defined, 5

and inflation, 267–68
interperiod, 82–83
rate (*See* Interest rate)
simple, 9–10
Interest period, 5
Interest rate. *See also* Effective interest
    rate; Rate of return
continuous compounding, 76–77
definition, 5
inflation-adjusted, 268
inflation free (real), 267, 275
market, 268
multiple, 156–65
nominal *versus* effective, 73–77
for public sector, 185
unknown, 21–23, 36, 57–58
on unrecovered balance, 143
Interest tables, 439–64
Internal rate of return. *See* Rate
    of return
Inventory turnover ratio, 423
Invested capital, cost of, 368
Investment(s). *See also* First cost
extra, 149–52
permanent, 104–08, 130–32
safe, 7, 268
IRR function, Excel, 21–22, 57, 146,
    166–67, 227, 410–11

L

Land, depreciation, 320
Lang factors, 301
Least common multiple (LCM)
and annual worth, 125
and future worth, 103
and incremental cash flow, 149–51
and incremental rate of return, 152,
    154–56
and independent projects, 109
and present worth, 101–02
Learning curve, 303–05
Leveraging, 362–63
Liabilities, 418–19

Life. *See* Lives
Life cycle costs, 103
Likert scale, 382
Lives
common multiple, 101–02, 125
equal, 98, 152
finite, 107
independent projects, 108–09
minimum cost, 244–46
perpetual, 104–08, 130–32, 183
and rate of return, 152–56
recovery, in depreciation, 319
unequal, 100–02, 107, 125
useful, 319
Logical IF, Excel, 162, 165, 409–10

M

MACRS (Modified Accelerated Cost
    Recovery System), 325–28
and Canadian depreciation system,
    329
depreciation rates, 326
and Excel functions, 338, 415
recovery period, 328
straight line alternative, 328
switching, 330
U.S., required, 321, 325
Maintenance and operating (M&O)
    costs, 3, 96, 346
Make-or-buy decisions, 127,
    207, 215
Manufacturing progress function, 304
Marginal tax rates, 347–48
Market interest rate. *See* Interest rate
Market value
and depreciation, 319
in ESL analysis, 245
estimating, 243, 245
in PW analysis, 100
in replacement analysis, 243,
    247–50
MARR. *See* Minimum attractive rate
    of return

Measure of worth, 5, 217, 219, 395
Minimum attractive rate of return
  after-tax *versus* before-tax, 350
  definition, 6
  establishing, 7, 360–61
  as hurdle rate, 7
  and independent projects, 108–10
  inflation-adjusted, 277
  in PW and AW analysis, 98, 127
  and rate of return, 152–56
  in sensitivity analysis, 216
  and WACC, 361
MIRR function, Excel, 162, 164, 411
M&O costs. *See* Annual operating
    costs; Maintenance and
    operating costs
Modified benefit/cost ratio, 189
Modified rate of return (MIRR)
    method, 161–65
Monte Carlo sampling, 394–98
Most likely estimate, 221–22, 385
Multiple alternatives
  benefit/cost analysis, 191–96
  independent, 108–10
  rate of return analysis, 152–56
Multiple attribute evaluation, 380–85
Multiple rates of return
  definition, 156
  determining, 156–60
  removing, 160–65
Municipal bonds, 99–100
Mutually exclusive alternatives
  annual worth analysis, 127–32
  benefit/cost analysis, 191–96
  capitalized cost analysis, 104–08
  formation of, 96–98
  future worth analysis, 103
  present worth analysis, 98–103
  rate of return analysis, 148–56

N

Net cash flow, 17, 223–24, 349
Net investment procedure, 162–65
Net operating income, 346

Net present value. *See* NPV function,
    Excel; Present worth
Net operating profit after taxes
    (NOPAT), 367–69
Net worth, 368, 418
NOMINAL function, Excel,
    84–86, 411
Nominal interest rate, 73–77
  and APR, 73
  and effective rate, 73–74
Nonconventional cash flow
    series, 156–57
Noneconomic factors, 4, 380–85
Non-owner's perspective,
    replacement, 243
No-return payback, 223–25
Norstrom's criterion, 157–59
NPER and *n* function, 21, 53, 412
  and unknown *n,* 21, 53–55
NPV function, Excel, 21, 412
  embedding in PMT, 55, 134, 254–56
  and present worth, 111–12
  in PW *vs. i* graphs, 168

O

One-year-later replacement analysis,
    247–50
Operating costs. *See* Annual operating
    costs; Cost estimation
Optimistic estimate, 221–22
Order of magnitude estimates, 293
Overhead rates. *See* Indirect costs
Owner's equity, 418

P, Q

P/A factor, 40–42, 98–99
Payback analysis, 223–25
  limitations, 224
  uses, 224, 230
Payment period
  of bonds, 99
  defined, 77–78
  relative to compounding
    period, 77–84

Payout period. *See* Payback analysis
Percentage depletion. *See* Depletion
Permanent investments. *See* Capitalized cost
Personal property, 320, 328
Pessimistic estimate, 221–22
P/F factor, 35–36
P/G factor, 44–45. *See also* Gradients, arithmetic
Planning horizon. *See* Study period
PMT function, Excel, 21–22, 41, 412–13
  and annual worth, 54, 127, 132–34
  and capital recovery, 132
  and economic service life, 254–57
  embedded with NPV function, 55, 134, 254–56
  and shifted series, 54
Point estimates, 17, 385
Power law and sizing model, 299
Present worth, 98–113
  and annual worth, 126
  arithmetic gradient, 44
  assumptions, 98, 101
  of bonds, 99–100
  of depreciation, 330
  for equal lives, 98
  geometric gradient, 46–47
  of income taxes, 356
  and independent projects, 108–10
  and inflation, 269–73
  and life cycle cost, 103
  and multiple interest rates, 156–65
  and rate of return, 152–56
  and sensitivity analysis, 216–19
  single-payment factor, 35
  for unequal lives, 100–03
  vs. i graph, 168
Present worth factors
  gradient, 44–45
  single-payment, 35
  uniform-series, 40–41
Probability
  defined, 386
  and expected value, 389–91

Probability distribution, 386–88
Productive hour rate method, 307
Profit, 209–11
Profitability ratios, 421
Profit-and-loss statement. *See* Income statement
Project net investment, 162–65
Property class, 328
Public sector projects, 183–87
  and annual worth, 130–32
  benefit/cost analysis, 188–96
  capitalized cost, 104–08
  and contract types, 187
  and ethics, 187–88
Purchasing power, 266, 274–75
PV function, Excel, 21, 41, 413
  and present worth, 52, 100, 111
  and single-payment, 36
  and uniform series present worth, 41
PW *vs. i* graph, 168
Quick ratio, 422

R

RAND function, Excel, 387, 413–14
Random samples, 386
Random variable, 386–98
  classification, 386–87
  expected value, 389
  independence of, 395
  probability distribution of, 386–87
  standard deviation, 391–94
Rank and rate technique, 383
RATE and *i* functions, 21–22, 146, 166–67, 414
Rate of depreciation
  declining balance, 323
  MACRS, 325–26
  straight line, 322
Rate of return (ROR), 143–69
  after-tax, 350–52
  and annual worth, 147, 156
  of bonds, 147

Rate of return (*continued*)
  cautions, 148
  on debt and equity capital, 362–63
  defined, 5, 143
  external, 160
  on extra investment, 152–56
  incremental, 148–52
  and independent projects, 153
  internal, 143, 160
  minimum attractive (*see* Minimum
    attractive rate of return)
  and MIRR method, 161, 163–64
  multiple, 156–60
  and mutually exclusive
    alternatives, 152–56
  ranking inconsistency, 149
  and reinvestment rate, 160–61
  and ROIC method, 162, 163–65
  and sensitivity analysis, 220–21
Ratios, business, 421–24
Real interest rate, 267, 275
Real property, 320, 326
Recaptured depreciation
  definition, 353
  in replacement study, 358–60
  and taxes, 353–55
Recovery period
  defined, 319
  effect on taxes, 356–58
  MACRS, 328
Recovery rate. *See* Rate of
    depreciation
Reinvestment rate, 160–61
Replacement analysis, 243–57
  after-tax, 358–60
  before-tax, 246–54
  cash flow approach, 250
  conventional approach, 250
  depreciation recapture, 359–60
  economic service life, 244–46
  equal lives, 250
  first costs, 243–44
  market value, 243, 248–49
  opportunity cost approach, 250
  outsider's viewpoint, 243
  and study periods, 247, 250–54
  sunk costs, 243
Replacement value, 250
Retained earnings, 419
Retirement life. *See* Economic
    service life
Return on assets ratio, 423
Return on invested capital (ROIC)
    method, 162–65
Return on investment (ROI), 5, 141.
    See also Rate of return
Return on sales ratio, 423
Revenue alternative, 96, 150,
    152, 191
Risk
  and decision making, 4, 385–86
  and sensitivity analysis, 216, 385
  and simulation, 394–98
  and variation in estimates, 385

 S

Safe investment, 7, 268
Sales, return on, 422
Salvage value
  in B/C analysis, 189
  and capital recovery, 126–27
  and depreciation, 320–25
  description, 17, 96
  and market value, 100, 128, 248
  in replacement analysis, 243, 248
Sample, 386
  average, 389
  expected value, 389
  standard deviation, 391
Sampling, Monte Carlo, 394–97
Savings, tax, 350
Scatter charts. *See x-y* charts, Excel
Section 179 deduction, 321, 354
Sensitivity analysis, 216. *See also*
    Breakeven analysis; GOAL
    SEEK, Excel
  of one parameter, 216–19
  of several parameters, 219–21

using simulation, 394–98
with three estimates, 221–23
Service life. *See* Economic service life
Shifted gradients, 50–52
Shifted series, 47–50
Simple cash flow series, 156–57
Simple interest, 9–10
Simulation, 394–98
Single-payment compound amount
    factor (F/P), 35–36
Single payment factors, 35–40
Single-payment present worth factor
    (P/F), 35–36
Sinking fund (A/F) factor, 42–43
SLN function, Excel, 322, 414
Solvency ratios, 421
SOLVER, Excel, 110, 219, 416
Spreadsheet analysis. *See also* Excel
    after-tax, 364–67
    annual worth, 132–34, 365–67
    B/C analysis, 197–98
    breakeven analysis, 225–26
    depreciation, 334–37
    effective interest, 84–86
    equivalency computations, 20–23,
        52–58
    EVA, 368–69
    inflation, 279–82
    layout, 406–07
    present worth, 110–12
    rate of return, 166–69
    replacement analysis, 254–57
    sampling and simulation, 397
    sensitivity analysis, 227–28
Standard deviation, 391–94
STDEV function, Excel, 392, 394
Stocks, 360, 362–63
Straight line alternative, in
    MACRS, 328
Straight line depreciation, 321–22
Straight line rate, 321–22
Study period
    AW evaluation, 125, 128
    equal service, 101, 109

FW analysis, 103
independent projects, 109
PW evaluation, 100–02
replacement analysis, 246–47,
    250–54
Sunk costs, 243
Switching, depreciation methods,
    330–31, 336

T

Tax depreciation, 319
Taxable income, 346–60
    and CFAT, 349–52
    and depreciation, 354, 356–58
    negative, 350
Taxes, 346–60. *See also* After-tax;
    Economic value added; Income
    tax; Taxable income
Terminology, 14–15
Time value of money, 4–5
    and financial calculator TVM
        functions, 22, 38, 41–42
Top down approach, 292–93
Total cost relation. *See* Breakeven
    analysis
Trade-in value, 244, 250, 320.
    See also Market value; Salvage
    value

U

Uniform distribution, 386–87
Uniform gradient. *See* Gradient,
    arithmetic
Uniform percentage method. *See*
    Declining balance
    depreciation
Uniform series
    compounding period and payment
        period, 77–84
    description, 14
    factors, 40–44
    functions, spreadsheet/calculator,
        21–22
    shifted, 47–50

Unit method, 294–95
Unrecovered balance, 143–44

V–Y

Value added analysis. *See* Economic
    value added (EVA)
Value engineering, 293
Variable. *See* Random variable
Variable cost, 209–10
VDB function, Excel, 336, 415
    for MACRS rates, 336, 415

VLOOKUP function, Excel, 397
Weighted attribute method, 380–85
Weighted average cost of capital
    (WACC), 361–62
x-y charts, Excel, 168, 227, 255, 337,
    405–06
Year(s). *See also* Half-year
    convention
    cash flow diagrams, 18–20
    end-of-period convention, 18
    symbols, 14–15

Glossary of Terms and Symbols

Term	Symbol	Description
Annual amount or worth	A or AW	Equivalent uniform annual worth of all cash inflows and outflows over estimated life.
Annual operating cost	AOC	Estimated annual costs to maintain and support an alternative.
Average	$\overline{X}$	Measure of central tendency; average of sample values.
Benefit/cost ratio	B/C	Ratio of a project's benefits to costs expressed in PW, AW, or FW terms.
Bond dividend	I	Dividend (interest) paid periodically on a bond.
Book value	BV	Remaining investment in an asset after depreciation is removed.
Borrowing rate	i_b	Interest rate on externally borrowed capital funds.
Breakeven point	Q_{BE}	Quantity at which revenues and costs are equal, or two alternatives are equivalent.
Capital recovery	CR	Equivalent annual cost of owning an asset plus the required return on the initial investment.
Capitalized cost	CC or P	Present worth of an alternative that will last forever (or a long time).
Cash flow	CF	Actual cash amounts which are receipts (inflow) and disbursements (outflow).
Cash flow before or after taxes	CFBT or CFAT	Cash flow amount before relevant taxes or after taxes are applied.
Compounding frequency	m	Number of times interest is compounded per period (year).
Cost estimating relationships	C_2 or C_T	Relations that use design variables and changing costs over time to estimate current and future costs.
Cost of capital	WACC	Interest rate paid for the use of capital funds; for debt and equity considered, it is weighted average cost of capital.
Debt-equity mix	D-E	Percentages of debt and equity investment capital used by a corporation to fund projects.
Depreciation	D	Reduction in the value of assets using specific models and rules; there are book and tax depreciation methods.
Depreciation rate	d_t	Annual rate for reducing the value of assets using depreciation.
Economic service life	ESL	Number of years at which the AW of costs is a minimum.
Expenses	E	All corporate costs incurred in transacting business.
First cost	P and B	Total initial cost—purchase, setup, etc.; also the basis for depreciation.
Future amount or worth	F or FW	Amount at some future date considering time value of money.
Gradient, arithmetic	G	Uniform change (+ or −) in cash flow each time period.
Gradient, geometric	g	Constant rate of change (+ or −) each time period.
Gross income	GI	Income from all sources for corporations or individuals.

(Continued)